AF771383

Herbicides and Weed Control

Herbicides and Weed Control

Editor: Bobby Jones

Callisto Reference
www.callistoreference.com

Callisto Reference,
118-35 Queens Blvd., Suite 400,
Forest Hills, NY 11375, USA

Visit us on the World Wide Web at:
www.callistoreference.com

© Callisto Reference, 2018

This book contains information obtained from authentic and highly regarded sources. Copyright for all individual chapters remain with the respective authors as indicated. All chapters are published with permission under the Creative Commons Attribution License or equivalent. A wide variety of references are listed. Permission and sources are indicated; for detailed attributions, please refer to the permissions page and list of contributors. Reasonable efforts have been made to publish reliable data and information, but the authors, editors and publisher cannot assume any responsibility for the validity of all materials or the consequences of their use.

ISBN: 978-1-63239-903-8 (Hardback)

The publisher's policy is to use permanent paper from mills that operate a sustainable forestry policy. Furthermore, the publisher ensures that the text paper and cover boards used have met acceptable environmental accreditation standards.

Trademark Notice: Registered trademark of products or corporate names are used only for explanation and identification without intent to infringe.

Printed in China.

Cataloging-in-Publication Data

Herbicides and weed control / edited by Bobby Jones.
p. cm.
Includes bibliographical references and index.
ISBN 978-1-63239-903-8
1. Herbicides. 2. Weeds--Control. I. Jones, Bobby.
SB951.4 .H47 2018
632.954--dc23

Table of Contents

Preface

Farming techniques require regular treatment and management of crops in order to protect them from harm caused by unwanted weeds and plants. Some of the popular weed control practices include seedling growth inhibitors, amino acid biosynthesis inhibitors, plant growth regulators, etc. This book is a valuable compilation of topics, ranging from the basic to the most complex advancements in the field of herbicides and weed control. This book on herbicides and weed control provides in-depth knowledge into the practices of weed eradication. The readers would gain knowledge that would broaden their perspective about this field.

The researches compiled throughout the book are authentic and of high quality, combining several disciplines and from very diverse regions from around the world. Drawing on the contributions of many researchers from diverse countries, the book's objective is to provide the readers with the latest achievements in the area of research. This book will surely be a source of knowledge to all interested and researching the field.

In the end, I would like to express my deep sense of gratitude to all the authors for meeting the set deadlines in completing and submitting their research chapters. I would also like to thank the publisher for the support offered to us throughout the course of the book. Finally, I extend my sincere thanks to my family for being a constant source of inspiration and encouragement.

Editor

1

Cell Suspension Culture-Mediated Incorporation of the Rice *Bel* Gene into Transgenic Cotton

Liping Ke[1], RuiE Liu[1], Bijue Chu[1], Xiushuang Yu[1], Jie Sun[2], Brian Jones[3], Gang Pan[4], Xiaofei Cheng[1], Huizhong Wang[1], Shuijin Zhu[4], Yuqiang Sun[1]*

1 College of Life and Environmental Science, Hangzhou Normal University, Hangzhou, Zhejiang, China, **2** College of Agronomy, Shihezi University, Shihezi, Xinjiang, China, **3** Faculty of Agriculture, Food and Natural Resources, The University of Sydney, Sydney, Australia, **4** College of Agriculture and Biotechnology, Zhejiang University, Hangzhou, Zhejiang, China

Abstract

Cotton plants engineered for resistance to the herbicides, glyphosate or glufosinate have made a considerable impact on the production of the crop worldwide. In this work, embryogenic cell cultures derived from *Gossypium hirsutum* L. cv Coker 312 hypocotyl callus were transformed via *Agrobacterium tumefaciens* with the rice cytochrome P450 gene, *CYP81A6* (*bel*). In rice, *bel* has been shown to confer resistance to both bentazon and sulfanylurea herbicides. Transformed cells were selected on a liquid medium supplemented alternately or simultaneously with kanamycin (50mg/L) and bentazon (4.2 μmol). A total of 17 transgenic cotton lines were recovered, based on the initial resistance to bentazon and on PCR detection of the *bel* transgene. *Bel* integration into the cotton genome was confirmed by Southern blot and expression of the transgene was verified by RT-PCR. In greenhouse and experimental plot trials, herbicide (bentazon) tolerance of up to 1250mg/L was demonstrated in the transgenic plants. Transgenic lines with a single copy of the *bel* gene showed normal Mendelian inheritance of the characteristic. Importantly, resistance to bentazon was shown to be stably incorporated in the T1, T2 and T3 generations of self-fertilised descendents and in plants outcrossed to another upland cotton cultivar. Engineering resistance to bentazon in cotton through the heterologous expression of *bel* opens the possibility of incorporating this trait into elite cultivars, a strategy that would give growers a more flexible alternative to weed management in cotton crops.

Editor: Sunghun Park, Kansas State University, United States of America

Funding: This research was funded by a Foundation for the Author of National Excellent Doctoral Dissertation of People's Republic of China (FANEDD, 201175), the National Natural Science foundation of China (Grant No. 31101417) and the National Program on Research and Development of Transgenic Plants. The funders had no role in study design, data collection and analysis, decision to publish, or preparation of the manuscript.

Competing Interests: The authors have declared that no competing interests exist.

* E-mail: yqsun@hznu.edu.cn

These authors contributed equally to this work.

Introduction

Cotton (*Gossypium hirsutum* L.) has long been the world's most important source of natural textile fiber. Although the cotton plant is susceptible to a wide range of environmental stressors, over 180 million people worldwide depend on it for their livelihood. Many advances have been made through traditional selection and breeding programs. Biotechnologies have now been introduced into these programs. Cotton was one of the first cultivated crops to include transgene-mediated herbicide tolerance (Ht). Weed control based on the application of herbicides to eliminate susceptible weeds within a herbicide-resistant transgenic crop has been widely adopted by growers [1,2,3]. Herbicide-resistant transgenic crops have now been generated in a number of species for non-selective herbicides such as glyphosate, bromoxynil, triazolopyrimidine sulfonanilides, imidazolinones and 2,4-dichlorophenoxyacetic acid (2,4-D) [1,3]. A flexible approach is vital for any weed management program and, amongst other benefits, the development of new herbicide-tolerant crops would minimize the potential for the evolution of tolerance to the herbicides currently used with Ht transgenic lines. In this paper, we describe the novel integration into cotton of a rice gene, *CYP81A6* (*bel*), which has been shown to confer resistance to the herbicides, bentazon and sulfonylurea.

Bentazon, known commercially as Basagran, is a benzothiadiazinone contact herbicide that controls broadleaf weeds and sedges by disrupting photosystem II electron transfer [4,5,6,7]. Resistance to bentazon in non-susceptible species is primarily based on metabolic detoxification, mediated by oxidative reactions that are catalyzed by cytochrome P450 mono-oxygenases [8,9,10]. Sulfonylurea herbicides act by inhibiting the essential enzyme, acetolactate synthase. The herbicide resistant, *CYP81A6*, *bel* gene was originally cloned from two rice bentazon lethal mutants, *bel*[a] and *bel*[b], that were more susceptible to the herbicide than wild-type plants. The bentazon- and sulfonylurea-lethal phenotypes of the mutants were the result of defects in the ability of the plants to metabolize these xenobiotics [11].

Cytochrome P450 monooxygenases are NADPH-dependent heme proteins that code for a large and diverse group of isozymes that mediate a wide range of oxidative reactions in plants, animals, and microorganisms [12,13,14,15]. In plants, cytochrome P450s are known, for example, to mediate the biosynthesis of lignins, terpenoids, alkaloids, sterols, fatty acids, and many plant defense-related secondary compounds [14,15,16]. Several plant P450s have been shown to be capable of metabolizing herbicides to harmless metabolites. These include the *bel* gene, the *CYP71A11* and *CYP81B2* genes from tobacco [17], *CYP71B1* from *Thlaspi arvensae* [18], *CYP71A10* from soybean [19], and *CYP73A1* [20], *CYP76B1* [21], and *CYP81B1* [22] from *Helianthus tuberosus*. All, apart from *bel*, are known to detoxify chlortoluron [17,18,19,23].

Traditional breeding programs for commercial cotton fiber production are based on the upland cotton species, *Gossypium hirsutum* L. These breeding programs have led to a steady improvement in agronomic traits. However, a paucity of useful economic characteristics in the breeding stock and the availability of potentially useful genes in other organisms have led to the incorporation of genetic transformation technologies into cotton breeding programs [24,25,26]. Although several genetically modified cotton lines have been widely adopted, difficulties with the transformation and regeneration of cotton remains an impediment to its widespread adoption. A number of improvements to the transformation procedure have been published that increase transformation efficiency and reduce somaclonal variation. A reliable and effective step-by-step protocol that incorporates numerous improvements and modifications was published by Wilkins et al. (2004) [25]. We have refined the protocol for cotton transformation further to improve its efficiency and reducing the time required to develop new transgenic lines. Through the use of our protocol, we have developed transgenic cotton lines heterologously expressing the rice *bel* gene, that confer resistance to the herbicide, bentazon. This provides the potential for further flexibility for weed management in cotton crops.

Materials and Methods

Plant Materials

G. hirsutum L. Coker 312 seeds were de-coated and sterilized by dipping in 70% (v/v) ethanol prior to a 10 min exposure to 0.1% (w/v) $HgCl_2$. They were subsequently rinsed in sterile distilled water and germinated on 1/2 MS (Murashige & Skoog medium) [27] medium with 10% (w/v) glucose and 0.25% (w/v) Phytagel (Sigma, USA) at 28°C in the dark for 3 d and transferred to the culture room (28°C, 14-h photoperiod, irradiance of 135 μmol $m^{-2}s^{-1}$ provided by cool-white fluorescent lamps) for 4 days.

Hypocotyls were excised from aseptic seedlings and cut into 5–7 mm segments. Callus induction was carried out on MSB medium [MS inorganic salts and B_5 vitamins [28]] supplemented with 3% (w/v) glucose, 0.25% (w/v) Phytagel, 0.1 mg/L 2,4-dichlorophenoxyacetic acid (2,4-D)(Sigma, USA), and 0.2 mg/L kinetin (Sigma, USA) for 4–6 weeks. Embryogenic callus was induced on MSB medium supplemented with 18.8 mM KNO_3, 13.7 mM glutamine, 7.6 mM asparagine, 3% (w/v) glucose, 0.25% (w/v) Phytagel, 0.5 mg/L Indole-3-butyric acid (IBA) (Sigma, USA), and 0.1 mg/L kinetin. After subculturing for 12 to 15 days, embryogenic calluses were collected and inoculated into 30 ml of liquid MSB medium in 100-ml Erlenmeyer flasks containing 18.8 mM KNO_3, 13.7 mM glutamine, 7.6 mM asparagine, and 3% (w/v) glucose, in order to establish suspension cultures for transformation (15 to 60 days). The liquid medium was removed every 7 days and replaced with 30 ml of fresh liquid medium.

Bacterial Strains and Plasmids

The *Agrobacterium tumefaciens*-disarmed helper strain LBA4404 [29], harbouring the plasmid pC450-2 that carries p35SCYP81A6 and the *npt*II gene as the selectable marker and was used for transformation and regeneration experiments. CYP81A6 was cloned from rice by Pan et al (2006) [11] (Figure 1) and had been shown to confer resistance to the bentazon and sulfonylurea herbicides. The binary vector was transferred to *Agrobacterium* by the heat-shock method [30]. The transformed *Agrobacterium* was grown on LB medium plates containing rifampicin (25 mg/l) and kanamycin (50 mg/l) (Sigma, USA) (LB medium [31]: Bacto Tryptone 10 g/L, Bacto Yeast Extract 5 g/L and NaCl 10 g/L). Five single colonies were inoculated individually in 20 ml LB medium containing rifampicin (25 mg/l) and kanamycin (50 mg/l) in flasks and these were grown for 36 h at 28°C with 200 rpm shaking. Another 10 ml of MGL medium (Tryptone 5 g/L, NaCl 5 g/L, $MgSO_4.7H_2O$ 0.1 g/L, $KH_2PO_4$0.25 g/L, Gly 1.0 g/L, pH 5.8) was added to the *Agrobacterium* suspension for *Agrobacterium* activation on a shaker at 200 rpm to $A6_{00}$ 1.0–1.5 for 2–3 h at 28°C in a 100 ml flask. Just prior to co-cultivation, the *Agrobacterium* were diluted by fresh MGL medium to OD_{600} to 0.8–1.0. Additional acetosyringone (AS) (pH 5.8) was added to the culture at a final concentration of 100 μM.

Co-cultivation

One, 15, 30, 45 and 60 day-old suspension cultures were prepared for infection by *Agrobacterium*. Friable embryogenic calluses was cultured in liquid MS medium for one day was also prepared for infection and used as a control. 3-4 day old post-subculture embryogenic cell suspension cultures growing on liquid MSB medium with IBA (0.5 mg/l) were used for transformations. The process of *Agrobacterium*-mediated transformation of suspension cultures is shown in Figure 2. Somatic cell suspension cultures were collected by sterile pipette and sieved through a 100 μm filter to remove large cell clumps and then desiccated for 15–20 min. The suspension cultures were incubated with *Agrobacterium* culture LBA4404 (OD_{600} 0.8–1.0) for 30 min. The suspension cultures were then transferred onto sterile filter paper and placed on the solid co-culture medium (CM medium, MSB medium, 0.25% (w/v) phytagel, 3% (w/v) Glucose) for 36 to 48 hours in the dark at 21°C. 1 ml of 50 mg/L As was added to the sterile filter paper. Three flasks of suspension cultures with different culture times (1, 15, 30, 45 and 60 day, respectively) were co-cultivated with *Agrobacterium* with OD_{600} value of 0.8–1.0. The co-cultivation with cultures of different ages was repeated three times and the number of bentazon-resistant colonies per flask was determined.

After co-cultivation, the cells along with the filters were transferred to sterile water supplemented with cefotaxime (200 mg/l) (Sigma, USA) and washed 2-3 times. The cells were

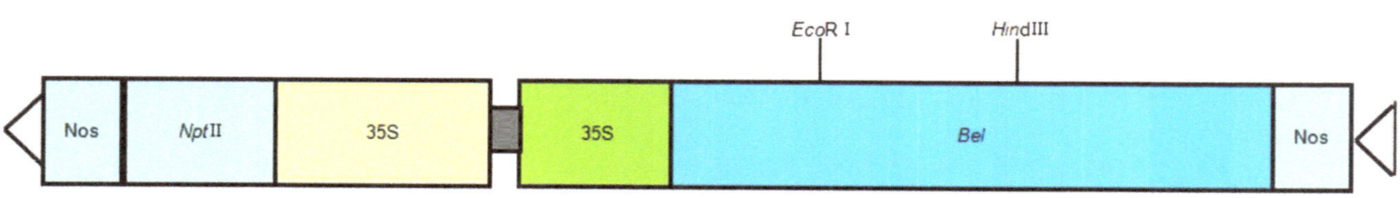

Figure 1. Schematic diagram of T-DNA incorporating the *Bel* expression cassette and the *npt* II selective marker gene.

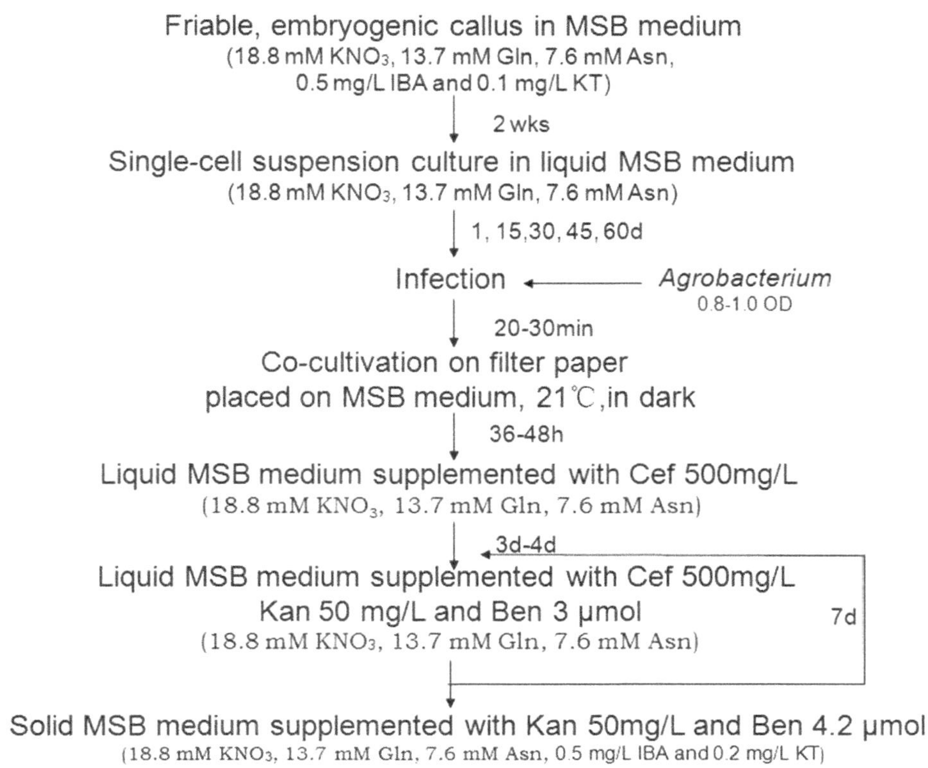

Figure 2. Scheme for *Agrobacterium*-mediated transformation of suspension cultures and plant regeneration in upland cotton.

then cultured in liquid MSB medium supplemented with cefotaxime (500 mg/l) for 3-4 days. The liquid medium was then removed and the same liquid medium supplemented with kanamycin (50 mg/l) and bentazon (3 μM) (Sigma, USA) was applied. The cultures were then incubated at 28°C at 110 rpm on a rotary shaker. The liquid medium was changed every 7 days for 4 times to eliminate the false positive cells and proliferate the resistant cells, when no more *Agrobacterium* contamination appeared, the cefotaxime was eliminated from the medium. The suspension cells were then transferred to solid MSB medium supplemented with kanamycin (50 mg/l) and bentazon (4.2 μM) for proliferation.

Inducing Kanamycin and Bentazon Resistant Callus

After the first selection in liquid suspension culture, cell suspension cultures were again transferred to a solid selection medium (MSB, 3% (w/v) Glucose, 400 mg/L Ticarcillin, 50 mg/L Kan or/and Ben 4.2 μmol, 0.5 mg/L IBA and 0.2 mg/L KT, 0.3 (w/v) % Phytagel) for 3-4 weeks. Proliferation of Kan and Ben-resistant calluses was subcultured on the same solid media.

Differentiation of Somatic Embryos and Plant Regeneration

Calluses resistant to Kanamycin and Bentazon were transferred onto solid medium (MSB, 3% (w/v) Glucose, 18.8 mM KNO_3, 0.25% (w/v) Phytagel, Gln 1.0 g/L, Asn 0.5 g/L, IBA 0.5 mg/L, KT 0.15 mg/L) to induce the embryogenic calluses, then embryogenic calluses were transferred onto embryo differentiation medium (MSB, 3% (w/v) Glucose, 0.25% (w/v) Phytagel, Gln 2.0 g/L, Asn 1.0 g/L, IBA 0.5 mg/L, KT 0.15 mg/L) for induction and development of somatic embryos every 4 weeks. Mature embryos (cotyledon embryo and torpedo embryo) were transferred onto the filter paper on the DM medium for germination for 30 days. After root and shoot elongation, the germinated embryos were transferred to solid medium (MSB, 3% (w/v) Glucose, 0.25% (w/v) Phytagel, IBA 0.5 mg/L, KT 0.1 mg/L) to conversion into plantlets with normal roots. Then the plantlets were planted into pots containing a 1:1:1 mixture of vermiculite, top soil, and peat, and covered to increase humidity. Plantlets were gradually exposed to ambient humidity over a two week period and then placed in the greenhouse. Regenerated plants with minimal root systems were grafted onto wild-type Coker 312 rootstock material.

Screening and Analysis of Transgenic Plants

PCR and southern blot analysis. Cotton genomic DNA was extracted from young leaves using methods described by Paterson et al. (1993) [32]. Polymerase chain reaction (PCR) and Southern blotting were used to confirm the presence of the *bel* gene and the copy number in transgenic plants. PCR analysis for detection of the *npt*II gene was performed using the primers 5′-TCG GCT ATG ACT GGG CAC AAC AGA-3′ (forward) and 5′-AAG AAG GCG ATA GAA GGC GAT GCG-3′ (reverse) that amplify a section of the *npt*II gene coding region. PCR analysis for detection of the *bel* gene was performed with the primers 5′-GAA GTT CAT GCC GGA GAG-3′ (forward) and 5′-ATT GCG GGA CTC TAA TCA TA-3′ (reverse). PCR was performed in 25 μl reaction mixtures consisting of 10 × reaction buffers, 50 ng DNA templates, 15 mM $MgCl_2$, 10 mM dNTPs, 50 ng of each primer and 1 unit Taq DNA polymerase in an ABI thermal cycler (Veriti 9902, USA) using the following conditions: initial denaturation at 94°C for 2 min followed by 35 cycles of denaturing at 94°C for 1 min, annealing at 55°C for *npt*II and

Figure 3. Formation of transformed callus resistant to Bentazon and Kanamycin and transgenic plant regeneration via somatic embryogenesis. A. Infected suspension cultures on solid medium, B. Cell masses browning and new callus formation, C. New somatic embryogenic callus formation, D–E. Callus proliferation and somatic embryogenesis from different colonies, F–H. Somatic embryo formation, maturity, I. Cotton plant regeneration on medium supplemented with Kanamycin and Bentazon.

Table 1. The percentage of transformed colonies resistant to Bentazon and Kanamycin according to duration of suspension culture.

time	1 day			15 days			30 days			45 days			60 days		
R-colonies1	4	2	3	13	[illegible]	8	11	9	8	15	22	19	14	16	21
R-colonies2	5	2	–	12	[illegible]	10	13	10	8	17	20	19	23	18	20
R-colonies3	7	–	3	7	[illegible]	–	–	14	12	18	16	21	19	17	15
Average	3.7			10.3			10.6			18.6			18.1		

R-colonies represent the number of colonies resistant to Bentazon and Kanamycin from the infected suspension cultures in one bottle, the transformation was repeated 3 times with different suspension cultures (1, 15, 30, 45 and 60 day-old); – represents callus colonies polluted by *Agrobacterium*; data were counted from each bottle.

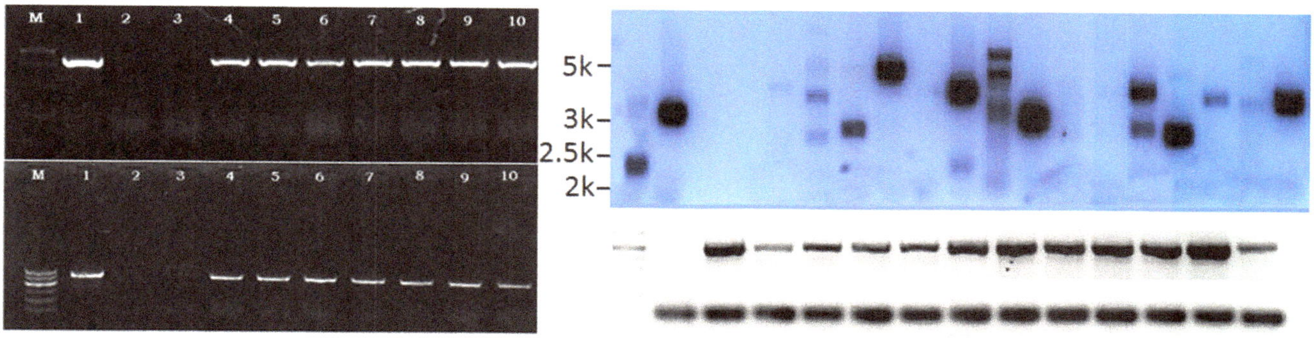

Figure 4. PCR and Southern blot analysis of transgenic cotton. A. PCR-based verification of stable insertion of the *npt II* gene into the genomic DNA of cotton using an *npt II* -specific probe; and B, of the *bel* gene with a *bel*-specific probe, C. Southern blot of T1 plants, Lane 1 shows positive control plasmid DNA, D. *Bel* expression analysis of RT-PCR of six plants of individual T2 and T3 plants, using *actin* as internal control. (Lane 1 showing Coker 312 as a negative control, Lane 2–7 showing T2 plants, Lane 8–13 showing T3 plants generated from T2 plants from plants in Lanes 2–7; *Actin* gene in cotton as reference gene).

annealing at 55°C for *bel* gene for 1 min, extension at 72°C for 3 min and final extension at 72°C for 5 min. PCR products were analyzed by gel electrophoresis on 1% (w/v) agarose gels.

Southern blot analysis was carried out according to Sambrook et al. (1989) [33] using the DIG DNA Labeling and Detection Kit (Roche, Switzerland) according to the manufacturer's recommendations. Plasmid DNA was used as a positive control and non-transgenic Coker 312 was used as a negative control. Approximately 10 µg of total genomic DNA was digested with *EcoR* I at 37°C overnight in a total volume of 100 µl and separated by electrophoresis on 0.8% (w/v) agarose gel at 30 V for 8 h. DNAs were blotted onto charged nylon membranes (Hybond N$^+$, Bedford, MA01730) and hybridized following manufacturer's instructions of Roche kit. PCR-amplified *bel* fragments were used as a probe with boitin-labelling with the DIG Detection Kit (DIG DNA Labeling and Detection Kit, Roche, Switzerland).

In 2007, T1 transgenic cotton plant families derived from self-fertilized T0 transgenic plants were grown in a greenhouse at the Huajiachi Campus of Zhejiang University in Hangzhou, China. Bentazon (1250 mg/L) was sprayed during the seedling stage, to test for herbicide resistance of the transgenic plants. Strongly herbicide resistant lines with normal gross morphological phenotypes were self-fertilised. The copy number of *bel* transgenes inserted was determined in the T1, herbicide tolerant lines using Southern blot analysis. Mature seeds were harvested from individual high-Bentazon-resistant plants with a single-copy insertion. Homozygous transgenic cotton lines were selected by molecular and genetic analysis of T2 lines in 2008 to get T3 seeds. T2 homozygous transgenic cotton plants were crossed with an elite cultivar for hybrid selection and testing with Bentazon sprays in the experimental plot.

RT-PCR analysis. In order to test for the expression level of the *bel* transgene, total RNA was isolated from young leaves of untransformed control and transformed single copy T2 and T3 plants from using the RNeasy Plant Mini Kit (Qiagen, Germany) following manufacturer's instructions. cDNA obtained from this RNA (Eppendorf, Germany) was then used as template for PCR amplification of *bel* transcripts using the primer pair: 5′ AGAAGAAGAGCATGATCGCC 3′ (forward), 5′ TGGTTCAGCAGTAGCGACAT 3′ (reverse).

PCR conditions were: 94°C for 5 min for initial melting followed by 35 cycles of amplification with each cycle consisting of the following steps: 94°C for 1 min, 56°C for 1 min and 72°C for 1 min 15 s with a final extension at 72°C for 10 min. Primers specific to the actin gene were used as internal control: 5′ CCGATGCCTTGATGAAGATT 3′ (forward), 5′ GCAGTCTCCAGTTCCTGCTC 3′ (reverse). PCR products were analyzed by gel electrophoresis on 1% (w/v) agarose gels.

Results and Discussion

Generation of Herbicide Resistant Transgenic Cotton Plants

Although glyphosate, glufonidase and a number of other herbicides have previously been used to effectively control dicot and monocot weeds in cotton crops, it is desirable to develop additional transgenic cotton varieties resistant to other herbicides, as no one herbicide is sufficient for an effective weed control program. In addition, the availability of transgenic lines with tolerance or resistance genes to different herbicides can reduce the danger of developing acquired herbicide resistance in weed species. We have used an improved transformation system to develop transgenic cotton lines resistant to the herbicide bentazon through the incorporation of the rice *Bel* gene into the cotton genome. *Bel* is a cytochrome P450 gene, originally identified from work conducted on bentazon and sulfonylurea-lethal rice mutants [11].

Using our protocol, embryogenic callus and embryos were formed from the cell suspension cultures after 4 weeks. After co-cultivation with *Agrobacterium*, transformed cells were selected and putative positive colonies were transferred onto solid medium (Figure 3 A) containing 50 mg/L kanamycin and 4.2 µM bentazon. Many of the cell masses proliferating on the kanamycin and bentazon selection media browned and died. However, in many instances, yellow living calluses formed from the browning cell mass. These were subcultured onto fresh selection media (Figure 3 B, C) and 5 to 6 weeks after the co-cultivation with *Agrobacterium*, multiple distinct living colonies were obtained. A total of 504 transformed colonies were obtained from four of the suspension culture treatments, with frequencies of 3.7% in the 1-day suspension cultures, 10.3% in the 15-day cultures, 10.6% in the 30-day cultures, 18.6% in the 45-day n cultures, and 18.1% in the 60-day cultures, indicating a significant improvement in transformation efficiency between 30 and 45 day cultures (Table 1). After 4 weeks of cell proliferation and differentiation on the selective media, many somatic embryos appeared on the surface of the resistant calluses (Figure 3 C, D, E). Mature embryos were transferred to fresh solid media for conversion into plantlets via

Figure 5. Transgenic cotton plants in pots, greenhouse and experimental plot. A. Transgenic plants grafted onto rootstock of Coker312 showing resistance to bentazon and rootstock showing sensitivity to bentazon (Arrow shows wilting 2 days after spraying with Bentazon), B. T0 transgenic cotton plant in greenhouse, C. T1 generation plants showing resistance to Bentazon, and some that are sensitive to Bentazon. D. T2 plants resistant to Bentazon and non-transgenic Coker 312 plants sensitive to Bentazon in the experimental plot.

somatic embryogenesis (Figure 3 F–I). In order to verify that the colonies were comprised of transgenic cells, PCR testing was undertaken for the presence of the the *npt*II and *bel* gene. Material from 17 of 29 lines tested was confirmed as containing the *npt* II and *bel* genes incorporated in the cotton genome (Figure 4 A, B). The 17 positive lines were subsequently carried forward through the cell proliferation, somatic embryogenesis and plant regeneration processes. Twenty-six primary (T0) transgenic cotton plantlets were obtained from the 17 lines and re-confirmed by PCR analysis as containing the *bel* and *npt*II transgenes. The regenerated, transgenic cotton plants were grafted as scions onto Coker 312 rootstocks. After the graft union had formed, the grafted plants were sprayed with the herbicide (Bentazon 500 mg/L) on the leaves of the scion and the cotyledons of the rootstock. The leaves of transgenic scion remained green, whereas the rootstock cotyledons wilted, indicating that the rootstocks retained sensitivity to the herbicide, while the transgenic scion maintained its acquired resistance (Figure 5 A). Grafted transgenic plants (T0) were transferred to pots in the greenhouse for further growth and self-fertilization (Figure 5 B).

The grafted T0 transgenic lines were allowed to self-fertilize and the resulting T1 plants were tested by spraying with bentazon (1250 mg/L). Surviving plants were classed as putatively bentazon-resistant. These were screened by PCR for the presence of the *npt*II and *bel* genes. The PCR screen confirmed that all of the tested plants were positive for the transgenes. Finally, the T1 plants were examined by Southern blot for determination of transgene copy number. The Southern blots indicated that 7 lines out of the 17 tested had one copy of the T-DNA inserted into the cotton genome. Other lines were shown to have two or three copies, and four of the lines showed no signals on the Southern blot (Figure 4 C). The 7 lines with a single copy of the *bel* gene were further self-fertilized to obtain T2 seeds. Resistant seedlings with one copy of the *bel* gene appeared at a ratio of approximately 3:1 (Figure 5 C). Similarly, the T2 plants were self-fertilized to obtain T3 seeds. Testing of T2 and T3 plants showed that the bentazon resistance was carried through to these subsequent generations. RT-PCR analysis was also conducted on T2 and T3 plants from the same lines. RT-PCR results showed that all lines had high levels of expression of the *bel* gene (Figure 4 D). Importantly, hybrids between the T2 transgenic cotton lines and an elite varieties retained resistance to bentazon. Seedlings of the T2 and T3 generations of transgenics and wild-type Coker 312 were sprayed with Bentazon (1250 mg/L) for two times 3 days later at the two true leaf stage of development in the field. All of the Coker 312 plants subsequently wilted and died. The T2 and T3 seedlings, however, continued to grow normally (Figure 5 D). Similarly, the F1 hybrid seedlings between the transgenic and the elite lines retained their resistance in the face of the same herbicide treatment used for the T2 and T3 transgenic cotton lines.

In this study, we further refined the procedure for *Agrobacterium*-mediated cotton transformation, using suspension cell cultures as explants to produce transgenic cotton lines with resistance to the herbicide, bentazon. Overall, the process of transformation, from suspension cultures to transgenic cotton plants took approximately 3-5 months. Previously published cotton transformation protocols take 6-10 months for cotton in order to obtain transgenic plants [25,34]. Wilkins et al. [25] published a detailed protocol using hypocotyls as explant material for co-cultivation with *Agrobacterium*, the protocol requires approximately 8-10 months for the production of transgenic plantlets [25]. Our improvements to the published protocols therefore make the process less time consuming. One of the difficulties has been that many of the previously published protocols have proven difficult to replicate in other laboratories. In our protocol, cell suspension cultures are used as explants material for transformation and for the selection of transgenic clones because it has the advantages of: improving the survival of transgenic colonies; decreasing the percentage of false positive resistant calluses; and reducing the occurrence of deleterious somaclonal variations that are common in material derived from solid selection media. Certainly, our method is efficient in terms of the number of transgenic lines produced. It is known that in cell suspension cultures, many of the single cells and small cell masses become growth stage synchronous. It is possible that the success of our method is at least partially due to a uniformity of cell physiology because of this synchrony [35,36,37]. We believe that widespread adoption of our cotton transformation protocol has the potential to significantly accelerate the generation of transgenic lines with other agronomical useful traits.

We also expect that the bentazon resistance incorporated in the lines developed in this study will be incorporated into elite line breeding programs and will eventually give growers a useful additional tool in their arsenal.

Author Contributions

Conceived and designed the experiments: YQS SJZ. Performed the experiments: LPK REL BJC XSY. Analyzed the data: GP HZW SJZ. Contributed reagents/materials/analysis tools: JS GP XFC. Wrote the paper: YQS BJ.

References

1. Bayley C, Trolinder N, Ray C, Morgan M, Quisenberry JE, et al. (1992) Engineering 2, 4-D resistance into cotton. Theor Appl Genet 83: 645–649.
2. Keller G, Spatola L, Mcabe D, Martinell B, Swain W, et al. (1997) Transgenic cotton resistant to herbicide bialaphos. Transgenic Res 6: 385–392.
3. Stalker DM, Kiser JA, Baldwin G, Coulombe B, Houck CM (1996) Cotton weed control using the BXN™ system. In: Duke, S.O. Ed. Herbicide-Resistant Crops: Agricultural, Environmental, Economic, Regulatory and Technical Aspects. New York, USA: Lewis Publishers. 93–105.
4. McFadden JJ, Gronwald JW, Eberlein CV (1990) In vitro hydroxylation of bentazon by microsomes from naphthalic anhydride-treated corn shoots. Biochem Biophys Res Commun 168: 206–213.
5. Leah JM, Worrall TL, Cobb AH (1991) A study of bentazon uptake and metabolism in the presence of cytochrome P450 and acethyl-coenzyme, a carboxylase inhibitor. Pestic Biochem Physiol 39: 232–239.
6. Burton JD, Maness EP (1992) Constitutive and inducible bentazon hydroxylation in shattercane (*Sorghum bicolor*) and johnsongrass (*Sorghum halepense*). Pestic Biochem Physiol 44: 40–49.
7. Forthoffer N, Helvig C, Dillon N, Benveniste I, Zimmerlin A, et al. (2001) Induction and inactivation of a cytochrome P450 conferring herbicide resistance in wheat seedlings. Eur J Drug Metab Pharmacokinet 26: 9–16.
8. Brown HM (1990) Mode of action, crop selectivity, and soil relations of the sulfonylurea herbicides. Pestic Sci 29: 263–281.
9. Koeppe MK, Brown HM (1995) Sulfonylurea herbicide plant metabolism and crop selectivity. Agro Food Ind Hi-Tech 6: 9–14.
10. Deng F, Hatzios KK (2003) Characterization of cytochrome P450-mediated bensulfuron-methyl *O*-demethylation in rice. Pestic Biochem Physiol 74: 102–115.
11. Pan G, Zhang XY, Liu KD, Zhang JW, Wu XZ, et al. (2006) Map-based cloning of a novel rice cytochrome P450 gene CYP81A6 that confers resistance to two different classes of herbicides. Plant Mol Biol 61: 933–943.
12. Bolwell GP, Bozak K, Zimmerlin A (1994) Plant cytochrome P450. Phytochemistry 37: 1491–1506.
13. Durst F, O'Keffe DP (1995) Plant cytochrome P450: an overview. Drug Metab Drug Interact 12: 171–187.
14. Schuler MA (1996) The role of cytochrome P450 monooxgenases in plant-interactions. Plant Physiol 112: 1411–1419.
15. Chapple C (1998) Molecular-genetic analysis of plant cytochrome P450-dependent monooxygenases. Annu Rev Plant Physiol Plant Mol Biol 49: 311–343.
16. Cou WW, Kutchan T (1998) Enzymatic oxidations in the biosynthesis of complex alkaloids. Plant J 15: 289–300.
17. Yamada T, Kambara Y, Imaishi H, Ohkawa H (2000) Molecular cloning of a novel cytochrome P450 species induced by chemical treatments in tobacco cells. Pestic Biochem Physiol 68: 11–25.
18. Lamb SB, Lamb DC, Kelly SL, Stuckey DC (1998) Cytochrome P450 immobilisation as a route to bioremediation/biocatalysis. FEBS Lett 431: 343–346.
19. Siminszky B, Corbin FT, Ward ER, Fleischmann TJ, Dewey RE (1999) Expression of a soybean P450 monooxygenase cDNA in yeast and tobacco enhances the metabolism of phenylurea herbicides. Proc Natl Acad Sci USA 96: 1750–1755.
20. Pierrel MA, Batard Y, Kazmaier M, Mignotte-Vieus C, Durst F, et al. (1994) Catalytic properties of the plant cytochrome P450 CYP73 expressed in yeast. Substrate specificity of a cinnamate hydroxylase. Eur J Biochem 224: 835–844.
21. Didierjean L, Gondet L, Perkins R, Lau SC, Schaller H, et al. (2002) Engineering herbicide metabolism in tobacco and *Arabidopsis* with CYP76B1, a cytochrome P450 enzyme from Jerusalem artichoke. Plant Physiol 130: 179–189.
22. Cabello-Hurtado F, Batard Y, Salaun JP, Durst F, Pinot F, et al. (1998) Cloning, expression in yeast, and functional characterization of CYP81B1, a plant cytochrome P450 that catalyzes in-chain hydroxylation of fatty acids. J Biol Chem 273: 7260–7267.
23. Robineau T, Batard Y, Nedelkina S, Cabello-Hurtado F, LeRet M, et al. (1998) The chemically-inducible plant cytochrome P450 CYP76B1 actively metabolizes phenylureas and other xenobiotics. Plant Physiol 118: 1049–1056.

24. Sawahel WA, Cove DJ (1992) Gene transfer strategies in plants. Biotechnology Advances 10 (3): 393–412.
25. Wilkins TA, Mishra R, Trolinder NL (2004) *Agrobacterium*-mediated transformation and regeneration of cotton. Food Agriculture Environment 2: 179–187.
26. Wilkins TA, Rajasekaran K, Anderson DM (2000) Cotton biotechnology. Crit Rev Plant Sci 19: 511–550.
27. Murashige T, Skoog F (1962) A revised medium for rapid growth and bioassays with tobacco tissue culture. Physiol Plant 15: 473–497.
28. Gamborg OL, Miller RA, Ojima K (1968) Nutrient requirements of suspension cultures of soybean root cells. Exp Cell Res 50: 151–158
29. Hooykaas PJJ (1988) *Agrobacterium* molecular genetics. *In* SB Gelvin, RA Schilperoort, eds, Plant Molecular Biology Manual. Kluwer Academic Publishers, Dordrecht, The Netherlands, A4–A13.
30. Bowyer P (2001) DNA-mediated transformation of fungi. In: Talbot N (ed) *Molecular and Cellular Biology of Filamentous Fungi*. Oxford Univ Press, Oxford, 33–46.
31. Bertani G (1951) Studies on lysogenesis. I. The mode of phage liberation by lysogenic Escherichia coli. J. Bacteriol. 62: 293–300.
32. Paterson AH, Brubaker CL, Wendel JF (1993) A rapid method for extraction of cotton (*Gossypium* spp.) genomic DNA suitable for RFLP or PCR analysis. Plant Mol Biol Rep 11: 122–127.
33. Sambrook J, Fritschi EF, Maniatis T (1989) Molecular cloning: a laboratory manual, Cold Spring Harbor Laboratory Press, New York.
34. Firoozabady E, DeBoer DL, Merlo DJ, Halk EL, Amerson LN, et al. (1987) Transformation of cotton (*Gossypium hirsutum* L.) by *Agrobacterium tumefaciens* and regeneration of transgenic plants. Plant Mol Biol 10: 105–116.
35. Sun YQ, Zhang XL, Huang C, Nie YC, Guo XP (2005) Plant regeneration via somatic embryogenesis from protoplasts of six explants in Coker 201 (*Gossypium hirsutum* L.). Plant Cell Tiss Organ Cult 82: 309–315.
36. Sun YQ, Zhang XL, Nie YC, Guo XP, Jin SX, et al. (2004) Production and characterization of somatic hybrids between upland cotton (*Gossypium hirsutum*) and wild cotton (*G. klotzschianum* Anderss) via electrofusion. Theor Appl Genet 2004, 109: 472–479.
37. Sun YQ, Zhang XL, Nie YC, Guo XP (2005) Production of fertile somatic hybrids of *Gossypium hirsutum* + *G. bickii* and *G. hirsutum* + *G. stockii* via protoplast fusion. Plant Cell Tiss Organ Cult 83: 303–310.

Quantifying Synergy: A Systematic Review of Mixture Toxicity Studies within Environmental Toxicology

Nina Cedergreen*

Department of Plant and Environmental Sciences, University of Copenhagen, Frederiksberg, Denmark

Abstract

Cocktail effects and synergistic interactions of chemicals in mixtures are an area of great concern to both the public and regulatory authorities. The main concern is whether some chemicals can enhance the effect of other chemicals, so that they jointly exert a larger effect than predicted. This phenomenon is called synergy. Here we present a review of the scientific literature on three main groups of environmentally relevant chemical toxicants: pesticides, metal ions and antifouling compounds. The aim of the review is to determine 1) the frequency of synergy, 2) the extent of synergy, 3) whether any particular groups or classes of chemicals tend to induce synergy, and 4) which physiological mechanisms might be responsible for this synergy. Synergy is here defined as mixtures with minimum two-fold difference between observed and predicted effect concentrations using Concentration Addition (CA) as a reference model and including both lethal and sub-lethal endpoints. The results showed that synergy occurred in 7%, 3% and 26% of the 194, 21 and 136 binary pesticide, metal and antifoulants mixtures included in the data compilation on frequency. The difference between observed and predicted effect concentrations was rarely more than 10-fold. For pesticides, synergistic mixtures included cholinesterase inhibitors or azole fungicides in 95% of 69 described cases. Both groups of pesticides are known to interfere with metabolic degradation of other xenobiotics. For the four synergistic metal and 47 synergistic antifoulant mixtures the pattern in terms of chemical groups inducing synergy was less clear. Hypotheses in terms of mechanisms governing these interactions are discussed. It was concluded that true synergistic interactions between chemicals are rare and often occur at high concentrations. Addressing the cumulative rather than synergistic effect of co-occurring chemicals, using standard models as CA, is therefore regarded as the most important step in the risk assessment of chemical cocktails.

Editor: Aamir Nazir, CSIR-Central Drug Research Institute, India

Funding: No funding provided apart from my own salary from the University of Copenhagen. The funders had no role in study design, data collection and analysis, decision to publish, or preparation of the manuscript.

Competing Interests: The author has declared that no competing interests exist.

* E-mail: ncf@plen.ku.dk

Introduction

Background

Cocktail effects and synergistic interactions of chemicals in mixtures are an area of great concern to both the public [1,2] and regulatory authorities in the US and Europe ([3] and references therein). There are two general aspects underlying this concern: The first is the uncertainty as to whether we are monitoring and regulating the most harmful chemicals? The second concerns whether the chemicals we regulate on a single compound basis, and deem "safe", potentiate or are being potentiated by other chemicals so that they jointly exert a larger effect than predicted? The latter is called synergy, and is one of the factors that create uncertainty around models proposed for the implementation in chemical risk assessment of mixtures. For those legislations where mixtures are considered, which are few, dose- or concentration additivity is proposed as the default model [3]. But are synergistic interactions really an area that should concern us? Earlier reviews have shown that synergistic interactions, at least within pesticide mixtures and realistic low-dose chemical mixtures in mammals, are a rather rare phenomenon, constituting approximately 5% of the tested mixture combinations [4–7]. This percentage is rather low given the fact that experiments are often designed to search for synergistic interactions, thereby biasing the databases towards synergistic interactions. If, however, these 5% are combinations that often co-occur in humans and the environment, they might nonetheless be of quantitative importance. Hence, if we could identify the groups of chemicals that are likely to induce synergistic interactions, special precautions could be taken in the risk assessment of these chemicals. Identifying the potential synergists would reduce the uncertainty of using the models proposed for risk assessment of mixtures of the remaining 95% of antagonistic or non-interacting chemicals [3].

The aim of this review is therefore to define which groups of chemicals are involved in well documented synergistic interactions, and if possible, to identify the mechanisms behind their synergistic effects. This will be done within three large groups of chemicals that often co-occur in the environment at measurable concentrations: The first group consists of pesticides, which is probably the most well studied chemical group within ecotoxicological mixtures studies. This is not only due to the use of chemical mixtures in pesticide formulations and tank mixtures and the resulting co-occurance in agricultural areas, but just as much because of the in depth knowledge of their physiological mode of action. This makes them ideal candidates for testing mixture models based on chemical mode of action and understanding the physiological mechanisms behind possible interactions [8,9]. The second group of chemicals are metals. Metals typically co-occur in potentially toxic concentration in relation to mining, smelting and other industrial activities and a substantial body of literature on metal mixtures is available [10]. The last group are antifouling biocides, which consist both of traditional organic biocides, organo-metals and metal ions [11], making this group a mixture of the two above thereby opening the possibility of finding other synergistic mechanisms. Antifoulants co-occur in harbour areas and marine and freshwater areas with substantial boat traffic [12,13].

Chemical mixtures from waste water treatment plants, oil spills, industrial effluents and other sources yielding very complex mixtures have not been included for two reasons: The first is that they are often chemically very poorly described; hence, we do often not know which chemicals cause the majority of the toxicity [14,15]. The second is that the probability for severe interactions decrease when the number of pollutants adding to the joint toxicity increase [16,17], hence, severe interactions are more likely to occur when a few chemicals dominate the overall toxicity, as is more often seen for e.g. pesticide pollution, in comparison to effluent pollution [18].

The Definition of Synergy

Defining synergy as two or more chemicals exerting a larger effect than predicted implies that we can predict joint effects of chemicals under certain assumptions. The aim of being able to do so, has been a research topic for more than a century [19], and the two major concepts underlying all valid assessments of joint chemical effects were framed already in the first part of the twentieth century by Loewe and Muischnek (1926) and Bliss (1939), respectively [20,21]. Loewe and Muischnek (1926) based their concept on the assumption that all chemicals in a mixture acted on the same biological target site and therefore could be viewed as being dilutions of each other, each having a different chemical potency. Hence, if the chemical potency of chemical A and B in a binary mixture was based on the Effect Concentration (EC) of each chemical causing x% effect on any endpoint in a test-system ($1/EC_{xA}$ and $1/EC_{xB}$), then the sum of the concentration of chemicals (c_A and c_B) multiplied with their respective potency in a mixture provoking x% effect would be equal to 1 [20]:

$$\frac{c_A}{EC_xA}+\frac{c_B}{EC_xB}=1 \quad (1)$$

The concept has been re-invented several times since 1926 and has received many names such as Loewe Additivity, Dose Addition, The Additive Dose Model or Concentration Addition, depending on whether it has been used within pharmaceutical, agricultural, toxicological or ecotoxicological sciences [22]. In this review we will use the term Concentration Addition (CA). Bliss (1939) worked with test-systems where mortality was the endpoint, and added another way of looking at mixtures, in the cases where the tested chemicals obviously did not affect the organisms through a similar molecular target. Different target sites were by Bliss defined by their concentration-response curves having different shapes [21]. Bliss viewed death by a chemical as a stochastic event. The probability of surviving or dying due to exposure to several chemicals acting on independent targets in the organism could therefore be calculated based on probabilities of surviving or dying from exposure to the individual chemicals [21]. Hence, the probability of surviving two independently acting chemicals (R_{mix}) would be equal to the probability of surviving the first chemical (R_1) multiplied by the probability of surviving the second chemical (R_2). Or, if assessing the probability of dying from two independently acting chemicals (E_{mix}), this is equal to the probability of dying from the first chemical (E_1) plus the probability of dying from the second chemical (E_2), minus the probability of dying from both chemicals ($E_1 \times E_2$) [21].

$$E_{mix}=E_A+E_B-E_AE_B \quad (2)$$

This concept has likewise been re-invented several times and has been named Bliss Independence, Response Multiplication, Response Addition, Effect Addition, Independent Action a.o. depending on the inventor and context [22]. In this review we will use the term Independent Action (IA). Both concepts can be extended to an infinite number of chemicals and can be used to predict mixture toxicity effects of all mixture ratios and effect levels, providing that entire dose- or concentration response relationships for the single chemicals in the desired test-system are available. Often such data are not available and reduced approaches must be used. A recent review of mixture models and their uses can be found in Cedergreen et al. (2013) [22]. How they are proposed to be used in different chemical legislation is reviewed by Backhaus and Faust (2010) [3]. Common for both concepts is also the assumption that the chemicals do not interact chemically or affect the toxicity of each other [20,21]. If the chemicals do interact, the joint effects might deviate from the predictions resulting either in the before mentioned synergistic effects or in antagonistic effects, which are defined as smaller effects than predicted [22].

Synergy can therefore be defined in relation to two basic concepts: CA and IA. Empirical evidence, however, shows that even mixture toxicity of dissimilarly acting compounds can be described with a high level of accuracy with CA, as well as with IA, despite their different underlying assumptions [6,23]. CA generally generate slightly more conservative predictions (predicting larger effects than IA), and as databases on chemicals often only provides EC_x data or No Observable Effect Concentrations (NOECs) or Lowest Observable Effect Concentrations (LOECs) which only makes CA predictions possible and not IA, CA is most often the recommended model for risk assessment purposes [3]. In this review, synergy is therefore defined in relation to CA predictions.

Experimental data are always determined with variance. For mixture studies this applies both to the toxicity data of the individual compounds used to make the model prediction, and to the tested mixture toxicity data. The consequence of this is that small deviations from the reference models can be difficult to detect statistically and repeat experimentally [24]. Biologically significant and reproducible synergy is therefore here defined as a more than two fold deviation from CA, as was also proposed by Belden et al (2007) [5]. That is, the concentration predicted to yield a certain effect is more than twice the concentration actually observed giving the proposed effect [5]. Belden calls the ratio of predicted versus observed effect concentrations for the Model Deviation Ratio (MDR) [5]. Many of the mixtures showing MDRs slightly below two, most likely also include true synergists. But to exclude false positives and to focus on combinations where the size of the synergistic interactions might be of quantitative importance, we have chosen to set the MDR limit defining synergy at two.

Materials and Methods

Identification of Experiments

A flow chart of the record selection for each of the three toxicant groups is presented in Figure 1. To evaluate the frequency of chemicals, chemical mixtures and species groups involved in synergistic (MDR>2), additive (0.5≤MDR≤2) and antagonistic

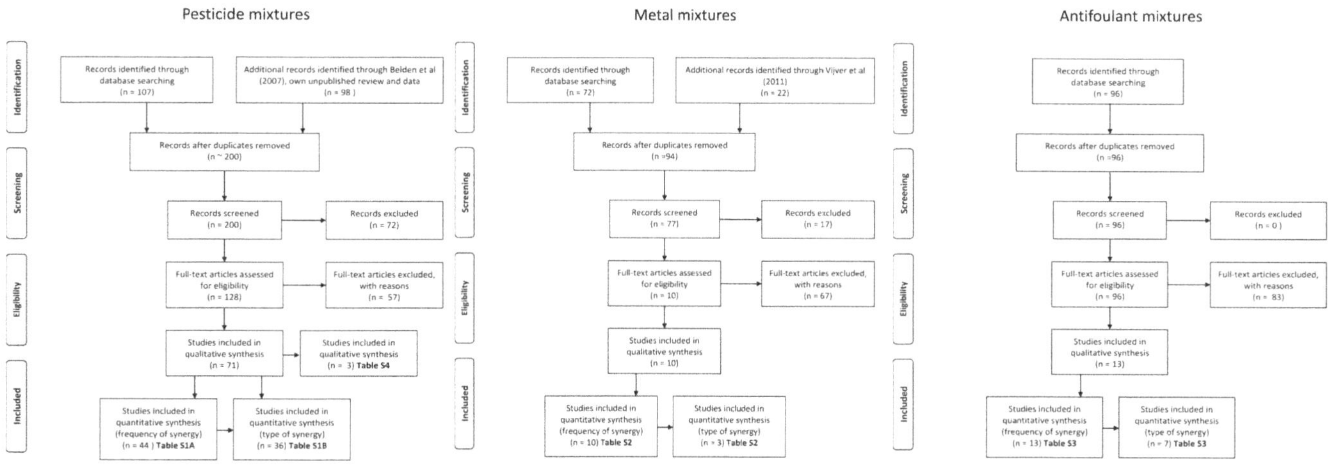

Figure 1. PRISMA 2009 Flow Diagram [90]**.** A flow diagram depicting the process of selection of records used in the review for the three main groups of toxicants: Pesticides, metals and antifoulants. Data selection has, for pesticides and metals, been built on previous reviews and data compilations, given in the top right text-box, supplemented with database searched using ISI Web of Science. Search criteria and criteria for selecting eligible records are given in the Material and Methods section. For each toxicant the search resulted in two types of databases: One to determine the frequency of synergy in a randomly selected number of mixtures studies, and another focussing only on defined synergistic mixtures. It should be noted that many records contain data on several independent mixtures studies; hence the number of records given in the figure does not match the number of selected studies reported in the results section. References to tables in supporting material giving the raw data on specific chemical mixtures, test species, endpoint and timecourse of the experiment, and the record providing the information are given in the figure.

(MDR<0.5) mixture experiments, the database of Belden et al. (2007) was used for the pesticides, the one of Vijver et al (2011) was used as a starting point for the metals, while our own data-collection was used for the antifoulants (Figure 1, Supporting information: Table 1A, 2 and 3).

For the pesticide mixtures, Belden et al (2007 and our own database on synergistic interactions was expanded with more recent studies screening the database ISI Web of Science using the search words "pesticide*", "mixture*" and "synerg*" in the period 2008–2013. For the metals, the review by Vijver et al (2011) [10] was supplemented by newer studies using ISI Web of Science and the search words "metal*", "mixture*", "synerg*" and "toxic*" for the period 2009–2013. The antifoulant mixture compilation using ISI Web of Science and the search words "antifoul*" and "mixture*" for the time period 1990–2013 to be able to detect the frequency of synergy in a similar way as had been done in the study by Belden et al. (2007).

Only studies complying with the criteria developed by Belden et al (2007) were used: Mixture studies should be conducted using only pure substances. Hence, studies using formulated pesticide or formulated antifouling biocides were excluded, as the formulation products could affect the results. Studies using metals in the form of nano-particles were likewise excluded. To avoid biasing the database with similar experiments, duplicated experiments using the same mixture and species presented in the same manuscript were entered in the database as one study, but giving the MDRs of each individual replicate. If multiple mixture ratios were tested in the same experiment, the MDR from the mixture ratio closest to the ratio where both chemicals contributed equally to the toxicity (equipotent ratio) was used in the cases of isobole designs, where several mixture ratios were tested. Otherwise the numerically larger MDR was used. Finally, the experiment had to be conducted in a way that an MDR could be calculated. That is, comparable ECx values or Toxic Units ($1/EC_x$) from individual compounds and their mixtures should be available either directly or from reading off graphs. From each study, the following information was collected: The chemicals involved, the species tested, the higher taxonomic group of the species, monitored endpoint and duration of the toxicity test, and the original reference where the raw data were reported. Studies on species communities were not included.

It should be noted that the published data does not represent a random selection of chemical mixtures tested on representative ecological species, but rather represent mixtures selected because of co-occurrence or suspicion of synergy tested on standard laboratory species. The choice of chemicals biases the database towards detecting synergies, while the choice of robust laboratory organisms, on the other hand, might give conservative estimates on synergies as they might not represent the most susceptible species.

All data treatments were done in excel.

Results

The Frequency of Synergy

Figure 1 presents the selection process of record for the study. Several of the records reported more than one mixture toxicity experiment. In the following the individual mixture toxicity experiments will be discussed. The records from where data has been retrieved can be found in the tables S1–S4 in File S1 in the supporting information. A PRISMA Checklist for reviews is given in Checklist S1.

Pesticide mixtures. The database of Belden et al (2007) provided data on 207 pesticide mixtures of which 194 were binary and another 13 consisted of more than two pesticides [5].

Metal mixtures. Evaluating the meta-analysis of Vijver et al (2011) on metal mixtures according to the criteria set by Belden et al (2007), reduced the number of usable studies from 22 to 6 studies reporting 10 experiments where MDR could be calculated and another 7 experiments, where data shown on graphs could be evaluated as being over or under-predicted by CA (Table S2 in File S1). Of the remaining 15 studies, five studies only allowed IA predictions and 10 studies reported metal tissue accumulations, but not effects. Since there is not always a straight forward

Table 1. The overall group, name and proposed Modes of Action (MoA) of the antifouling compounds.

Group	Name	IUPAC name	Mode of Action
Photosystem II inhibitors	Atrazin	6-chloro-N^2-ethyl-N^4-isopropyl-1,3,5-triazine-2,4-diamine	Inhibits the electron transport in photosystem II
	Irgarol1051	2-tert-butylamino)-4-(cyclopropylamino)-6-(methylthio)-1,3,5-triazine	Inhibits the electron transport in photosystem II
	Seanine211	4,5-dichloro-2-n-octyl-4-isothiazoline-3-one	Inhibits the electron transport in photosystem II
	Diuron	3-(3,4.dichlorophenyl)-1,1-dimethylurea	Inhibits the electron transport in photosystem II
Metals and organometals	Cd	Cadmium ion	General toxicant, interacts with enzymes[a]
	Cu	Copper ion	General toxicant, interacts with enzymes[a]
	CuPT	Copper 2-pyridinethiol-1-oxide	General toxicant, interacts with enzymes[a]
	Zn	Zink ion	General toxicant, interacts with enzymes[a]
	ZnPT	Zinc 2-pyridinethiol-oxide	General toxicant, interacts with enzymes[a]
	Ziram	Zinc bis(N,N'-dimethyl)-dithiocarbamate	Dimethyldithiocarbamate fungicide with Zn. Inhibitor of enzymes containing copper ions or sulfhydryl groups, including P450 monooxygenases of the CYP 2A6 group[b]
	TBT	tri-butyl-tin-chloride	PSII inhibitor (with tin), endocrine disruptor[c]
Fungicides	Chlorothalonil	Tetrachloroisophthalonitrile	Conjugation with, and depletion of, thiols (particularly glutathione) from germinating fungal cells, leading to disruption of glycolysis and energy production, fungistasis and fungicidal action.
	Dichlofluanid	*N*-dichlorofluoromethylthio-*N'*,*N'*-dimethyl-*N*-phenylsulfamide	Multi-site mode of action, non-specific thiol reactant, inhibiting respiration.
	IPBC	3-iodo-2-propynyl butylcarbamate	AChE inhibitor and fungicide and bactericide[d]
	PTPB	Pyridine triphenylboron	Fungicide[d]
	TCMTB	2-thio cyano methyl thio benzothiazole	Fungicide, Inhibitor of mitochondrial electron transport[c]
	Tolylfluanid	*N*-dichlorofluoromethylthio-*N'*,*N'*-dimethyl-*N*-*p*-tolylsulfamide	Multi-site mode of action, non-specific thiol reactant, inhibiting respiration.

[a] Altenburger, 2011 [88].
[b] Walker, 2009 [49].
[c] Fernandez-Alba et al, 2002 [89].
[d] Zhou et al, 2006 [40].
Particularly for the fungicides, which have multiple and often undefined modes of action, different target sites are given in different references. For herbicides and fungicides used as pesticides we use the definition of Tomlin 2002 [29]. For the remaining compounds, the source of the MoA are given as footnotes.

correlation between tissue accumulation and toxic effect [25,26] these were disregarded. A recent paper of Xu et al (2011) [27] added another 11 metal mixtures where MDR-values could be calculated, making it a total of 28 mixtures from 8 studies tested on 7 species. Of these, 21 mixtures were binary while the remaining 7 mixtures consisted of more than two metals.

Antifoulants mixtures. For antifoulants 136 mixtures where MDR-values could be calculated were found. These were presented in 14 studies comprising mixtures of 20 chemicals tested on 15 different species (Table S3 in File S1). There were 103 binary mixtures and 33 mixtures with more than two chemicals. The frequencies of synergy in the binary mixtures were 7%, 3% and 26% for pesticides, metals and antifoulants, respectively, while 88%, 86% and 64% was within two fold of the CA prediction (Figure 2). For the 13 pesticide mixtures where more than two chemicals were included, only one was synergistic [illegible], while for the 33 antifoulant mixtures with more than two chemicals 61% showed severe synergy (Table S3 in File S1).

Types of Synergy

Pesticide mixtures. In addition to Belden et al (2007) [5] and the review by Cedergreen et al (2008) [28] another 84 papers were reviewed for synergy where the MDR ratios were >2. This resulted in a database on synergistic interactions including 73 cases of synergy from both Belden et al (2007) and the data search compiled from 36 studies. These studies tested the effect of combinations of 54 pesticides on 27 different species. Of all the mixture combinations, 69 were binary mixtures while the remaining four mixtures consisted of combinations of three or five organophosphate insecticides or eight chloroacetamide herbicide safeners (Table S1B in File S1). Dividing the pesticides into groups with common modes of action according to Tomlin

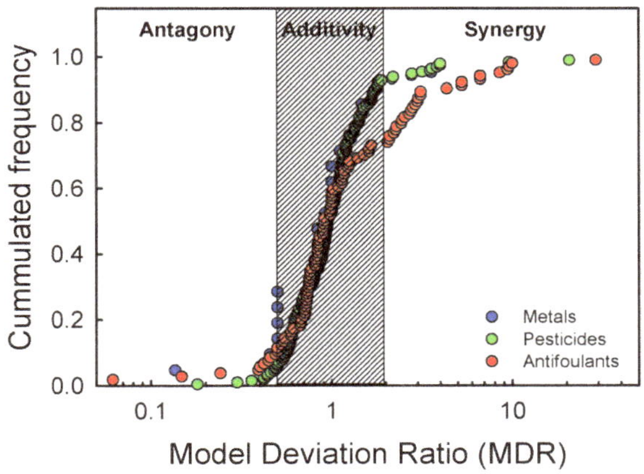

Figure 2. Cummulated frequency of Model Deviation Ratios. Cummulated frequency of Model Deviation Ratios. (MDR) of binary mixtures of pesticides ($n = 195$), metals ($n = 20$), and antifoulants ($n = 103$). The hatched interval where $0.5 \leq MDR \leq 2$ defines the mixtures that deviates less than two-fold from a Concentration Addition predictions. Mixtures having MDR values<0.5 are termed antagonistic, while mixtures with MDR values>2 are synergistic.

(2002) [29] showed that particularly five groups of pesticides were overrepresented in the synergistic mixtures. These were the organophosphate and carbamate insecticides (Cholinesterase inhibitors), azole fungicides (Ergosterol biosynthesis inhibitors), triazine herbicides (Photosystem II inhibitors) and pyrethroid insecticides (interferes with sodium channels in nerve cells) (Figure 3A). Grouping the cholinesterase inhibitors together and looking at which of the binary combinations of the above pesticide groups induced synergy in auto-trophic organisms (plants and algae) and hetero-trophic organisms (microorganisms and animals) showed no cases of synergy within the autotrophic organisms (Figure 3B). In the group of hetero-trophic organisms 69 of the 73 synergistic mixtures (95%) contained either cholinesterase inhibitors (organophosphates or carbamates) or azole fungicides (Figure 3C). The remaining four mixtures were the before mentioned mixture of 8 herbicide safeners, a mixture of a pyrethroid with an organochloride insecticide, a pyrethroid insecticide and a piperidine fungicide and a photosystem II (PSII) inhibiting herbicide and a growth regulator (Table S1B in File S1). Of the 69 binary mixtures 76% contained a cholinesterase inhibitor and another 24% an azole fungicide (Figure 3C). The triazines only entered in synergistic mixtures together with either chlorpyriphos, diazinon, malathion, methidathion, methyl-parathion, which belong to the phosphorothioate and phosphorodithioates class of organophosphates, or trichlorfon, a phosphate class organophosphate. Pyrethroids, on the other hand, only entered in synergistic mixtures together with azole fungicides.

An evaluation of which types of the pesticides from the review of Belden et al (2007) were dominant in the antagonistic mixtures and those conforming to CA, showed that cholinesterase inhibitors and azole fungicides made up 29% of the antagonistic mixtures and 48% of the mixtures conforming to CA (Figure 3B and C), which is considerably less than the 95% of the synergistic mixtures. Hence, though these modes of action were present in all types of mixtures, they were clearly overrepresented in the mixtures displaying synergistic interactions. The triazines occurred in 1% of the antagonistic mixtures, 22% of the concentration additive mixtures and in 12% of the synergistic mixtures. Hence, triazines did not seem to occur particularly frequently in the synergistic mixtures, and when they did, only in mixtures with the before

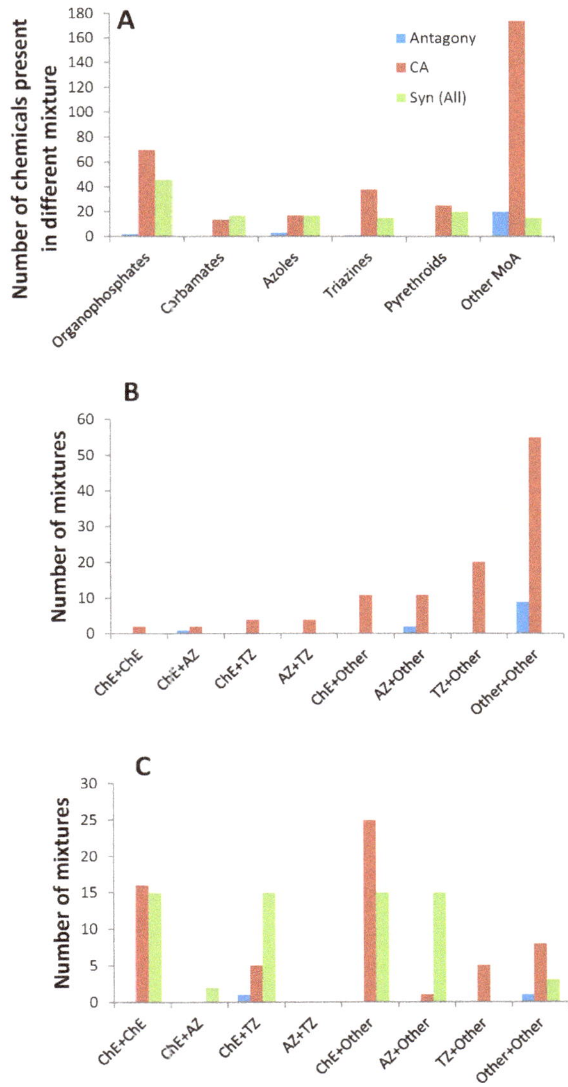

Figure 3. Frequency of pesticide antagony, additivity and synergy. Figure 2A shows the number of times a pesticide belonging to the group of organophosphates, carbamates, azoles, triazines, pyrethroids or some other Mode of Action (other MoA) occur in a binary mixture resulting in antagony (blue bars), concentration additivity (CA) (red bars) or synergy (green bars). In figure B and C, the number of binary combinations of cholinesterase inhibitors (ChE) (The organophosphates and carbamates), azoles (AZ), triazines (TZ) and other Modes of Action (Other) resulting in either antagony, concentration additivity or synergy are shown for mixtures tested on B) autotropic organisms (plants and algae, $n = 120$) or C) heterotrophic organisms (microorganisms and animals, $n = 128$).

mentioned organophosphates. The 19 triazine mixtures with an MDR<1 were dominated by Auxin transport inhibitors, branched chain- and aromatic amino acid synthesis inhibitors, while the 19 triazine mixtures with MDR values between 1 and 2 were dominated by organophosphates, PSII inhibitors and cell division inhibiting herbicides. All 22 additive mixtures including pyrethroids, were mixtures with organophosphates, carbamates or other pyrethroids (Table S1A in File S1).

Metal mixtures. Going through the 55 selected potential papers found on ISI Web-of-Science using the key-words given

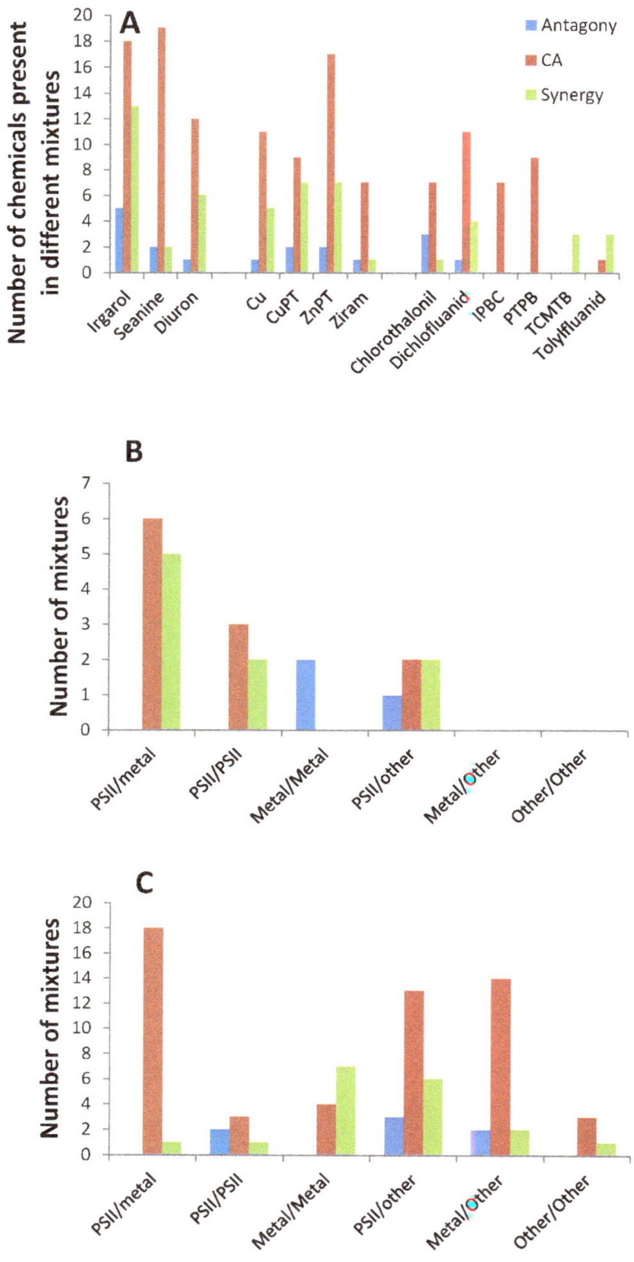

Figure 4. Frequency of antifoulant antagony, additivity and synergy. Figure 3A shows the number of times each of the antifoulants occur in a binary mixture resulting in antagony (blue bars), concentration additivity (CA) (red bars) or synergy (green bars). Antifoulants occurring in less than 1% of the mixtures were excluded. In figure B and C, the number of binary combinations of photosystem II herbicides (PSII) metal ions or metal containing compounds (Metal) and other organic compounds (Other) resulting in either antagony, concentration additivity or synergy are shown for mixtures tested on B) auto-tropic organisms (plants and algae, $n = 23$) or C) heterotrophic organisms (microorganisms and animals, $n = 80$).

above only revealed two additional studies with three experiments where MDR>2 could be estimated from figures (Table S2 in File S1). Hence, despite the large numbers of studies made on metal mixtures, well documented severe synergistic metal-metal interactions seem to be rare. The four binary mixtures giving synergy were, Cd+Zn, Cu+Zn, Cu+Cd and Cd+As tested on the shrimp *Penaeus setiferus*, the fish *Gobiocypris rarus* and the water-flee *Daphnia magna* (Table S2 in File S1).

Antifoulants mixtures. In the antifoulants database (Table S3 in File S1), we found 47 cases of synergy from 8 studies, testing the effect of mixtures of 12 chemicals on 9 organisms. Another 7 chemicals were tested that did not occur in any of the synergistic mixtures. The antifoulants were more difficult to categorise according to physiological mode of action compared to the pesticides, as this information is not required for registration. While pesticides are often developed to act physiologically very specifically in specific target organisms, antifoulants are selected to be toxic to the wide range of organisms settling on ship hulls. Hence, their physiological mode of action is more likely to be general, targeting physiological pathways important for a broad range of species. Hence, the analysis of the frequency of chemicals in synergistic, additive and antagonistic mixtures were done on the individual chemicals roughly divided into three groups: Herbicides (2,4-D, atrazine, irgarol 1051, Seanine 211, and diuron), metals and metal containing organic compounds (Cd, Cu, Cu Pyrethrione (PT), Zn, ZnPT, Ziram and Tributhyltin (TBT)), and other organic compounds (chlorothalonil, dichlorfluanid, 3-iodo-2-propynyl butylcarbamate (IPBC), pyridine triphenylboron (PTPB), 2-thio cyano methyl thio benzothiazole (TCMTB) and tolylfluanid). Chemical class and proposed modes of action are given in Table 1. For the 103 binary mixtures the frequency of synergy was markedly higher than the frequency of antagony for mixtures containing either irgarol or diuron, Cu, CuPT or ZnPT, TCMTB, dichlorofluanid or tolyfluanid (Mixtures including Cd, Zn, or TBT were excluded in this analysis as they were included in <3 binary mixtures each) (Figure 4A). Analysing the frequency of binary mixtures combined of the above defined three overall groups for the 23 binary studies on plants and algae and the 80 studies on animals and microorganisms separately, showed that all synergistic mixtures tested on plants or algae contained a PSII inhibiting herbicide either in combination with another PSII inhibiting herbicide, or metal or an organic antifoulant (Figure 4B). This is contrary to the pesticide study, where no synergy was found in studies on auto-trophic organisms at all (Figure 3B and C, Table S1B in File S1). For the autotrophic organisms the PSII inhibitors were present in slightly more additive than synergistic mixtures (Figure 4B). For the group of heterothrophs, PSII inhibiting herbicides also caused synergy, particularly together with dichlorfluanid, tolyfluanid and TCMTB (Figure 4C, Table S3 in File S1). Combinations of two metal containing compounds induced synergy in seven of 11 cases for this group (64%). Hence, contrary to mixtures of metal ions (Table S2 in File S1), mixtures of organically bound metals seem to be much more potent in inducing synergy.

Of the 23 ternary mixtures of antifoulants and the ten quarternary mixtures, four mixtures were antagonistic, nine additive and the remaining 20 mixtures had an MDR>2 (Table S2 in File S1). The frequency of occurrence of the different antifoulants in antagonistic, additive and synergistic mixtures is shown in figure 5, confirming that particularly irgarol, Seanine, CuPT, dichlofluanid and tolylfluanid often occur in synergistic mixtures.

Synergistic mixtures of metals and organic compounds. In the search for synergistic mixtures, some mixtures showed up that could not be categorised as either pesticides or metals, as they contained both. Synergistic interactions between metals and pesticides seemed to be quite frequent compared to synergistic mixtures of metals alone, but since a comprehensive database on mixture experiments of metals and

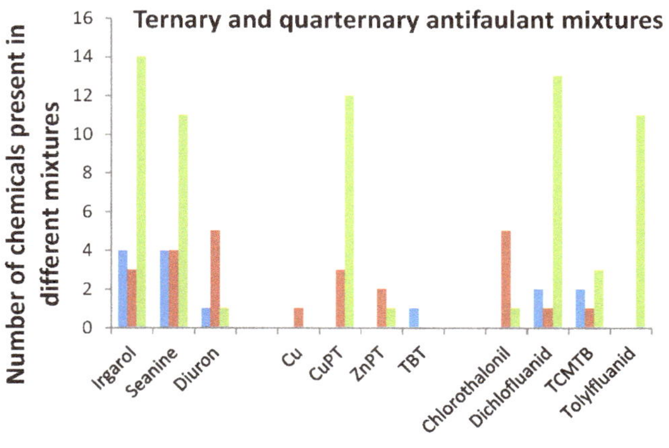

Figure 5. Frequency of antifoulant interactions in ternary and quaternary mixtures. The number of times each of the antifoulants occur in a ternary ($n=23$) or quaternary ($n=10$) mixture resulting in antagony (blue bars), concentration additivity (CA) (red bars) or synergy (green bars).

pesticides has not been made, this cannot be tested. Table S4 in File S1 in the Supplementary material show 11 mixtures of metals and pesticides from three studies, of which eight had a MDR value >2.

Discussion

Which are the Chemicals Causing Synergy?

The review showed that for pesticides, the combinations causing synergy were not random but included either cholinesterase inhibitors or azole fungicides in 95% of the described cases. The proposed mechanisms behind these synergies are relatively well investigated, as discussed below. The synergy frequency for metal ion mixtures was very low, hence no general conclusion in terms of which compounds caused synergy could be made. When metal ion synergy occurred, it was in the mg L^{-1} concentration range for three of the four cases [30,31]. These concentrations are high, compared to the concentrations normally found in metal polluted waters being in the lower ng to μg L^{-1} range [32,33]. For the antifouling compounds synergistic interactions were also related to specific chemical groups, though more synergistic combinations of different chemical groups were involved than seen for the pesticides (Figure 3, 4). The high frequency of synergistic interactions observed for the antifoulants, particularly in the mixtures with more than two active ingredients, is most likely due to the selection for compounds able to induce synergy in antifouling products, which most often are composed of more than one active ingredient [34]. The mechanisms behind the synergistic interactions of the antifoulants are, contrary to what is seen for the pesticides, rarely investigated. In the following, the proposed mechanisms behind the synergistic interactions of pesticide, metal ions and antifouling mixtures are discussed.

Mechanisms Causing Synergistic Interactions

Interactions between chemicals can basically affect six processes that are important for the resultant toxicity of a chemical towards an organism: bioavailability, uptake, internal transportation, metabolization, binding at the target site and excretion. The synergistic interactions identified in the present study are most likely caused by interactions around one or more of these processes. In the following, known mechanisms behind the identified synergistic interactions will be discussed in terms of which processes are most likely affected by the interactions.

Bioavailability. Interactions between chemicals can take place outside the organism, with one chemical affecting the availability of the other. This is commonly seen for metal ions, where ion speciation and competition for binding sites to organic matter in soil, sediments and the water phase can change free ion availability and composition [35–37]. If a less toxic ion replaces a bound or chelated ion with a higher toxicity, this will lead to apparent synergistic interactions, if the toxicities are estimated based on total metal concentrations rather than bioavailable concentrations. These types of interactions, however, most often occur when binding sites are limited. Hence, either the metal ion concentrations are high, or the binding site density low, as would for example be the case with ions in water with low concentrations of dissolved organic matter or mineral ions as calcium carbonate and other salts [37]. None of the four metal-metal ion synergies found in this study (Table S4 in File S1) allow for an assessment of whether the synergistic interactions occurred due to changed ion availability, as only total metal concentrations were given. Changes in speciation outside the organism as a cause of synergistic interaction has, however, been well documented for mixtures of pyrithione antifoulants [38]. When ZnPT and Cu ions are mixed together the more toxic CuPT complex is formed, making the mixture more toxic than predicted from the toxicities of ZnPT and Cu alone [38–40]. As the affinity of pyrithione for Cu is higher than for Zn, then the equilibrium between the metal-pyrithione complexes and free pyrithione will be shifted in favour of CuPT [41]. If there is a metal ion surplus in the ZnPT and CuPT synergistic mixture observed by Koutsaftis and Aoyama (2006) [42], a shift towards a larger proportion of CuPT might be taking place. It could therefore be hypothesised that changes in speciation outside the organism is a main mechanism behind all the reported synergistic interactions of metal/metal mixtures among the antifoulants (Figure 4C), apart from the mixture of Cu and Ziram [40].

Uptake rates and transport to the target site. One chemical can affect the uptake rate of the other by for example competition at biological ligands or competitive inhibition of transport proteins, as is often observed for interactions on metal uptake ([35,36] and references herein); though not all studies explicitly describe external ion availabilities, making it difficult to determine whether the interactions measured on internal concentration stem from interactions on bioavailability or on uptake.

Interactions on uptake rates have, however, also been measured for combinations of organic contaminants. When Belden and Lydy (2000) investigated the synergistic interactions between the herbicide atrazine and the organophosphate insecticide chlorpyriphos, they found that the addition of atrazine increased chlorpyriphos uptake by 40% [43]. This increase in contaminant uptake was proposed to be caused by an increased oxygen consumption, leading to higher ventilation rates and thereby higher uptake rates of a contaminant as chlorpyriphos, which is predominantly taken up over the gills. Increased ventilation alone could not explain the observed four-fold increase in toxicity. Hence, though it is likely that many contaminants will increase ventilation rates when the organisms start spending energy metabolizing them, thereby increasing uptake of other contaminants taken up over gills, lungs and tracheid's, the quantitative importance of this extra uptake is most likely of little importance for the more severe synergistic cases reported in the literature.

Potential important effects on uptake was also proposed by Kennaugh et al (1993) in a study on the effect of the known synergist Piperonyl Butoxide (PBO) on the cytochrome P450 mediated metabolic rate of the pyrethroid insecticide Permethrin in permethrin resistant and wildtype *Helicoverpa armigera* [44]. The ability of PBO to break the 20-fold resistance could not be explained by differences in P450 monooxygenase mediated permethrin detoxification rates, since they were identical for the resistant and non-resistant genotypes. Hence, it was proposed that PBO instead affected a P450 mediated "penetration resistance" developed by the resistant strain, making the resistant strain take up less pyrethroid. The proposed effect on uptake rates was, however, never confirmed by actual studies of permethrin uptake. Hence, the P450 mediated "penetration resistance" is still a hypothesis.

Many of the synergists known to enhance uptake belong to the large group of surfactants and other additives added to formulated pesticides with the exact purpose of enhancing the uptake of the active compounds [45]. As this review has excluded all studies with formulated compounds and surfactants, the database does not include examples on surfactant synergies, despite of their frequent use. Though there is a proven effect of the surfactants on uptake of active compounds when hitting their target at high concentrations, it is likely that most lose their "uptake enhancing" potency when diluted in environmental matrices, even though they might still act as dilute pollutants adding to the overall toxicity according to concentration addition. This is supported by a toxicity study on formulated versus technical herbicides on aquatic plants and algae showing no difference in potency for nine out of ten herbicides [46].

The transport rate of one chemical towards its molecular target can be affected by the presence of another chemical, as is for example the proposed mechanisms behind the strong antagonistic responses often seen in plants when a rapidly acting photosynthetic inhibitor is mixed with a slower acting systemic herbicide [47]. No studies have, to my knowledge, shown that one chemical can actively increase the transport of another chemical to their target.Thist is, nonetheless, the proposed mechanism behind many hypotheses regarding nano-particle facilitated increase in chemical toxicity [48], which we will not touch upon here, and therefore cannot be excluded either for chemical/chemical interactions.

Metabolic enzyme activities. Alternations of metabolic activity that are the most well investigated mechanisms behind observed synergistic patterns. A chemical can either increase or decrease the metabolization rate of another chemical. Decreased metabolization will typically lead to a higher toxicity than expected, when the toxic effect is caused by the unchanged parent compound. In contrast, increased metabolization will increase the toxicity of chemicals which are metabolically activated.

Synergistic interactions involving azole fungicides are most likely all examples of cases where the metabolization of the pesticides is inhibited by the azole. Azole fungicides are known inhibitors of a wide range of P450 monooxygenases, which are enzymes responsible for the phase I metabolization of lipophilic compounds [49], together with a range of biosynthesis processes in both plants and animals [50,51]. Hence, the toxicity of lipophilic insecticides such as pyrethroids are often severely enhanced when mixed with azole fungicides [49,52–55].

The synergistic cases involving cholinesterase inhibitors, which made up 76% of all the synergistic pesticide mixtures, most likely all also involve interactions on metabolism. The dominant mechanisms are, however, different depending on which compounds are involved. Basically three mechanisms can be involved:

First, besides the target enzyme acethylcholinesterase (AChE), organophosphate and carbamate insecticides can also inhibit esterases, which are responsible for phase II metabolization of other xenobiotics, including organophosphates and carbamates themselves [49]. Although having the same mode of action and therefore supposedly following concentration addition, mixtures of some organophosphates and carbamates do act synergistically [56–61] (Figure 3).

Second, organophosphates, from the phosphorothioate and phosphorodithioates class of organophosphates, must be metabolically activated to their more active oxon form in order to inhibit the target site AChE [43]. This means that compounds that can induce the production of P450 monooxygenases, will increase the rate of oxon formation and hence increase the toxicity of the organophosphates. This mechanism has been proposed as being the main mechanism responsible for the cases of synergy between triazine herbicides and organophosphates [43,62]. Belden and Lydy (2000) elegantly showed how the amount of polar metabolites of chlorpyriphos increased in *Chironomus tentans* in the presence of atrazine [43], explaining the majority of the observed synergy. Triazine herbicides have also been shown to induce P450 activity in fish [63,64]. The fact that all the cases of pesticide synergy between triazines and organophosphates include organophosphates belonging to the class of phosphorothioate and phosphorodithioates (chlorpyriphos, diazinon, malathion, methidathion, methyl-parathion) [43,62,62,65–68], or being transformed into one [69,29], indicate that triazine induced P450 induction is the main cause of this synergistic interaction. However, not only triazines induce P450 activity. Many xenobiotics, ranging from polyaromatic hydrocarbons, dioxins and ethanol [49,70] to natural substances in honey and metal ions [71,72] are proven P450 inducers. The synergistic interactions between organo(thio-)phosphates and neonicotenoids in the nematode *Caenorhabditis elegans* were also proposed to stem from P450 induction of neonicotenoids [73]. It even seems as if compounds that inhibit P450 activity at high concentrations induce activity at low concentrations or on a longer time-scale. Azole fungicides have, for example, shown to give protective effects against pyrethroid toxicity at low doses in bees [52] and the aquatic invertebrate *Daphnia magna* (pers.obs.), and have been shown to induce synergy together with organophosphates in birds pre-treated with prochloraz [74–76]. In the bird studies increased metabolism of the organophosphates was measured, strongly indicating P450 induction [75,76].

Third and finally, phosphorothioate organophosphates are known to inhibit some types of P450 monooxygenases, thereby not only affecting phase II but also affecting phase I metabolism of xenobiotics [49]. New studies have shown that the inhibitions and activations of different P450 genes are compound specific [77], as are the xenobiotics affinities for the different monooxygenases [49]. Hence, it is likely that the majority of severe synergistic interactions can be explained by interactions on metabolism. Which types of interactions plays the largest role for specific chemical combinations, and at which concentrations and time-scales the interactions are most severe for different species, is, however, still largely unexplored.

Excretion. As the ability of an organism to excrete a compound is mainly related to its ability to transform xenobiotis to an excretable form, excretion is closely related to metabolization. One exception is active excretion of essential metals as Cu and other ions, for which specific transporters or other excretion systems exist, aiding in keeping internal concentrations within a non-toxic range [78,79]. It could be hypothesised that interactions on these excretion processes could lead to synergistic interactions if they were in some way inhibited, though none of the synergistic mixtures included in this study have proven these mechanisms to be important.

Synergistic Interactions where the Mechanisms are Unknown

Apart from the pesticide and metal examples given above, where mechanisms causing synergistic interactions are, if not proven, then at least suggested, the review also revealed synergistic chemical combinations where the mechanisms are unknown. These were mainly the interactions between PSII herbicides with other PSII herbicides, metals or non-azole fungicides in the antifouling mixtures together with the mixtures of metals and organo-metals (Table S3 in File S1) or simply metals and organic pesticides (Table S4 in File S1).

Photosystem II herbicides did not induce synergy in any of the 33 mixtures performed on plants or algae in the pesticide database (Table S1 in File S1). Hence, it was surprising to find that significant synergy was found in nine of the 21 antifoulant mixtures including PSII herbicides when tested on plants or algae. Five of these nine mixtures were with the metals Cd, Cu and Zn, which were not part of any of the PSII mixture in the pesticide database. A proposed synergistic mechanism between metals and PSII inhibitors in autotrophs could be that metals might prevent the repair of not only damaged PSII complexes, which are constantly repaired during photosynthesis [80], but also the damage caused by the reactive oxygen species (ROS) created by the PSII inhibition and the metals themselves, by interacting with enzymes responsible for the repair. The two synergistic PSII/PSII mixtures were between irgarol and diuron, while the remaining two were between irgarol and chorothalonil or TCMTB. The synergies between irgarol and the two general fungicides, chorothalonil and TCMTB, could be similar to the mechanism proposed for the PSII/metal interactions, as both fungicides create ROS [81] and additionally chlorothalonil conjugates with gluthatione [29], an important ROS scavenger. These hypotheses, however, need to be tested.

The mode of action of PSII inhibitors in heterotrophs is largely unknown, as these organisms lack photosystems. The studies on pesticides revealed that triazines, such as irgarol, can induce P450 activity in heterotrophs, thereby enhancing the effect of the organophosphates which needed to be metabolically activated. A study by Suzuki et al (2004) show that also dichlofluanid and chlorothalonil need activation by P450 monooxygenases to reach their full lipid-oxidation potential, with dichlofluanid being far more potent than chlorothalonil [81]. As tolylfluanid is chemically related to dichlofluanid, it might also have to be oxidised by P450 to be fully activated. A study on fish have shown that also the metal ions Cu^{+} and Pb^{+} can induce P450 activity [72]. Five of the seven synergistic mixtures involving PSII inhibitors or metals together with organics are mixtures of the triazine irgarol or Cu and either dichlofluanid or tolylfluanid. In addition, for the twenty synergistic mixtures with more than two antifoulants, these combinations were present in all but three mixtures. It could therefore be hypothesised that the main mechanism behind the synergy between irgarol or Cu (or CuPT), and the fungicides dichlofluanid and tolylfluanid were irgarol and Cu mediated P450 induction leading to faster activation of dichlofluanid and tolylfluanid. But this hypothesis would have to be tested.diuron and TCMTB also induced synergy together with irgarol in two cases each, in the heterotrophic organisms, but information in terms of possible P450 induced activation of these two compounds has not been found.

Metal induced P450 activity could possibly also play a role in the synergies with the phosphorothioate organophosphates malathion, chlorpyriphos and dimethoate [56,82](Sejerøe 2011) (Table S4 in File S1), but as synergies between metal ions an phosphate organophosphates as dichlorvos, the carbamate carbofuran and the azole penconazole [56]Sejerøe 2011) was also found, other mechanisms are most likely also of importance. Lister et al (2011) found an increased uptake and metabolization rate of chlorpyriphos in the presence of Ni, but data were two variable to say anything definite [83]. However, as both P450 monooxygenases and esterases are important enzymes in many biochemical processes, changes in their activity could also affect uptake, excretion or possibly some of the mechanisms used for inactivation of metals in different organisms, though these hypotheses must be subjected to experimental scrutiny.

Is Synergy of any Importance in Nature?

For synergistic interactions to take place in the environment, interacting chemicals have both to co-occur and to be present at levels high enough to induce the synergy. Co-occurrence does happen, as has been shown for both pesticides and antifoulants [11,59,84]. Looking at the cases presented in this review, however, most experiments showing significant synergy use chemical concentrations in the high μg L^{-1} to mg L^{-1} range, which is considerably above the concentrations most often monitored in the environment (pg L^{-1} to the low μg L^{-1} range) [11,59,84]. Very few studies though use realistic concentration ranges, Laetz et al (2009) being such an exception. It is, however, likely that a threshold for synergistic interactions exists for most synergists, and that only a few proven synergists will act as synergists at any endpoint when diluted down to realistic environmental levels. This loss of efficiency as a synergist has for example been shown for piperonyl butoxice (PBO), a known P450 inhibitor, when used to formulate pyrethroid insecticides for mosquito control [85]. In this case, adding the synergist to the aquatic environment did not increase the efficacy of the insecticide towards an aquatic crustacean. Another case, however, showed PBO to enhance pyrethroid toxicity down to concentrations as low as 25 μg L^{-1} [86]. Hence, more data is needed to determine if a lower threshold for synergists interfering with metabolic processes do exist. In these studies it will be important to include sub-lethal endpoints such as growth and reproduction so that true long term effects on population growth can be estimated.

Conclusion

From the present review of possible mechanisms causing the observed synergies, it can be concluded that interactions on metabolic processes affecting the transformation of xenobiotics seem to be far the most common mechanism of synergy, though interactions on availability and uptake might play an important role for metal/metal synergies. For the synergistic interactions between pesticides, with cholinesterase inhibitors and azole fungicides being present in 95% of the described synergistic cases, the chemical groups causing synergy can be well defined. For the antifoulants the pattern was less clear, primarily due to the lack of knowledge on the interference of the compounds with metabolic processes. However, knowing that most synergistic interactions most likely stem from interactions on metabolic processes, it would be possible to screen for potential synergists using either *in vitro* assays on P450 monooxygenase or esterase inhibition potential, or by investigating metabolization kinetics *in vivo* in representative test species; though the latter is quite labour intensive.

In the introduction it was stated that if we could identify the groups of chemicals that are likely to induce synergistic interactions, special precautions could be taken in the risk assessment of these chemicals. The present review shows that some groups of potential synergists can indeed be identified, while others need more research to be specifically defined as synergists.

That said, considering the generally high chemical concentrations needed to induce synergistic interactions, their importance as synergists within naturally occurring exposure scenarios is most likely of a relatively small importance compared to the additive effect of many co-occuring pollutants. Even if one compound enhances the effect of another compound four-fold, it only takes another three compounds of a similar strength to arrive at the same joint toxicity. And considering the complex pollution patterns monitored [13,33,87], the additive effect of the many co-occurring pollutants might likely project a larger hazard than those of the presence of a few synergist. Hence, in a regulatory perspective addressing the cumulative effect of co-occurring chemicals is the first and most important step in providing a more realistic hazard assessment of chemical cocktails in both man and environment.

Supporting Information

Checklist S1 PRISMA 2009 checklist for systematic reviews and meta-analyses.

File S1 Table S1A. Antagonistic and additive pesticide mixtures. All binary antagonistic and concentration additive pesticide mixtures from Belden et al (2007) sorted with increasing Model Deviation ration (MDR). The synergistic mixtures from Belden et al (2007) are included in Table S1B. For information on species tested, endpoint and original references, please see Belden et al (2007), Supplementary material, Table 1. Table S1B. Synergistic pesticide mixtures. The mixtures are sorted with increasing MDR and including information on the test species, its phylum, sub-phylum or class, the endpoint tested and the reference of the original study. The synergistic mixtures also included in Belden et al (2007) are given in bold. In the cases where the same mixtures were repeated on the same organism in independent experiments, MDR-values are given for all experiments and are sorted according to the highest MDR value. One full ray-design is defined as one experiment, even though several mixture ratios were tested. Table S2. Metal mitures. Antagonistic and concentration additive mixtures of metal ions from Vijvers et al (2011) and Xu et al (2011) from which MDR-values could be calculated, sorted with the binary mixtures first and then with increasing MDR. Below are the four synergistic mixtures found of which one mixture, given in bold, was obtained from Vijvers et al (2011). The table includes information on the test species, its phylum, sub-phylum or class, the endpoint tested and the reference of the original study. The last three entries are the three extra synergistic mixtures found by the additional database study. Table S3. Mixtures of antifoulants. All mixtures of antifoulants (Antif) from which MDR-values could be calculated, sorted with the binary mixtures first and then with increasing MDR. The table includes information on the test species, its phylum, sub-phylum or class, the endpoint tested and the reference of the original study. For full chemical names and chemical class and mode of action of the antifaulants, please consult Table 1 in the manuscript. The following names are abbreviated: Irgarol1051 (Irgarol), Sea-nine211 (Seanine), Chlorothalonil (Chlorot.), Dichlofluanid (Dichlo), Tolylfuanid (Tolyl). Table S4. Additional synergistic mixtures. Synergistic mixtures between metals and organic compounds which did not fit into any of the three categories; pesticides, metals or antifoulants, sorted with increasing MDR. The table includes information on the test species, its phylum, sub-phylum or class, the endpoint tested and the reference of the original study.

Acknowledgments

I wish to thank Christian Neerstrøm and Maj-Britt A. Bjergager for starting the compilation of literature on pesticides and Andreas Kretschmann, Michele Gottardi, Claus Svendsen, David Spurgeon and Helle Marcussen for fruitful discussions on modes of synergistic interactions, and comments on an earlier version of the manuscript. No additional funding was received for this study apart from my salary provided by the University of Copenhagen.

Author Contributions

Conceived and designed the experiments: NC. Wrote the paper: NC.

References

1. Colborn T, Dumandski D, and Myers JP (1996) Our Stolen Future: Are We Threatening Our Fertility, Intelligence, and Survival? - Scientific Detective STory. New York: Dutton. 306 p.
2. Kepner J (2004) Synergy: The Big Unknowns of Pesticide Exposure. Pesticides and You 23: 17–20.
3. Backhaus T, Blanck H, Faust M (2010) Hazard and Risk Assessment of Chemical Mixtures under REACH: State of the Art, Gaps and Options for Improvement. 3/10: 1–75.
4. Deneer JW (2000) Toxicity of mixtures of pesticides in aquatic systems. Pest Manage Sci 56: 520.
5. Belden JB, Gilliom RJ, Lydy MJ (2007) How well can we predict the toxicity of pesticide mixtures to aquatic life? Int Environ Assess Manage 3: 364–372.
6. Cedergreen N, Christensen AM, Kamper A, Kudsk P, Mathiasen S, et al.(2008) A review of independent action as a reference model for binary mixtures of compounds with different molecular target sites. Environ Toxicol Chem 27: 1621–1632.
7. Boobis A, Budinsky R, Collie S, Crofton K, Embry M et al.(2011) Critical analysis of literature on low-dose synergy for use in screening chemical mixtures for risk assessment. Crit Rev Toxicol 41: 369–383.
8. Altenburger R, Backhaus T, Boedeker W, Faust M, Scholze M, et al. (2000) Predictability of the toxicity of multiple chemical mixtures to Vibrio fischeri: Mixtures composed of similarly acting chemicals. Environ Toxicol Chem19: 2341–2347.
9. Backhaus T, Altenburger R, Boedeker W, Faust M, Scholze M, et al. (2000) Predictability of the toxicity of multiple mixture of dissimilarly acting chemicals to *Vibrio Fischeri*. Environ Toxicol Chem19: 2348–2356.
10. Vijver MG, Elliott EG, Peijnenburg WJGM, De Snoo GR (2011) Response Predictions for Organisms Water-Exposed to Metal Mixtures: A Meta-Analysis. Environ Toxicol Chem30: 1482–1487.
11. Readman JW (2006) Development, occurance and regulation of antifouling paint biocides: Historical review and future trends. In: Antifouling Paint Biocides. Berlin: Springer. 1–15.
12. Mohr S, Schroder H, Feibicke M, Berghahn R, Arp W, et al. (2008) Long-term effects of the antifouling booster biocide Irgarol 1051 on periphyton, plankton and ecosystem function in freshwater pond mesocosms. Aquat Toxicol 90: 109–120.
13. Cassi R, Tolosa I, de Mora S (2008) A survey of antifoulants in sediments from Ports and Marinas along the French Mediterranean coast. Mar Pollut Bull 56: 1943–1948.
14. Vermeirssen ELM, Hollender J, Bramaz N, van der Voet J, Escher BI (2010) Linking Toxicity in Algal and Bacterial Assays with Chemical Analysis in Passive Samplers Deployed in 21 Treated Sewage Effluents. Environ Toxicol Chem29: 2575–2582.
15. Grung M, Naes K, Fogelberg O, Nilsen AJ, Brack W, et al.(2011) Effects-Directed Analysis of Sediments From Polluted Marine Sites in Norway. J Toxicol Environ Health-Part A-Current Issues 74: 439–454.
16. Warne MSJ, Hawker DW (1995) The Number of Components in A Mixture Determines Whether Synergistic and Antagonistic Or Additive Toxicity Predominate - the Funnel Hypothesis. Ecotoxicol Environ Saf 31: 23–28.
17. Cedergreen N, Sørensen H, Svendsen C (2012) Can the effect of ternary mixtures be predicted from binary mixture toxicity results? Sci Tot Environ 427–428: 229–237.
18. Alabaster JS, Lloyd R (1980) Water Quality Criteria for Freshwater Fish. London: Butterworth. 200 p.
19. Fraser TR (1892) Lecture on The Antagonism Between the Actions of Active Substances. Britt Med J Nov. 2: 485–487.

20. Loewe S, Muischnek H (1926) Über kombinationswirkungen. 1. Mitteilung: Hilfsmittel der Fragestellung. Naunyn-Schmiedebergs Arch Exp Pathol Pharmakol 114: 313–326.
21. Bliss CI (1939) The toxicity of poisons applied jointly. Ann App Biol 26: 585–615.
22. Cedergreen N, Svendsen C, Backhaus T (2013) Chemical Mixtures: Concepts for Predicting Toxicity. In: Jorgensen SE, editors. Encyclopedia of Environmental Management. New York: Taylor & Francis. 2601–2610.
23. Altenburger R, Boedeker W, Faust M, Grimme LH (1996) Regulations for combined effects of pollutants: Consequences from risk assessment in aquatic toxicology. Food Chem Toxicol 34: 1155–1157.
24. Cedergreen N, Kudsk P, Mathiassen SK, Sørensen H, Streibig JC (2007) The reproducability of binary mixture toxicity studies. Environ Toxicol Chem26: 149–156.
25. Lock K, Janssen CR (2001) Zinc and cadmium body burdens in terrestrial oligochaetes: Use and significance in environmental risk assessment. Environ Toxicol Chem 20: 2067–2072.
26. Cedergreen N, Johansson HKL, Nielsen JK, Nørhave N, Spurgeon D, et al. (2013) Low temperatures enhance the chronic toxicity of Cd and Cu through different mechanisms. Environ Toxicol Chem. 32: 2274–2283.
27. Xu X, Li Y, Wang YA, Wang YH (2011) Assessment of toxic interactions of heavy metals in multi-component mixtures using sea urchin embryo-larval bioassay. Toxicol in Vitro 25: 294–300.
28. Cedergreen N, Streibig JC (2004) Independent action and apparent synergistic effects between toxicants. SETAC Globe 5: 33–34.
29. Tomlin CDS (2002) The e-Pesticide Manual.
30. Vanegas C, Espina S, Botello AV, Villanueva S (1997) Acute toxicity and synergism of cadmium and zinc in white shrimp, Penaeus setiferus, juveniles. Bull Environ Contam Toxicol 58: 87–92.
31. Zhu B, Wu ZF, Li J, Wang GX (2011) Single and joint action toxicity of heavy metals on early developmental stages of Chinese rare minnow (Gobiocypris rarus). Ecotoxicol Environ Saf 74: 2193–2202.
32. Rahman MS, Saha N, Molla AH, Al-Reza SM (2014) Assessment of Anthropogenic Influence on Heavy Metals Contamination in the Aquatic Ecosystem Components: Water, Sediment, and Fish. Soil Sediment Contam 23: 353–373.
33. Nriagu JO, Wong HKT, Lawson G, Daniel P (1998) Saturation of ecosystems with toxic metals in Sudbury basin, Ontario, Canada. Sci Tot Environ 223: 99–117.
34. Okamura H, Mieno H (2006) Present status of antifouling systems in Japan: tributyltin substitutes in Japan. In: Antifouling paint biocides. Berlin: Springer. 201–212.
35. Posthuma L, Baerselman R, Van Veen RPM, Dirven-Van Breemen EM (1997) Single and joint toxic effects of copper and zinc on reproduction of Enchytraeus crypticus in relation to sorption of metals in soils. Ecotoxicol Environ Saf 38: 108–121.
36. Elliott NG, Swain R, Ritz DA (1986) Metal Interaction During Accumulation by the Mussel Mytilus-Edulis-Planulatus. Mar Biol 93: 395–399.
37. Meyer JS, Ranville JF, Santore RC, Ryan AC, Gorsuch JW, et al.(2013) Toxicity of metal mixtures to *Daphnia magna*: Competitive binding to dissolved organic matter and biotic ligand can explain non-additive interactions. SETAC Europe 23rd Annual Meeting, Glasgow, UK, 12–16 May 2013.
38. Mochida K, Ito K, Harino H, Kakuno A, Fujii K (2006) Acute toxicity of pyrithione antifouling biocides and joint toxicity with copper to red sea bream (Pagrus major) and toy shrimp (Heptacarpus futilirostris). Environ Toxicol Chem25: 3058–3064.
39. Bao VWW, Leung KMY, Kwok KWH, Zhang AQ, Lui GCS (2008) Synergistic toxic effects of zinc pyrithione and copper to three marine species: Implications on setting appropriate water quality criteria. Mar Pollut Bull 57: 616–623.
40. Zhou XJ, Okamura H, Nagata S (2006) Remarkable synergistic effects in antifouling chemicals against Vibrio fischeri in a bioluminescent assay. J Health Sci 52: 243–251.
41. Nakajima K, Yasuda T, Nakazawa H (1990) High-Performance Liquid-Chromatographic Determination of Zinc Pyrithione in Antidandruff Preparations Based on Copper Chelate Formation. J Chromatography 502: 379–384.
42. Koutsaftis A, Aoyama I (2006) The interactive effects of binary mixtures of three antifouling biocides and three heavy metals against the marine algae Chaetoceros gracilis. Environ Toxicol 21: 432–439.
43. Belden JB, Lydy MJ (2000) Impact of atrazine on organophosphate insecticide toxicity. Environ Toxicol Chem19: 2266–2274.
44. Kennaugh L, Pearce D, Daly JC, Hobbs AA (1993) A piperonyl butoxide synergizable resistance to permethrin in *Helocoverpa armigera* which is not due to increased detoxification by cytochrome P450. Pest Biochem Physiol 45: 234–241.
45. Stephenson GR, Solomon KR (2007) Pesticide selection, formulation and application. In: Stephenson GR, Solomon KR, editors. Pesticides and the Environment. Guelph: Canadian Network of Toxicology Centres Press. 57–68.
46. Cedergreen N, Streibig JC (2005) The toxicity of herbicides to non-target aquatic plants and algae: Assessment of predictive factors and hazard. Pest Manage Sci 61: 1152–1160.
47. Cedergreen N, Kudsk P, Matthiasen S, Streibig JC (2007) Combination effects of herbicides: Do species and test system matter? Pest Manage Sci 63: 282–295.
48. Hartmann NB, Von der Kammer F, Hofmann T, Baalousha M, Ottofuelling S, Baun A (2010) Algal testing of titanium dioxide nanoparticles-Testing considerations, inhibitory effects and modification of cadmium bioavailability. Toxicol 269: 190–197.
49. Walker CH (2009) Factors determining the toxicity of organic pollutants to animals and plants. In: Organic Pollutants. London: CRC Press. 17–66.
50. Bolwell GP, Bozak K, Zimmerlin A (1994) Plant Cytochrome-P450. Phytochemistry 37: 1491–1506.
51. Guengerich FP (2008) Cytochrome P450 and chemical toxicology. Chem Res Toxicol 21: 70–83.
52. Johnson RM, Dahlgren L, Siegfried BD, Ellis MD (2013) Acaricide, Fungicide and Drug Interactions in Honey Bees (Apis mellifera). Plos One: 8.
53. Nørgaard KB, Cedergreen N (2010) Pesticide cocktails can interact synergistically on aquatic crustaceans. Environ Sci Pollut Res 17: 957–967.
54. Pilling ED, Jepson PC (1993) Synergism Between EBI Fungicides and A Pyrethroid Insecticide in the Honeybee (Apis-Mellifera). Pest Sci 39: 293–297.
55. Pilling ED, Bromley-Challanor KAC, Walker CH, Jepson PC (1995) Mechanism of synergism between the pyrethroid insecticide lamda-cyhalothrin and the imidazole fungicide prochloraz in the honeybee (*Apis mellifera* L.). Pest Biochem Physiol 51: 1–11.
56. Forget J, Pavillon JF, Beliaeff B, Bocquene G (1999) Joint action of pollutant combinations (pesticides and metals) on survival (LC50 values) and acetylcholinesterase activity of Tigriopus brevicornis (Copepoda, Harpacticoida). Environ Toxicol Chem18 912–918.
57. Marking LL, Dawson VK (1975) Method for Assessment of Toxicity Or Efficacy of Mixtures of Chemicals. US Fish and Wildlife Service Investigations in Fish Control 67: 1–8.
58. Cacciatore LC, Kristoff G, Guerrero NRV, Cochon AC (2012) Binary mixtures of azinphos-methyl oxon and chlorpyrifos oxon produce in vitro synergistic cholinesterase inhibition in Planorbarius corneus. Chemosphere 88: 450–458.
59. Laetz CA, Baldwin DH, Collier TK, Hebert V, Stark JD, et al. (2009) The Synergistic Toxicity of Pesticide Mixtures: Implications for Risk Assessment and the Conservation of Endangered Pacific Salmon. Environ Health Perspec 117: 348–353.
60. Mcleese DW, Metcalfe CD (1979) Toxicity of Mixtures of Phosphamidon and Methidathion to Lobsters (Homarus-Americanus). Chemosphere 8: 59–62.
61. Woods M, Kumar A, Correll R (2002) Acute toxicity of mixtures of chlorpyrifos, profenofos, and endosulfan to Ceriodaphnia dubia. Bull Environ Contam Toxicol 68: 801–808.
62. Schuler LJ, Trimble AJ, Belden JB, Lydy MJ (2005) Joint toxicity of triazine herbicides and organophosphate insecticides to the midge Chironomus tentans. Arch Environ Contam Toxicol 49: 173–177.
63. Fu Y, Li M, Liu C, Qu J, Zhu W, et al. (2013) Effect of atrazine and chlorpyrifos exposure on ctochrome P450 contents and enzyme activities in common carp gills. Ecotoxicol Environ Saf 94: 28–36.
64. Dong XL, Zhu LS, Wang JH, Wang J, Xie H, et al. (2009) Effects of atrazine on cytochrome P450 enzymes of zebrafish (Danio rerio). Chemosphere 77: 404–412.
65. Jin-Clark Y, Lydy MJ, Zhu KY (2002) Effects of atrazine and cyanazine on chlorpyrifos toxicity in Chironomus tentans (Diptera: Chironomidae). Environ Toxicol Chem21: 598–603.
66. Lydy MJ, Austin KR (2004) Toxicity assessment of pesticide mixtures typical of the Sacramento-San Joaquin Delta using Chironomus tentans. Arch Environ Contam Toxicol 48: 49–55.
67. Anderson TD, Lydy MJ (2002) Increased toxicity to invertebrates associated with a mixture of atrazine and organophosphate insecticides. Environ Toxicol Chem21: 1507–1514.
68. Lydy MJ, Linck SL (2003) Assessing the impact of triazine herbicides on organophosphate insecticide toxicity to the earthworm Eisenia fetida. Arch Environ Contam Toxicol 45: 343–349.
69. Pape-Lindstrom PA, Lydy MJ (1997) Synergistic toxicity of atrazine and organophosphate insecticides contravenes the response addition mixture model. Environ Toxicol Chem16: 2415–2420.
70. Vanderweiden MEJ, Tibosch HJH, Bleumink R, Sinnige TL, Vandeguchte C, et al. (1993) Cytochrome-P450-1A Induction in the Common Carp (Cyprinus-Carpio) Following Exposure to Contaminated Sediments with Halogenated Polyaromatics. Chemosphere 27: 1297–1309.
71. Johnson RM, Mao WF, Pollock HS, Niu GD, Schuler MA, et al. (2012) Ecologically Appropriate Xenobiotics Induce Cytochrome P450s in Apis mellifera. Plos One 7.
72. Henczova M, Deer AK, Filla A, Komlosi V, Mink J (2008) Effects of Cu2+ and Pb2+ on different fish species: Liver cytochrome P450-dependent monooxygenase activities and FTIR spectra. Comparative Biochemistry and Physiology C-Toxicology & Pharmacology 148: 53–60.
73. Svendsen C, Siang P, Lister LJ, Rice A, Spurgeon DJ (2010) Similarity, Independence, Or Interaction for Binary Mixture Effects of Nerve Toxicants for the Nematode Caenorhabditis Elegans. Environ Toxicol Chem29: 1182–1191.
74. Snegaroff J, Bach J (1989) Effects of the fungicide prochloraz on xenobiotic metabolism in rainbow trout - Inhibition invitro and time course of induction invivo. Xenobiotica 19: 255–267.
75. Johnston G, Walker CH, Dawson A (1994) Interactive Effects Between Ebi Fungicides (Prochloraz, Propiconazole and Penconazole) and Op Insecticides (Dimethoate, Chlorpyrifos, Diazinon and Malathion) in the Hybrid Red-Legged Partridge. Environ Toxicol Chem13: 615–620.

76. Johnston G, Walker CH, Dawson A (1994) Interactive Effects of Prochloraz and Malathion in Pigeon, Starling and Hybrid Red-Legged Partridge. Environ Toxicol Chem13: 115–120.
77. Mao WF, Schuler MA, Berenbaum MR (2011) CYP9Q-mediated detoxification of acaricides in the honey bee (Apis mellifera). Proceedings of the National Academy of Sciences of the United States of America 108: 12657–12662.
78. Burkhead JL, Reynolds KAG, Abdel-Ghany SE, Cohu CM, Pilon M (2009) Copper homeostasis. New Phytologist 182: 799–816.
79. Buchwalter DB, Cain DJ, Martin CA, Xie L, Luoma SN, et al. (2008) Aquatic insect ecophysiological traits reveal phylogenetically based differences in dissolved cadmium susceptibility. Proc Natl Acad Sci USA 105: 8321–8326.
80. Taiz L, Zeman F (2010) Photosynthesis: The light reaction. In: Plant Physiology. Sunderland, USA: Sinauer Associates Inc. 163–198.
81. Suzuki T, Nojiri H, Isono H, Ochi T (2004) Oxidative damages in isolated rat hepatocytes treated with the organochlorine fungicides captan, dichlofluanid and chlorothalonil. Toxicol 204: 97–107.
82. Amorim MJB, Pereira C, Menezes-Oliveira VB, Campos B, Soares AMVM, et al.(2012) Assessing single and joint effects of chemicals on the survival and reproduction of Folsomia candida (Collembola) in soil. Environ Pollut 160: 145–152.
83. Lister LJ, Svendsen C, Wright J, Hooper HL, Spurgeon DJ (2011) Modelling the joint effects of a metal and a pesticide on reproduction and toxicokinetics in Lumbricid earthworms. Environ International 37: 663–670.
84. Wang H, Li Y, Huang HH, Xu X, Wang YH (2011) Toxicity Evaluation of Single and Mixed Antifouling Biocides Using the Strongylocentrotus Intermedius Sea Urchin Embryo Test. Environ Toxicol Chem30: 692–703.
85. Lawler SP, Dritz DA, Johnson CS, Wolder M (2008) Does synergized pyrethrin applied over wetlands for mosquito control affect Daphnia magna zooplankton or Callibaetis californicus mayflies? Pest Manage Sci 64: 843–847.
86. Brander SM, Werner I, White JW, Deanovic LA (2009) Toxicity of a dissolved pyrethroid mixture to *Hyalella azteca* at environmentally relevant concentrations. Environmental Toxicology and Chemistry 28: 1493–1499.
87. Martin JD, Crawford CG, Larson SJ (2003) Pesticides in streams. 1–5.
88. Altenburger R (2011) Understanding combined effects for metal co-exposure in ecotoxicology. In: Sigel A, Sigel H, Sigel KO, editors. Metal Ions in Life Sciences. London: Royal Society of Chemistry. 1–26.
89. Fernandez-Alba AR, Hernando MD, Piedra L, Chisti Y (2002) Toxicity evaluation of single and mixed antifouling biocides measured with acute toxicity bioassays. Anal Chim Acta 456: 303–312.
90. Moher D, Liberati A, Tetzlaff J, Altman DG (2009) Preferred Reporting Items for Systematic Reviews and Meta-Analyses: The PRISMA Statement. Plos Medicine 6: 1–6.

3

Reference Genes to Study Herbicide Stress Response in *Lolium* sp.: Up-Regulation of P450 Genes in Plants Resistant to Acetolactate-Synthase Inhibitors

Arnaud Duhoux, Christophe Délye*

INRA, UMR1347 Agroécologie, Dijon, France

Abstract

Variation in the expression of numerous genes is at the basis of plant response to environmental stresses. Non-target-site-based resistance to herbicides (NTSR), the major threat to grass weed chemical control, is governed by a subset of the genes involved in herbicide stress response. Quantitative PCR assays allowing reliable comparison of gene expression are thus key to identify genes governing NTSR. This work aimed at identifying a set of reference genes with a stable expression to be used as an internal standard for the normalisation of quantitative PCR data in studies investigating NTSR to herbicides inhibiting acetolactate synthase (ALS) in the major grass weed *Lolium* sp. Gene expression stability was assessed in plants resistant or sensitive to two ALS inhibitors, subjected or not to herbicide stress Using three complementary approaches implemented in the programs BestKeeper, NormFinder and geNorm, cap-binding protein, glyceraldehyde-3-phosphate-dehydrogenase and ubiquitin were identified as the most suitable reference genes. This reference gene set can probably be used to study herbicide response in other weed species. It was used to compare the expression of the genes encoding two herbicide target enzymes (ALS and acetyl-coenzyme A carboxylase) and five cytochromes P450 (CYP) with potential herbicide-degrading activity between plants resistant or sensitive to ALS inhibitors. Overall, herbicide application enhanced *CYP* gene expression. Constitutive up-regulation of all *CYP* genes observed in resistant plants compared to sensitive plants suggested enhanced secondary metabolism in the resistant plants. Comprehensive transcriptome studies associated to gene expression analyses using the reference gene set validated here are required to unravel NTSR genetic determinants.

Editor: Christian Schönbach, Kyushu Institute of Technology, Japan

Funding: This work was supported by funding from Arvalis - Institut du Végétal (http://www.arvalisinstitutduvegetal.fr) to AD. The funder had no role in study design, data collection and analysis, decision to publish, or preparation of the manuscript.

Competing Interests: The authors have declared that no competing interests exist.

* E-mail: delye@dijon.inra.fr

Introduction

Plant response to environmental stresses is mediated by the regulation of gene expression. A major abiotic stress encountered by arable weeds infesting agricultural fields is herbicide applications. Herbicide applications therefore trigger stress response pathways in weed plants [1]. Due to inherent intraspecific genetic variation, these pathways can differ among individual weed plants. In some plants, some of the stress response pathways triggered by herbicide applications can enable plants to survive herbicide applications. These particular pathways are at the basis of non-target-site based resistance (NTSR) to herbicides, an adaptive response [1]. NTSR is the major cause for herbicide resistance in grass weeds, and is thus agronomically and economically very important [1]. As a part of plant stress response pathways, NTSR is under a complex genetic control that is still poorly understood, but involves changes in the regulation of a range of genes in resistant plants compared to sensitive plants. In particular, an increase in glutathione-S-transferase, cytochrome P450 (CYP) or glycosyl-transferase enzyme activities leading to an acceleration of herbicide degradation in herbicide-resistant weed plants has often been observed, but hardly any data is available regarding the genes involved [1]. Yet, identifying NTSR genes is crucial for understanding, diagnosing and managing herbicide resistance. As NTSR seems largely endowed by differences in gene expression between resistant and sensitive plants, identifying NTSR genes requires to reliably be able to quantify differences in gene expression.

Quantitative reverse transcription-polymerase chain reaction (RT-qPCR) is the most accurate tool to date to accurately determine differences in gene expression [2]. For this purpose, it is necessary to normalise qPCR data using a set of reference genes with a constant expression level in the system studied [2,3]. In plants, suitable reference genes have mostly been identified in species with associated genomic resources, such as crop species (e.g. [4]), or model species (*Arabidopsis thaliana* and *Brachypodium distachyon*; e.g. [5,6]). Few sets of reference genes have been validated for weeds, because most weeds are species without associated genomic resources. Extrapolation from other species, even in closely related taxa, can be tricky: several reports have shown that the expression of many widely used reference genes can vary considerably with the experimental conditions, tissues and species [7,8]. As a consequence, no "universal" reference genes have been identified to date.

To date, only one set of reference genes has been validated under herbicide stress in a study considering the grass weed *Alopecurus myosuroides* and herbicides inhibiting acetyl-CoA carboxylase (ACCase) [9]. Here, we considered the two herbicides inhibiting acetolactate-synthase (ALS) that are most broadly used against the grass weed *Lolium sp.* (rye-grass). *Lolium sp.* is one of the world's worst weeds, with a major economic and agronomic negative impact [10]. It is ranked first among the most herbicide-resistance-prone weeds worldwide [10]. ALS is a key enzyme in the biosynthesis of branched amino-acids. ALS inhibitors are the second most broadly used herbicide class worldwide, and are also the herbicide class that is most prone to select for resistance in weeds [10,11]. NTSR to herbicides is frequently observed in *Lolium* sp. [12]. While CYP activity has been shown to play a role in NTSR of *Lolium sp.* to ALS inhibitors [12], none of the genes involved has been identified to date. This is in part due to the absence of relevant reference genes enabling reliable measurement of gene expression levels under herbicide stress in *Lolium sp.*

The first aim of this study was therefore to validate a set of reference genes with a stable expression under ALS-inhibiting herbicide stress in *Lolium sp.* This set was used for a secondary objective: assessing the effect of ALS inhibitor application on the expression of two genes encoding herbicide target proteins and of five genes encoding CYPs with potential herbicide-degrading activity in plants resistant or sensitive to ALS inhibitors, in order to identify candidate NTSR genes.

Materials and Methods

Lolium sp. Populations

Seeds of four distinct *Lolium sp.* populations (RG08-994, RG08-914, RG08-068 and RG07-043) were collected in French fields where control of *Lolium sp.* using ALS inhibitors failed. No specific permits were required for the collection of *Lolium sp.* seeds. *Lolium sp.* is not an endangered or protected species, and can freely be collected in agricultural fields by governmental institutions such as INRA. Seedlings from each population were grown in individual 2L-pots in a glasshouse at 22°C/18°C day/night with 14-hour photoperiod. To ensure resistant plants were resistant because of NTSR, the occurrence of mutations known to endow herbicide resistance within the gene encoding ALS was excluded by genotyping all seedlings at the *ALS* locus as described [13] prior to herbicide application.

When each plant had developed at least a dozen tillers, the individual tillers were separated and transplanted into individual pots to obtain individual one-tiller plants. The one-tiller plants issued from a same plant were clones, i.e., genetically identical plants at the same growth stage (3–4 leaves). This allowed to use a given plant in different experimental modalities.

Plant Material Production for the Validation of a Reference Gene Set

A batch of samples was produced to assess the stability of expression of candidate reference genes. A time-course experiment consisting of six modalities was conducted for each herbicide studied. Modalities were: before treatment (BT), 2 hours after treatment (2HAT), 6 hours after treatment (6HAT), 24 hours after treatment (24HAT), unsprayed control and sprayed control. Two clones were used par plant and per modality, i.e., a total of 12 clones par plant studied. A sample consisted of the above-ground part of the two clones used for a given plant and a given modality that was cut, immediately frozen in liquid nitrogen, and stored at −80°C prior to RNA extraction. The two ALS-inhibiting herbicides most frequently sprayed against *Lolium sp.* were used for treatment: a mixture of two sulfonylureas (iodosulfuron+mesosulfuron; Archipel WG; 3% w/w iodosulfuron and mesosulfuron; Bayer CropScience, applied at the recommended field rate: 250 g formulated herbicide per ha) and a triazolopyrimidine (pyroxsulam; Abak; 7.5% w/w pyroxsulam; Dow Agroscience, applied at the recommended field rate: 250 g formulated herbicide per ha). An adjuvant enhancing the herbicide penetration into the plant tissues was added to the mixture as recommended (Actirob B; Bayer CropScience, applied at the recommended field rate: 1 L per ha). Herbicide application using a custom-built, single-nozzle sprayer was as described [9]. At least two clones from a reference herbicide-sensitive plant were included in each spraying experiment to check herbicide application efficacy.

The sprayed control modality was used to determine the phenotype (resistant or sensitive) of the source plant by observing clone survival four weeks after herbicide application. If the two clones were hardly damaged and extended new, green leaves, the source plant was considered resistant. If the two clones were dead, the source plant was considered sensitive. If the two clones were damaged but extended new, green leaves, the source plant was considered moderately resistant. The samplings produced using (iodosulfuron+mesosulfuron) and pyroxsulam will be referred to as sampling IM1 and sampling P1, respectively (Table 1).

Plant Material Production to Assess the Effect of Herbicide Application on the Expression of Seven Genes of Interest

A second batch of samples was produced to measure the expression of genes of interest in the presence or absence of herbicide. For this purpose, a second time-course experiment was conducted as before for both herbicides studied. The modalities consisted of clones collected before treatment (BT) and 24 hours after treatment (24HAT), unsprayed control and sprayed control. The samplings produced using (iodosulfuron+mesosulfuron) and pyroxsulam in this experiment will be referred to as sampling IM2 and sampling P2, respectively (Table 2).

cDNA Synthesis

Total RNA extraction and DNA contamination removal were performed using the RNeasy plant mini kit and the RNase free DNase set (Qiagen, Courtaboeuf, France) according to the manufacturer's instructions. Nucleic acid concentration was measured at 260 nm using a NanoDrop ND-1000 spectrophotometer (LABTECH, Luton, UK). RNA quality was considered adequate for A_{260}/A_{280} and A_{260}/A_{230} absorption ratio values between 1.8 and 2.2, and 2 and 2.2, respectively. cDNA synthesis was performed from 5 µg total RNA for each sample using the Masterscript RT-PCR System (5 PRIME, Hamburg, Germany). For each RNA sample used, two independent reverse-transcription reactions were performed to obtain technical replicates, as recommended [3].

Primer Design and PCR Conditions

The candidate reference genes tested were reference genes which expression had been proven stable under various stresses in *Lolium* sp. [14,15] or in other grasses [4,16,17]. Primer sequences are given in Table 3. They were designed using the available genes in *Lolium perenne*, *Lolium rigidum* or *Lolium temulentum* deposited in GenBank/EMBL. The web interface Primer3Plus [18] was used to design primers using a primer length of 21±3 nucleotides, a melting temperature (Tm) of 60°C ±3°C, a guanine-cytosine content between 40% and 60% and an expected amplicon size between 90 and 150 base pairs. Primers targeting *TUB* and *GAPDH* were designed to amplify an intron-containing amplicon to enable the detection of gDNA contaminations.

Table 1. RNA samples used to assess the stability of candidate reference genes (samplings IM1 and P1).

Sampling IM1. Herbicide: (iodosulfuron+mesosulfuron)

Herbicide application[1]	**Populations**									
	RG08-994		**RG08-914**		**RG08-068**		**RG07-043**		**Total**	
	S[2]	**R**	**S**	**R**	**S**	**R**	**S**	**R**	**S**	**R**
BT	2	1	1	3	1	1	1	0	5	5
2HAT	2	1	1	3	1	1	1	0	5	5
6HAT	2	2	1	3	1	0	1	0	5	5
24HAT	2	2	1	1	1	2	1	0	5	5
Total	8	6	4	10	4	4	4	0	20	20

Sampling P1. Herbicide: pyroxsulam.

Herbicide application[1]	**Populations**									
	RG08-994		**RG08-914**		**RG08-068**		**RG07-043**		**Total**	
	S[2]	**R**	**S**	**R**	**S**	**R**	**S**	**R**	**S**	**R**
BT	0	1	3	2	0	2	2	0	5	5
2HAT	0	1	3	2	0	1	2	1	5	5
6HAT	2	2	2	2	0	1	1	0	5	5
24HAT	0	1	3	1	0	3	2	0	5	5
Total	2	5	11	7	0	7	7	1	20	20

[1]BT, before treatment; xHAT, x hours after treatment.
[2]S, sensitive; R, resistant.

Specific amplification of each candidate reference gene from cDNA was first tested by PCR followed by electrophoresis on 2% (w/v) agarose gels. Amplicons were sequenced on both strands to confirm the gene targeted was amplified.

qPCR Experiments

Validation of the candidate reference genes complied with the criteria proposed by [3]. qPCR reactions were as described [4], using the StepOnePlus™ Real-Time PCR System Thermal Cycling Block (Applied Biosystems, Foster City, USA). Amplicon sizes were checked on 3% (w/v) agarose gels. The absence of gDNA contamination was checked for every sample by seeking the intron-containing amplicons of TUB and GAPDH. Assessment of qPCR efficiency was performed using five point, five-fold cDNA dilution series as described elsewhere [9]. To assess candidate reference gene stability, all cDNA samples in samplings IM1 and P1 were diluted 125-fold prior to qPCR. Every diluted cDNA sample was used in two independent qPCRs. As two independent reverse transcriptions had been performed per RNA sample, this yielded four technical replicates per RNA sample.

Data were analyzed with the StepOne Software v2.2 (Applied Biosystems, Foster City, USA). For all genes, the threshold for fluorescence detection was manually set at 0.2 ΔRn, i.e., above the background fluorescence and at the beginning of the region of exponential amplification of the amplification curve.

Selection of a Set of Reference Genes with Stable Expression Under Herbicide Action in *Lolium sp.*

cDNA samples obtained from samplings IM1 and P1 (Table 1) were used to assess the stability of the candidate reference genes. qPCR data were analysed separately for each sampling. The reference genes with the most stable expression were therefore identified independently for each herbicide. This enabled to check

Table 2. Number of plants in the samplings used for target gene expression quantification BT and 24HAT (samplings IM2 and P2).

Sampling	**Populations**									
	RG08-068		**RG08-994**			**RG08-914**			**RG07-043**	
	S[2]	**R**	**S**	**R**	**r**	**S**	**R**	**r**	**S**	**R**
IM2[1]	5	5	4	4	1	3	2	1	3	0
P2[1]	0	0	0	0	0	5	2	1	0	0

[1]IM2, sprayed with iodosulfuron+mesosulfuron; P2, sprayed with pyroxsulam.
[2]S, sensitive; R, resistant, r, moderately resistant. All plants were analysed before treatment (BT) and 24 hours after treatment (24HAT).

Table 3. Candidate reference genes tested and primer information.

Code[1] (GenBank accession)	Primer sequence (5′-3′)[2]	Amplicon length (bp)	Tm (°C)	PCR efficiency (%)[3]	Regression coefficient[3]	Cq[4]
TUB	(F) ATACAATGCCACTCTCTCAGTC	140	60	IM1:101	0,996	23.6
(AY742902)	(R) GAGATGAGATGGTTCAAATCAC			P1:101	0.999	
CAP	(F) CTCCAGGGAAGATGCTGAAG	95	57	IM1:96	0.996	25.7
(EU328530)	(R) CTTGAAAGCCCCAATCAAAA			P1:96	0.993	
EF1	(F) CACTGGTCACCTGATCTACAA	100	60	IM1:102	0.990	20.1
(EU168438)	(R) GTACTTGAAGGACCTCTTGTTCA			P1:93	0.999	
GAPDH	(F) AGGTTATCAATGACAAGTTTGG	83	60	IM1:103	0.999	20.8
(EU328533)	(R) ATCAACAGTCTTCTGGGTAGC			P1:104	0.994	
RUB	(F) GGAGTATGAAACCAAGGATACTG	123	57	IM1:97	0.993	16.8
(HM850132)	(R) GTTGTCCATGTACCAGTAGAAGA			P1:99	0.990	
UBQ	(F) CAAGAAGAAGACGTACACCAAG	85	57	IM1:95	0.997	22.2
(EF470423)	(R) GACCTTGTAGAACTGGAGGAG			P1:101	0.990	
18S	(F) GTGACGGAGAATTAGGGTTC	98	57	IM1:98	0.998	18.4
(AY346367)	(R) TGTCAGGATTGGGTAATTTG			P1:93	0.990	
25S	(F) GGATTAACGAGATTCCCACT	115	57	IM1:92	0.996	18.1
(EF463062)	(R) CGGACTAGAGTCAAGCTCAA			P1:94	0.991	

[1]*TUB*, beta tubulin; *CAP*, Capsine phosphatase; *EF1*, Elongation factor 1; *GADPH*, Glyceraldehyde 3-phosphate dehydrogenase; *RUB*, Ribulose-1,5-bisphosphate carboxylase oxygenase; *UBQ*, Ubiquitin; *18S*, ribosomal RNA 18S; *25S*, ribosomal RNA 25S.
[2]F, forward primer; R, reverse primer.
[3]Values obtained using samplings IM1 or P1 (Table 1).
[4]Average Cq value computed over all samples in samplings IM1 and P1.

whether a same set of reference genes was suitable for both ALS inhibitors. For every candidate reference gene, qPCR data used for stability analysis consisted into the mean quantification cycle (Cq) value computed for the two RT replicates of every RNA sample in each of the two samplings analysed. The Cq values were exported for each sample and each gene after positioning the fluorescence detection threshold manually at 0.2 ΔRn.

The stability of the expression of the candidate reference genes expressed as Cq values was assessed using BestKeeper [19], geNorm [20] and NormFinder [21]. Each of these programs provides a particular statistical framework to estimate the variation in the expression of the candidate reference genes. BestKeeper was used as a first approach. It computes the variation in Cq values and its Standard Deviation (SD) for each gene. Genes with SD <1 are considered stable and ranked on the basis of pairwise correlations between their Cq value and the geometric mean of Cq values of all candidate genes with SD <1 (BestKeeper Index). Candidate genes with the strongest correlations with the BestKeeper Index are considered the most stable.

NormFinder and geNorm require the transformation of Cq values using the $2^{-\Delta\Delta C_q}$ method [20,21]. NormFinder allows creating subsamples of the sample set, which in this work corresponded to the experimental modalities. NormFinder ranks the candidate reference genes within and among subsamples after their Stability Value (SV) that is based on the variation of their respective transformed Cq values within and among subsamples. Candidate genes with the lowest SVs are considered the most stable. GeNorm computes all possible pairwise variations between the transformed Cq values of the candidate genes. It then provides an estimate of the expression stability (M) of each reference gene tested. M <1.5 identifies candidate genes with a stable expression, and genes with the lowest M value are considered the most stable.

As a single reference gene rarely allows adequate normalisation, geNorm also computes the optimal number of genes necessary for reliable normalisation based on the pairwise variation between normalisation factors NF_n and NF_{n+1}, where n and n+1 are the number of genes considered. NF_n is estimated from the geometric mean of the transformed Cq values of the n most stable candidate genes.

Effect of ALS-inhibiting Herbicides on the Expression Levels of Two Herbicide Target Genes and Five CYPs

CAP, *GAPDH* and *UBQ* were used to normalise the expression data of the seven genes investigated. The primers used in qPCR and the expected amplicon sizes for each gene are given in Table 4. qPCR data was generated from samplings IM2 and P2 (Table 2). Expression of the seven genes was compared within each population for each herbicide separately. In each population, gene expression was compared in resistant plants *versus* sensitive plants before treatment (BT) to identify constitutive differences in expression, and in resistant plants *versus* sensitive plants 24 hours after treatment (24HAT), in resistant plants BT *versus* 24HAT, and in sensitive plants BT *versus* 24HAT to identify genes differentially regulated in resistant plants by herbicide application. qPCR results were analysed using REST-MCS version 2 [22] that compares the expression level of a target gene between two groups of samples, taking into account the respective amplification efficiencies of the target gene and of reference genes. Gene expression data consisted into the average Cq values computed from two replicates (two independent qPCR runs, each performed from one of two independent reverse transcription reactions). Gene expression comparison was performed for the seven genes investigated using relative expression ratios computed for each gene and each plant after equation 1 in [23], using three herbicide-sensitive plants in

Table 4. Target genes investigated, primer and qPCR information.

Original organism[1] (accession)	Target gene in *Lolium* sp.[2] Code (accession, % identity)	Primer sequence (5′-3′)[3]	Amplicon length (bp)	Tm (°C)	PCR efficiency (%) (regression coefficient)
Lolium multiflorum	*ALS*	(F) GCAATCAAGAAGATGCTTGAGAC	132	60	93
(AF310684)	(AF310684, 100%)	(R) TCCTGCCATCACCTTCCATGAT			(0.983)
Lolium rigidum	*ACCase*	(F) CACAAGACACAGCTAGATAGTGGCG	174	60	87
(AF359513)	(AF359513, 100%)	(R) TTCCAACAGTTCGTCCAGTCACAAAT			(0.994)
Lolium rigidum	*CYP71R4*	(F) GAACATGATGTACCATTTCGACTG	101	60	99
(AB097496)	(AB097496, 100%)	(R) CAGACTTAAGGCCAGGCGACAAC			(0.997)
Zea mays	*CYP72A*	(F) CAGTGATGACTTGCTAGGATTG	173	63	97
(AF465265)	(AF321869, 81%)	(R) CATGCTGAGCAGAATTAGTGTC			(0.987)
Helianthus tuberosus	*CYP81B1*	(F) GTCTGTTCATGATACCGTTCG	146	63	94
(AJ000477)	(AB159037, 45%)	(R) CTCCGTCATGTCCACCTTC			(0.992)
Oryza sativa	CYP81A	(F) AGGGGAGACGGGATGCTGG	179	60	101
(ABC69856)	(HF565053, 78%)	(R) TTGGGCATGGTGATCCCTGG			(0.989)
Homo sapiens	*CYP92A*	(F) CTCGATGTTAAGGGGCAGGATTA	176	60	91
(NM_000771)	(HF565054, 26%)	(R) GATCTCCTCCATGTTCAGCTCC			(0.988)

[1]Organism from which the gene of interest had originally been isolated.
[2]*Lolium* sp. genes investigated in this work. *ALS*, Acetolactate synthase; *ACCase*, Acetyl-coenzyme A carboxylase; *CYP*: cytochrome P450. Accession numbers are given for the *Lolium CYP* genes available in GenBank/EMBL that display the highest amino-acid identity with the *CYP* gene of interest. % identity, % of amino-acid identity between the *CYP* gene of interest and the *Lolium CYP* gene studied.
[3]F, forward primer; R, reverse primer.

population RG07-063 collected BT as the reference (control) sample.

The genes investigated consisted of *ACCase*, *ALS* and five *CYP* genes. *ACCase* and *ALS* encode herbicide target enzymes. *ACCase* encodes acetyl-coenzyme a carboxylase, a crucial enzyme in lipid biosynthesis and an herbicide target enzyme [24]. *ACCase* is not the target of the herbicides used herein, and is therefore expected to be stably expressed BT in our experiments. *ALS* encodes the target of the herbicides used herein, and was included in our experiments to investigate whether *ALS* over-expression could play a role in resistance in the populations we analysed.

To identify genes potentially involved in NTSR to ALS inhibitors, we considered six genes encoding CYPs known to have an herbicide-degrading activity. *CYP71R4* had been identified in herbicide-resistant *L. rigidum*. *CYP71R4* had been shown to catalyse the degradation of the herbicide chlorotoluron that inhibits photosynthesis [25]. *CYP81B1* had been identified in Jerusalem artichoke (*Helianthus tuberosus*). *CYP81B1* had also been shown to catalyse the degradation of the herbicide chlorotoluron [26]. *CYP72A* and *CYP81A6* had been identified in maize and rice, respectively. *CYP72A* and *CYP81A6* had been shown to catalyse the degradation of several pesticides, including herbicides inhibiting ALS [26,27]. *CYP2C9* is a human gene that had been introduced into transgenic rice plants where it had been shown to catalyse the degradation of chlorsulfuron and imazosulfuron, two herbicides inhibiting ALS [28].

Except *CYP71R4*, all *CYPs* of interest had been identified in plant species different from *Lolium* sp. For each *CYP* gene of interest, primers (Table 4) were designed using the sequence of the *Lolium sp. CYP* gene available in GenBank/EMBL that displayed the highest identity in amino-acids with the *CYP* gene considered. The *Lolium* sp. *CYP* gene in GenBank/EMBL most similar to *CYP81B1* had also been reported as *CYP81B1*. The *Lolium* sp. *CYP* genes most similar to *CYP72A*, *CYP81A6* and *CYP2C9* had not been assigned to a *CYP* family. We used a *CYP*-specific Blast tool [29] to identify the family to which each gene could be assigned. To sum up, in this study, *Lolium* sp. genes corresponding to *CYP72A*, *CYP81A6*, *CYP81B1* and *CYP2C9* will be referred to as *CYP72A*, *CYP81A*, *CYP81B1* and *CYP92A*, respectively (Table 4). Amplicons were sequenced on both strands to confirm the amplification of the targeted *Lolium* sp. *CYP* gene.

Results

Specificity and Efficiency of Amplification of Candidate Reference Genes

Samplings IM1 and P1 (Table 1) were used to identify reference genes with a stable expression under herbicide action. Each sampling comprised 40 RNA samples (Table 1), and enabled to assess the effect of one herbicide. As two RT reactions were performed per RNA sample, a total of 160 cDNA samples were analysed.

Eight genes were tested as candidate reference genes (Table 3). Agarose gel electrophoresis and the single-peak melting curves obtained for every gene confirmed the absence of primer dimers and of non-specific amplicons (Fig. S1). The absence of gDNA contamination was confirmed by the single-peak melting curves obtained with primers targeting *TUB* and *GAPDH*, which flank an intron-containing region. Specific amplification of the targeted amplicons was confirmed by amplicon sequencing (Table S1).

The non-template controls yielded Cq values that were always above 32 cycles for all primer pairs tested. The corresponding melting curves showed a single small peak that was always located before the position of the single amplicon peak that was obtained in all template samples. As no amplicon peak was detected in the non-template controls, the positive Cq values obtained for non-template controls were considered to be due to primer dimer formation, and were therefore ignored.

qPCR efficiency together with the corresponding correlation coefficients computed from the five-fold dilution series are given in Table 4. Amplification efficiency values obtained for the eight candidate genes were comprised between 90% and 110% and were therefore considered acceptable and comparable [30].

Stability of Candidate Reference Genes

For each of the two samplings, expression data was subdivided in 10 subsets according to the phenotype (2 subsets), the time after herbicide application (4 subsets) and the *Lolium sp.* population of origin (4 subsets) (Table 1).

a) BestKeeper analyses. Gene stability was assessed in each sampling separately using all samples as a single set where all possible combinations of the tested effects (phenotype, time after herbicide application and population) were present. In sampling IM1, all candidate genes were found suitable for normalisation (SD <1 or very close to 1, Table 5). The three most stable genes were by ranking order *GAPDH*, *CAP* and *UBQ*. In sampling P1, five genes were found suitable for normalisation (SD <1 or very close to 1, Table 5). The three most stable genes were by ranking order *GAPDH*, *CAP* and *UBQ*. *18S* and *25S* were the least stable genes in both samplings.

b) NormFinder analyses. This algorithm was first used on each sampling without subset assignment. The most stable genes were by ranking order *GAPDH*, *CAP* and *UBQ* in sampling IM1, and *CAP*, *UBQ* and *GAPDH* in sampling P1 (Table 5). To assess the phenotype effect, every sample in each sampling was assigned to one of the two subsets "sensitive" or "resistant". The genes most stable between resistant and sensitive plants were by ranking order *GAPDH*, *CAP* and *UBQ* in sampling IM1, and *CAP*, *UBQ* and *GAPDH* in sampling P1 (Table 5).

To assess the effect of time after herbicide application, every sample in each sampling was assigned to one of the four subsets "BT", "2HAT", "6HAT" and "24HAT". The genes most stable under herbicide effect were by ranking order *CAP*, *GAPDH* and *UBQ* in sampling IM1, and *CAP*, *UBQ* and *RUB* in sampling P1 (Table 5). *GAPDH* was ranked the fourth most stable gene under herbicide action in sampling P1.

To assess the population effect, every sample in each sampling was assigned to the corresponding population subset. The genes most stable among populations were by ranking order *GAPDH*, *CAP* and *UBQ* in sampling IM1, and *GAPDH*, *CAP* and *EF1* in sampling P1 (Table 5). *UBQ* was ranked fourth in sampling P1.

Overall, *CAP*, *GAPDH* and *UBQ* were identified as the most stable genes.

c) geNorm analyses. Analysis showed that all candidate genes were suitable for normalisation in sampling IM1 (M<1.5). In sampling P1, all genes except *18S* were found suitable for normalisation. *CAP* was identified as the most stable gene in both samplings, with *GAPDH* and *UBQ* being the next two most stable genes. *18S* and *25S* were the least stable genes in both samplings (Fig. 1A).

The optimal number of reference genes required for accurate normalisation was computed in each sampling. *18S* was excluded from the analyses conducted in sampling P1. All values obtained for the pairwise variation between consecutive normalisation factors (Vi/j values, Fig. 1B) were higher than the proposed 0.15 cutoff threshold [20]. As recommended in this case [20], we considered the change in the Vi/j values when including additional reference genes, and we used the lowest Vi/j value to determine the number of reference genes adequate for normalisation. For both samplings, the inclusion of the third most stably expressed gene yielded the lowest variation of the normalisation factor (V2/3; Fig. 1B). Using the three most stable reference genes (*CAP*, *GAPDH* and *UBQ*) as reference genes is considered to be a valid normalisation strategy in this case [20].

Variation in the Expression of Two Genes Encoding Herbicide Targets and of Five *CYP* Genes

Samplings IM2 and P2 (Table 2) were analysed using *CAP*, *GAPDH* and *UBQ* for normalisation. Relative expression ratio values ranged from 0.011 (*CYP81A* BT in a sensitive plant in population RG08-994, Fig. 2G) to 56.0 (*CYP81B1* 24HAT in a resistant plant in population RG08-994, Fig. 2D). Overall, considerable variation in the relative expression ratio values was observed among genes, but also among populations, among individual plants in a given population, and among experimental modalities for a given plant (Fig. 2). For instance, *CYP72A* was significantly up-regulated (7.6-fold) in the resistant plants in population RG08-068 compared to the resistant plants in population RG08-994. *CYP72A* was also significantly up-regulated (6.0-fold) in the sensitive plants in population RG08-068 compared to the sensitive plants in population RG08-994 (Figure 2C). Within a given population, substantial differences in relative expression ratios could be observed between plants with a same phenotype (resistant or sensitive). For instance, differences in the relative expression ratios of *CYP71R4* among resistant plants in population RG08-068 could be up to 100-fold (Fig. 2E). As a consequence, no gene showed differences in expression enabling to discriminate all the resistant plants from all the sensitive plants.

No significant differences in the expression of *ALS* were observed between resistant and sensitive plants for all conditions tested (Fig. 2A, Fig. S2A). The highest differences were observed BT in population RG08-068, where the average *ALS* expression in resistant plants was two-fold higher than in sensitive plants. No significant differences in the expression of *ACCase* were observed between resistant and sensitive plants for all conditions tested, except in population RG08-068 in sampling IM2 (Fig. 2B, Fig. S2B). In this population, *ACCase* was found to be significantly up-regulated overall (2.2-fold) in resistant plants compared to sensitive plants 24HAT. This was because *ACCase* expression did not significantly vary 24HAT compared to BT in the resistant plants, while it was significantly down-regulated 24HAT in four of the five sensitive plants in this population (2.4-fold compared to BT) (Fig. 2B, Fig. S2B).

CYP71R4 was significantly up-regulated BT in the resistant plants in population RG08-068 only (3.3-fold compared to the sensitive plants in this population) (Fig. 2C, Fig. S2C). *CYP71R4* was significantly up-regulated 24HAT in the sensitive plants in population RG08-994 in sampling IM2 and in the resistant plants in population RG08-914 in sampling IM2 (6.4 and 5.8-fold, respectively, compared to the same plants BT).

CYP72A was significantly up-regulated BT in resistant plants overall and in populations RG08-068, RG08-994 and RG08-914 in sampling IM2 (8.3, 6.0 and 6.2-fold up-regulated compared to the sensitive plants in the corresponding populations, respectively) (Fig. 2D, Fig. S2D). At 24HAT, no significant up-regulation of *CYP72A* was detected in the resistant plants compared to the sensitive plants in any population. This was due to *CYP72A* being more strongly up-regulated 24HAT compared to BT in the sensitive plants than in the resistant plants in all populations in both samplings (e.g., 15-fold up-regulation on average in the resistant plants in population RG08-994 in sampling IM2 *versus* above 50-fold in the sensitive plants in this population).

CYP81B1 was significantly up-regulated BT in the resistant plants overall and in each population in sampling IM2 (Fig. 2E, Fig. S2E). It was 8.3, 3.9 and 4.7-fold up-regulated in the resistant plants compared to the sensitive plants in populations RG08-068,

Table 5. Ranking of the eight candidate reference genes according to the stability of their expression in samplings IM1 (top) and P1 (bottom) computed using BestKeeper and NormFinder.

Sampling IM1[1]

Candidate genes[2]	BestKeeper			NormFinder							
				All samples		Phenotype effect		Herbicide effect		Population effect	
						Sensitive/ Resistant		Before /2HAT, 6HAT, 24HAT			
	SD[3]	r[4]	rank	SV[5]	rank	SV	rank	SV	rank	SV	rank
TUB	1.039	0.796*	6	0.723	5	0.199	5	0.463	4	0.237	5
UBQ	**0.870**	**0.863***	**3**	**0.504**	**3**	**0.151**	**3**	**0.309**	**3**	**0.153**	**3**
GAPDH	**0.565**	**0.857***	**1**	**0.274**	**1**	**0.103**	**1**	**0.277**	**2**	**0.095**	**1**
EF1	0.935	0.626*	5	0.672	4	0.181	4	0.484	7	0.211	4
RUB	0.769	0.310	4	0.889	7	0.269	8	0.463	4	0.263	6
18S	1.085	0.439*	8	0.974	8	0.261	7	0.477	6	0.316	8
25S	1.030	0.498*	7	0.864	6	0.247	6	0.502	8	0.277	7
CAP	**0.628**	**0.901***	**2**	**0.324**	**2**	**0.137**	**2**	**0.221**	**1**	**0.103**	**2**

Sampling P1[1]

Candidate genes[2]	BestKeeper			NormFinder							
				All samples		Phenotype effect		Herbicide effect		Population effect	
						Sensitive/ Resistant		Before /2HAT, 6HAT, 24HAT			
	SD[3]	r[4]	rank	SV[5]	rank	SV	rank	SV	rank	SV	rank
TUB	1.053	0.356*	5	0.971	7	0.242	6	0.620	8	0.620	7
UBQ	**0.628**	**0.679***	**3**	**0.58**	**2**	**0.161**	**2**	**0.385**	**2**	**0.426**	**4**
GAPDH	**0.579**	**0.869***	**1**	**0.594**	**3**	**0.178**	**3**	**0.455**	**4**	**0.380**	**1**
EF1	1.143	0.411*	6	0.724	4	0.192	4	0.561	6	0.388	3
RUB	1.025	0.651*	4	0.784	5	0.239	5	0.381	3	0.521	5
18S	1.873	0.556*	8	1.273	8	0.314	8	0.602	7	0.684	8
25S	1.407	0.387*	7	0.941	6	0.241	6	0.541	5	0.554	6
CAP	**0.620**	**0.877***	**2**	**0.512**	**1**	**0.143**	**1**	**0.337**	**1**	**0.381**	**2**

[1]Sampling composition given in Table 1.
[2]The three reference genes selected are in bold. *TUB*, beta tubulin; *CAP*, Capsine phosphatase; *EF1*, Elongation factor 1; *GADPH*, Glyceraldehyde 3-phosphate dehydrogenase; *RUB*, Ribulose-1,5-bisphosphate carboxylase oxygenase; *UBQ*, Ubiquitin; *18S*, ribosomal RNA 18S; *25S*, ribosomal RNA 25S.
[3]Standard deviation of Cq values. SD values higher than the threshold value (1.00) are underlined.
[4]Pearson's coefficient of correlation.
[5]Stability value.
*associated p-value <0.001.

RG08-994 and RG08-914 in sampling IM2, respectively. At 24HAT, *CYP81B1* was only significantly up-regulated (4.5-fold) in the resistant plants compared to the sensitive plants in population RG08-994 in sampling IM2. Overall, *CYP81B1* was significantly up-regulated 24HAT in all the plants in populations RG08-994 and RG08-914 and in the sensitive plants in population RG08-068 in sampling IM2, and in the sensitive plants in population RG08-914 in sampling P2 (from 5.5-fold in the sensitive plants in population RG08-994 in sampling P2 up to 15-fold in both the resistant and the sensitive plants in population RG08-994 in sampling IM2, compared to the same plants BT).

CYP81A was significantly up-regulated in the resistant plants in populations RG08-068 and RG08-994 in sampling IM2 BT (3.5 and 4.8-fold, respectively, compared to the sensitive plants) and 24HAT (3.9 and 10.5-fold, respectively, compared to the sensitive plants) (Fig. 2F, Fig. S2F).

CYP92A was significantly up-regulated BT and 24HAT only in the resistant plants in population RG08-068 in sampling IM2 (11.9 and 6.1-fold compared to the sensitive plants in this population BT and 24HAT, respectively) (Fig. 2G, Fig. S2G). *CYP92A* was also significantly up-regulated BT in the resistant plants in population RG08-914 in sampling P2 (3.6-fold compared to the sensitive plants in this population).

All other differences in expression observed among experimental modalities were not significant. In particular, in the three moderately resistant plants analysed, the relative expression ratios of all the genes investigated were always in range of values observed for sensitive plants (Fig. 2).

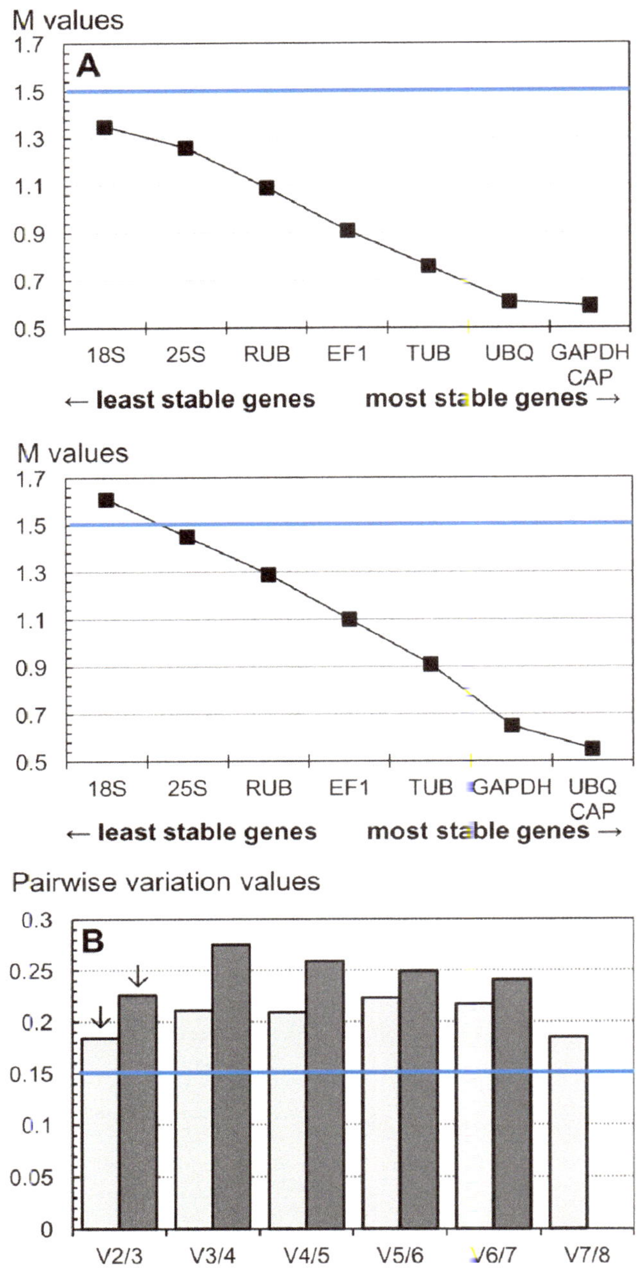

Figure 1. geNorm ranking of the eight candidate reference genes according to their average expression stability value M (A) and determination of the optimal number of reference genes for accurate normalisation (B). A, ranking was performed for all RNA samples in samplings IM1 (top) and P1 (bottom). M-values of the remaining genes at each step during stepwise exclusion of the least stable gene are shown. The genes are ranked according to increasing expression stability (i.e., decreasing M-value). B, pairwise variation analysis to determine the optimal number of reference genes for accurate normalisation in samplings IM1 (open boxes) and P1 (solid boxes). The pairwise variation between consecutive normalisation factors (Vi/i+1) indicating the optimal number of reference genes is arrowed.

Spearman's rank correlation tests performed on the relative expression ratios of each gene in all resistant *versus* all sensitive plants BT and 24HAT showed a significant correlation between the relative expression ratios of *CYP72A* and *CYP81B1* BT ($\rho = 0.83$; $p < 0.001$). All other tests were non-significant. When considering the expression data for all five *CYP* genes in each plant individually, it appeared that the plants resistant to (iodosulfuron+mesosulfuron, sampling IM2) generally displayed high relative expression ratio values for most, or all, genes. This was particularly true for two resistant plants in population RG08-068 (Fig. 2C to G). Sensitive plants could also display high relative expression ratio values, but, for a given plant, this was only observed for one or two of the five *CYP* genes. For instance, one sensitive plant in population RG08-068 in sampling IM2 displayed high relative expression ratio values for *CYP72A* (5th and 4th highest expression ratio in population RG08-068 BT and 24HAT, respectively) and *CYP81B*1 (4th and 3rd highest expression ratio in this population BT and 24HAT, respectively), but low relative expression ratio values for the other three genes (8th to 10th highest expression ratios in this population) (Fig. 2C to G). Regarding the two pyroxsulam-resistant plants investigated (sampling P2), one displayed high relative expression ration values for all *CYP* genes except for *CYP81A*. The other displayed relative expression ratio values for all five *CYP* genes that were much lower, and similar to those observed for most sensitive plants (Fig. 2C to G).

Discussion

Validation of a Set of Reference Genes Suitable for Gene Expression Studies Under the Action of Herbicides Inhibiting ALS

Eight candidate reference genes were tested using two very heterogeneous samplings (Table 1). As no single method is generally accepted to identify the most stably expressed genes, we used three different algorithms in our analyses. All three identified *CAP*, *GAPDH* and *UBQ* as the genes most stably expressed overall among populations, between phenotypes and under herbicide action for the two herbicides studied. *18S* and *25S* were the genes least stably expressed overall (Fig. 1A and Table 5). The optimal number of reference genes required for adequate expression data normalisation computed by geNorm was three (Fig. 1B). We therefore selected *CAP*, *GAPDH* and *UBQ* as the set of reference genes suitable for normalisation of qPCR data in *Lolium sp.* under herbicide inhibiting ALS stress.

The only other study reporting validated reference genes for the normalisation of qPCR data obtained from plants under herbicide stress considered the grass weed *Alopecurus myosuroides Huds.* and one herbicide inhibiting ACCase. *TUB*, *GAPDH* and *UBQ* were identified as the most stable genes [9]. Our results obtained in a different grass species using two herbicides with a different target enzyme confirmed the stability of expression of *GAPDH* and *UBQ* under herbicide stress. *TUB* was ranked among the three genes most stable under herbicide action in the previous study [9]. Here, *TUB* was ranked among the least stable genes by BestKeeper and NormFinder, and was ranked the fourth most stable gene by geNorm only (Fig. 1A, Table 5). This may be due to the different modes of action of the herbicides used in both studies, which could have a different impact on plant gene regulation. Alternatively, as plants contain several different genes encoding β-tubulin (e.g., [31]), it may be that the *TUB* genes used in each study are not orthologs, and thus do not have the same regulation. From both studies, it appears that *GAPDH* and *UBQ* could be suitable reference genes for the normalisation of qPCR data obtained from grasses under herbicide stress. *CAP* was not tested in the previous study. It was consistently ranked among the two most stable genes in our study. Thus, *CAP*, *GAPDH* and *UBQ* could be suitable

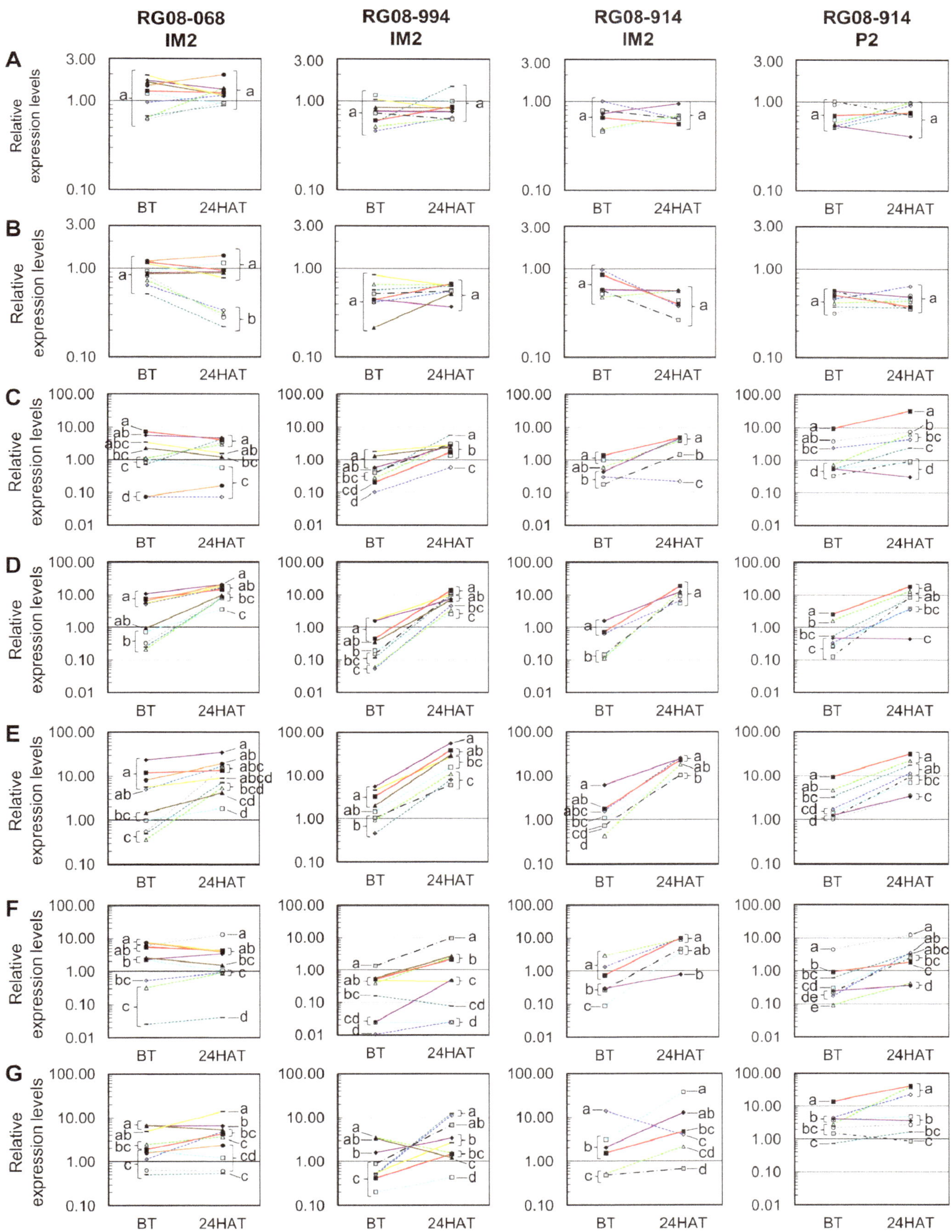
RG08-068
IM2
RG08-994
IM2
RG08-914
IM2
RG08-914
P2
A
B
C
D
E
F
G
Relative expression levels
BT
24HAT

Figure 2. Expression of *ALS, ACCase* and five *CYP* genes in individual *Lolium* sp. plants in four groups of samples. A group of samples consists of the plants from the same population that have been used to assess the effect of the same herbicide in sampling IM2 or P2 (Table 2). For every individual plant, the relative expression levels of each gene measured in the absence of herbicide (BT) and 24 hours after herbicide application (24HAT) are connected by a solid line for herbicide-resistant plants, a dotted line for herbicide-sensitive plants or a dashed-and-dotted line for plants with a moderately resistant phenotype. A given color in all panels in a column (i.e., in a given group of samples) indicates the same individual plant. Panels on lanes A to G, relative expression levels for *ALS, ACCase, CYP71R4, CYP72A, CYP81B1, CYP81A* and *CYP92A*, respectively. The herbicide considered is indicated below the population code using the sampling code (IM2, iodosulfuron+mesosulfuron; P2, pyroxsulam, see Table 2). Identical letters in a given panel and in a given experimental modality (BT or 24HAT) indicate relative expression levels that are not significantly different.

reference genes for the normalisation of qPCR data obtained from grasses under herbicide stress.

Lolium sp. not being a model species, the reference genes we identified were obtained using the few candidate genes that could be tested based on their stable expression under stress in grasses [4,14–17]. With the rise of next-generation sequencing, genomic resources are now becoming available for a growing number of plant species, including *Lolium sp.* [32]. This will greatly facilitate the identification of genes that are orthologous between closely related model and non-model species. Thus, obtaining a broad range of candidate reference genes for non-model species based on the stability data available for model species will be possible in the near future. In the case of *Lolium sp.*, this could result in the identification of yet more stable reference genes for stress response studies.

Variation in the Expression of *ACCase*, *ALS* and Five CYP-encoding Genes

The second aim of this study was to investigate potential transcriptional changes of two genes encoding herbicide target enzymes and of five genes encoding CYPs with potential herbicide-degrading activity in herbicide-resistant and sensitive plants in response to the application of herbicides inhibiting ALS.

ACCase encodes an enzyme not targeted by the herbicides studied here. As expected, *ACCase* expression did not vary between resistant and sensitive plants or under herbicide action, except in four sensitive plants in one population where *ACCase* was down-regulated after herbicide application. This could be attributed to a redirection of the metabolism of these plants towards the compensation of the herbicide effects [1]. Observing this effect only in one population clearly illustrates the variety of stress responses among populations from a same species.

ALS encodes the target enzyme of the herbicides used in this study. Although the resistant plants in population RG08-068 tended to show an up-regulation of *ALS* expression (Fig. 2A), *ALS* expression did not significantly vary between resistant and sensitive plants, or under herbicide action. This is consistent with the literature, where over-expression of the gene encoding the herbicide target had very rarely been reported, and seems an uncommon resistance mechanism in plants [12].

CYPs belong to plant secondary metabolism, and the regulation of their expression is involved in plant response to stresses [33]. Subsets of the *CYP* genes present in a plant genome are involved in different stress response pathways, including herbicide stress response [1]. Among the *CYP* genes involved in the overall response of plants to herbicide stress, a subset can be involved in NTSR. A high expression of the genes in this subset is expected to enhance herbicide degradation in resistant plants compared to sensitive plants [1,33]. The expression of *CYP* genes involved in NTSR to herbicides can therefore be expected to be up-regulated in resistant plants compared to sensitive plants, either BT and after herbicide application (constitutive up-regulation), or only after herbicide application (herbicide-induced up-regulation) [1,25,27,28]. Thus, comparing the expression of *CYP* genes potentially catalysing herbicide degradation between herbicide-resistant and sensitive plants could allow identification of CYPs involved in NTSR. The expression patterns of each of the five *CYP* genes studied herein varied broadly from plant to plant. Constitutive and herbicide-induced up-regulation was observed both in resistant and in sensitive plants for every gene. No gene displayed expression patterns that were specific to resistant plants. Thus, our data provide no direct proof of the role of any of the five *CYP* genes in NTSR to the two herbicides inhibiting ALS studied here. Two explanations can be proposed.

The first, obvious, explanation is that none of the *CYP* genes investigated is involved in NTSR in the plants studied. Among the five *CYP* genes, only *CYP71R4* had originally been isolated in *Lolium* sp. Yet, its activity on the herbicides investigated here, if any, had not been assessed [25]. The other four *CYP* genes considered had originally been isolated in species different from *Lolium* sp. [26,27,28]. The *Lolium* sp. genes studied here are the genes available in GenBank/EMBL that displayed the highest homology with the genes of interest. However, as *Lolium* sp. are not model species, only 50 annotated *CYP* genes have been identified in this genus. In plants, hundreds of *CYP* genes are present that belong to multigenic families, and substrate specificity varies greatly among CYP isoforms [33,34]. It may be that the genes studied here were not the true homologues of the genes of interest, or that the genes of interest have no homologues with an activity on the herbicides studied in *Lolium* sp. This underlines the need to screen the whole CYP complement (CYPome) of *Lolium* sp. to identify *CYPs* genes involved in NTSR.

A second explanation can be suggested. NTSR had been proposed to evolve by the accumulation of different genes, each conferring a moderate level of resistance. When a combination of genes conferring a high enough resistance level is present in a plant, this plant shows a resistant phenotype [1]. The few studies that investigated the genetic control of NTSR demonstrated that NTSR was governed by a set of genes present in resistant plants. This set varied among plants, even in a given population [35,36]. The variability in the *CYP* gene expression patterns we observed among plants is consistent with these results. It may thus be that, in the plants studied, some of the *CYP* genes investigated could be involved in NTSR, but none was expressed at a level sufficient to confer NTSR alone (i.e., a combination of several up-regulated genes may be necessary to confer a resistant phenotype). Under this hypothesis, observing similar, high relative expression ratios for a given *CYP* gene in resistant and sensitive plants does not enable to exclude that the CYP in question is not involved in NTSR: additional gene(s) not investigated here can also be present and up-regulated in the resistant plants, but not in the sensitive plants.

Overall, the five *CYP* genes studied were all constitutively up-regulated in resistant plants compared to sensitive plants. Our results thus suggest that at least some secondary metabolism pathways are enhanced in *Lolium* sp. plants resistant to ALS inhibitors compared to sensitive plants, as previously shown in *Arabidopsis thaliana* [37]. Our results are also consistent with a previous study which demonstrated constitutive or herbicide-induced enhanced secondary metabolism and stress oxidative response in *A. myosuroides* plants showing NTSR to herbicides compared to herbicide-sensitive plants [38]. Thus, our study is a further confirmation that NTSR mechanisms belong to a subset of the mechanisms involved in the plant response to herbicide stress [1].

Conclusions

Using a validated reference gene set to assess the possible role of five *CYP* genes in NTSR to herbicides inhibiting ALS, we illustrated the complexity of investigating the genetic bases of NTSR. NTSR is a quantitative trait, and resistant phenotypes clearly seem to be endowed by several genes. Thus, approaches such as the one conducted herein based on heterologous candidate genes have little chance to be fruitful. As proposed elsewhere [1], an approach enabling to compare the full transcriptome of resistant and sensitive plants appears to be the most straightforward way towards the identification of candidate NTSR genes. In such approaches, the set of reference genes validated herein will be key for the measurement of differences in the expression of candidate NTSR genes identified using transcriptome-scanning methods. Beyond NTSR, our reference gene set can also be used to investigate the genetic bases of the overall response of *Lolium* sp. to herbicide stress. It is very likely that this gene set can also be used in herbicide studies addressing other grass weeds, subject to preliminary validation.

Supporting Information

Figure S1 Primer specificity test. Melting curves generated for *TUB* (**A**), *CAP* (**B**), *EF1* (**C**), *GAPDH* (**D**), *RUB* (**E**), *UBQ* (**F**), *18S* (**G**), *25S* (**H**), *ALS* (**I**), *ACCase* (**J**), *CYP71R4* (**K**), *CYP72A* (**L**), *CYP81B1* (**M**), *CYP81A* (**N**), *CYP92A* (**O**).

Figure S2 Expression of *ALS*, *ACCase* and five *CYP* genes according to the phenotypes of *Lolium* sp. plants in four groups of samples. A group of samples consists of the plants from the same population that have been used to assess the effect of the same herbicide in sampling IM2 or P2 (Table 2). Red, orange and yellow boxes indicate the plants from populations RG08-068, RG08-994 and RG08-914 in sampling IM2, respectively. Blue boxes indicate the plants from population RG08-914 in sampling P2. Panels **A** to **G**, relative expression levels for *ALS*, *ACCase*, *CYP71R4*, *CYP72A*, *CYP81B1*, *CYP81A* and *CYP92A*, respectively. The relative expression levels of each gene are computed for each group of samples in the resistant plants in the absence of herbicide (BT) using the sensitive plants in the same group as the reference condition (R/S BT), in the resistant plants 24 hours after herbicide application (24HAT) using the sensitive plants in the same group as the reference condition (R/S 24HAT), in the sensitive plants 24HAT using the same plants BT as the reference condition (S BT/24HAT), and in the resistant plants 24HAT using the same plants BT as the reference condition (R BT/24HAT).

Table S1 Sequences of the amplicons obtained in RT-qPCR. Primers binding sites are underlined.

Acknowledgments

The authors are grateful to Séverine Michel (INRA Dijon) for skilful assistance in the experiments.

Author Contributions

Conceived and designed the experiments: AD CD. Performed the experiments: AD. Analyzed the data: AD CD. Contributed reagents/materials/analysis tools: AD CD. Wrote the paper: AD CD.

References

1. Délye C (2013) Unravelling the genetic bases of non-target-site-based resistance (NTSR) to herbicides: a major challenge for weed science in the forthcoming decade. Pest Manag Sci, 69: 176–187.
2. Bustin SA, Benes V, Garson JA, Hellemans J, Huggett J, et al. (2009) The MIQE guidelines: minimum information for publication of quantitative realtime PCR experiments. Clin Chem, 55: 611–622.
3. Bustin SA, Beaulieu JF, Huggett J, Jaggi R, Kibenge FSB, et al. (2010) MIQE precis: Practical implementation of minimum standard guidelines for fluorescence-based quantitative real-time PCR experiments. BMC Mol Biol 11: 74.
4. Jarošová J, Kundu JK (2010) Validation of reference genes as internal control for studying viral infections in cereals by quantitative real-time RT-PCR. BMC Plant Biol 10: 146.
5. Czechowski T, Stitt M, Altmann T, Udvardi MK, Scheible WR (2005) Genome-wide identification and testing of superior reference genes for transcript normalization in *Arabidopsis thaliana*. Plant Physiol 139: 5–17.
6. Hong S, Seo PJ, Yang MS, Xiang F, Park CM (2008) Exploring valid reference genes for gene expression studies in *Brachypodium distachyon* by realtime PCR. BMC Plant Biol 8: 112.
7. Thellin O, Zorzi W, Lakaye B, De Borman B, Coumans B, et al. (1999) Housekeeping genes as internal standards: use and limits. J Biotechnol 75: 291–295.
8. Stürzenbaum SR, Kille P (2001) Control genes in quantitative molecular biological techniques: the variability of invariance. Comp Biochem Physiol B Biochem Mol Biol 130: 281–289.
9. Petit C, Pernin F, Heydel JM, Délye C (2012) Validation of a set of reference genes to study response to herbicide stress in grasses. BMC Res Notes 5: 18.
10. Heap I (2013) The international survey of herbicide resistant weeds. [http://www.weedscience.org/In.asp].
11. Tranel PJ, Wright TR (2002) Resistance of weeds to ALS-inhibiting herbicides: what have we learned? Weed Sci 50(6): 700–712.
12. Powles S, Yu Q (2010) Evolution in action: Plants resistant to herbicides. Annu Rev Plant Biol 61: 317–347.
13. Délye C, Boucansaud K, Pernin F, Le Corre V (2009) Variation in the gene encoding acetolactate-synthase in *Lolium* species and proactive detection of mutant, herbicide-resistant alleles. Weed Res 49(3): 326–336.
14. Lee J, Roche JR, Donaghy DJ, Thrush A, Sathish P (2010) Validation of reference genes for quantitative RT-PCR studies of gene expression in perennial ryegrass (*Lolium perenne* L.). BMC Mol Biol 11: 8.
15. Dombrowski JE, Martin RC (2009) Evaluation of reference genes for quantitative RT-PCR in *Lolium temulentum* under abiotic stress. Plant Sci 176: 390–396.
16. Silveira ED, Alves-Ferreira M, Guimarães LA, Rodrigues da Silva F, Tavares de Campos Carneiro V (2009) Selection of reference genes for quantitative real-time PCR expression studies in the apomictic and sexual grass *Brachiaria brizantha*. BMC Plant Biol 9: 84.
17. Ovesná J, Kučera L, Vaculová K, Štrymplová K, Svobodová I, et al. (2012) Validation of the *β-amy1* transcription profiling assay and selection of reference genes suited for a RT-qPCR assay in developing barley caryopsis. PLoS ONE 7: e41886.
18. Primer3Plus. [http://www.bioinformatics.nl/cgi-bin/primer3plus/primer3plus.cgi].
19. Pfaffl MW, Tichopad A, Prgomet C, Neuvians TP (2004) Determination of stable housekeeping genes, differentially regulated target genes and sample integrity: BestKeeper-Excel-based tool using pair-wise correlations. Biotechnol Lett 26: 509–515.
20. Vandesompele J, De Preter K, Pattyn F, Poppe B, Van Roy N, et al. (2002) Accurate normalization of real-time quantitative RT-PCR data by geometric averaging of multiple internal control genes. Genome Biol 3: research0034.1–0s034.11.
21. Andersen L, Jensen JL, Ørntoft TF (2004) Normalization of real-time quantitative reverse transcription-PCR data: a model-based variance estimation approach to identify genes suited for normalization applied to bladder and colon cancer data sets. Cancer Res 64: 5245–5250.
22. Pfaffl MW, Horgan GW, Dempfle L (2002) Relative expression software tool (REST©) for group-wise comparison and statistical analysis of relative expression results in real-time PCR. Nucleic Acids Res 30(9): e36.
23. Pfaffl MW (2001 A new mathematical model relative quantification in real-time RT-PCR. Nucleic Acids Res 29 (9): e45.
24. Délye C (2005) Weed resistance to acetyl coenzyme A carboxylase inhibitors: an update. Weed Sci 53: 728–746.

25. Tsujii H, Dillon N, Ohkawa H (2004) Molecular functions of cytochrome P450 species involved in herbicide resistance in *Lolium rigidum* biotypeWLR2. *Int. Symp. Cytochrome P450: Biodiversity and Biotechnology, 7th, Hyogo, Japan*, p. 69 (Abstr.).
26. Werck-Reichhart D, Hehn A, Didierjean L (2000) Cytochromes P450 for engineering herbicide tolerance. Trends Plant Sci 5: 116–123.
27. Lei Z, Qing L, Honggao C, Gang P, Shushu X, et al. (2007) Identification of a cytochrome P450 hydroxylase, CYP81A6, as the candidate for the bentazon and sulfonylurea herbicide resistance gene, Bel, in rice. Mol Breed 68: 19–59.
28. Inui H, Shiota N, Ido Y, Inoue T, Hirose S, et al. (2001) Herbicide Metabolism and Tolerance in the Transgenic Rice Plants Expressing Human CYP2C9 and CYP2C19. Pestic Biochem Physiol 71: 156–169.
29. P450 Blast Server. [http://blast.uthsc.edu/].
30. Livak KJ, Schmittgen TD (2001) Analysis of Relative Gene Expression Data Using Real-Time Quantitative PCR and the $2^{-\Delta\Delta CT}$ Method. Methods 25: 402–408.
31. Snustad DP, Haas NA, Kopczak SD, Silflow CD (1992) The small genome of *Arabidopsis* contains at least nine expressed b-tubulin genes. Plant Cell 4: 549–556.
32. Pfeifer M, Martis M, Asp T, Mayer KFX, Lübberstedt T, et al. (2013) The perennial ryegrass GenomeZipper: targeted use of genome resources for comparative grass genomics. Plant Physiol 161: 571–582.
33. Werk-Reichart D, Feyereisen R (2000) Cytochromes P450: a success story. Genome Biol 1(6): 3003.1–3003.9.
34. Nelson D, Werck-Reichart D (2011) A P450-centric view of plant evolution. Plant J 66: 194–211.
35. Petit C, Duhieu B, Boucansaud K, Délye C (2010) Complex genetic control of non-target-site-based resistance to herbicides inhibiting acetyl-coenzyme A carboxylase and acetolactate-synthase in *Alopecurus myosuroides* Huds. Plant Sci 178: 501–509.
36. Busi R, Vila-Aiub MM, Powles SB (2010) Genetic control of P450 metabolism-based herbicide resistance mechanism in *Lolium rigidum*. Heredity 106(5): 817–24.
37. Glombitza S, Dubuis P-H, Thulke O, Welzl G, Bovet L, et al. (2004) Crosstalk and differential response to abiotic and biotic stressors reflected at the transcriptional level of effector genes from secondary metabolism. Plant Mol Biol 54: 817–835.
38. Cummins I, Bryant DN, Edwards R (2009) Safener responsiveness and multiple herbicide resistance in the weed black-grass (*Alopecurus myosuroides*). Plant Biotechnol J 7: 807–820.

4

Role of a Novel I1781T Mutation and Other Mechanisms in Conferring Resistance to Acetyl-CoA Carboxylase Inhibiting Herbicides in a Black-Grass Population

Shiv Shankhar Kaundun*, Sarah-Jane Hutchings, Richard P. Dale, Eddie McIndoe

Syngenta, Jealott's Hill International Research Centre, Biological Sciences, Bracknell, United Kingom

Abstract

Background: Knowledge of the mechanisms of herbicide resistance is important for designing long term sustainable weed management strategies. Here, we have used an integrated biology and molecular approach to investigate the mechanisms of resistance to acetyl-CoA carboxylase inhibiting herbicides in a UK black-grass population (BG2).

Methodology/Principal Findings: Comparison between BG2 phenotypes using single discriminant rates of herbicides and genotypes based on ACCase gene sequencing showed that the I1781L, a novel I1781T, but not the W2027C mutations, were associated with resistance to cycloxydim. All plants were killed with clethodim and a few individuals containing the I1781L mutation were partially resistant to tepraloxydim. Whole plant dose response assays demonstrated that a single copy of the mutant T1781 allele conferred fourfold resistance levels to cycloxydim and clodinafop-propargyl. In contrast, the impact of the I1781T mutation was low (Rf = 1.6) and non-significant on pinoxaden. BG2 was also characterised by high levels of resistance, very likely non-target site based, to the two cereal selective herbicides clodinafop-propargyl and pinoxaden and not to the poorly metabolisable cyclohexanedione herbicides. Analysis of 480 plants from 40 cycloxydim resistant black grass populations from the UK using two very effective and high throughput dCAPS assays established for detecting any amino acid changes at the 1781 ACCase codon and for positively identifying the threonine residue, showed that the occurrence of the T1781 is extremely rare compared to the L1781 allele.

Conclusion/Significance: This study revealed a novel mutation at ACCase codon position 1781 and adequately assessed target site and non-target site mechanisms in conferring resistance to several ACCase herbicides in a black-grass population. It highlights that over time the level of suspected non-target site resistance to some cereal selective ACCase herbicides have in some instances surpassed that of target site resistance, including the one endowed by the most commonly encountered I1781L mutation.

Editor: Tianzhen Zhang, Nanjing Agricultural University, China

Funding: The authors have no support or funding to report.

Competing Interests: All authors are employed by Syngenta, Jealott's Hill International Research Centre. There are no patents, products in development or marketed products to declare.

* E-mail: deepak.kaundun@syngenta.com

Introduction

Black-grass (*Alopecurus myosuroides* Huds.) is one of the most problematic weeds in North Western Europe [1]. It is a particularly challenging pest in autumn sown crops as most seeds germinate during this period [2]. Once established it is difficult to eliminate as it sheds seeds before the harvest, thus constantly replenishing the seed bank [3]. Though relatively less competitive than species such as wild oats, it can occur at high densities causing important losses in crop yield. On average 12 black grass plants per m^2 can cause a yield reduction of 5% in winter cereals in the UK [4].

The most preferred method of black-grass control in arable crops is via the use of herbicides such as inhibitors of acetolactate synthase and acetyl-CoA carboxylase (ACCase). Of interest here, the ACCase herbicides are classified into aryloxyphenoxypropionates (FOPs), cyclohexanediones (DIMs) and phenylpyrazolin (DEN) based on their chemical structures [5,6]. In all cases they bind to the carboxyltransferase domain of chloroplastic ACCase of most grass weeds with little to no action on the cytoplasmic isoform [7]. In so doing they inhibit the formation of malonyl-CoA, depleting the cells of important fatty acids, leading to rapid necrosis and plant death [8,9]. Their selective action on grass weeds is attributable to the dissimilar structures of plastidic ACCase in the different plant groups. In Poaceae, acetyl-CoA carboxylase is homomeric with the biotin carboxyl carrier protein (BCCP), biotin carboxylase (BC) and carboxyl transferase α and β subdomains located on a single polypeptide. In contrast, the subunits are encoded by four different genes that are co-ordinately expressed to form a functional enzyme in other plant species [10].

Over time, the extensive use of ACCase herbicides has resulted in the evolution of resistance in the major grass weeds worldwide [11]. The first case of resistance to ACCase herbicides in black-grass was reported as early as 1982 in the UK. To date, resistance is widespread and affects most arable farms infested with black-grass [12]. Black-grass resistance to ACCase herbicides is also relatively common in France and is a growing problem in Germany, Belgium, Turkey, the Netherlands and Italy [13,14].

Two different mechanisms account for resistance to ACCase herbicides. Resistance can be due to subtle amino acid changes in the ACCase, resulting in reduced sensitivities to the toxophores. In black-grass the I1781L, W2027C, I2041N, D2078G and G2096A mutations have been reported to confer varying levels of resistance to ACCase inhibiting herbicides [15,16]. Recently, a comprehensive survey across its geographical range has shown that non-target site resistance to ACCase herbicides is more widespread than target site mechanisms [13]. The genetic basis of non-target site resistance appears to be very complex and governed by up to seven different loci [17]. Two sets of proteins involved in xenobiotic metabolism are associated with non-target site resistance to ACCase herbicides, namely cytochrome p450 and gluthathione *S*-transferase (*GST*) enzymes [18]. In addition to their herbicide degrading properties, *GST* enzymes from some multiple herbicide resistant black-grass populations have been shown to exhibit peroxidase activities, protecting the plants from the noxious effects of reactive oxygen radicals following herbicide treatment [19]. The complexity associated with *GST* based metabolism is further exacerbated as the expression of the detoxifying enzymes varies with plant growth stages and environmental conditions [20].

In the UK, whilst black-grass resistance to extensively used cereal selective ACCase herbicides is widespread, the poorly metabolisable DIM herbicides which are often employed in rotational dicotyledonous crops are less affected and estimated to occur in around half of the populations resistant to FOP herbicides [21]. Here we have investigated the molecular basis of cycloxydim resistance in a black-grass population from the UK and determined its cross resistance to a number of commonly used ACCase herbicides, paying particular attention to estimating the level of resistance conferred by the different target site mutations and non-target site resistance mechanisms contained in this population. Additionally, we have developed two robust DNA markers following the identification of a novel mutation at ACCase codon position 1781.

Materials and Methods

Materials

Seeds from the suspected resistant black-grass population BG2 were sampled from a wheat field located in Buckinghamshire, UK. Over the years the field was treated with a range of ACCase herbicides applied in small grain cereal and rotational dicotyledonous crops. A standard sensitive population (STD1), used for comparison in all biological and molecular studies, was sourced from Herbiseed (Twyford, UK).

Additionally, 40 cycloxydim resistant black-grass populations collected from across the UK between 2002 and 2009 were analysed to estimate the relative frequency of a novel mutation identified at ACCase codon position 1781.

"No specific permissions were required for the location where the ryegrass seeds were collected. This study did not involve any endangered or protected species".

Methods

Confirmation of resistance to a DIM, FOP and DEN herbicides. Around 50 seeds each of BG2 and STD1 were sown and grown at the 2-leaf stage in 1 litre pot containing peat and compost in 1:1 ratio. The pots were maintained and fertilised as necessary in controlled glasshouse conditions at 24°C/16 hr day, 18°C night, 65% relative humidity, and a photon flux density of approximately 250 µmol quanta m^{-2} s^{-1}. Three replicate pots each of suspected resistant (BG2) and sensitive (STD1) populations were sprayed with the recommended field rates of cycloxydim (200 g ai/ha), clodinafop-propargyl (60 g ai/ha) and pinoxaden (60 g ai/ha) (a.i.: active ingredient).

Selection of cycloxydim sensitive and resistant plants and ACCase gene sequencing. The BG2 and STD1 plants were sown and grown as described above. At the one-leaf stage 48 BG2 and 16 STD1 plants were individually transplanted in single pots and grown to the two-leaf stage. The plants were sprayed with the recommended field rate of cycloxydim and survival was recorded 21 days after treatment. Prior to herbicide application, a 1-cm leaf tip segment was harvested from each of the 48 BG2 and 16 STD1 plants for sequencing a DNA fragment containing the carboxyl-transferase domain of *ACCase*. The RT-PCR methodology and primers were the same as reported in Kaundun (2010) [22]. All 64 plants were assembled into sensitive and resistant subgroups and their DNA sequences were examined with the Seqman software (Lasergene, USA) in view of identifying the ACCase mutation(s) responsible for resistance to cycloxydim in BG2.

To formally analyse the influence of the threonine and leucine mutations at ACCase codon 1781 on the efficacy of cycloxydim, only plants that were wild at codon 2027 were considered. For each mutation, heterozygous and homozygous plants were grouped. Survival data from the co-segregation experiment was arranged as a 2×2 contingency table for each of the II1781 (homozygous wild type) vs IT1781+TT1781 (heterozygous and homozygous threonine mutants) and II1781 (homozygous wild type) vs IL1781+LL1781 (heterozygous and homozygous leucine mutants) combinations and analysed using Fisher's Exact test. A p-value of less than 0.05 indicates a statistically significant result at the 5% probability level.

Cross resistance to two other DIM herbicides. In order to investigate the cross resistance to two other DIM herbicides commonly used for black-grass control, 48 randomly selected and individually potted BG2 plants each were treated with the recommended field rates of tepraloxydim (100 g ai/ha) and clethodim (120 g ai/ha). For comparison, 16 plants from the standard sensitive population STD1 were treated with the two DIM herbicides. Survival to the herbicides was assessed 21 days after treatment. Prior to herbicide treatment a 1-cm leaf segment was sampled from each plant for DNA analysis at the 1781 and 2027 ACCase codon positions for phenotype to genotype correlation. The resulting data was statistically analysed with the Fisher Exact test in the same way as described for cycloxydim.

Level of resistance conferred by mutations at ACCase codon position 1781 and non-target site mechanisms. Sixteen plants with pre-determined wild and mutant genotypes were used to assess the resistance factors associated with non-target site and target site resistance due to the threonine and leucine mutations at ACCase codon position 1781. All plants were wild type at ACCase codon position 2027. The plants consisted of four each of homozygous wild type BG2-II1781, heterozygous mutant BG2-IT1781 and heterozygous mutant BG2-IL1781 from population BG2, and four homozygous wild type STD1-II1781 genotypes from the standard sensitive population STD1. These 16 mother plants were tiller propagated for a period of one year to produce identical tillers for use in the whole plant dose response assays. Tillering was aided by fertilising and trimming the plants regularly. Two genetically identical tillers of each mother plant were sprayed with the following rates of cycloxydim: 1.56, 3.125, 6.125, 12.5, 25, 50, 100, 200, 400, 800 g ai/ha; clodinafop-propargyl: 1.875, 3.75, 7.5, 15, 30, 60, 120, 240, 480 and 960 g ai/ha; pinoxaden: 1.875, 3.75, 7.5, 15, 30, 60, 120, 240, 480 and 960 g ai/ha.

The pots were arranged in a randomized complete block design after herbicide application (blocking per mother plant). Twenty-one days after treatment the tillers were cut above ground level, put into paper envelopes and dried in an oven at 70°C for three days. The dry mass for each individual was recorded to the nearest mg.

Since the mother plants represent the independent biological replicates in the experiment, the dry weight measurements were averaged across technical replicates (tillers) separately for each compound, rate and mother plant, and divided by the average dry weight of the untreated tillers from the corresponding mother plant. GR_{50} estimates were derived from the resulting percent of untreated values separately for each mother plant by a non-linear least-squares regression model described by the equation:

$$P = \frac{100}{1+e^{-\beta(x-\mu_{ij})}}$$

where x denotes log(Rate); μ_{ij} denotes the $\log GR_{50}$ for mother plant i of genotype j and β denotes the slope of the rate-response.

The estimated resistance factor (Rf) between a pair of genotypes was calculated as the ratio of the geometric means of the estimated GR_{50} values across mother plants. Confidence intervals around the estimated resistance factors were calculated using the error term derived from an analysis of variance carried out on the $\log GR_{50}$ values. All statistical analyses were carried out using SAS software, version 9.2.

Establishment of two dCAPS assays for the identification of mutations at ACCase codon position 1781. The dCAPS (derived Cleaved Amplified Polymorphic Sequence) method operates in a 3-step procedure consisting of PCR, restriction digest and gel electrophoresis [23]. It allows selective digestion and differentiation between wild and mutant DNA by the forced introduction of nucleotide change(s) in the dCAPS primer. The dCAPS primers and corresponding restriction enzymes were selected using the freeware developed by Neff et al. (2002) [24]. The first assay consisted of detecting any known mutations at ACCase codon 1781 using the enzyme *Nsi*I whilst the second assay employed the enzyme *Nde*I for the positive identification of a threonine at this position. The recognition sites for *Nsi*I and *Nde*I are ATGCAT and CATATG respectively. The reverse dCAPS primers used in the *Nsi*I and *Nde*I assays were AGAATA**C**G-CACTGGCAATAGCAGCACTTCCATG<u>C</u> and CCCTAGAA-TAGGCACTGGCAATAGCAGCACTTC<u>C</u>AT<u>A</u>T respectively. The forced mutations on the primers are underlined. Both assays utilised the same forward primer: ATGACTGACGAAGACCAT-GATCG. With the first assay, changes in any one of the first two bases of the 1781 codon would disrupt the *Nsi*I restriction site (ATGCAT) and result in an uncut DNA fragment on the agarose gel. In contrast, a restriction site (CATATG) is present for the enzyme *Nde*I when the isoleucine allele is changed into threonine at codon position 1781. Consequently, the presence of a positive threonine allele is manifested as a restricted band on the agarose gel. It is noteworthy that a second forced mutation at position N+31 (in bold) with respect to the first base of the 1781 codon was included on the *Nsi*I primer to eliminate a second non-specific site for this enzyme in black-grass plants. The 48 BG2 and 16 STD1 plants previously sequenced for the entire ACCase CT domain were used to develop the two dCAPS assays. The procedure for the dCAPS assays was as follows: 1 cm leaf segment from each plant was ground on a Spex Certiprep (Metuchen, NJ, USA) 2000 model Genogrinder; DNA was extracted from the ground material on a Beckman Coulter Biomek FX robot (High Wycombe, Buckinghamshire, UK) using a Wizard Magnetic 96 DNA Plant System kit (Promega, Madison, WI, USA). Polymerase chain reactions were carried out using PuReTaq Ready-To-Go PCR beads (Amersham Biosciences, Bucks, UK) in a total volume of 25 µL containing 0.8 µM of each primer and from 10–50 ng genomic DNA and run on an Eppendorf Master Cycle Gradient Thermocycler Model 96 programmed for an initial denaturation step of 94°C of 2 min followed by 40 cycles of 30 s at 94°C, 30 s at 64°C and 1 min at 72°C. A final extension step for 10 min at 72°C was also included. For the positive identification of the wild type isoleucine residue, 8 µl of the PCR product was then digested with 1 µL (5 units) of *Nsi*I at 37°C for 2 hours in a 40 µL reaction. Ten µL of the *Nsi*I treated PCR product was then migrated on a 1×TBE 2% agarose gel containing 0.5 µg mL^{-1} ethidium bromide. The same procedure was employed for the positive identification of the mutant threonine allele, except that *Nsi*I was replaced by the enzyme *Nde*I.

Frequency of the T1781 allele in a range of cycloxydim resistant black-grass populations. Forty UK black-grass populations from our cycloxydim resistant collection were screened with 200 g ai/ha cycloxydim in a whole plant assay. Around 50 seedlings were tested per population. Twenty one days after treatment, twelve survivors per population were assessed with the established dCAPS markers described above in order to estimate the relative frequency of T1781 allele among black-grass plants mutated at the corresponding codon. The 480 cycloxydim survivors in all were assayed in a two-step procedure. All plants were screened with the *Nsi*I based dCAPS marker method. Upon non-digestion of the PCR fragment with *Nsi*I, and thus indication of a mutated allele at codon position 1781, the plants were further tested with the *Nde*I based T1781 specific assay.

Results

Resistance Profile of Black-grass Population BG2

All BG2 plants survived the pinoxaden and clodinafop-propargyl treatments while 50±5.8% biomass reduction was achieved with cycloxydim with distinct live and dead plants identified in the pots. As expected, the three ACCase herbicides fully controlled the standard sensitive population. Given that cycloxydim is a poorly metabolisable herbicide, an insensitive target rather than enhanced herbicide detoxification was suspected as the likely mechanism of resistance to this DIM herbicide.

Selection of Sensitive and Cycloxydim RESISTANT Plants and Identification of Target Site Resistance in BG2

The BG2 population treated with a single rate of cycloxydim at 200 g ai/ha segregated into 21 dead and 27 live plants while all the sensitive STD1 were killed. Sequencing of a PCR fragment encompassing the whole CT domain revealed gene sequences showing over 97% homology with published black-grass ACCase sequences. The sequences were particularly conserved among the 48 BG2 and 16 STD1 plants with only seven nucleotide changes recorded at six ACCase codon positions. Four base changes at codon positions 1801, 1833, 1923 and 1988 were silent. The non-synonymous changes consisted of the commonly reported I1781L and W2027C resistance mutations and a novel I1781T mutation. The I1781L mutation involved a transversion of an adenine to thymine change on the first base of the codon while the I1781T mutation resulted from a transition of a thymine to a cytosine at the second base of the codon. In all, eight different combined 1781/2027 genotypes could be distinguished among the BG2 plants (Table 1). The most common genotypes consisted of wild type sequences for both the 1781 and 2027 codon positions followed by heterozygous mutant IT1781 and IL1781 plants. As expected with highly outcrossing species, compound heterozygotes LT1781 genotypes and plants containing one each of the L1781 or T1781 and C2027 mutant alleles were also identified. It is noteworthy that only a few plants were homozygous for any mutations, thus indicating that the population was very probably at a relatively early stage of resistance selection with ACCase herbicides. Importantly, when the plant genotypes and phenotypes were compared, there was clear evidence that either the T1781 or L1781 alleles were associated with cycloxydim resistance ($P<0.001$) irrespective of their amino acid status at ACCase codon position 2027. When present alone, a single copy of the mutant cytosine 2027 allele did not confer resistance to cycloxydim.

Cross Resistance with Two other DIM Herbicides

ACCase gene analysis of the two additional sets of 48 BG2 plants, each treated with tepraloxydim and clethodim revealed genotypic frequencies comparable to the one observed with the cycloxydim treated plants (Table 1) but with two additional combined IL1781/WC2027 and II1781/CC2027 genotypes identified. All 48 plants sprayed with clethodim were controlled regardless of their genotypes at codon positions 1781 and 2027. Similarly, all BG2 individuals with either the T1781 or C2027 alleles were killed with tepraloxydim either alone or in combination with another mutation. One out of 12 IL1781-WW2027 and one out of two LL1781-WW2027 plants survived the tepraloxydim treatment. Formal statistical analysis with Fisher Exact test did not identify a significant difference between wild type and mutant L1781 plants for tepraloxydim ($P = 0.493$).

Level of Resistance Conferred to a FOP a DIM and DEN Herbicides

As all BG2 plants were found to be resistant to pinoxaden and clodinafop-propargyl in the initial resistance confirmation tests, dose response assays were conducted on pre-determined wild and mutant genotypes at ACCase codon position 1781 in order to assess the relative importance of target site versus non-target site resistance to these two herbicides as well as cycloxydim. Dose responses could not be obtained or fully exploited for the IL1781 genotypes as the leucine allele conferred very high levels of resistance to all three herbicides (Table 2). This was particularly the case for cycloxydim with a low average biomass reduction of 15% for leucine mutant plants compared to untreated controls at the highest rate tested (Figure 1). Similarly, clodinafop-propargyl and pinoxaden provided a maximum of 35% biomass reduction at the highest rate of 960 g ai/ha (Figure 2 and 3). Comparison of wild II1781 and IT1781 genotypes from population BG2 yielded estimated resistance factors of 4.7 and 4.0 for clodinafop-propargyl and cycloxydim respectively but only 1.6 for pinoxaden (Table 3). As the confidence intervals around this latter estimate contained the value 1, it can be concluded that a single copy of the T1781 allele does not confer resistance to the DEN herbicide. There was no evidence of a shift in the dose response between the wild type BG2-II1781 and STD1-II1781 genotypes for cycloxydim confirming the former co-segregation study which showed that the DIM herbicide is not affected by non-target site resistance in BG2. Significantly, very high levels of non-target site resistance were identified for pinoxaden (estimated $Rf = 28$) and clodinafop-propargyl (estimated $Rf = 18$) as determined by comparing the wild type BG2-II1781 and STD1-II1781 genotypes for these two herbicides (Table 3). It is noteworthy that the effects of target site

Table 1. Correlation between cycloxydim phenotypes and combined genotypes at ACCase codon positions 1781 and 2027.

Combined 1781 and 2027 ACCase genotypes	cycloxydim 200 g ai/ha number of plants			clethodim 120 g ai/ha number of plants			tepraloxydim 100 g ai/ha number of plants		
	Total	live	Dead	Total	live	Dead	Total	live	Dead
II1781/WW2027	16	0	16	11	0	11	10	0	10
IL1781/WW2027	11	11	0	10	0	10	12	1	11
IT1781/WW2027	9	9	0	6	0	6	5	0	5
TT1781/WW2027	2	2	0	1	0	1	0	0	0
LL1781/WW2027	1	1	0	0	0	0	2	1	1
LT1781/WW2027	2	2	0	1	0	1	3	0	0
II1781/WC2027	5	0	5	12	0	12	7	0	7
IT1781/WC2027	2	2	0	2	0	2	3	0	3
IL1781/WC2027	0	0	0	5	0	5	4	0	4
II1781/CC2027	0	0	0	0	0	0	2	0	2

and non-target site mechanisms were additive in negatively affecting these two cereal selective ACCase herbicides. This is illustrated by a resistance factor of 45 for pinoxaden (comparison of wild type STD1-II1781 and mutant BG2-IT1781 genotypes), which is equivalent to the product of the individual target site (Rf = 1.6) and non-target site resistance (Rf = 27.9) contributions.

dCAPS Assays for Identifying the I1781 and Specifically the T1781 Allele

Two dCAPS assays were developed to address the discovery of a novel mutation at ACCase codon position 1781. One of the assays using the enzyme *Nsi*I consisted of detecting any mutation at codon position 1781 and a second *Nde*I assay was employed to specifically identify a threonine residue at this position. Typical dCAPS profiles are provided on figures 4a and 4b. PCR generated fragments of 165 bp and 169 bp for the *Nsi*I and *Nde*I based assays respectively. Upon restriction with the *Nsi*I enzyme, a shorter band of 134 bp was generated for plants containing a wild type isoleucine allele or the AT dinucleotide at the first and second bases of the 1781 codon. Non-digested and undistinguishable fragments were present for either the L1781 or T1781 alleles characterised by the TT and AC dinucleotides respectively. Likewise, with the *Nde*I assay a restricted and shorter band of 132 bp was present for plants that specifically contained the T1781 allele, whilst either the I1781 wild type or L1781 mutant alleles were manifested as non-digested PCR fragments. When used in combination, the two *Nsi*I and *Nde*I assays allowed unambiguous discrimination of all types of wild and mutant 1781 alleles in BG2. Overall, the results from the dCAPS assay was totally corroborated with sequencing data available for 48 BG2 plants and 16 STD1 used in the cycloxydim co-segregation study, thus testifying the robustness of the two dCAPS assays developed here.

Frequency of the T1781 Mutation in a Wide Range of Cycloxydim Resistant Black-grass Populations from the UK

All the 40 black-grass populations screened showed some levels of cycloxydim resistance ranging from 0% to 75% biomass reduction relative to the untreated control. Testing of 12 survivors per population with the *Nsi*I assay identified at least three 1781 mutant plants per population. Overall the genotypic frequencies were 0.06, 0.66 and 0.28 for homozygous wild type, heterozygous mutant and homozygous mutant types at ACCase position 1781 among the cycloxydim resistant plants. Analysis of the mutant plants with the threonine specific assays showed that only two plants contained the T1781 allele at the heterozygous state. This amounted to a very low T1781 allelic frequency of 0.005 among the 1781 ACCase mutant plants. That an L1781 allele was predominantly present in the samples that were not digested with the *Nsi*I assay was further confirmed via sequencing analysis of a few populations.

Discussion

Black-grass is a growing threat to arable crop production in the UK. Field cultivation, delayed sowing, higher seeding rate and competitive crop varieties are reactively being practiced due to the evolution of resistance to the most effective single site post-emergence herbicides [25]. Once resistance has evolved, however, long term field experiments have shown that, non-chemical weed control measures, though important for managing the seed bank, will contribute very little to the overall percentage of resistant individuals in the population [26]. Therefore, it is imperative to better understand the basis of herbicide resistance evolution in order to make the best use of existing products and other cultural methods for sustainable black-grass control.

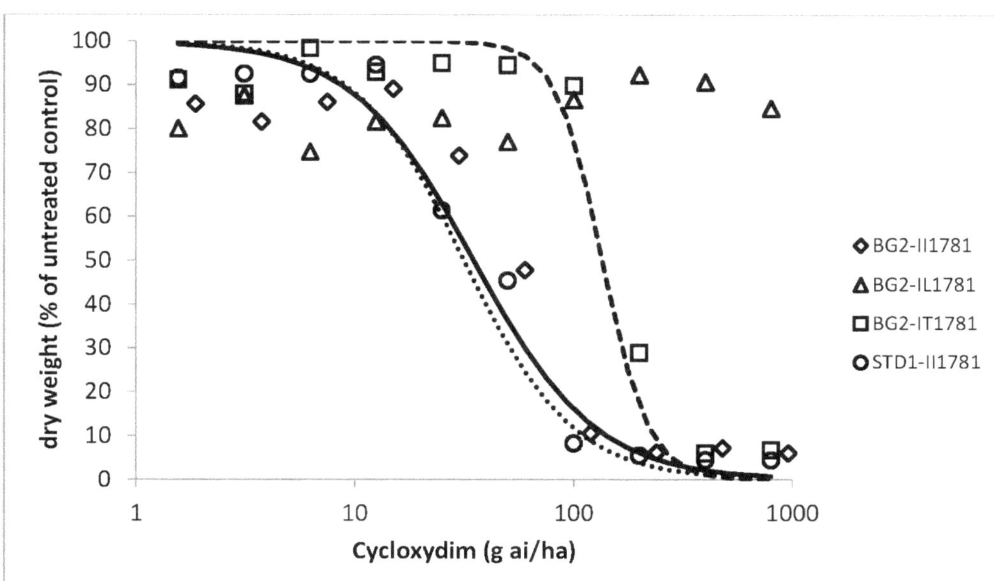

Figure 1. Cycloxydim whole plant dose response tests on four plant groups. Homozygous wild type STD1-II1781 from the standard sensitive population; homozygous wild type BG2-II1781; heterozygous mutant type BG2-IT1781 and heterozygous mutant type BG2-IL1781 from population BG2.

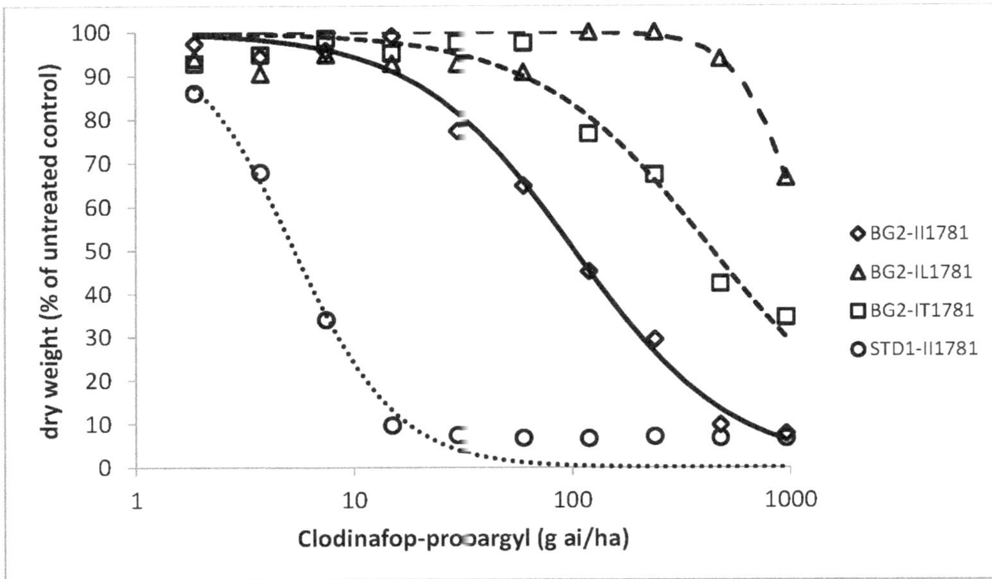

Figure 2. Clodinafop-propargyl whole plant dose response tests on four plant groups. Homozygous wild type STD1-II1781 from the standard sensitive population; homozygous wild type BG2-II1781; heterozygous mutant type BG2-IT1781 and heterozygous mutant type BG2-IL1781 from population BG2.

Target Site Resistance Mutations in BG2

Sequencing experiments showed that the ACCase carboxyl-transferase domain is very much conserved in black-grass. This agrees with previous sequence data from three French populations and in which only four non-synonymous ACCase mutations were identified [16]. Of these, three were found to be associated with ACCase target site resistance. The relative conservation of ACCase in black-grass contrasts sharply with *Lolium* species in which as many as seven non-synonymous mutations were found among 12 plants sequenced from a single population [22] and of these only one resulted in resistance to ACCase inhibiting herbicides. Similar observations were made in other ryegrass populations from the UK [27] and Australia [28]. Thus, black-grass appears to be a very suitable species for discovering new ACCase target site resistance mutations in grass weeds. In this respect, four out of the seven ACCase codon positions reported to be involved in an insensitive target were initially identified in

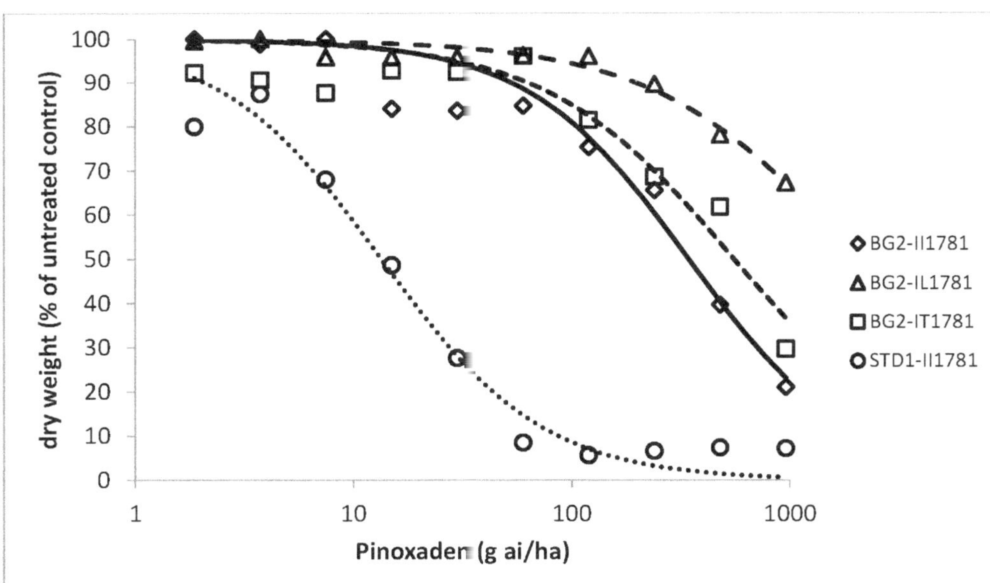

Figure 3. Pinoxaden whole plant dose response tests on four plant groups. Homozygous wild type STD1-II1781 from the standard sensitive population; homozygous wild type BG2-II1781; heterozygous mutant type BG2-IT1781 and heterozygous mutant type BG2-IL1781 from population BG2.

Table 2. Estimated GR50 values for three ACCase herbicides and different genotypes at ACCase codon position 1781.

	STD1-II1781	BG2-II1781	BG2-IT1781	BG2-IL1781
Cycloxydim	39.3 (24.6–62.8)	40.5 (25.3–64.8)	163 (102–261)	>800
Clodinafop-propargyl	5.15 (1.72–15.4)	91.5 (30.6–273.4)	433 (145–1292)	>959
Pinoxaden	11.0 (4.4–27.5)	307 (123–767)	500 (200–1248)	>1091

95% confidence limits in brackets.

black-grass before being found to endow resistance in other grass weed species [29,30,31].

Three target site mutations in all were identified in population BG2. These included the W2027C mutation which was detected in *Alopecurus myosuroides* [16] *Avena sterilis* [32], *Lolium* [28], *Phalaris minor* [33] and *Sorghum bicolor* [34]. In black-grass the W2027C mutation was found to confer resistance to the FOP herbicides clodinafop-propargyl, haloxyfop-P-methyl and fenoxaprop-P-ethyl but not to the DIM herbicides cycloxydim and clethodim [35]. Here we observed that the W2027C mutation did not confer resistance to tepraloxydim either. The influence of the W2027C mutation on clodinafop-propargyl and pinoxaden could not be investigated due to the small number of plants that contained this mutation and also potential confounding non-target site resistance present in BG2. Crossing of several heterozygous BG2-WC2027 plants to produce homozygous wild type WW2027 and CC2027 genotypes and subsequent whole plant dose response assays will allow precise estimation of the level of resistance conferred by the W2027C mutation on these two cereal selective ACCase herbicides.

Twenty five percent of BG2 individuals contained at least one copy of the mutant L1781 allele. The role of the I1781L mutation in conferring resistance to a range of ACCase herbicides is amply demonstrated via yeast gene replacement [36], enzyme [37] and molecular [38] assays since the first identification in a *Setaria viridis* population [39], and subsequently in four other major grass weeds including *Lolium* [36], *Avena* [40], *Alopecurus* [41] and *Echinochloa* [42] species. Given the conservation of the ACCase gene among grass species and the absence of a fitness penalty [43], it would not be surprising to find that the I1781L mutation is also involved in the additional 30 or so weed species that have evolved resistance to ACCase herbicides [11].

Table 3. Estimated resistance factors for three ACCase herbicides and four different 1781 ACCase plant groups.

Genotypes	Cycloxydim	Clodinafop-propargyl	Pinoxaden
BG2-II1781 v STD1-II1781	1.0 (0.5–2.0)	17.8 (3.8–83.5)	27.9 (5.9–131)
BG2-IT1781 v STD1-II1781	4.2 (2.1–8.1)	84.0 (17.9–395)	45 (9.6–213)
BG2-IL1781 v STD1-II1781	>20.4	>186	>99
BG2-IT1781 v BG2-II1781	4.0 (2.1–7.8)	4.7 (1.0–22.2)	1.6 (0.4–7.7)
BG2-IL1781 v BG2-II1781	>19.8	>10	>3.6
BG2-IL1781 v BG2-IT1781	>4.9	>2.2	>2.2

BG2- II1781, BG2-IT1781 and BG2- IT1781 originates from the same BG2 population and thus shares the same genetic background except for their amino acids at ACCase position 1781.

STD1-II1781 originates from standard sensitive population STD1 used for comparison.

This study has also identified a novel I1781T mutation found to confer comparatively lower levels of resistance to all three FOP, DIM and DEN herbicides tested. The I1781T present in one copy was found to confer dominant resistance to cycloxydim and clodinafop-propargyl but only low and non-significant levels of resistance to pinoxaden. It is noteworthy that the dose responses were conducted on heterozygous IT1781 plants and that when present at the homozygous state, the mutant threonine allele may confer significant levels of resistance to pinoxaden.

Overall three allelic variants have been identified at position 1781 to date, including the I1781V mutation reported to confer at best relatively low levels of resistance to some ACCase herbicides in a *Phalaris paradoxa* population [44]. However, the precise impact of the valine mutant allele on ACCase herbicides remains to be ascertained as non-target site resistance was overlooked in this population. The lower resistance factors associated with the threonine and valine as compared to the leucine mutant alleles is paradoxical given that leucine amounts to a smaller amino acid change relative to the valine and threonine substitutions. While recent crystallography studies have provided a better understanding of the precise binding of FOP, DIM and DEN herbicides to its target enzyme [45,46,47], more remains to be done to better appreciate how subtle amino acid changes can have a profound influence on the efficacies of ACCase herbicides.

a

b

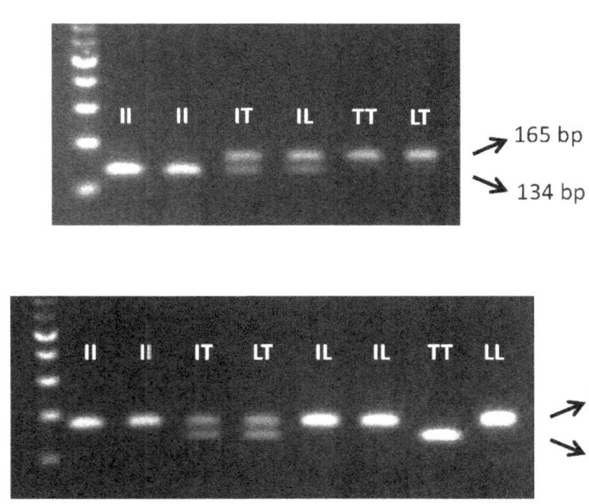

Figure 4. dCAPS procedures for the detection of the wild type isoleucine and mutant threonine amino acid residues at ACCase codon position 1781: (4a) ***NsiI*** **restricted (134 bp) correspond to the I1781 allele and unrestricted fragments (165 bp) correspond to either T1781 or L1781 alleles.** (4b) *NdeI* restricted fragments (132 bp) correspond to the mutant threonine allele while the undigested band (169 bp) corresponds to either the wild type isoleucine or leucine alleles. Heterozygous plants show one each of the restricted and unrestricted PCR fragments in the two assays.

Detection of Mutations at Codon Position 1781

Several DNA based methods have been developed for the quick identification of the I1781L target site ACCase mutation in grass weeds. These include the PASA [38], dCAPS [48,49], qPCR [50], Taqman [13], Maldi-Tof [51] and SNaPshot [15] assays. These methodologies altogether have contributed towards identifying other mechanisms of resistance to ACCase herbicides by quickly disregarding populations characterised by the I1781L mutation. Additionally, they have permitted the investigation of gene flow [52] and mode of evolution of ACCase target site resistance [53]. However, they are all ambiguous in the light of the present discovery of the I1781T mutation in that they have been designed to detect a nucleotide change on the first base of the 1781 nucleotide triplet only. Thus, in so doing they would have missed other mutations resulting from a nucleotide change on the second base of the 1781 triplet. The assay developed here is very similar to the one published previously [48] except that the dCAPS primer ends on the third base of the 1781 codon to take into account the possible nucleotide changes that can occur on the first and also second base of the codon triplet. It is to be noted, however, that further confirmation and differentiation between all the possible allelic variants would require other assays such as the one developed here for positively identifying the threonine mutant at codon position 1781.

Rarity of the I1781T Mutation

The I1781L mutation has been widely identified in several hundred grass weed populations worldwide [13,15,29]. In contrast, the I1781T has only recently been discovered in population BG2. The rarity of the threonine 1781 allele could be due to the lower level of resistance it confers to ACCase herbicides as compared to the L1781 mutation, making it less probable to be selected under ACCase herbicide selection pressure. However, the T1781 allele did confer dominant resistance to clodinafop-propargyl and cycloxydim, two widely used herbicides in the UK. Therefore both L1781 and T1781 alleles should have been selected to equal degrees in black-grass populations. The seldom encounter of the T1781 allele could be explained by the fact that all high throughput DNA based assays employed to date were ambiguous and may have missed the second nucleotide change that results in the substitution of an isoleucine to the threonine allele. The analysis of a large number of cycloxydim resistant black-grass samples using an unambiguous assay developed here has revealed that the T1781 allele is indeed a very rare event accounting for only 0.5% of mutations that occur at position 1781. Alternatively, the rarity of the I1781T could be due to a fitness cost associated with the threonine allele as compared to the leucine 1781 allele [43]. That dissimilar amino acids can confer different levels of fitness at the same codon position is illustrated by changes at the 2088 ACCase codon. The C2088R mutation appeared to be affected by a fitness cost in *Lolium* species based on initial enzyme data [28] whilst the C2088F is fixed in several species such as *Bromus* and even *Hordeum* and by definition is not affected by a fitness penalty [54].

Non-target Site Resistance Mechanisms in BG2

Population BG2 was completely resistant to clodinafop-propargyl and pinoxaden. Yet 25% of the plants did not have a mutation in the herbicide binding carboxyltransferable domain, allowing indirect inference of the involvement of non-target site based mechanisms in BG2. Non-target site resistance has previously been found to affect the metabolisable FOP and DEN herbicides in black-grass [55]. In a survey of over 200 French black-grass populations, an overwhelming 75% of resistant plants were found not to contain a known ACCase mutation [51]. The widespread occurrence of non-target site versus target site resistance to ACCase herbicides was upheld when the whole range of the species was investigated [13]. The frequency of non-target site resistance established by genotyping plants with SNP markers at known resistance codon positions should be taken with caution. DNA markers used to screen resistant populations are sometimes ambiguous, as demonstrated here, and novel mutations yet to be identified may be present in populations classified as non-target site resistant. Also, it has long been assumed that non-target site resistance and in particular, metabolic resistance, confers lower levels of resistance compared target site resistance. Here, we showed that the levels of resistance to clodinafop-propargyl and pinoxaden due to non-target site mechanisms were higher than the I1781T and, potentially, the I1781L mutations. Precise assessment of the level of non-target site resistance was possible by comparing wild type plants from BG2 and the standard sensitive population. If the target site mutant IT1781 subpopulation was directly compared to the standard sensitive population, as with many previous studies, then it would have been wrongly concluded the I1781T mutation conferred 45-fold resistance to pinoxaden. As established, this target site mutation contributed very little to pinoxaden resistance and that failure to control BG2 is largely attributed to non-target site mechanism(s).

Differential Efficacies of ACCase Herbicides and Management of Black Grass in Rotational Broadleaved Crops

Resistance to ACCase herbicides has been classified as FOP or DIM specific [16] though differences exist among compounds from the same chemical sub-class [15]. In a survey of black-grass populations from France for instance, increasing levels of control were observed from fenoxaprop-P-ethyl to pinoxaden through clodinafop-propargyl [55]. Similarly in this study, the non-metabolisable DIM herbicides showed different levels of efficacies on the black-grass population BG2. Cycloxydim killed all plants characterised by non-target site resistance and the W2027C ACCase mutation. Tepraloxydim was very effective at controlling all plants characterised by non-target site resistance, the I1781T and W2027C mutations and most individuals with the commonly encountered I1781L mutation. The competitive edge of tepraloxydim over cycloxydim may explain why the former is gradually supplanting the latter for controlling ACCase resistant black-grass populations in the UK. However, a few plants containing the I1781L mutation were resistant to tepraloxydim in such a way that over time increased use of this herbicide will select for homozygous LL1781 resistant plants and result in product failure. Clethodim was by far the best herbicide, providing 100% kill on all wild and mutant ACCase genotypes. Black-grass plants containing the D2078G ACCase mutation [16] and C2088R mutation (yet to be identified in this species), are known to affect clethodim [28,56]. Consequently its impending use for black-grass control in dicotyledonous rotational crops in the UK could lead to these mutations being favourably selected. It is imperative, therefore, that grass weed control should not only be carried out in conjunction with diverse herbicide modes of action but should also proactively include non-chemical weed control methods for prolonging the sustainability of DIM herbicides that are still very effective at controlling a significant proportion of black-grass populations.

Conclusion

This study highlights the importance of segregating grass weed populations into individual components for establishing the role of target site and non-target site mechanisms in conferring resistance to ACCase inhibiting herbicides. In addition to two known target site resistance mutations, a novel I1781T mutant ACCase allelic variant was identified and its importance on ACCase herbicide efficacies was precisely assessed in a black-grass population. The study also showed that contrary to common belief, accumulation of non-target site resistance in single plants and populations over the years has, in some cases, generated levels of resistance that are higher than the most common target site mechanisms.

Acknowledgments

The authors are very grateful to the plant production and screening teams at Syngenta, Jealott's Hill, UK for their contribution to the glasshouse tests, Suzanne Harris and Jonathan Garratt for maintaining the plants used in the dose response assays and Karthik Putta for his help at genotyping some of the black-grass populations. Our thanks also go to several Syngenta managers for their encouragement and support in conducting this research.

Author Contributions

Conceived and designed the experiments: SSK. Performed the experiments: SJH RPD. Analyzed the data: EM. Contributed reagents/materials/analysis tools: SSK SJH RPD EM. Wrote the paper: SSK.

References

1. Marechal P, Henriet F, Vancutsem F, Bodson B (2012) Ecological review of black-grass (Alopecurus myosuroides Huds.) propagation abilities in relationship with herbicide resistance. Biotechnol Agron Soc Environ 16: 103–113.
2. Colbach N, Dürr C, Roger-Estrade J, Chauvel B, Caneill J (2006) AlomySys: Modelling black-grass (*Alopecurus myosuroides* Huds.) germination and emergence, in interaction with seed characteristics, tillage and soil climate: I. Construction. Eur J Agron 24: 95–112.
3. Moss SR (1983) The production and shedding of *Alopecurus myosuroides* Huds. seeds in winter cereal crops. Weed Res 23: 45–51.
4. Blair AM, Cussans JW, Lutman PJW (1999)A biological framework for developing a weed management support system for weed control in winter wheat: weed competition and time of weed control.
5. Hofer U, Muehlebach M, Hole S, Zoschke A (2006) Pinoxaden – for broad spectrum grass weed management in cereal crops. J Plant Dis Prot 113: 989–995.
6. Price LJ, Herbert D, Moss SR, Cole DJ, Harwood JL (2003) Graminicide insensitivity correlates with herbicide-binding co-operativity on acetyl-CoA carboxylase isoforms. Biochem J 375: 415–423.
7. Konishi T, Shinohara K, Yamada K, Sasaki Y (1996) Acetyl-CoA carboxylase in higher plants: most plants other than Gramineae have both the prokaryotic and the eukaryotic forms of this enzyme. Plant and Cell Physiology 37: 2.
8. Shorrosh BS, Dixon RA, Ohlrogge JB (1994) Molecular cloning, characterization, and elicitation of acetyl-CoA carboxylase from alfalfa. Proc Natl Acad Sci USA 91: 4323–4327.
9. Herbert D, Walker K, Price L, Cole D, Pallett K, et al. (1997) Acetyl-CoA carboxylase: a graminicide target site. Pestic Sci 50: 67–71.
10. Incledon BJ, Hall CJ (1997) Acetyl-coenzyme A carboxylase: quaternary structure and inhibition by graminicidal herbicides. Pestic Biochem Physiol 57: 255–271.
11. Heap IM (2012) International Survey of Herbicide-Resistant Weeds. Available: http://www.weedscience.com.
12. Moss SR, Marshall R, Hull R, Alarcon-Reverte R (2011) Current status of herbicide-resistant weeds in the United Kingdom; Crop Protection in Southern Britain.
13. Délye C, Michel S, Berard A, Chauvel B, Brunel B, et al. (2010) Geographical variation in resistance to acetyl-coenzyme A carboxylase-inhibiting herbicides across the range of the arable weed *Alopecurus myosuroides* (black-grass). New Phytologist 186: 1005–1017.
14. Henriet F, Marechal PY (2009) Black-grass resistance to herbicides: three years of monitoring in Belgium. Commun Agric Appl Biol Sci 74: 471–478.
15. Marshall R, Hanley SJ, Hull R, Moss SR (2012) The presence of two different target-site resistance mechanisms in individual plants of *Alopecurus myosuroides* Huds., identified using a quick molecular test for the characterisation of six and seven ACCase SNPs. Pest Manag Sci In press.
16. Délye C, Zhang XQ, Michel S, Matéjicek A, Powles SB (2005) Molecular bases for sensitivity to acetyl-coenzyme A carboxylase inhibitors in black-grass. Plant Physiol 137: 794–806.
17. Petit C, Duhieu B, Boucansaud K, Délye D (2010) Complex genetic control of non-target-site-based resistance to herbicides inhibiting acetyl-coenzyme A carboxylase and acetolactate-synthase in Alopecurus myosuroides Huds. Plant Science 178: 501–509.
18. Hall LM, Moss SR, Powles SB (1997) Mechanisms of resistance to arryloxyphenoxypropionate herbicides in two resistant biotypes of *Alopecurus myosurides* (blackgrass): herbicide metabolism as a cross-resistance mechanism. Pest biochem and physiol 57: 87–98.
19. Cummins I, Cole DJ, Edwards R (1999) A role for glutathione transferases functioning as glutathione peroxidases in resistance to multiple herbicides in black-grass. Plant J 18: 285–292.
20. Milner LJ, Reade JP, Cobb AH (2001) Developmental changes in glutathione *S*-transferase activity in herbicide-resistant populations of *Alopecurus myosuroides* Huds (black-grass) in the field. Pest Manag Sci 57: 1100–1106.
21. Moss SR (2012) Developing and promoting more sustainable grass-weed control strategies to combat herbicide resistance. DEFRA Report. Project reference code: PS2714. http://randd.defra.gov.uk.
22. Kaundun SS (2010) An aspartate to glycine change in the carboxyl transferase domain of acetyl-CoA carboxylase is in part associated with resistance to ACCase inhibitor herbicides in a Lolium multiflorum population. Pest Manag Sci 66: 1249–1256.
23. Neff MM, Neff JD, Chory J, Pepper AE (1998) dCAPS, a simple technique for the genetic analysis of single nucleotide polymorphisms: Experimental applications in Arabidposis thaliana genetics. Plant Journal 14: 387–392.
24. Neff MM, Turk E, Kalishman M (2002) Web-based primer design for single nucleotide polymorphism analysis. Trends Genet 18: 613–615.
25. Moss SR, Tatnell LV, Hull R, Clarke JH, Wynn S, et al. (2010) Integrated Management of Herbicide Resistance. HGCA project report: 466 wwwhgca-com/publinkaspx?id = 7492.
26. Chauvel B, Guillemin JP, Colbach N (2009) Evolution of a herbicide-resistant population of *Alopecurus myosuroides* Huds. in a long term cropping system experiment. Crop prot 28: 343–349.
27. White GM, Moss SR, Karp A (2005) Differences in the molecular basis of resistance to the cyclohexanedione herbicide sethoxydim in *Lolium multiflorum.* Weed Res 45: 440–448.
28. Yu Q, Collavo A, Zheng MQ, Owen M, Sattin M, et al. (2007) Diversity of Acetyl-Coenzyme A Carboxylase Mutations in Resistant Lolium Populations: Evaluation Using Clethodim. Plant Physiol 145: 547–558.
29. Beckie H, Tardif FJ (2012) Herbicide cross resistance in weeds. Crop prot 35: 15–28.
30. Powles S, Yu Q (2010) Evolution in action: plants reistant to herbicides. Annu Rev Plant Biol 61: 317–347.
31. Délye C (2005) Weed resistance to acetyl coenzyme A carboxylase inhibitors: an update. Weed Sci 53: 728–746.
32. Liu W, Harrison DK, Chalupska D, Gornicki P, O'Donnell CC, et al. (2007) Single-site mutations in the carboxyltransferase domain of plastid acetyl-CoA carboxylase confer resistance to grass-specific herbicides. Proc Natl Acad Sci USA 104: 3627–3632.
33. Gherekhloo J, Osuna MD, De Prado R (2012) Biochemical and molecular basis of resistance to ACCase-inhibiting herbicides in Iranian Phalaris minor populations. Weed Res 52: 367–372.
34. Kershner KS, Al-Khatib K, Krothapalli K, Tuinstra MR (2012) Genetic resistance to acetyl-coenzyme A carboxylase-inhibiting herbicides in grain sorghum. Crop sci 52: 64–73.
35. Délye C, Matéjicek A, Michel S (2008) Cross-resistance patterns to ACCase-inhibiting herbicides conferred by mutant ACCase isoforms in *Alopecurus myosuroides* Huds. (black-grass), re-examined at the recommended herbicide field rate. Pest Manag Sci 64: 1179–1186.
36. Zagnitko O, Jelenska J, Tevzadze G, Haselkorn R, Gornicki P (2001) An isoleucine/leucine residue in the carboxyltransferase domain of acetyl-CoA carboxylase is critical for interaction with aryloxyphenoxypropionate and cyclohexanedione inhibitors. Proc Natl Acad Sci USA 98: 6617–6622.
37. Moss SR, Cocker KM, Brown AC, Hall L, Field LM (2003) Characterisation of target-site resistance to ACCase-inhibiting herbicides in the weed *Alopecurus myosuroides* (black-grass). Pest Manag Sci 59: 190–201.
38. Délye C, Calmès É, Matéjicek A (2002) SNP markers for black-grass (*Alopecurus myosuroides* Huds.) genotypes resistant to acetyl CoA-carboxylase inhibiting herbicides. Theor Appl Genet 104: 1114–1120.
39. Zhang X, Devine D (2000) A possible point mutation in the plastidic ACCase gene conferring resistance to sethoxydim in green foxtail (*Setaria viridis*). Weed Sci Soc Am 40: 33.
40. Christoffers MJ, Berg ML, Messersmith CG (2002) An isoleucine to leucine mutation in acetyl-CoA carboxylase confers herbicide resistance in wild oat. Genome 45: 1049–1056.
41. Brown AC, Moss SR, Wilson ZA, Field LM (2002) An isoleucine to leucine substitution in the ACCase of *Alopecurus myosuroides* (black-grass) is associated with resistance to the herbicide sethoxydim. Pest biochem and physiol 72: 160–168.
42. Huan ZB, Tao J, Zhang SY, Wang JX, Huan ZB, et al. (2011) Cloning and sequence analysis of plastid acetyl-CoA carboxylase cDNA from two *Echinochloa crus-galli* biotypes. Journal of Pestic Sci 36: 461–466.

43. Menchari Y, Chauvel B, Darmency H, Delye C (2008) Fitness costs associated with three mutant acetyl-coenzyme A carboxylase alleles endowing herbicide resistance in black-grass *Alopecurus myosuroides*. J appl ecol 45: 939–947.
44. Collavo A, Panozzo S, Lucchesi G, Scarabel L, Sattin M (2011) Characterisation and management of Phalaris paradoxa resistant to ACCase-inhibitors. Crop protection 30: 293–299.
45. Yu LPC, Kim YS, Tong L (2010) Mechanism for the inhibition of the carboxyltransferase domain of acetyl-coenzyme A carboxylase by pinoxaden. Proc Natl Acad Sci USA 107: 22072–22077.
46. Zhang H, Tweel B, Tong L (2004) Molecular basis for the inhibition of the carboxyltransferase domain of acetyl-coenzyme-A carboxylase by haloxyfop and diclofop. Proc Natl Acad Sci USA 101: 5910–5915.
47. Xiang S, Callaghan MM, Watson KG, Tong L (2009) A different mechanism for the inhibition of the carboxyltransferase domain of acetyl-coenzyme A carboxylase by tepraloxydim. Proc Natl Acad Sci USA 106: 20723–20727.
48. Kaundun SS, Windass JD (2006) Derived cleaved amplified polymorphic sequence, a simple method to detect a key point mutation conferring acetyl CoA carboxylase inhibitor herbicide resistance in grass weeds. Weed Res 46: 34–39.
49. Délye C, Pernin F, Michel S (2011) Universal' PCR assays detecting mutations in acetyl-coenzyme A carboxylase or acetolactate synthase that endow herbicide resistance in grass weeds. Weed Res 51: 353–362.
50. Kaundun SS, Cleere SM, Stanger CP, Burbidge JM, Windass JD (2006) Real-time quantitative PCR assays for quantification of L1781 ACCase inhibitor resistance allele in leaf and seed pools of *Lolium* populations. Pest Manag Sci 62: 1082–1091.
51. Delye C, Menchari Y, Guillemin JP, Matéjicek A, Michel S, et al. (2007) Status of black-grass (*Alopecurus myosuroides*) resistance to acetyl coenzyme A carboxylase inhibitors in France. Weed Res 47: 95–105.
52. Busi R, Michel S, Powles SB, Delye C (2011) Gene flow increases the initial frequency of herbicide resistance alleles in unselected *Lolium rigidum* populations. Agr Ecosyst Environ 142: 403–409.
53. Delye C, Straub C, Matejicek A, Michel S (2003) Multiple origins for black-grass (*Alopecurus myosuroides* Huds) target-site-based resistance to herbicides inhibiting acetyl-CoA carboxylase. Pest Manag Sci 60: 35–41.
54. Délye C, Michel S (2005) 'Universal' primers for PCR-sequencing of grass chloroplastic acetyl-CoA carboxylase domains involved in resistance to herbicides. Weed Res 45: 323–330.
55. Petit C, Bay G, Pernin F, Délye C (2010) Prevalence of cross- or multiple resistance to the acetyl-coenzyme A carboxylase inhibitors fenoxaprop, clodinafop and pinoxaden in black-grass (*Alopecurus myosuroides* Huds.) in France. Pest Manag Sci 66: 168–177.
56. Kaundun SS, Hutchings SJ, Dale R, McIndoe E (2012) Broad resistance to ACCase inhibiting herbicides in a ryegrass population Is due only to a cysteine to arginine mutation in the target enzyme. Plosone 7(6): e39759. doi:10.1371/journal.pone.0039759.

5

Increasing Seriousness of Plant Invasions in Croplands of Eastern China in Relation to Changing Farming Practices

Guo-Qi Chen[1], **Yun-He He**[1,2], **Sheng Qiang**[1]*

1 Weed Research Laboratory, Nanjing Agricultural University; Nanjing, China, 2 Department of Landscape Architecture, Zhejiang Agricultural and Forestry University, Hangzhou, China

Abstract

Arable areas are commonly susceptible to alien plant invasion because they experience dramatic environmental influences and intense anthropogenic activity. However, the limited reports on relevant factors in plant invasion of croplands have addressed single or a few invasive species and environmental factors. To elucidate key factors affecting plant invasions in croplands, we analyzed the relationship between 11 effective factors and changes in composition of alien plants, using field surveys of crop fields in Anhui Province conducted during 1987–1990 (historical dataset) and 2005–2010 (recent dataset), when rapid urbanization was occurring in China. We found that in the past few decades, the dominance and richness of alien plant populations approximately doubled, despite differences among the 4 regions of Anhui Province. Among the 38 alien invasive plant species observed in the sites, the dominance values of 11 species increased significantly, while the dominance of 4 species decreased significantly. The quantity of chemical fertilizer and herbicide applied, population density, agricultural machinery use, traffic frequency, and annual mean temperature were significantly related to increased richness and annual dominance values of alien plant species. Our findings suggest that the increase in alien plant invasions during the past few decades is primarily a result of increased application of chemical fertilizer and herbicides.

Editor: Fei-Hai Yu, Beijing Forestry University, China

Funding: This research was supported by National Basic Research and Development Program (2009CB1192), China National Natural Science Foundation (31070482), Graduate Student Research and Innovation Program of Jiangsu Province (CXZZ11.0647) and the 111 Project (B07030). The funders had no role in study design, data collection and analysis, decision to publish, or preparation of the manuscript.

Competing Interests: The authors have declared that no competing interests exist.

* E-mail: wrl@njau.edu.cn

Introduction

With their high levels of available resources and anthropogenic disturbance, agricultural areas are particularly susceptible to alien plant invasions [1,2].The continuous introduction and expansion of invasive plants into arable areas makes the management of alien crop weeds increasingly challenging. It is critically important to measure the seriousness and potential trends of alien weed invasions in arable lands, in order to form strategies for the management of these species [3].

Invasive plant species are generally capable of rapid adaptation to altered environments, climate, and human disturbance [1,2], and shifts in farming practices may facilitate plant invasions in arable areas [4]. For example, overuse of herbicides may lead to outbreaks of resistant or tolerant invasive weed populations; 346 herbicide-resistant biotypes of 194 weed species have been described [5], of which many (e.g., *Lolium multiflorum*, *Sorghum halepense*, and *Conyza canadensis*) are highly invasive. Long-term application of chemical fertilizer may promote alien plant invasions in agricultural areas, as habitats with higher levels of nutrient resources tend to be more susceptible to plant invasions [6,7]. Kovacs-Hostyanszki et al. [8] found that fertilizer had a negative impact on the richness of weed species with lower nitrogen preference, and on the coverage of native weeds. In addition, use of agricultural machinery may disperse seeds or other propagules (e.g., rhizomes) of invasive plant species over great distances [9,10]. Climate change also facilitates range expansion of many invasive plant species [3], such as *Sorghum halepense* [11] and *Carduus nutans* [12]. Comparative studies examining the relative influence of the wide variety of factors involved in plant invasion in arable areas are scarce.

In China, large numbers of invasive plant species have established and spread rapidly in arable areas since the 1980s [13]. Changes in the dynamics of plant invasions have occurred concomitantly with changes in agrochemical inputs, land use [13], farming methods [14], and climate [15]. Before 1980, weed control practices in China were limited to manual removal, tillage, and adjustments in crop rotation. Herbicide application began during the 1980s, and expanded rapidly to become the primary weed management strategy after the middle 1990s [14]. In addition, rapid urbanization was accompanied by abandonment of rural areas and farming by large numbers of people, leading to increased mechanization and fertilizer use. The effects of climate change have also become more apparent since the 1980s [15]. The combination of these diverse changes in environmental factors and farming practices may lead to profound changes in plant community dynamics. However, few studies have addressed the invasion of crops by alien weeds in China. In this study, we assessed weed species richness and dominance in croplands in Anhui Province, a typical agricultural province in eastern China.

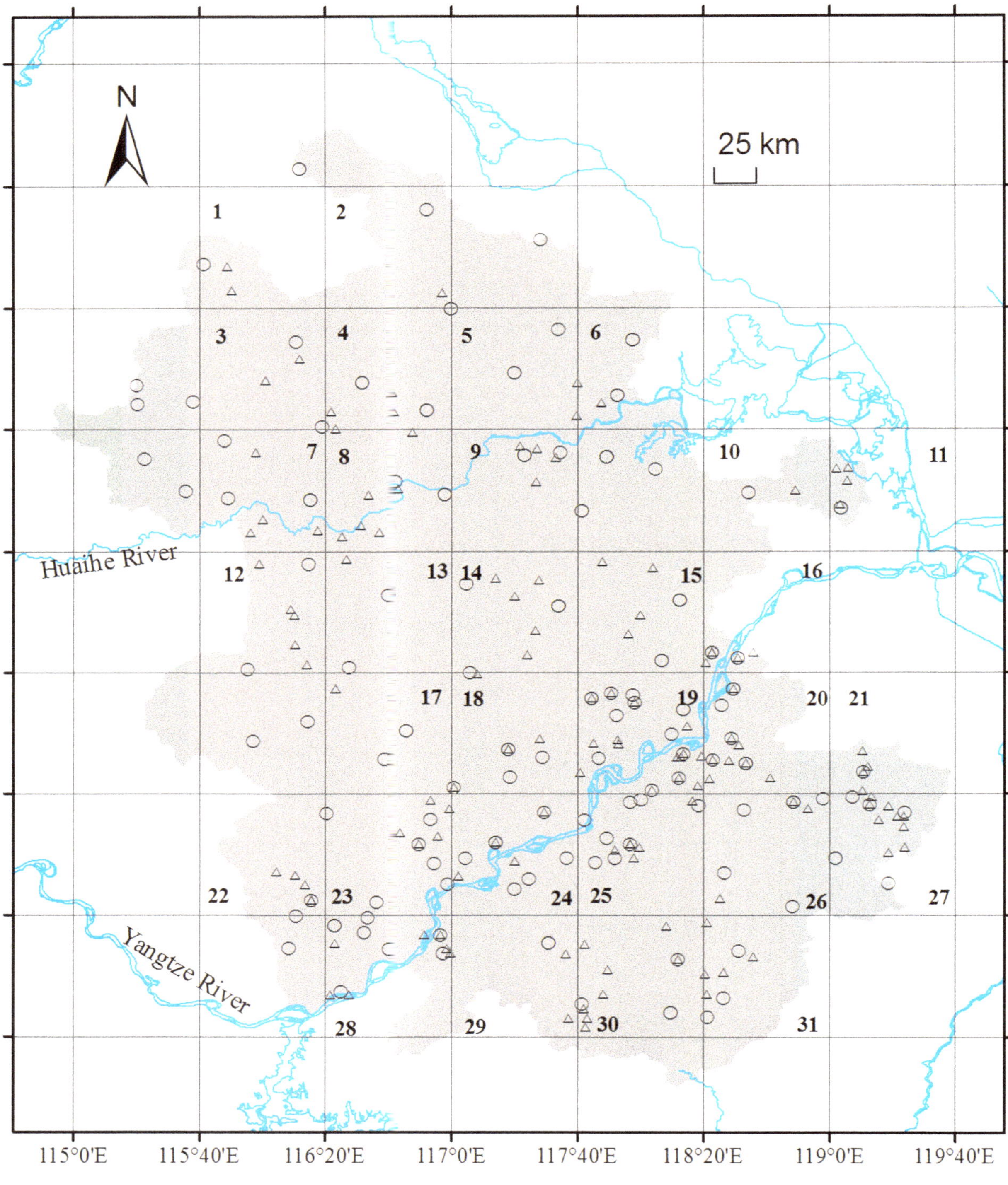

Figure 1. Sites surveyed during 1987–1990 (Δ) and 2005–2010 (○) in summer crop fields in Anhui Province, China.

We hypothesized that increasing applications of chemical herbicide and fertilizer, traffic frequency, and population density may promote alien plant invasions in croplands. Climate change may also have a strong influence on alien weed invasions in cropland. To test this hypothesis, we analyzed 2 datasets obtained from field surveys of summer crops (wheat and oilseed rape) conducted during 1987–1990and 2005–2010, and explored the key factors responsible for changes in alien plant species richness and dominance.

Materials and Methods

Ethics Statement

Here, by conducting field surveys, we studied weed communities in croplands in Anhui Province, China. No specific

Table 1. Visual scoring method for weed dominance value in crop fields.

	Maximum height in field		
Code	**>80 cm**	**20cm–80 cm**	**<20cm**
0.1	1–3 stems or total coverage <0.1%	<10 stems or total coverage <1%	<15 stems or total coverage <2%
0.5	4–10 stems or total coverage 0.2%–0.9%	11–15 stems or total coverage 1%–2%	16–30 stems or total coverage 3%–5%
1	11–15 stems or total coverage 1%–2%	16–30 stems or total coverage 3%–5%	31–60 stems or total coverage 6%–10%
2	16–30 stems or total coverage 3%–5%	31–60 stems or total coverage 6%–10%	61–100 stems or total coverage 11%–25%
3	31–60 stems or total coverage 6%–10%	61–100 stems or total coverage 11%–25%	101–200 stems or total coverage 25%–50%
4	61–100 stems or total coverage 11%–25%	101–200 stems or total coverage 25%–50%	201–500 stems or total coverage 50%–90%
5	>100 stems or total coverage >25%	>200 stems or total coverage >50%	>500 stems or total coverage >90%

permissions were required and the field studies did not involve endangered or protected species.

Study Area

Anhui Province (29°24′ to 34°57′N lat, 114°53′ to 119°39′E long) is located in eastern China with a total area of approximately 1.4×10^5 km^2 (Fig. 1). The province differs in climate from north to south. Average annual temperature ranges from 14 to 17°C, average rainfall from 800 to 1800 mm y^{-1}, and average frost-free period from 200 to 250 d [16].

Data Collection

Two datasets from field surveys made approximately 15 years apart in Anhui province (Fig. 1) were used to estimate changes in weed species richness and dominance. Field surveys were conducted from 1987 to 1990 ("historical dataset," by Sheng Qiang [17]) and from 2005 to 2010 ("recent dataset," this study, by Guo-Qi Chen and Yun-He He) using the same methods. All surveys were conducted after crops (wheat, *Triticum aestvum* L. or oilseed rape, *Brassica napus* L.) flowered, but before harvest. Eighty-three sites cultivated with oilseed rape and 47 sites with wheat were surveyed from 1987 to 1990, and 78 sites cultivated with oilseed rape and 69 sites with wheat were surveyed between 2005 and 2010. In Anhui Province, the amount of arable land per capita is approximately 0.06 ha [16], and most croplands are divided into small units with clear margins. At each survey site, we established ten 0.06-ha quadrats (approximately 666 m^2) [17] in which we recorded all weed species and their dominance values. Dominance values were divided into 7 categories according to relative coverage, abundance, and height of each weed species (Table 1). This method is one of the most common protocols employed in arable weed field surveys in China [17].

In order to determine the key factors affecting the increasing dominance and richness of invasive plant species in the surveyed summer croplands, we analyzed 11 environmental factors relating to crop type, crop rotation, climate, and human disturbance and farming mode (Table 2). We obtained these data for each site and survey year from records in the Anhui Statistical Yearbook [16]for the cities in which field sites were located. We obtained data on herbicide application from the Yearbooks and from local Plant Protection Stations, which were government organizations, which were responsible for consisting farmers cultivating crops, as well as introducing pesticides to farmers. Two types of crop rotations were employed in the surveyed croplands: (1) wet-dry rotation consisting of rice followed by wheat (oilseed rape); and (2) dry crop rotation, which was just2 dry crops in succession, such as corn (soybean) – wheat (oilseed rape) rotation.

Table 2. Mean values of environmental factors of 130 historical sites and 147 recent sites examined in this study.

Factor	Historic	Recent
Crop type (oilseed rape or wheat)	–	–
Crop rotation (wet-dry crop or just dry crops) ^{A}	–	–
Mean temperature of the coldest month (January, °C)	3.79	3.09*
Mean temperature of the hottest month (July, °C)	28.45	28.52 ^{NS}
Annual mean temperature (°C)	15.88	16.47*
Annual mean precipitation (mm)	1671.02	1239.29*
Population density (people/km^2)	329.54	485.30*
Traffic frequency (Freight turnover (10^4 ton km/km^2))	7.02	341.74*
Net cropland agricultural machinery power (kw/ha.)	4.01	12.11*
Net cropland chemical fertilizer applied (kg/ha.)	374.66	782.20*
Net cropland herbicide applied (kg/ha.)	0.53	9.77*

For each environmental factor, mean values of historical and recent datasets in 31 grids (see Figure 1) were compared with paired-sample t-tests. Note: ^{A}: the preceding crops in fields with wet-dry crop rotation were wet rice, while those in lands with dry crop rotation were dry crops such as soybean, corn, and cotton. "*": $P<0.05$ and "^{NS}": not significant.

Data Analysis

Data matrices for the historical and recent datasets included alien weed richness and dominance value and environmental

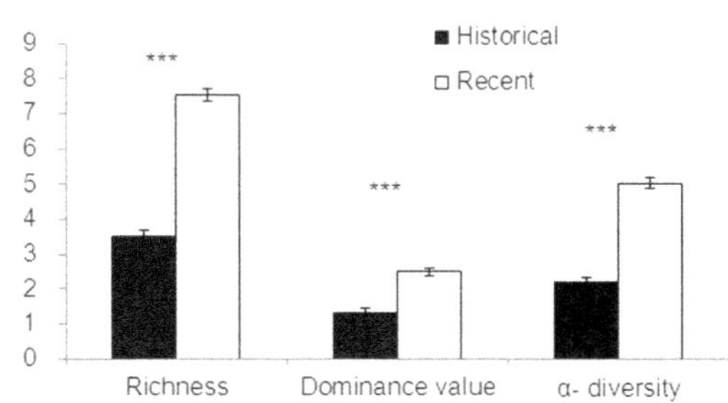

Figure 2. Comparisons between the historic and recent datasets in richness (number of species per site), dominance value, and α-diversity of overall alien weed species in summer crop fields in Anhui Province, China. Note: "***": $P<0.001$.

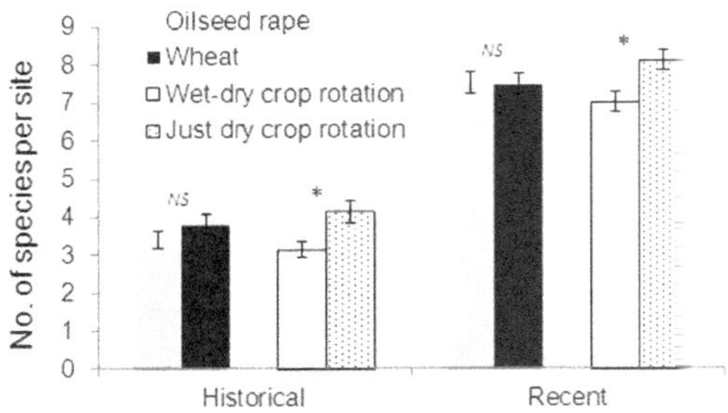

Figure 3. Overall dominance values of alien crop weeds in different groups of summer croplands surveyed in Anhui Province, China.

factors for each site. Mean dominance values by species were calculated for each site from the 10 quadrats to obtain a site-species data matrix. Shannon-Weiner α-diversity [18] was then calculated for both datasets. For each site, the number of alien weed species was estimated as the richness value for each site, and the sum of the dominance values of all alien weed species was calculated as overall dominance value.

To better test the relationship between changes in alien weed invasions and changes in environmental factors, we divided the surveyed area (Anhui Province) into small grids, each of which covered a geographic area of 40′×40′. The surveyed area included31 grids that contained both the historical and recent survey sites (see Fig. 1), which were used as the sample units for the analyses. For each index in a given grid, we used the average value of the sites located in that grid. We then calculated the change in each index for a given grid by the value in the recent dataset minus that for the related historical data. For each environmental factor, mean values of historical and recent datasets in the 31 grids were compared using paired-sample t-tests. On the basis of the data from these grids, we employed redundancy analysis (RDA) to test the relationship between changes in environmental factors and changes in dominance values of alien weed species, using the "vegan" add-on package in the R 2.12.1 Language and Environment for Statistical Computing [19].

Stepwise regression models [20,21] were used to explore relationships between changes in environmental factors and changes in total richness and dominance of alien weeds. The best-fit models were selected using Akaike's information criterion (AIC) [22,23]. Changes in total richness and dominance values, amount of herbicide and chemical fertilizer applied, and agricultural machinery power were calculated as ratios of recent to historical data. Changes in precipitation and temperature were calculated by subtracting historical values from recent values.

Results

Human disturbance factors and mean annual temperature increased significantly ($P<0.05$) from 1987 to 2010, while mean temperature of the coldest month and annual precipitation decreased significantly (Table 2). Compared with the historical dataset, the amount of chemical fertilizer applied doubled, the amount of chemical herbicide applied increased by 18 times, and traffic frequency was 49 times greater in the recent dataset. Moreover, both the overall richness and dominance values of alien weeds doubled.

In the historical dataset, 24 alien weed species from 17 genera and 10 families were recorded (Table 3). Among these, 9 species (37.5%) had frequencies higher than 10% and 2 (8.3%) had frequencies higher than 50%. The average number of alien weed species per site was 3.65 (Fig. 2). In the recent dataset, 35 alien weed species from 26 genera and 13 families were recorded (Table 3). Among these, 17 (48.57%) had a frequency higher than 10% and 6 (17.14%) had a frequency higher than 60%.Overall richness, dominance value, and α-diversity of alien weeds in the recent dataset were significantly higher ($P<0.001$) than the relative values in the historical dataset (Fig. 2). Wheat fields had a significantly higher ($P<0.01$) occurrence of alien weeds than oilseed rape fields in the historical dataset, but not in the recent dataset (Fig. 3). Croplands with wet-dry crop rotation (rice as the preceding crop) showed significantly lower dominance of alien weeds than those with dry crop rotation (another dry crop as the preceding crop) for both datasets, while the difference in the recent dataset was significantly lower ($P<0.01$).

According to the RDA results, the 31 grids could be organized into 4 groups (Fig. 4), and increasing dominance of alien weed species was significantly related to changes in 8 environmental factors (Figs. 4 and 5). These factors included annual precipitation and mean temperature, mean temperature of the hottest (July) and coldest (January) month, amount of herbicide applied, amount of chemical fertilizer applied, population density, and traffic frequency. Among these8 factors, increases in population density and traffic frequency showed the greatest influences on changes in alien species invasions in north Anhui (grids 1 to 5); the dominance values of several alien weed species, including *Avena fatua* and *Euphorbia helioscopia*, increased significantly ($P<0.05$) with increased population density and traffic frequency(Fig. 5). Increased annual precipitation and mean January temperature facilitated invasion of alien weeds, including *Veronica persica*, *Alternanthera philoxeroides*, and *Conyza* spp., in south Anhui (grids 29 to 31).The increase in mean July temperature showed a negative influence on *A. fatua*, *V. sativa*, and *E. helioscopia*. A few species (e.g., *Veronica persica* and *Conyza canadensis*) benefited from increases in annual mean temperature and herbicide application, while species with higher invasiveness (*Geranium caroliniamum*, *A. fatua*, *V. specie*, *A. philoxeroides*, *Erigeron annuus*, *C. bonariensis*, and *C. sumatrensis*) benefited from increased fertilizer application.

Stepwise regression models revealed relationships between changes in alien weed richness and dominance and environmental factors. Six of the 8 environmental factors were significantly related to an increased richness of alien weeds (Table 4); the factor with the greatest influence was the increase in the amount of chemical fertilizer applied. The model identified 4 environmental factors that affected change in overall weed dominance, among which 3 (increased herbicide application, traffic frequency, and mean temperature of the hottest month) showed significant effects (Table 4). The greatest (positive) effect was due to the amount of herbicide applied.

Discussion

Plant invasion is a major economic problem in agricultural fields, and a variety of factors may influence the expansion of alien plants [24]. We collected data from field surveys that measured changes in diversity and dominance of alien agricultural weeds in relation to a historical survey. The most important factors identified as being correlated with increased diversity and dominance of alien weeds were the amount of chemical fertilizer and herbicide applied in agricultural areas. This finding has implications for the ways in which further plant invasions might be managed.

Table 3. Alien weed species, and their frequencies and change in dominance value (DV), among all sites surveyed between 1987–1992 (historical, 130 sites) and 2005–2010 (recent, 147 sites).

Species	Frequency recent (%)	Frequency historical (%)	Change in DV
Geranium caroliniamum	90.79	22.96	0.554*
Conyza canadensis	85.53	59.26	0.152*
Alternanthera philoxeroides	69.74	8.39	0.210*
Vicia sativa	65.79	65.93	−0.041 ^{NS}
Veronica persica	61.84	15.33	0.022 ^{NS}
Erigeron annuus	61.18	26.67	0.129*
Avena fatua	58.55	47.41	−0.029*
Euphorbia helioscopia	39.47	18.52	0.152*
Sonchus asper	28.95	0.74	0.014*
Daucus carota	26.97	4.44	0.044*
Conyza bonariensis	26.97	1.48	0.086*
Bidens frondosa	25.00	0	0.045*
Conyza sumatrensis	20.39	0.74	0.060*
Aster subulatus	14.47	1.48	0.011*
Veronica polita	12.50	34.81	−0.208*
Coronopus didymus	11.18	0	0.007 ^{NS}
Cyperus rotundus	11.18	5.19	0.005 ^{NS}
Veronica arvensis	7.24	17.04	0.019 ^{NS}
Bidens pilosa	4.61	0.74	0.014 ^{NS}
Lepidium virginicum	3.29	0	0.005 ^{NS}
Aeschynomene indica	3.29	0	0.003 ^{NS}
Sonchus oleraceus	3.29	1.48	0.004 ^{NS}
Phytolacca americana	3.29	0	0.004 ^{NS}
Plantago virginica	2.63	0	0.006 ^{NS}
Bromus catharticus	2.63	0	0.012 ^{NS}
Ambrosia artemisiifolia	2.63	0	0.003 ^{NS}
Lolium temulentum	1.97	7.41	−0.006 ^{NS}
Thlaspi arvense	1.97	8.89	−0.040 ^{NS}
Amaranthus retroflexus	1.97	0	0.000 ^{NS}
Euphorbia maculata	1.32	0	0.000 ^{NS}
Crassocephalum crepidioides	1.32	0.74	0.003 ^{NS}
Lolium multiflorum	0.66	0	0.002 ^{NS}
Solidago canadensis	0.66	0	0.001 ^{NS}
Chenopodium ambrosioides	0.66	0	0.000 ^{NS}
Amaranthus tricolor	0.66	0	0.000 ^{NS}
Veronica hederaefolia	0	2.96	−0.007 ^{NS}
Coreopsis drummondii	0	0.74	0.000 ^{NS}
Veronica peregrina	0	9.63	−0.005*

Note: "*": $P<0.05$ and "^{NS}": not significant.
Note: Change in DV for each species in each grid was calculated by the DV in the recent dataset minus that in the corresponding historic dataset.

Human Factors

Factors associated with human activities are frequently found to facilitate plant invasions [24]. Together, our results implied that increasing traffic frequency and population density, as well as changed farming practices, have significant influences on the occurrence of alien crop weeds.

Firstly, increased application of herbicides negatively affected the richness but positively affected the dominance of alien weeds in the surveyed area. Herbicide application can be an effective means of addressing invasion by the most serious weed species, and chemical weed control has greatly improved the efficiency of weed management in agricultural areas [14]. Many agricultural weed species are controlled simultaneously by herbicides, resulting in a decrease in species richness with increased herbicide application [25], as found in our study. However, many invasive plant species are capable of adapting to herbicides [1], such that chemical weed control may promote the dominance of invasive plants.

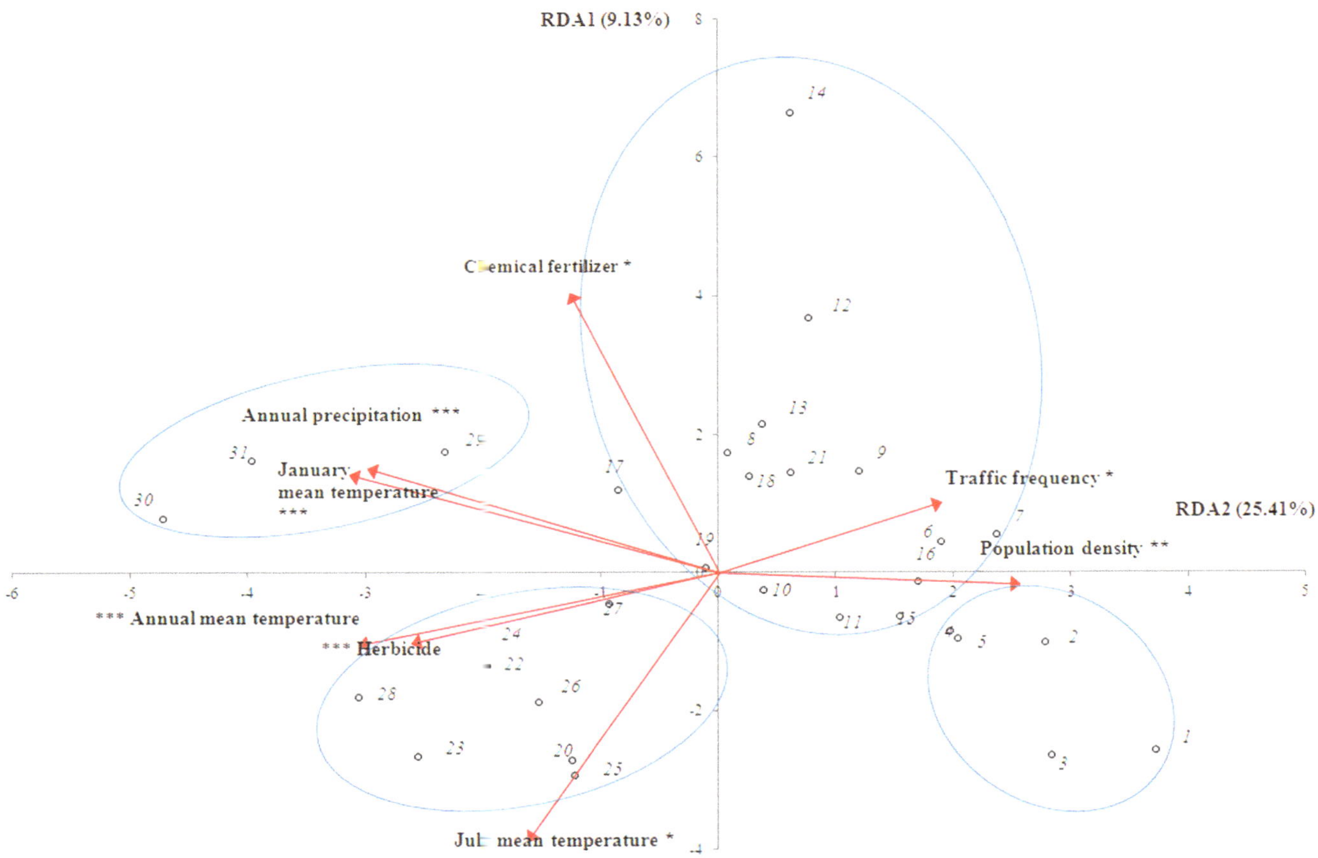

Figure 4. Redundancy analysis (RDA) showing the 8 significant environmental factors and the 31 geographic grids in Anhui Province, China (see Figure 1). RDA was conducted to analyze the relationship between changes in environmental factors and changes in the dominance values of alien weed species in croplands.

Many invasive plant species are also less sensitive to herbicides than are native plants or crop species. The active ingredients contained in the herbicides applied in our survey area primarily include 2, 4–D (2, 4-dicholrophenoxyacetic acid), MCPA (2-methyl-4-chlorophenoxy acetic acid), metsulfuron methyl, tribenuron-methyl, chlorsulfuron, isoproturon, fenoxaprop-P-ethyl, and quizalofop-p-ethyl [26–28]. *G. caroliniamum* showed higher tolerance than other species to most of the herbicides, [29,30]. In the surveyed areas, *G. caroliniamum* increased in frequency from 23% to 91%, and its dominance value per site increased by 6 times from the historical to the current survey period. Similarly, *E. helioscopia* is reported to be insensitive to 2, 4–D, MCPA, and tribenuron-methyl [31,32]; although increased herbicide application negatively impacted the dominance of this species, *E. helioscopia* maintained a high dominance value.

Long-term application of herbicides can result in selection and rapid distribution of herbicide-resistant weed species [33]. Our results suggested that the increased dominance of *C. canadensis* and *V. persica* was positively correlated with increased herbicide application (Fig. 5). Many invasive agricultural weeds show high genetic or phenotypic plasticity, and thus can become herbicide resistant in a short time, as has been reported for *C. canadensis* [34], *A. fatua* [35], *V. sativa* [36], and *Lolium multiflorum*. Ten years of intense glyphosate use (3.7 kg ha^{-1} y^{-1}) in fruit orchards in Chile led to selection of glyphosate-resistant *L. multiflorum* populations [37]. The most-serious herbicide-resistant weed species in South America, such as *S. halepense*, *C. canadensis*, and *L. multiflorum*, are all invasive species [38]. Further, many serious herbicide-resistant weed species in North America, including *Ambrosia artemisiifolia*, *Ambrosia trifida*, *S. halepense*, *Amaranthus retroflexus*, and *C. canadensis* [33,39] are serious invaders in countries outside of their native ranges [40]. Therefore, developing ecological weed management that is less reliant on herbicide application is of high importance [4,41]; integrated agricultural practices such as rice-fish [42] and rice-duck [43] co-culture systems, and well-designed crop rotation and intercropping [44], could be part of the solution to this challenge.

The application of chemical fertilizer and use of agricultural machinery may promote alien plant invasions. Our results suggested that the increasing amount of chemical fertilizer applied was the most significant factor related to the increase of alien plant species richness (Table 4), and that increased quantities of chemical fertilizers applied to crops was positively related to increased dominance of several invasive plant species in the surveyed croplands. Combine harvesters can disperse weed seeds and other propagules over great distances, as shown for *A. fatua* [10] and *A. philoxeroides* (according to our field observation). Agricultural mechanization in Anhui Province is still at a low level, but is developing quickly [16]; hence, the potential influence of agricultural machinery on alien plant invasions deserves more attention.

In addition, our study suggested that alien plant invasions increased more rapidly in oilseed rape fields than in wheat fields. This could be the result of changes in the composition of alien

Table 4. Results of the stepwise regression models used to test the relationships between changes in alien weed species richness and dominance, and changes in environmental factors between the 2 datasets surveyed in different time periods.

Parameter	Estimate	SE	*t*-value	*P*
Alien weed species richness				
Net cropland chemical fertilizer applied	0.008	0.001	7.349	<0.001
Net cropland herbicide applied	−0.004	0.001	−5.660	<0.001
Population density	0.844	0.258	3.274	0.004
Net cropland agricultural machinery power	0.023	0.007	3.057	0.006
Annual mean temperature	−0.899	0.347	−2.592	0.017
Traffic frequency	0.950	0.448	2.119	0.046
Mean temperature of the coldest month	−0.180	0.089	−2.013	0.057
Mean temperature of the hottest month	0.048	0.030	1.606	0.123
ΔAIC with null model	−60.34			
Overall dominance value of alien weeds				
Net cropland herbicide applied	0.002	0.000	4.168	<0.001
Mean temperature of the hottest month	−0.137	0.054	−2.526	0.018
Traffic frequency	−0.446	0.188	−2.366	0.026
Mean temperature of the coldest month	0.328	0.169	1.948	0.063
ΔAIC with null model	−17.28			

The changes in Akaike's information criteria between final and null models are also shown.

plant species in the surveyed croplands. In the historical dataset, the most dominant alien species was *A. fatua*, followed by *V. sativa*, *Veronica polita*, and *V. persica*, all of which are archaeophytes (introduced before 1840).These 4 alien species account for 86% of total alien dominance value, and each of these species tended to occur in wheat fields. Nevertheless, several neophytes (introduced after 1840) in the surveyed areas spread quickly, particularly *G. caroliniamum*, *A. philoxeroides*, and *C. canadensis*, all of which tended to occur in oilseed rape fields. The continuous introduction and expansion of alien neophytes have resulted in plant invasions in oilseed rape areas becoming as serious as those in wheat fields. Other types of cropland that currently have lower rates of plant invasion may increasingly be faced with similar problems.

Alien plant invasions in lands with wet-dry crop rotation were less serious than those in lands with dry crop rotation, but increased faster. Alien invasions in areas with wet-dry crop rotation were very low in the historical dataset. One explanation may be that the annual shift in soil moisture between wet and dry planting periods caused a barrier against maintenance of weed seed banks [45], particularly for alien species that lack a long history of co-evolution with crop cultivation in China. Recently, some perennial alien plant species with rapid vegetative reproduction have invaded croplands and spread quickly, such as *A. philoxeroides* [46], which is highly adapted to different moisture conditions [47]. Seeds of some alien plant species, including *G. caroliniamum*, *V. sativa*, and *V. persica* [48] can remain viable in wet-dry crop rotation systems, and frequently infest this type of cropland. Moreover, some invasive weeds, such as *C. canadensis* and *E. annuus* [49], produce large amounts of small seeds and maintain large seed banks from which they readily disperse into croplands. Well-designed crop rotation systems could help to manage weed invasions. If possible, wet-dry crop rotation system showed higher resistance to alien plant invasions. Moreover, invasive weed species that are well adapted to both moist and dry soil should be more carefully monitored and controlled, and regular field investigations on crop weed communities should be conducted every few decades, as well the distribution of serious invasive plants should be monitored.

In addition, increases in human population density and traffic frequency both showed positive influences on the richness of alien plant species, consistent with studies that have shown correlations between anthropogenic disturbance and plant invasion [50–52]. Our results also suggested that the increased dominance of alien plants in areas with greater increases in traffic frequency was less pronounced in north Anhui Province. In north Anhui Province, several archaeophytes have been major agricultural weeds over long time periods, and the occurrence of these species did not increase significantly. For example, the most dominant alien weed species in the historical dataset were *A. fatua*, *V. sativa*, *V. persica*, *V. polita*, and *Thlaspi arvense*. Among these species, the dominance of *A. fatua*, *V. polita*, and *T. arvense* decreased significantly and those of *V. sativa* and *V. persica* did not change significantly.

Physical Factors

Our results suggested that climatic change plays an important role in promoting the invasivity of alien plants in agricultural areas. Temperature and precipitation are widely known to be key factors in determining the distribution of plant species [3,53,54]. The increase in mean annual temperature and annual precipitation were positively correlated with increased dominance of several common invasive plants. However, the increase in mean temperature of the hottest month was negatively associated with dominance of most alien plant species. Located in the subtropical and temperate zones, Anhui Province has a climate with high precipitation, warm winters, and mild summers [16], which favor the establishment of a broad range of invasive plant species. Thus, the threat of plant invasions in this area may worsen in the future. The temperature in Anhui Province has shown a clear warming trend in recent decades [55], and annual precipitation has been highly variable with more frequent rainstorms [56]. A warming climate may enable the distribution of many invasive plants to expand [3]. Thus, cropping systems in many regions are likely to experience new vulnerabilities to exotic plant invasions in the future [11]. Additionally, more frequent rainstorms may result in increased flooding, which may further promote plant invasions by dispersing large numbers of seeds across a large area, as well as causing habitat fragmentation [57,58].

Conclusion

With increasing applications of herbicide and chemical fertilizer, higher population density, an upgraded traffic system, and the influence of climate change, plant invasions in crop areas have approximately doubled during the past few decades in Anhui Province, China. Differences in the seriousness of plant invasions among different types of cropping systems are fading. Much more attention should be focused on studying the current and potential distributions of invasive plants in agricultural areas, and to assess the risks of plant invasions in various croplands, particularly in areas that currently experience low rates of invasion. Considering that current patterns of biological invasion may better reflect

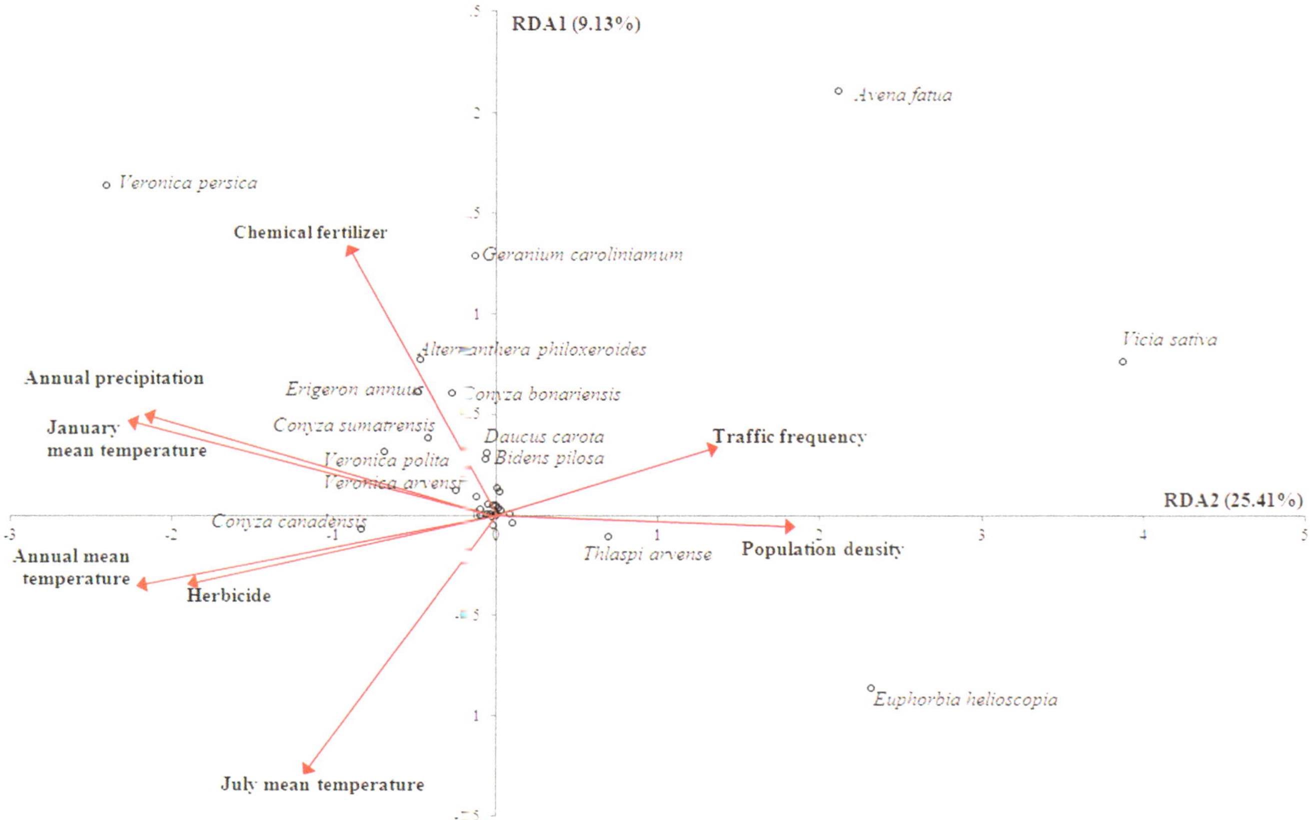

Figure 5. Redundancy analysis (RDA) showing the 8 significant environmental factors and alien weed species in croplands in Anhui Province, China. Note: species with lower correlations with RDA axes are not shown.

historical than recent human activities [59], the potential for crop weed invasions in this area could be a serious threat in the next several decades and beyond. Integrative weed management with reduced application of chemical fertilizers and herbicides deserves much more attention.

Acknowledgments

We thank Kerin Bentley, Joan West and Sandra Hoffberg from the University of Georgia (USA) for their help with preparing the manuscript.

Author Contributions

Conceived and designed the experiments: SQ Y-HH G-QC. Performed the experiments: SQ Y-HH G-QC. Analyzed the data: G-QC. Contributed reagents/materials/analysis tools: G-QC. Wrote the paper: G-QC Y-HH SQ.

References

1. Clements DR, DiTommaso A, Jordan N, Booth BD, Cardina J, et al. (2004) Adaptability of plants invading North American cropland. Agriculture, ecosystems & environment 104: 379–398.
2. Seastedt T (2007) Plant ecology - Resourceful invaders. Nature 446: 985–986.
3. Clements DR, Ditommaso A (2011) Climate change and weed adaptation: can evolution of invasive plants lead to greater range expansion than forecasted? Weed Research 51: 227–240.
4. Petit S, Boursault A, Le Guilloux M, Munier-Jolain N, Reboud X (2011) Weeds in agricultural landscapes. A review. Agronomy for sustainable development 31: 309–317.
5. HRAC (2012) International Survey of Herbicide Resistant Weeds.
6. DeGasperis BG, Motzkin G (2007) Windows of opportunity: Historical and ecological controls on *Berberis thunbergii* invasions. Ecology 88: 3115–3125.
7. Richardson D, Pyšek P (2006) Plant invasions: merging the concepts of species invasiveness and community invasibility. Progress in Physical Geography 30: 409.
8. Kovacs-Hostyanszki A, Batary P, Baldi A, Harnos A (2011) Interaction of local and landscape features in the conservation of Hungarian arable weed diversity. Applied Vegetation Science 14: 40–48.
9. Blanco-Moreno J, Chamorro L, Masalles R, Recasens J, Sans F (2004) Spatial distribution of *Lolium rigidum* seedlings following seed dispersal by combine harvesters. Weed Research 44: 375–387.
10. Shirtliffe SJ, Entz MH (2005) Chaff collection reduces seed dispersal of wild oat (*Avena fatua*) by a combine harvester. Weed Science 53: 465–470.
11. McDonald A, Riha S, DiTommaso A, DeGaetano A (2009) Climate change and the geography of weed damage: Analysis of U.S. maize systems suggests the potential for significant range transformations. Agriculture, Ecosystems & Environment 130: 131–140.
12. Zhang R, Jongejans E, Shea K (2011) Warming increases the spread of an invasive thistle. PLoS One 6.
13. Ding JQ, Mack RN, Lu P, Ren MX, Huang HW (2008) China's booming economy is sparking and accelerating biological invasions. BIoscience 58: 317–324.
14. Zhang ZP (2003) Development of chemical weed control and integrated weed management in China. Weed Biology and Management 3: 197–203.
15. Wang Z, Ding Y, He J, Yu J (2004) An updating analysis of the climate change in China in recent 50 years. Acta Meteorologica Sinica 62: 228–236 (in Chinese with English Abstract).
16. Statistics Bureau of Anhui Province (1991–2011) Anhui statistical yearbook. Beijing: China Statistics Press. (in Chinese)

17. Qiang S (2005) Multivariate Analysis, Description, and Ecological Interpretation of Weed Vegetation in the Summer Crop Fields of Anhui Province, China. Journal of Integrative Plant Biology 47: 1193–1210.
18. Ponce C, Bravo C, de León DG, Magaña M, Alonso JC (2011) Effects of organic farming on plant and arthropod communities: A case study in Mediterranean dryland cereal. Agriculture, Ecosystems & Environment 141: 193–201.
19. R Development Core Team (2011) R: A language and environment for statistical computing. Vienna, Austria: R Foundation for Statistical Computing.
20. Whittingham MJ, Stephens PA, Bradbury RB, Freckleton RP (2006) Why do we still use stepwise modelling in ecology and behaviour? Journal of Animal Ecology 75: 1182–1189.
21. Mundry R, Nunn CL (2009) Stepwise model fitting and statistical inference: turning noise into signal pollution. The American Naturalist 173: 119–123.
22. José-María L, Armengot L, Blanco-Moreno JM, Bassa M, Sans FX (2010) Effects of agricultural intensification on plant diversity in Mediterranean dryland cereal fields. Journal of Applied Ecology 47: 832–840.
23. Burnham KP, Anderson DR (2002) Model selection and multimodel inference: a practical information-theoretic approach. New York: Springer.
24. Wang R, Wang JF, Qiu ZJ, Meng B, Wan FH, Wang YZ (2011). Multiple mechanisms underlie rapid expansion of an invasive alien plant. New Phytologist 191: 828–839.
25. Dekker J (1997) Weed diversity and weed management. Weed Science: 357–363.
26. Yu EL (2010) Investigation and strategies of control of weed in wheat and rapeseed in Anhui Province. Anhui Agri. Sci. Bull. 16: 172–173.(in Chinese with English abstract)
27. Zhang BW, Fang WH (2002) Research and application of weed control strategies in oilseed rape fields in Anqing City. Pesticide 41: 12–14. (in Chinese)
28. Zhang QY, Xu J, Zhou QF, Hu M (2003) Research progress on chemical weed control in wheat fields in Anhui Province. Plant Protection Technology and Extension 23: 34–35. (in Chinese)
29. Hu YZ (2010) Experiments on controlling broadleaved weeds such as *Geranium caroliniamum* in wheat fields with herbicides. Anhui Agri. Sci. Bull. 16: 143–143. (in Chinese)
30. Wei YB (2012) A study on chemical control against *Geranium caroliniamum* - a serious weed species in wheat fields. Anhui Agri. Sci. Bull. 18: 68–69. (in Chinese)
31. He JR, Han BR, Zhen ZZ, Xu W, Du JH (2008) Compositions and integrative management of weed communities in winter wheat fields in Suqian City. Anhui Agri. Sci. Bull. 14: 41–43.(in Chinese)
32. Song AY (2011) A preliminary study on the succession of weed community of wheat fields in Xiaoxian county.Anhui Agri. Sci. Bull. 17: 33–35. (in Chinese)
33. Powles SB (2008) Evolved glyphosate-resistant weeds around the world: lessons to be learnt. Pest Management Science 64: 360–365.
34. Mueller TC, Massey JH, Hayes RM, Chris L, Stewart CN Jr (2003) Shikimate accumulates in both glyphosate-sensitive and glyphosate-resistant horseweed (*Conyza canadensis* L. Cronq.). Journal of Agricultural and Food Chemistry 51: 680–684.
35. Beckie HJ, Thomas AG, Légère A, Kelner DJ, Van Acker RC, et al. (1999) Nature, occurrence, and cost of herbicide-resistant wild oat (*Avena fatua*) in small-grain production areas. Weed Technology: 612–625.
36. Nandula VK, Foy CL, Orcutt DM (1999) Glyphosate for Orobanche aegyptiaca Control in Vicia sativa and Brassica napus Weed Science 47: 486–491.
37. Perez A, Kogan M (2003) Glyphosate-resistant *Lolium multiflorum* in Chilean orchards. Weed Research 43: 12–19.
38. Vila-Aiub MM, Vidal RA, Balbi MC, Gundel PE, Trucco F, et al. (2008) Glyphosate-resistant weeds of South American cropping systems: an overview. Pest Management Science 64: 366–371.
39. Powles SB, Preston C (2006) Evolved Glyphosate Resistance in Plants: Biochemical and Genetic Basis of Resistance 1. Weed Technology 20: 282–289.
40. Holm LG, Pancho JV, Herberger JP (1979) A geographical atlas of world weeds. New York: John Wiley and Sons.
41. Thomas AG, Légere A, Leeson JY, Stevenson FC, Holm FA, et al. (2011) Weed community response to contrasting integrated weed management systems for cool dryland annual crops. Weed Research 51: 41–50.
42. Xie J, Hu L, Tang J, Wu X, Li N, et al. (2011) Ecological mechanisms underlying the sustainability of the agricultural heritage rice–fish coculture system. Proceedings of the National Academy of Sciences 108: E1381–E1387.
43. Huang Y, Wang H, Huang H, Feng Z, Yang Z, et al. (2005) Characteristics of methane emission from wetland rice–duck complex ecosystem. Agriculture, ecosystems & environment 105: 181–193.
44. Zhu Y, Chen H, Fan J, Wang Y, Li Y, et al. (2000) Genetic diversity and disease control in rice. Nature 406: 718–722.
45. Wu J, Zhou H (2000) Seed bank of weeds in paddy fields. Chinese Journal of Rice Science 14: 37–42.
46. Wang N, Yu FH, Li PX, He WM, Liu J, et al. (2009) Clonal integration supports the expansion from terrestrial to aquatic environments of the amphibious stoloniferous herb *Alternanthera philoxeroides*. Plant Biology 11: 483–489.
47. Pan XY, Geng YP, Zhang WJ, Li B, Chen JK (2006) The influence of abiotic stress and phenotypic plasticity on the distribution of invasive *Alternanthera philoxeroides* along a riparian zone. Acta Oecologica 30: 333–341.
48. Zuo RL, Qiang S, Li RH (2007) Relationship between weed seeds dispersed by irrigation water and soil weed seedbank of paddy field in rice-growing region. Chinese Journal of Rice Science 21: 417–424 (in Chinese with English Abstract).
49. Li YH (1998) Weed flora in China. Beijing: China Agriculture Press. (in Chinese)
50. Huang QQ, Wang GX, Hou YP, Peng SL (2011) Distribution of invasive plants in China in relation to geographical origin and life cycle. Weed Research 51: 534–542.
51. Gavier-Pizarro GI, Radeloff VC, Stewart SI, Huebner CD, Keuler NS (2010) Housing is positively associated with invasive exotic plant species richness in New England, USA. Ecol Appl 20: 1913–1925.
52. Seipel T, Kueffer C, Rew LJ, Daehler CC, Pauchard A, et al. (2012) Processes at multiple scales affect richness and similarity of non-native plant species in mountains around the world. Global Ecology and Biogeography 21: 236–246.
53. Freckleton RP, Stephens PA (2009) Predictive models of weed population dynamics. Weed Research 49: 225–232.
54. Graziani A, Steinmaus SJ (2009) Hydrothermal and thermal time models for the invasive grass, *Arundo donax*. Aquatic Botany 90: 78–84.
55. Xu XW, Sun MY, Fang YY, He XQ, Xue F, et al. (2011) Impact of climatic change on rice production and response strategies in Anhui Province. Journal of Agro-Environment Science 30: 1755–1763 (in Chinese with English abstract).
56. Xie WS, Tian H (2011) Characteristics of rainstorm in Anhui in the past 50 years. Meteorological Science and Technology: 160–164 (in Chinese).
57. Price JN, Berney PJ, Ryder D, Whalley RDB, Gross CL (2011) Disturbance governs dominance of an invasive forb in a temporary wetland. Oecologia 167: 759–769.
58. Touchette B, Romanello G (2010) Growth and water relations in a central North Carolina population of *Microstegium vimineum* (Trin.) A. Camus. Biological Invasions 12: 893–903.
59. Essl F, Dullinger S, Rabitsch W, Hulme PE, Hulber K, et al. (2011) Socioeconomic legacy yields an invasion debt. Proc Natl Acad Sci U S A 108: 203–207.

6

Risk and Prognostic Factors of Inpatient Mortality Associated with Unintentional Insecticide and Herbicide Poisonings: A Retrospective Cohort Study

Wu-Chien Chien[1], Chi-Hsiang Chung[2], Jouni J. K. Jaakkola[3], Chi-Ming Chu[1], Senyeong Kao[1], Sui-Lung Su[1], Ching-Huang Lai[1]*

1 Department of Public Health, National Defense Medical Center, No. 161, Section 6, Neihu District, Taipei City, Taiwan, Republic of China, **2** Graduate Institute of Life Sciences, National Defense Medical Center, No. 161, Section 6, Neihu District, Taipei City, Taiwan, Republic of China, **3** Center for Environmental and Respiratory Health Research, Institute of Health Sciences, University of Oulu, Oulu, Finland

Abstract

Introduction: Pesticide poisoning is an important public health problem worldwide. The study aimed to determine the risk of all-cause and cause-specific inpatient mortality and to identify prognostic factors for inpatient mortality associated with unintentional insecticide and herbicide pesticide poisonings.

Methods: We performed a retrospective cohort study of 3,986 inpatients recruited at hospitalization between 1999 and 2008 in Taiwan. We used the International Classification of Disease, 9th ed., Clinical Modification external causes of injury codes to classify poisoning agents into accidental poisoning by insecticides and herbicides. Comparisons in mortality rates were made between insecticide poisoning patients and herbicide poisoning patients by using the Cox proportional hazards models to estimate multivariable-adjusted hazard ratios (HRs) and their 95% confidence intervals (CIs).

Results: There were 168 deaths during 2 ,583 person-days of follow-up evaluation (7.8 per 1,000 person-days). The major causes of mortality for insecticide poisonings were the toxic effect of organophosphate and coma, and the major causes of mortality for herbicide poisonings were the toxic effect of other pesticides and the toxic effect of organophosphate. The mortality for herbicide exposure was fourfold higher than that for insecticide exposure. The factors associated with inpatient mortality were herbicide poisonings (HR = 4.58, 95% CI 3.29 to 6.37) and receiving mechanical ventilation treatment (HR = 3.85, 95% CI 2.73 to 5.42).

Conclusions: We demonstrated that herbicides stand out as the dominant agent for poisoning-related fatalities. The control of and limiting access to herbicide agents and developing appropriate therapeutic regimens, including emergency care, should be priorities.

Editor: Devendra Amre, University of Montreal, Canada

Funding: JJ was supported by Academy of Finland (http://www.aka.fi/eng), grants no. 129419 (SALVE program) and 138691. The funders had no role in study design, data collection and analysis, decision to publish, or preparation of the manuscript.

Competing Interests: The authors have declared that no competing interests exist.

* E-mail: lgh@mail.ndmctsgh.edu.tw

Introduction

Pesticide poisoning is a significant public health problem worldwide. With the development of modern agriculture, patients are increasingly exposed to new types of insecticides and herbicides. Currently, the most common type of pesticide in Taiwan is organophosphate pesticides (OP) [1], and glyphosphate and paraquat are the two types of herbicides that are used most frequently [2].

Pesticide exposure in Taiwan was recently evaluated using the nationwide registry maintained by the network of Taiwan's Poison Control Centers (PCC) [3]. Over the course of eight years (1985–1993), 23,436 telephone calls concerning human poisoning exposure were recorded. The most frequent cause of poisoning was pesticides (29.3%). Among all fatal cases, paraquat and organophosphate exposures were the two leading causes of death [4]. Another study reported that there were 4799 organophosphate pesticide (OP) exposures from July 1985 through December 2006 [1,5]. The fatality rate was 12.7%. These studies were based on information collected in telephone interviews on the topic of poisoning exposures. The source population was poorly specified, and therefore, these figures could not be used to estimate the incidence of poisoning. Furthermore, previous studies were of limited sample size and were largely based on one or few health-care facilities.

The National Health Insurance (NHI) in Taiwan was launched on March 1, 1995. Since then, medical institutions have been required by law to make inpatient claims to the NHI Bureau. Given that Taiwan's coverage rate is over 99%, the NHI Research Database provides a representative and evidence-based source of data for studies in this field. Our understanding of the fatality rate of pesticide poisoning has improved [6]. However, patients experiencing pesticide poisoning may experience multiple severe episodes during the same hospitalization or after discharge. Data on the contribution of subsequent episodes to the disease burden are still limited because most past studies used overall fatality as the measured outcome. Furthermore, little information is available on the inpatient hospital stay of pesticide poisoning patients and on the survival rates of patients hospitalized for pesticide poisoning.

During the 10-year period from 1999 to 2008, unintentional pesticide poisonings accounted for 6,328 hospitalizations in Taiwan [6]. To address the extent and major agents of exposure of acute pesticide poisoning in Taiwan resulting in hospitalization, we obtained data from a nationwide hospitalization database to determine the rate of all-cause and cause-specific inpatient mortality and to identify prognostic factors of inpatient mortality among insecticide and herbicide pesticide poisonings.

Methods

The hospitalization data were retrieved from the NHI Research Database (NHIRD). Categorization of the poisoning rates, the causes of hospitalization, and the cause of death after hospitalization occurring between 1999 and 2008 was carried out using the International Classification of Disease, 9th ed., Clinical Modification (ICD-9-CM). This study was conducted using a data base without patient identifications, thereby conforming to the Declaration of Helsinki.

Patients were categorized into six age groups: infants and toddlers aged 0–4, children aged 5–14, young adults aged 15–24, mature adults aged 25–44, middle-aged adults aged 45–64 and elderly aged 65 or older. We defined unintentional poisoning by using the ICD-9-CM external causes of injury codes (E-Codes) to classify poisoning types into accidental poisoning by insecticides of organophosphate (E863.1) and accidental poisoning by herbicides (E863.5). Because the most common type of insecticide in Taiwan is organophosphate insecticide (OP) [1], we excluded patients who presented with poisoning with organochlorine compounds (E863.0) (n = 27) or carbamates (E863.2) (n = 97), poisoning by a mixture of insecticides (E863.3) (n = 170), or poisoning by other and unspecified insecticides (E863.4) (n = 1197), fungicides (E863.6) (n = 9), rodenticides (E863.7) (n = 238), fumigants (E863.8) and other and unspecified agricultural and horticultural chemicals and pharmaceutical preparations other than plant foods and fertilizers (E863.9) (n = 508).

The NHI database recorded the status of prognosis after each hospitalization. Information on death after hospitalization was obtained from the variable of Trans_code from Inpatient expenditures by admissions (DD) data file. The operational definition of death after hospitalization is that the patient dies during hospitalization or immediately after leaving a hospital in a critical condition voluntarily. All cause and cause-specific mortality of inpatients due to unintentional insecticide and herbicide poisonings were defined by using ICD-9-CM N-Code. Cause-specific mortality of inpatients due to unintentional insecticide and herbicide poisonings were further categorized as toxic effect of organophosphates (989.3)(n = 64), toxic effect of other pesticides, not elsewhere classified (989.4)(n = 93), anoxic brain damage (348.1)(n = 1), cardiac arrest (427.5) (n = 1), congestive heart failure (428.0)(n = 1), pneumonitis due to inhalation of food or vomiting (507.0)(n = 1), acute respiratory failure (518.81)(n = 1), alcoholic liver damage, unspecified (571.3)(n = 1), acute renal failure (584.9)(n = 1), coma (780.01)(n = 2), and toxic effect of unspecified lead compound (984.9)(n = 2). Overall mortality analyses used time-to-event methods and were based on 21,583 person-days of inpatient follow-up evaluation. The person-days of follow-up evaluation were calculated from the date of hospitalization to inpatient death or the last date of the inpatient stay. Because the duration of hospitalization varies, the person-time approach is a useful way of expressing mortality taking into account differences in the hospitalization. For patients with multiple hospitalizations, all hospitalization days were summed. Mortality rates, expressed in no of deaths per 1000 person-days, were calculated by dividing the number of deaths by corresponding person time, and survival curves based on person time were derived by the Kaplan-Meier method. The log-rank test was used to compare the survival curves.

Comparisons in mortality rates were made between insecticide poisoning patients and herbicide poisoning patients by using the Cox proportional hazards models to estimate multivariable-adjusted hazard ratios (HRs) and their 95% confidence intervals (CIs). The covariates used in the analyses included the following: type of pesticide poisoning (insecticides/herbicides), age (0–4, 5–14,15–24, 25–44, 45–64, 65 or older), sex (male/female), Charlson Comorbidity Index (CCI), living area (northern, central, southern, eastern, offshore), level of care (medical center, regional hospital, local hospital), mechanical ventilation treatment (yes/no). "Catastrophic illness" meant that patients had one or more following diseases or injuries: malignant neoplasm, congenital deficiency of clotting factors, hemolytic or aplastic anemia, renal disease with renal failure, systemic lupus erythematosus, systemic sclerosis, rheumatoid arthritis, polymyositis, dermatomyositis, vasculitis, pemphigus, Sjogern's syndrome, Crohn's syndrome, ulcerative colitis, schizophrenic disorders, disorders of metabolism, congenital anomalies, burn of >20% of total body surface, complications of transplant, acute poliomyelitis with paralysis, injury severity score ≥16, used respirator 6 hours per day continue 30 days, a fully intravenous diet for 30 days, decompression sickness, air embolism, myasthenia gravis, disorder of immune mechanism, fracture of vertebral column with spinal cord injury, occupational disease, cerebrovascular disease with acute stage, multiple sclerosis, congenital muscular dystrophy, congenital anomalies integument, leprosy, liver cirrhosis with complications, premature infants determined to have medium impairments 3 months after birth, toxic effect of arsenic and its compounds (black foot), motor neurone disease, Creutzfeldt-Jakob disease, and rare diseases [7].

To calculate a Charlson Comorbidity Index (CCI), the first five diagnostic codes (N-Code) of a patient were each multiplied respectively by the scores assigned to 19 different diseases put forth by Charlson et al., and the five resultant figures were then totaled to reach a final index. A higher index was representative of more, or more serious, accompanying diseases [8]. Statistical significance was determined by 2-tailed tests ($P<0.05$). SPSS 18.0 software was used to conduct the statistical analyses.

Results

Demographics

Of the 3,986 patients who were hospitalized because of acute poisoning by insecticides (E863.1) (2,886/3,986) and herbicides (E863.5) (1,100/3,986) between 1999 and 2008 in Taiwan, there were 168 pesticide poisoning deaths during hospitalization.

Table 1 provides demographic information. Males, the dominant group, accounted for 75.4% of pesticide poisoning hospitalizations. The overall mortality rate was higher for males than females.

With respect to age, individuals 15 to 24 years old who experienced herbicide poisoning had a higher mortality rate than

Table 1. Demographics of the study subjects, 1999–2008.

	Insecticides (E863.1) (n = 2,886; 15,884 person-days)						Herbicides (E863.5) (n = 1,100; 5,699 person-days)						Total (n = 3,986; 21,583 person-days)					
	N	%	Mean ± SD	No. of deaths	Mortality rate per 1,000 person-days	Fatality (%)	N	%	Mean ± SD	No. of deaths	Mortality rate per 1,000 person-days	Fatality (%)	N	%	Mean ± SD	No. of deaths	Mortality rate per 1,000 person-days	Fatality (%)
Sex																		
Male	2185	75.71		45	3.82	2.06	821	74.64		84	19.52	10.23	3,006	75.41		129	8.02	4.29
Female	701	24.29		19	4.62	2.71	279	25.36		20	14.34	7.17	980	24.59		39	7.09	3.98
Age group			54.69±16.48						49.34±19.98						53.22±17.07			
0–4	29	1.00		0	0	0	17	1.55		0	0	0	46	1.15		0	0	0
5–14	12	0.42		0	0	0	5	0.45		0	0	0	17	0.43		0	0	0
15–24	90	3.12		2	3.84	2.22	56	5.09		8	26.49	14.29	146	3.66		10	12.15	6.85
25–44	659	22.83		15	4.46	2.28	389	35.36		43	19.57	11.05	1,048	26.29		58	10.44	5.53
45–64	1218	42.20		17	2.75	1.40	383	34.82		28	16.15	7.31	1,601	40.17		45	5.68	2.81
65+	878	30.42		30	5.27	3.42	250	22.73		25	17.72	10.00	1,128	28.30		55	7.75	4.88
Charlson Comorbidity Index, CCI			0.29±0.79	64	4.03	2.22			0.33±0.90	104	18.25	9.45			0.31±0.83	168	7.78	4.23
0	2,309	80.01		43	3.72	1.86	888	80.73		83	18.90	9.35	3,197	80.21		126	7.90	3.94
1	427	14.80		12	3.92	2.81	136	12.36		14	16.70	10.29	563	14.12		26	6.67	4.62
2	84	2.91		5	7.85	5.95	37	3.36		4	16.74	10.81	121	3.04		9	10.27	7.44
3+	66	2.29		4	6.30	6.06	39	3.55		3	13.04	7.69	105	2.63		7	8.09	6.67
Living area																		
Northern Taiwan	266	9.22		13	6.70	4.89	117	10.64		15	18.36	12.82	383	9.61		28	10.16	7.31
Central Taiwan	1839	63.72		32	3.23	1.74	698	63.45		68	20.51	9.74	2,537	63.65		100	7.56	3.94
Southern Taiwan	632	21.90		13	4.04	2.06	155	14.09		14	15.68	9.03	787	19.74		27	6.56	3.43
Eastern Taiwan	144	4.99		6	7.54	4.17	129	11.73		7	10.45	5.43	273	6.85		13	8.87	4.76
Offshore Islands	5	0.17		0	0	0	1	0.09		0	0	0		0.15		0	0	0
Level of care																		
Medical center	602	20.86		15	3.31	2.49	434	39.45		56	20.79	12.90	1,036	25.99		71	9.83	6.85
Regional hospital	1176	40.75		34	4.72	2.89	438	39.82		42	20.52	9.59	1,614	40.49		76	8.21	4.71
Local hospital	1108	38.39		15	3.61	1.35	228	20.73		6	6.26	2.63	1,336	33.52		21	4.11	1.57
Surgical treatment																		
Yes	787	27.27		49	6.28	6.23	409	37.18		69	24.71	16.87	1,196	30.01		118	11.14	9.87
No	2099	72.73		15	1.85	0.71	691	62.82		35	12.04	5.07	2,790	69.99		50	4.55	1.79
Mechanical ventilation																		

Table 1. Cont.

	Insecticides (E863.1)						Herbicides (E863.5)						Total					
	(n = 2,886; 15,884 person-days)						(n = 1,100; 5,699 person-days)						(n = 3,986; 21,583 person-days)					
	N	%	Mean ± SD	No. of deaths	Mortality rate per 1,000 person-days	Fatality (%)	N	%	Mean ± SD	No. of deaths	Mortality rate per 1,000 person-days	Fatality (%)	N	%	Mean ± SD	No. of deaths	Mortality rate per 1,000 person-days	Fatality (%)
Yes	286	9.91		35	10.21	12.24	85	7.73		22	37.80	25.88	371	9.31		57	14.21	15.36
No	2600	90.09		29	2.33	1.12	1015	92.27		82	16.03	8.08	3,615	90.69		111	6.32	3.07
Hemodialysis																		
Yes	24	0.83		3	10.42	12.50	104	9.45		26	47.62	25.00	128	3.21		29	34.77	22.66
No	2,862	99.17		61	3.91	2.13	996	90.55		78	15.14	7.83	3,858	96.79		139	6.70	3.60
Length of hospital stay (Mean ± SD)			5.50±8.50	Death: 4.30±6.72					5.18±6.71	Death: 5.58±8.27					5.41±8.04	Death: 5.09±7.72		

other age groups The mortality rate for herbicide poisoning was consistently higher than that for insecticide poisoning. The proportion of patients hospitalized in Central Taiwan was higher than that of other areas. For patients hospitalized in medical centers or regional hospitals, the mortality rate was higher than that of local hospitals. Of the patients, 9.91% (286/2,886) and 7.73% (85/1,100) were treated with mechanical ventilation among insecticide and herbicide poisonings, respectively. There was a higher proportion of herbicide poisoning patients (9.45%) who underwent hemodialysis treatment compared with insecticide poisoning patients (0.83%). The average hospital stays were 5.5 days and 5.2 days for the insecticide and herbicide poisoning groups, respectively.

All-cause and cause-specific mortality rates

The all-cause and cause-specific mortality rates for the entire inpatient cohort separated into the two poisoning groups are presented in Table 2. There were 168 deaths over 21,583 person-days of follow-up for an overall death rate, corresponding to 7.8 deaths per 1000 person-days. The all-cause mortality rate for the herbicide poisoning patients was higher than that of insecticide patients. The top two causes of mortality for insecticide poisoning patients were toxic effects of the organophosphates (3.5 per 1000 person-days) and coma (0.1per 1000 person-days); the major causes of mortality for herbicide poisoning patients were the toxic effects of other pesticides (16.3 per 1000 person-days) and the toxic effects of organophosphates (1.4 per 1000 person-days). There were 89% (93 out of 104) of those who died after herbicide poisoning, the cause of death is stated as 'toxic effects of other pesticides, not elsewhere classified' that deaths due to herbicides in general (and including paraquat and glyphosate) are classified in this category.

Prognostic factors of mortality in pesticide poisoning patients

As shown in Figure 1, the overall survival for herbicide inpatients was significantly worse than that for insecticide poisoning patients ($p<0.001$). The median survival time of length of stay for herbicide was 60 days. Table 3 shows the Cox regression hazard model evaluating cause of poisoning, sex, age, CCI, living area, level of care, mechanical ventilation treatment and their association with mortality risk.

Overall, the significant prognostic factors of all-cause mortality in this population were herbicide poisoning (HR = 4.58, 95% CI 3.29 to 6.37), and receiving mechanical ventilation treatment (HR = 3.85, 95% CI 2.93 to 6.00). With regard to the level of medical care, patients hospitalized in medical centers or regional hospitals had hazards ratios of 1.65 (95% CI 1.00–2.75) times and 1.79 (95% CI 1.09–2.91) times compared with patients hospitalized in local hospitals.

Discussion

This was a nationwide population-based study, allowing us to trace the medical services received by all inpatients after unintentional poisoning. The cumulative survival rate for herbicide inpatients was significantly worse than that of insecticide poisoning patients. The mortality rate for herbicide exposure was almost fivefold higher than that of insecticide exposure. The factors associated with inpatient mortality were herbicide poisonings, level of medical care, receiving ventilation and hemodialysis treatment.

Table 2. All-Cause and Cause-Specific Mortality (n = 3,986).

Cause of death	Insecticides		Herbicides	
	(E863.1) (n = 2,886; 15,884 person-days)		(E863.5) (n = 1,100; 5,699 person-days)	
	No. of deaths	Mortality rate (per 1,000 person-days)	No. of deaths	Mortality rate (per 1,000 person-days)
All Causes	64	4.03	104	18.25
Toxic effect of organophosphates (989.3)	56	3.53	8	1.40
Toxic effect of other pesticides, not elsewhere classified (989.4)	0	0	93	16.32
Anoxic brain damage (348.1)	1	0.06	0	0
Cardiac arrest (427.5)	1	0.06	0	0
Congestive heart failure (428.0)	0	0	1	0.18
Pneumonitis due to inhalation of food or vomiting (507.0)	1	0.06	0	0
Acute respiratory failure (518.81)	1	0.06	0	0
Alcoholic liver damage, unspecified (571.3)	0	0	1	0.18
Acute renal failure (584.9)	1	0.06	0	0
Coma (780.01)	2	0.13	0	0
Toxic effect of unspecified lead compound (984.9)	1	0.06	1	0.18

Cause-specific mortality and prognostic factors

The high mortality rate for herbicide poisoning was in concordance with previous reports [3,9]. One study that recruited various poisoning patients through emergency departments in Southern Taiwan showed that herbicides caused the highest fatality rate [9]. Paraquat and glyphosate were found to be the most frequent herbicides involved in fatalities.

Hemodialysis is one among the treatments for paraquat poisoning [10], as a consequence, much more patients suffered from severe herbicide poisoning in this study received hemodialysis therapy. It has been shown that patients exposed to glyphosate through oral ingestion died more frequently than patients exposed via other pathways. Shock and respiratory failure accounted for most fatalities [4]. Intubation and mechanical ventilation are essential treatment for the most severe poisoning cases [11]. Clinical features such as the development of respiratory distress, impaired consciousness, pulmonary edema, shock, arrhythmias, renal failure requiring hemodialysis and the presence of infiltrate on chest x-ray were some parameters associated with poisoning-related fatality for glyphosate poisoning patients [11]. In addition, hemodialysis and hemoperfusion are two elimination methods used to treat paraquat poisoning patients, but these treatments are unlikely to change the clinical course. The case-fatality rate is quite

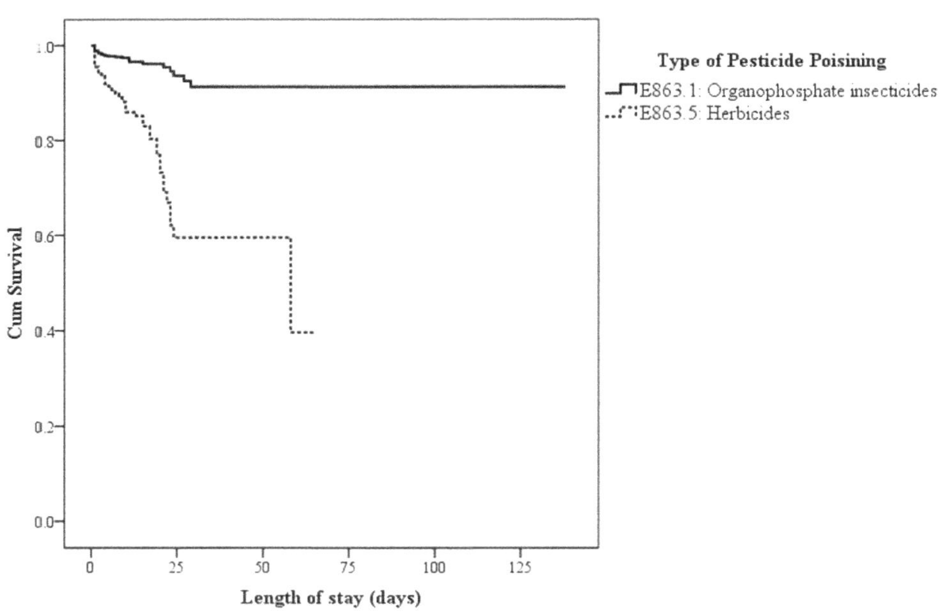

Figure 1. Kaplan-Meier analysis of overall survival for 3,986 inpatients divided into organophosphate insecticides and herbicides poisoning groups (log-rank $p<0.001$).

Table 3. Predictors of mortality for pesticide inpatients in Taiwan, 1999–2008[a].

Variables[b]	B	Adjusted HR	95% CI	p-Value
Type of pesticide poisonings				
E863.1: insecticides		Reference[c]		
E863.5: herbicides	1.52	4.58	3.29–6.37	<0.001
Sex				
Male	0.07	1.08	0.75–1.54	0.70
Female		Reference		
Age group				
0–4		NA		
5–14		NA		
15–24	0.42	1.53	0.76–3.08	0.24
25–44	0.10	1.11	0.75–1.64	0.61
45–64	−0.35	0.70	0.47–1.05	0.09
65+		Reference		
Charlson Comorbidity Index (CCI)				
0		Reference		
1	0.04	1.04	0.67–1.59	0.87
2	0.28	1.33	0.66–2.67	0.43
3+	0.05	1.05	0.48–2.29	0.90
Living area				
Northern Taiwan		Reference		
Central Taiwan	−0.33	0.72	0.47–1.10	0.13
Southern Taiwan	−0.30	0.74	0.43–1.27	0.27
Eastern Taiwan	−0.22	0.80	0.41–1.58	0.52
Offshore islands		NA[d]		
Level of care				
Medical center	0.50	1.65	1.00–2.75	0.04
Regional hospital	0.58	1.79	1.09–2.91	0.02
Local hospital		Reference		
Mechanical Ventilation				
Yes	1.35	3.85	2.73–5.42	<0.001
No		Reference		

[a]Using Cox proportional hazards models.
[b]Type of pesticide poisoning (insecticides/herbicides), age (0–4, 5–14,15–24, 25–44, 45–64, 65 or older), sex (male/female), Charlson Comorbidity Index (CCI), living area (northern, central, southern, eastern, offshore), level of care (medical center, regional hospital, local hospital), mechanical ventilation treatment (yes/no).
[c]Reference group.
[d]NA: no mortality cases.

high despite different variations in treatment [10]. Our finding is consistent with above reports that hemodialysis treatment is associated with a high herbicide poisoning fatality rate. Death among organophosphate poisoning patients is chiefly related to respiratory failure, central nervous system depression, seizures or ventricular arrhythmias. The pathogenesis of respiratory distress and failure was related to excessive bronchial secretions, bronchospasm, pulmonary edema, aspiration pneumonia and paralysis of respiratory muscles. This association might explain that ventilation treatment is another indicator of serious complications of organophosphate and herbicide poisonings.

With regard to the level of medical care, patients hospitalized in medical centers or regional hospitals had hazards ratios 1.65 times or 1.79 times compared with patients hospitalized in local hospitals. We found that only 29% and 50% of patients were transferred to a higher level of medical service among insecticide and herbicide poisonings, respectively. The primary hospital might have lacked of medication, equipment, or staff [12]. The level of medical care used in our study represented only the most recent hospitalization. Our results could be influenced by the fact that higher medical care facilities are prone to an over-representation of severe poisoning cases. For poisoning by highly toxic pesticides such as paraquat, patients transferred to a higher level of medical service might not experience better outcomes.

It is critically important to know exactly the type of chemical involved in a pesticide poisoning. However, further information on the specific type of pesticide within the categorization of organophosphate or herbicide was not available. Currently, organophosphate insecticides are the most common type of insecticides in Taiwan including Disulfoton,Terbufos, Azinphos-Methyl, Mevinphos, Methidathion, Demeton-S-Methyl, Methamidophos, Phorate, Dicrotophos, Trichlorfon. There were categorized on the basis of the WHO pesticide hazard classification [13] as moderately hazardous to extremely hazardous classes. Paraquat and glyphosphate are two types of herbicides that are used most frequently in Taiwan [14].

Intentional vs. non-intentional

Lin et al. (2008) reported that adults experienced more intentional OP exposures (67%) and that ingestion was the most common route of exposure (74.5%) based on the network of Taiwan's Poison Control Centers (PCC). Lee et al. (2008) reported a similar proportion (66.1%) of poisoning exposures involving suicidal intent. However, our study included only pesticide poisoning hospitalizations in which the poisoning was non-intentional.

There was a large variation in the case fatality rate among pesticide poisonings, ranging from 0% to 42%. Even for the same chemical and/or World Health Organization (WHO) toxicity classification of pesticides and for those used for similar agricultural indications, a marked variation in lethality was observed [15]. Patients can come into contact with toxic substances intentionally or by accident. Unlike the situation in which a substance is deliberately administered, in accidental exposure cases, there is usually only a poor understanding of the pesticides to which a subject is exposed, both in a qualitative and a quantitative sense.

In addition, the occurrence and magnitude of harmful effects and the part of the patients' bodies in which these effects arise depend on the form in which the patients are exposed to the pesticides. The form of exposure is determined by the environmental compartment in which the victim exists, the physical state of the pesticide (gas or liquid), the physicochemical prosperities of the pesticide (soluble in water or not) and the nature of contact (single or prolonged). Furthermore, whether a pesticide is absorbed, and at what rate, is determined by the barriers that are encountered along the route of entry into the body.

For 8 persons categorized as herbicide poisoning, the cause of death was indicated as 'toxic effects of organophosphates for 93 (out of the 104 deaths) as 'toxic effects of other pesticides, not elsewhere classified'. This could be explained either by misclassification of the type of poisoning and/or cause of death as well as by the presence of two types of substances, a herbicide and organophosphate It has been previously reported that glyphosate herbicide poisoning patients received atropine or pralidoxime therapy [4,16] even though atropine and pralidoxime are not antidotes for glyphosate poisoning. These reports imply that some physicians confuse glyphosates with organophosphates.

Limitations

Several limitations need to be considered in the interpretation of our findings. First, no further information on the specific type of pesticide within the categorization of organophosphate or herbicide was available, and no information was available about the exposure route and the amount of exposure. Second, hospitalization data have an over-representation of severe poisonings and severe symptoms because of self-selection to the hospital policlinics as well as admission to the hospital care. The quality of diagnostic practice and criteria for hospitalization may vary and it is possible that pesticide poisoning is under-reported in Taiwan.

Third, the comorbidity diagnoses, which rely completely on the claims data reported by the physicians or hospitals, may be less accurate than if all individuals had been assessed using a single standardized procedure [17]. Health professionals may not receive much training in environmental toxicology or pesticide poisoning. The signs and symptoms of pesticide poisoning often resemble those of more common conditions, which may be diagnosed preferentially. The NHIRD used discharge diagnoses provided by treating physicians; no standardized criteria were used to define hospitalization cases. This lack of standardization increases the probability for case misclassification. The NHIRD, designed as an administrative dataset, does not include some important individual characteristics for further analyses such as smoking and alcohol consumption, all of which may contribute to death or hospitalization. The mortality and hospitalization rates may be underestimates because of the possible under-reporting of cases in the databases used in the analyses. Administrative databases are known to be subject to possible undercoding and overcoding errors [18]. The difficulty of receiving reimbursement through workers' compensation may also bias health care providers' diagnoses and the reporting of episodes of unintentional pesticide poisoning. The health care professionals may fear that their patients may be subject to retaliation.

Strengths of this study

Litchfield [19] categorized studies on acute pesticide poisoning in agriculture into three categories: clinical case reports, descriptive epidemiology studies, and cross-sectional studies. Several studies conducted in China, India, and Taiwan have been centered on hospital-based case reports [1,3–9,20,21]. These reports included insufficient information on the source population to estimate the poisoning rates. A particular strength of the current study is the use of a nationwide population-based dataset, allowing us to trace the medical services received by all patients after poisoning. Large computerized databases derived from this system by the Bureau of National Health Insurance, Taiwan (BNHI) and maintained by the National Health Research Institutes, Taiwan, are provided to scientists in Taiwan for research purposes.

NHIB established a uniform system to control the quality of medical services and coding. Chou et al. (2003) reported the accuracy of the comorbidity in the claims data using medical records as the gold standard [22]. The kappa values for all of the demographic data between two abstractors were greater than 95%. In addition, the kappa values for 18 comorbidities were more than 84% in inter-abstractor reliability. The agreement of kappa value of comorbidities between medical records and claims data were in a range between 0.31 and 0.86. Authors concluded the validity of comorbidity data in NIH claims data was in the range of moderate and substantial agreement. Several studies have been published also supported the strong validity of the NHIRD [23,24].Using the same ICD-9-CM codes over the study period maintained the internal validity of the temporal trend analyses. Except for the limited information of the route of exposure of pesticide poisoning, the present study is the most complete nationwide population-based study conducted to assess the risk of inpatient mortality and to identify some prognostic factors of inpatient mortality among insecticide and herbicide pesticide poisonings.

Conclusions

The major causes of mortality for insecticide poisonings were the toxic effect of organophosphates and coma, whereas the major causes of mortality for herbicide poisonings were the toxic effect of other pesticides and the toxic effect of organophosphates. The factors associated with inpatient mortality were herbicide poisoning and receiving further medical treatment; however, we could not further classify the different substances in the organophosphate insecticide and herbicide categories. We demonstrated that herbicides stand out as the dominant agent for poisoning-related fatalities. From public health perspective, there is an obvious need to strengthen the efforts to get more information on the details of herbicides for this database. This would allow identification of the most hazardous herbicides and could lead to better treatment and prevention of accidents in the future. Furthermore, the control of and limiting access to herbicide agents and developing appropriate therapeutic regimens, including emergency care, should be priorities. Physicians and general practitioners should be aware and capable of recognizing these types of poisonings early because such poisonings are rapidly fatal unless correctly diagnosed and treated. More research on chronic pesticide exposure is also needed because there is still relatively little known about the health effects of these types of poisonings.

Acknowledgments

This study is based in part on data from the National Health Insurance Research Database provided by the Bureau of National Health Insurance, Department of Health and managed by National Health Research Institutes. The interpretation and conclusions contained herein do not represent those of Bureau of National Health Insurance, Department of Health or National Health Research Institutes.

Author Contributions

Conceived and designed the experiments: CHL WCC. Analyzed the data: CHC CMC SLS. Wrote the paper: CHC CWC. Suggestions for revision of the manuscript: JJ SK.

References

1. Lin TJ, Walter FG, Hung DZ, Tsai JL, Hu SC, et al. (2008) Epidemiology of organophosphate pesticide poisoning in Taiwan. Clin Toxicol (Phila) 46: 794–801.
2. Lee CH, Shih CP, Hsu KH, Hung DZ, Lin CC (2008) The early prognostic factors of glyphosate-surfactant intoxication. Am J Emerg Med 26: 275–281.
3. Yang CC, Wu JF, Ong HC, Hung SC, Kuo YP, et al. (1996) Taiwan National Poison Center: epidemiologic data 1985–1993. J Toxicol Clin Toxicol 34: 651–663.
4. Chen YJ, Wu ML, Deng JF, Yang CC (2009) The epidemiology of glyphosate-surfactant herbicide poisoning in Taiwan, 1986–2007: a poison center study. Clin Toxicol (Phila) 47: 670–677.

5. Alarcon WA, Calvert GM, Blondell JM, Mehler LN, Sievert J, et al. (2005) Acute illnesses associated with pesticide exposure at schools. JAMA 294: 455–465.
6. Chien (2011) Characteristics and trends with respect to unintentional pesticide poisoning mortality and hospitalization in Taiwan, 1999–2008. Pesticides in the Modern World: InTech - Open Access Publisher.
7. National Health Research Institutes TR (2011) National health insurance research database- Introduction of service.
8. Charlson ME, Pompei P, Ales KL, MacKenzie CR (1987) A new method of classifying prognostic comorbidity in longitudinal studies: development and validation. J Chronic Dis 40: 373–383.
9. Lee HL, Lin HJ, Yeh ST, Chi CH, Guo HR (2008) Presentations of patients of poisoning and predictors of poisoning-related fatality: findings from a hospital-based prospective study. BMC Public Health 8: 7.
10. Gawarammana IB, Buckley NA (2011) Medical management of paraquat ingestion. Br J Clin Pharmacol.
11. Bradberry SM, Proudfoot AT, Vale JA (2004) Glyphosate poisoning. Toxicol Rev 23: 159–167.
12. Senarathna L, Adams J, De Silva D, Buckley NA, Dawson AH (2008) Personal and professional challenges in the management of deliberate self-poisoning patients in rural Sri Lanka: a qualitative study of rural hospital doctors' experiences and perceptions. BMC Public Health 8: 373.
13. WHO (2005) The WHO recommended classification of pesticides by hazard and guidelines to classification 2004. Geneva: World Health Organization.
14. Bureau of Animal and Plant Health Inspection and Quarantine EY, Taiwan (ROC) (2012) Pesticides Statistic Analysis.
15. Dawson AH, Eddleston M, Senarathna L, Mohamed F, Gawarammana I, et al. (2010) Acute human lethal toxicity of agricultural pesticides: a prospective cohort study. PLoS Med 7: e1000357.
16. Tominack RL, Yang GY, Tsai WJ, Chung HM, Deng JF (1991) Taiwan National Poison Center survey of glyphosate–surfactant herbicide ingestions. J Toxicol Clin Toxicol 29: 91–109.
17. Lien YC, Chen CH, Lin HC (2009) Risk factors for 24-hour mortality after traumatic rib fractures owing to motor vehicle accidents: a nationwide population-based study. Ann Thorac Surg 88: 1124–1130.
18. Shen HN, Lu CL, Yang HH (2010) Epidemiologic trend of severe sepsis in Taiwan from 1997 through 2006. Chest 138: 298–304.
19. Litchfield MH (2005) Estimates of acute pesticide poisoning in agricultural workers in less developed countries. Toxicol Rev 24: 271–278.
20. Liu Q, Zhou L, Zheng N, Zhuo L, Liu Y, et al. (2009) Poisoning deaths in China: type and prevalence detected at the Tongji Forensic Medical Center in Hubei. Forensic Sci Int 193: 88–94.
21. Jaiprakash H, Sarala N, Venkatarathnamma PN, Kumar TN (2011) Analysis of different types of poisoning in a tertiary care hospital in rural South India. Food Chem Toxicol 49: 248–250.
22. Chou P (2003) Validation of comorbidity derived from Health Insurance data.
23. Cheng CL, Kao YH, Lin SJ, Lee CH, Lai ML (2011) Validation of the National Health Insurance Research Database with ischemic stroke cases in Taiwan. Pharmacoepidemiol Drug Saf 20: 236–242.
24. Chen YH, Chen KY, Lin HC (2011) Non-alcoholic cirrhosis and the risk of stroke: a 5-year follow-up study. Liver Int 31: 354–360.

7

Novel Bioassay for the Discovery of Inhibitors of the 2-*C*-Methyl-*D*-erythritol 4-Phosphate (MEP) and Terpenoid Pathways Leading to Carotenoid Biosynthesis

Natália Corniani[1], Edivaldo D. Velini[1], Ferdinando M. L. Silva[1], N. P. Dhammika Nanayakkara[2], Matthias Witschel[3], Franck E. Dayan[4]*

1 São Paulo State University, Faculty of Agronomic Sciences, Botucatu, SP, Brazil, **2** National Center for Natural Products Research, School of Pharmacy, University of Mississippi, University, MS, United States of America, **3** BASF SE, GVA/HC-B009, Ludwigshafen, Germany, **4** USDA-ARS Natural Products Utilization Research Unit, University, MS, United States of America

Abstract

The 2-*C*-methyl-*D*-erythritol 4-phosphate (MEP) pathway leads to the synthesis of isopentenyl diphosphate in plastids. It is a major branch point providing precursors for the synthesis of carotenoids, tocopherols, plastoquinone and the phytyl chain of chlorophylls, as well as the hormones abscisic acid and gibberellins. Consequently, disruption of this pathway is harmful to plants. We developed an *in vivo* bioassay that can measure the carbon flow through the carotenoid pathway. Leaf cuttings are incubated in the presence of a phytoene desaturase inhibitor to induce phytoene accumulation. Any compound reducing the level of phytoene accumulation is likely to interfere with either one of the steps in the MEP pathway or the synthesis of geranylgeranyl diphosphate. This concept was tested with known inhibitors of steps of the MEP pathway. The specificity of this *in vivo* bioassay was also verified by testing representative herbicides known to target processes outside of the MEP and carotenoid pathways. This assay enables the rapid screen of new inhibitors of enzymes preceding the synthesis of phytoene, though there are some limitations related to the non-specific effect of some inhibitors on this assay.

Editor: Manfred Jung, Albert-Ludwigs-University, Germany

Funding: Grant# BEX 0226/12-2 from the Brazilian Agency CAPES (Coordination for the Improvement of Higher Education Personnel) that supported the research of N. Corniani in a USDA laboratory. The funders had no role in study design, data collection and analysis, decision to publish, or preparation of the manuscript.

Competing Interests: The authors from BASF SE, the University of Sao Paulo, The University of Mississippi and the USDA ARS declare that no competing interests exist.

* Email: franck.dayan@ars.usda.gov

Introduction

The terms isoprenoid, terpenoid, and terpene are used interchangeably in the literature to refer to a broad class of natural products derived from C5 isopentenyl diphosphate (IPP) [1,2]. Plants produce a myriad of isoprenoids that are functionally important in many physiological and biochemical processes [3,4]. Carotenoids comprise a large isoprenoid family that are derived from the C40 tetraterpenoid phytoene [5] and produced by all photosynthetic organisms (plants, algae and cyanobacteria) as well as certain non-photosynthetic bacteria and fungi [6]. In plants, carotenoids participate in photosynthetic processes including light harvesting, energy conversion, electron transfer, and quenching of excited chlorophyll triplets [7] in addition to a number of other functions.

Through evolution, two independent biosynthetic routes have been selected for the synthesis of these two basic building blocks [8]. In the cytosol and mitochondria, IPP and dimethylallyl diphosphate (DMAPP) are assembled from three molecules of acetyl-CoA by the mevalonate (MVA) pathway. This pathway was first described in the early work of Bloch and Lynen [9,10], and was thought to be the sole source of all terpenoids. However, it is now known that it is responsible for the synthesis of sterols and ubiquinone. The MVA pathway is the subject of several reviews [5,11], and is not the focus of this paper.

The existence of an alternative pathway was suggested based on the observation that genes encoding enzymes catalyzing the late steps of the MVA pathway are absent in some archaeal genomes [12]. Furthermore, plants treated with the herbicide clomazone had reduced carotenoid levels but their levels of sterols were not affected [13–15]. This plastid-localized independent pathway, called the 2-*C*-methyl-*D*-erythritol 4-phosphate (MEP) pathway (also non-mevalonate or 1-deoxy-*D*-xylulose 5-phosphate (DOXP) pathway), was reported in 1997 (Figure 1) [9,16–19]. The MEP pathway is a major branch point providing precursors for the synthesis of plastidic monoterpenes, diterpenes, carotenoids, the phytyl chain of chlorophylls, tocopherols, plastoquinone as well as the hormones abscisic acid and gibberellins. There is limited crossover between the two pathways [20,21].

The MEP and carotenoid pathways are well characterized and have been reviewed extensively [1,5,12,22]. Briefly, carotenoid

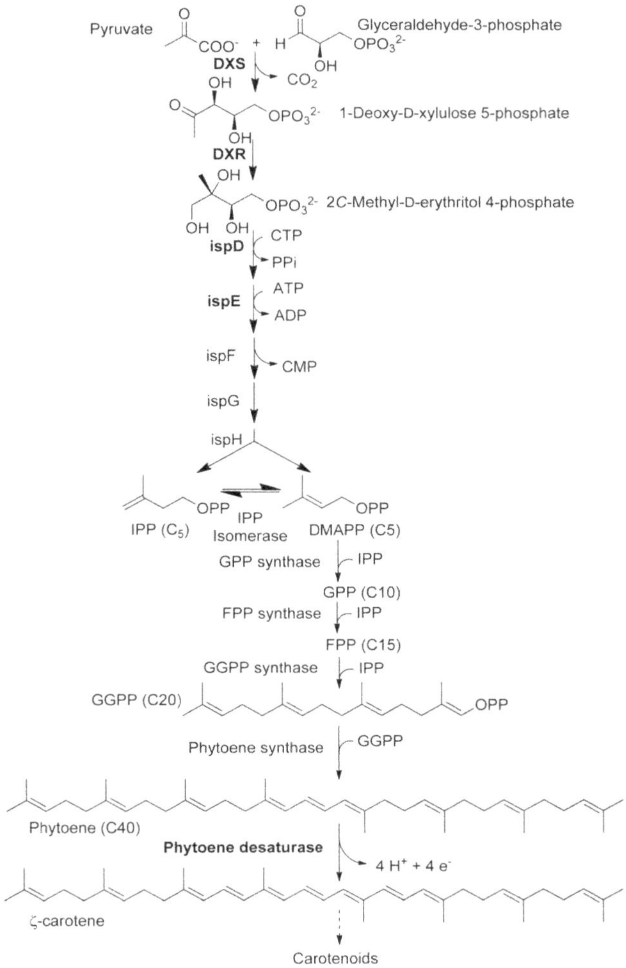

Figure 1. Biosynthesis of carotenoids starts with the 2-*C*-methyl-*D*-erythritol 4-phosphate (MEP) pathway leading to the formation of IPP, continues with the isoprenoid pathway to obtain GGPP. The first committed step to the synthesis of carotenoids consists of the head to head condensation of 2 GGPP to form phytoene. The enzymes in bold letters denote enzyme targets that were tested in this study.

biosynthesis can be divided into three phases (Figure 1). Phase I includes the formation of IPP and DMAPP via the plastid-localized MEP pathway. The first step of the MEP pathway is catalyzed by 1-deoxy-*D*-xylulose 5-phosphate synthase (DXS, EC 2.2.1.7), converting pyruvate and glyceraldehyde-3-phosphate to 1-deoxy-*D*-xylulose 5-phosphate (DOXP). The intramolecular rearrangement and reduction of DOXP to 2-*C*-methyl-*D*-erythritol 4-phosphate (MEP) is catalyzed by 1-deoxy-*D*-xylulose 5-phosphate reductoisomerase (DXR, EC 1.1.1.267). Diverse experimental evidence demonstrates that DXS and DXR represent potential regulatory control points in the MEP pathway [23,24]. Phase I concludes with the formation of the C5 building blocks IPP and DMAPP. In Phase II, a single DMAPP serves as the substrate for successive head-to-tail condensations of IPP units to ultimately form the C20 geranylgeranyl diphosphate (GGPP) [25]. Phase III begins with the head to head condensation of two GGPP molecules to produce phytoene (Figure 1) by the enzyme phytoene synthase (PSY, EC 2.5.1.32). Subsequently, phytoene desaturase (PDS, EC 1.3.5.5) and ζ-carotene desaturase (ZDS, EC 1.3.5.6) catalyze similar dehydrogenation reactions introducing four double bonds in phytoene to form lycopene. Desaturation requires a plastid-localized terminal oxidase and plastoquinone in photosynthetic tissues [26,27].

Carotenoids are important for plant survival, especially in their role as protection from photooxidation [4,28]. Several important bleaching herbicides inhibit carotenoid synthesis [29]. PDS is the target site for several herbicides such as norflurazon, fluridone and flurochloridone [30]. When sensitive plants are exposed to these herbicides, PDS activity is inhibited, resulting in a rapid accumulation of phytoene and cessation of carotenoid biosynthesis [31]. In plants, inhibition of *p*-hydroxyphenylpyruvate dioxygenase (HPPD, EC 1.13.11.27) by triketone, isoxazole and pyrazole herbicides affects the formation of homogentisic acid [32], which is a key precursor for the biosynthesis of plastoquinone, a critical cofactor of PDS [26] and leads to inhibition of its enzymatic activity. This class of herbicides represents the last herbicide mode of action to have been commercialized in the last twenty years [33].

Blockage of any of the steps preceding the formation of lycopene inhibits carotenoid synthesis. However, only the herbicide clomazone targets the MEP pathway by inhibiting DXS [34], although it does so indirectly [35]. It has been postulated that, once absorbed into the plant, clomazone is oxidized by cytochrome P450 monooxygenases (P450s) to form ketoclomazone (Figure 2), which is the putative active herbicidal form [34]. Early evidences of this requirement for metabolic activation was observed in plants treated with phorate or other P450s inhibitors being protected from the herbicidal effect of clomazone [36,37]. The antibiotic fosmidomycin is not used as an herbicide, but this compound inhibits DXR, the second step of the MEP pathway [38,39].

The MEP and carotenoid pathways are not present in animals, and thus, their enzymes are preferred targets for new herbicides [40]. In spite of the potential relevance of all other enzymes of this pathway, no herbicide targeting the early steps of carotenoid synthesis, other than clomazone, has been developed [41].

The emergence of resistance to herbicides is an increasing problem facing agriculture [42,43], and there do not appear to be any herbicides with novel mechanisms of action being developed [33]. The pathways leading to carotenoids (MEP and isoprenoid pathways) offer several attractive targets for new molecules discovery efforts [44]. Indeed, the unique target sites inhibited by clomazone [37] and fosmidomycin [45] illustrate the potential benefits of developing new herbicides which interfere with the early steps of carotenoid synthesis.

One approach to discover novel inhibitors has been high throughput *in vitro* assays that are most likely to identify

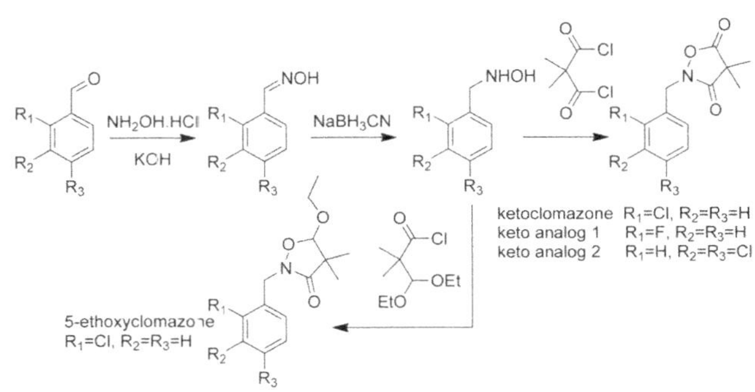

Figure 2. Schematics of the synthesis of 5-ethoxyclomazone, ketoclomazone and keto analogs 1 and 2.

mechanism-based inhibitors of the various steps in the MEP pathway [46]. The aim of our research was to develop a simple, fast, and inexpensive, *in vivo* assay to identify inhibitors of the early steps in carotenoid synthesis by measuring the carbon flux through the MEP and isoprenoid pathways using phytoene as a biomarker.

Materials and Methods

Chemicals and supplies

Phorate, *O,O*-diethyl *S*-[(ethylthio]methyl) phosphorodithioate; fosmidomycin (3-(formylhydroxyamino)propyl)-phosphonic acid sodium salt; FR-900098, *p*-[3-(acetylhydroxyamino)propyl]-phosphonic acid; amitrol, 1,2,4-triazol-3-amine; dinoterb, 2-(1,1-dimethylethyl)-4,6-dinitro-phenol; endothall monohydrate, 7- oxabicyclo[2.2.1] heptane-2, 3-dicarboxylic acid monohydrate; oryzalin, 4-(dipropylamino)-3,5-dinitro-benzenesulfonamide; sulcotrione, 2-[2-chloro-4-(methylsulfonyl)benzoyl]-1,3-cyclohexanedione were purchased from Sigma-Aldrich (St. Louis, MO 63103) and dichlobenil (2,6-dichlorobenzonitrile); glufosinate-ammonium, 2-amino-4-(hydroxymethylphosphinyl)butyric acid ammonium salt were purchased from Allied-Signal Inc., (Morristown, NJ 07960).

Clomazone, 2-[(2-chlorophenyl)methyl]-4,4-dimethyl-3-isoxazolidinone; imazapyr, 2-[4,5-dihydro-4-methyl-4-(1-methylethyl)-5-oxo-1*H*-imidazol-2-yl]-3-pyridinecarboxylic acid; quinclorac, 3,7-dichloro-8-quinolinecarboxylic acid; atrazine, 6-chloro-*N*2-ethyl-*N*4-(1-methylethyl)-1,3,5-triazine-2,4-diamine; paraquat CL tetrahydrate, 1,1′-dimethyl-4,4′-bipyridinium dichloride; imazethapyr, 2-(4,5-dihydro-4-methyl-4-(1-methylethyl)-5-oxo-1*H*-imidazol-2-yl)-5-ethyl-3-pyridinecarboxylic acid; alachlor, 2-chloro-*N*-(2,6-diethylphenyl)-*N*-(methoxymethyl) acetamide; asulam, *N*-[(4-aminophenil)sulfonyl]-carbamic acid methyl ester; sulfentrazone, *N*-[2,4-dichloro-5-[4-(difluoromethyl)-4,5-dihydro-3-methyl-5-oxo-1*H*-1,2,4-triazol-1-yl]-phenyl] methanesulfonamide; glyphosate-isopropylammonium, isopropylammonium *N*-(phosphonomethyl) glycine; diclofop-methyl, 2-[4-(2,4-dichlorophenoxy)phenoxy]-propanoic acid methyl ester were purchased from Chem-Service (West Chester, PA 19381).

Norflurazon, 4-chloro-5-(methylamino)-2-[3-(trifluoromethyl)-phenyl]-3(2*H*)-pyridazinone was provided by Sandoz, Inc. Crop Protection (now Syngenta, Greensboro, NC 27419), and experimental inhibitors for some of the enzymes of the MEP pathway were provided by BASF-SE (Ludwigshafen, Germany) [41].

Synthesis of halogenated analogs

Reagents and solvents were purchased from Sigma-Aldrich Chemical Co. (St Louis, MO, USA) and Fisher Scientific (Pittsburgh, PA, USA). NMR spectra were recorded on a Varian-Mercury-plus-400 or Varian Unity-Inova-600 spectrometer using $CDCl_3$ and methanol-d_4 unless otherwise stated. MS data were obtained from an Agilent Series 1100 SL equipped with an ESI source (Agilent Technologies, Palo Alto, CA, USA). Column chromatography and preparative TLC were performed on Merck silica gel 60 (230–400 mesh) and silica gel GF plates (20×20 cm, thickness 0.25 mm), respectively. General synthesis of 5-ethoxyclomazone, ketoclomazone and analogs is shown in Figure 2.

Synthesis of halogenated-benzaldehyde oximes

Oximes of 2-chloro-, 2-fluoro, and 3,4-dichlorobenzaldehyde were prepared by the procedure reported by Zamponi et al. [47].

2-Chlorobenzaldehyde oxime: ^{1}H NMR δ(CDCl3): 7.26 (1H, brt, J = 7.2 Hz), 7.31 (1H, td, J = 7.6, 1.6 Hz), 7.38 (1H, dd, J = 8.0, 1.2 Hz), 8.62 (1H, s).

2-Fluorobenzaldehyde oxime: ^{1}H NMR δ(CDCl3): 7.10 (1H, brt, J = 9.2 Hz), 7.66 (1H, brt, J = 7.6 Hz), 7.37 (1H, brq, J = 7.2 Hz), 7.71 (1H, t, J = 7.6 Hz), 8.38 (1H, s).

3,4-Dichlorobenzaldehyde oxime: ^{1}H NMR δ(CDCl3): 7.40 (1H, dd, J = 8.4, 2.0 Hz), 7.46 (1H, d, J = 8.4 Hz), 7.67 (1H, d, J = 2.0 Hz), 8.01 (1H, s).

Synthesis of halogenated N-(benzyl)hydroxylamine

A mixture of oxime (500 mg) in glacial acetic acid (10 ml) was treated with sodium cyanoborohydride ($NaCNBH_3$) portion wise under stirring while maintaining the temperature below 20°C until the reaction was complete as evidenced by TLC. The reaction mixture was basified with ice-cold NaOH and extracted with ethyl acetate. The organic layer was washed with water, dried and evaporated to afford a white solid. This product was chromatographed over silica gel and eluted with ethyl acetate:hexane 3:7 to yield a pure product which was crystallized from $CHCl_2$/hexanes.

N(2-Chlorobenzyl)hydroxylamine: ^{1}H NMR δ(CDCl3): 4.1 (2H, s), 7.21–7.26 (2H, m), 7.35–7.39 (2H, m).

N(2-Fluorobenzyl)hydroxylamine: ^{1}H NMR δ(CDCl3): 4.06 (2H, s, CH_2), 7.05 (1H, brt, J = 8.8 Hz), 7.13 (1H, brt, J = 7.2 Hz), 7.27 (1H, brq, J = 6.8 Hz), 7.33 (1H, brt, J = 7.6 Hz).

N(3,4-Dichlorobenzyl)hydroxylamine (**6**): ^{1}H NMR δ(CDCl3): 3.93 (2H, s), 7.15 (1H, dd, J = 8.0, 2.0 Hz), 7.40 (1H, d, J = 8.0 Hz), 7.44 (1H, d, J = 2.0 Hz).

Synthesis of halogenated ketoclomazone

Dimethylmalonyl dichloride (1.25 mM) in CH_2Cl_2 (1 ml) was slowly added to a solution of hydroxylamine (1 mM) and triethylamine (0.3 ml) in CH_2Cl_2 (5 ml) at 10°C. After addition was complete, the reaction mixture was stirred for 30 min poured onto ice and extracted with CH_2Cl_2. The organic layer was washed with aqueous Na_2CO_3, 1 N HCl and saturated aqueous NaCl. The resulting solution was then dried and the gummy product obtained was chromatographed over silica gel. The product was eluted with ethyl acetate:hexane 5:95 and crystallized in CH_2Cl_2/hexanes.

Ketoclomazone: ^{1}H NMR δ(CDCl3): 1.44 (6H, s), 5.05 (2H, s), 7.24–7.31 (2H, m), 7.34 (1H, m), 7.39 (1H, m). 13C NMR δ(CDCl3): 21.4 (CH_3), 41.9 (C), 47.3 (CH_2), 127.3 (CH), 130.0 (CH), 130.1 (CH), 130.3 (CH), 131.4 (C), 133.8 (C), 172.1 (C), 173.8 (C); HRESIMS $[M+H]^+$ m/z 254.0600 (calcd for $(C_{12}H_{12}ClNO_3+H)^+$254.0584).

2-(2-Fluorobenzyl)-4,4-dimethylisoxazolidine-3,5-dione (keto analog **1**): ^{1}H NMR δ($CDCl_3$): 1.41 (6H, s), 4.98 (2H, s), 7.07 (1H, brt, J = 9.2 Hz), 7.13 (1H, brt, J = 7.6 Hz), 7.25–7.34 (2H, m). 13C NMR δ(CDCl3): 21.2 (CH_3), 41.8 (C), 47.3 (CH_2, J_{CF} = 4.4 Hz), 115.8 (CH, J_{CF} = 21.3 Hz), 120.7 (CH, J_{CF} = 14.0 Hz), 124.5 (CH, J_{CF} = 3.6 Hz), 130.5 (CH), 130.6 (CH, J_{CF} = 5.9 Hz), 160.9 (C, J_{CF} = 248 Hz), 172.4 (C), 173.6 (C); HRESIMS $[M+H]^+$ m/z 237.0811 (calcd for $(C_{12}H_{12}FNO_3+H)^+$237.0801).

2-(3,4-Dichlorobenzyl)-4,4-dimethylisoxazolidine-3,5-dione (keto analog **2**): ^{1}H NMR δ(CDCl3): 1.40 (6H, s), 4.83 (2H, s), 7.17 (1H, dd, J = 8.0, 2.0 Hz), 7.41 (1H, brs), 7.42 (1H, d, J = 2 Hz), 7.43 (1H, d, J = 8.0 Hz). 13C NMR δ(CDCl3): 21.3 (CH_3), 42.0 (C), 48.8 (CH_2), 128.0 (CH), 130.6 (CH), 131.1 (CH), 133.1 (C), 133.2 (C), 134.0 (C), 172.9 (C), 173.5 (C); HRESIMS $[M+H]^+$ m/z 288.0207 (calcd for $(C_{12}H_{11}Cl_2NO_3+H)^+$288.0194).

Synthesis of 5-ethoxyclomazone

A mixture of ethyl 3,3-diethoxy-2,2-dimethylpropionate [48] and KOH (2.2 g) in 10% aqueous ethanol (40 ml) was refluxed for 2 hours and the solvent was evaporated under vacuum. The residue was dissolved in water (40 ml), neutralized with succinic acid and extracted with CH_2Cl_2. The organic layer was washed with saturated aqueous NaCl, dried over anhydrous Na_2SO_4, and evaporated to give 3-diethoxy-2-dimethylpropanoic acid as a thick oil. Upon storage at 4°C this oil formed a white crystalline solid.

1H-NMR (400 MHz) δ 1.17 (6H, s, CH_3), 1.17 (3H, t, J = 7.0 Hz, CH3), 3.55 (2H, dq, J = 16.0, 7.2 Hz, CH_2), 3.82 (2H, dq, J = 16.0, 7.2 Hz, CH_2), 4.54 (1H, s, CH); 13C-NMR (100 MHz) δ 15.3 (CH_3), 19.6 (CH_3), 48.4 (C), 66.5 (CH_2), 107.3 (CH), 181.9 (CO); HRESIMS $[M+Na]^+$ m/z 213.1099 (calcd for $(C_9H_{18}O_4+Na)^+$213.1103).

Oxalyl chloride (260 mg, 4 mm) was added to a solution of 3-diethoxy-2-dimethylpropanoic acid (380 mg) in toluene (3 ml) and the reaction mixture was stirred at 60°C for 30 minutes. The solvent was evaporated under vacuum to afford oil. This oil was dissolved in toluene 5 ml and evaporated under vacuum. This product was used in the next reaction immediately without further purification.

Triethylamine (0.5 ml) and 2-chloro-*N*-hydroxybenzylamine (200 mg 1.27 mm) were added sequentially to a solution of 3-diethoxy-2-dimethylpropanoic acid chloride (2 mm) in CH_2Cl_2 (4 ml) at 0°C under stirring. The reaction mixture was stirred over night at room temperature and partitioned between water and CH_2Cl_2. The organic layer was washed with water, dried over Na_2SO_4, and evaporated. The products were chromatographed on silica gel and elution with ethyl acetate/hexane (1:99) gave the least polar product, 5-ethoxyclomazone, as an colorless oil (32 mg).

1H-NMR (400 MHz) δ 1.06 (3H, t, J = 7.2 Hz, CH_3), 1.16 (3H, s, CH_3), 1.16 (3H, s, CH_3), 1.24 (3H, s, CH_3), 3.38 (1H, dq, J = 16.0, 7.2 Hz, CH) 3.53 (1H, dq, J = 16.0, 7.2 Hz, CH), 4.73 (1H, d, J = 16.8 Hz, CH_2), 4.83 (1H, s, CH), 4.73 (1H, d, J = 16.8 Hz, CH), 4.89 (1H, d, J = 16.8 Hz, CH), 7.16–7.25 (2H, m, CH), 7.29–7.35 (2H, m, CH); 13C-NMR (100 MHz) δ 14.8 (CH_3), 16.6 (CH_3), 22.4 (CH_3), 46.1 (CH_2), 46.3 (C), 64.1 (CH_2), 108.0 (CH), 126.8 (CH), 129.0 (CH), 129.2 (CH), 129.4 (CH), 132.9 (C), 133.2 (C), 172.8 (CO); HRESIMS $[M+H]^+$ m/z 283.0969 (calcd for $(C_{14}H_{18}ClNO_3+H)^+$283.0975).

Plant material

Barley (*Hordeum vulgare* L.) seeds were purchased from Johnny's Selected Seeds (Waterville, Maine 04903). Seeds were sown in moist commercial Metromix potting soil and grown either in a dark growth chamber set at 25°C or in the greenhouse under natural light for 4 days.

Bioassays

Approximately 0.1 g of fresh young barley leaves were weighed, cut in 3 mm sections with a razor blade, and incubated (60×15 mm Petri dishes) in 5 ml of 5 mM 2-[*N*-morpholino]ethanesulfonic acid buffer (MES, pH 6.5) containing 200 μM of norflurazon for 24 h in a growth chamber with the 16/8 light/dark cycle at 25°C. The herbicides and other test compounds (see section *Chemicals and Supplies*) were tested either at a fixed concentration or at different concentrations(dose-response curves ranged from 0.1 to 100 μM inhibitor in acetone). Control tissues were exposed to the same amount of acetone as the treated tissues but without the test compounds. All experiments had 3–5 replicates and were repeated in time. The effect of 50 μM phorate, a cytochrome P450 monooxygenase inhibitor, was tested in some of the assays with clomazone to determine the requirement for the metabolic activation of this herbicide.

The usefulness of this bioassay in identifying potential novel inhibitors of the early steps of carotenoid biosynthesis was tested with a number of experimental compounds provided by BASF or synthesized in our laboratory. These compounds were tested at a 100 μM final concentration on greening etiolated barley leaf cuttings and their activity is expressed as inhibition of phytoene accumulation relative to the amount of phytoene accumulating in the norflurazon alone treatment. Finally, the specificity of this assay was evaluated by testing representative compounds inhibiting all the known target sites of commercial herbicides. These compounds were tested at a 100 μM final concentration on etiolated barley leaf cutting either exposed to light or maintained in total darkness for 24 h. Herbicides causing a 10% or less reduction of phytoene level were considered not active. Those causing 10 to 50% inhibition and those causing more than 50% inhibition were considered slightly and highly active, respectively.

Phytoene extraction and determination

Phytoene was extracted and quantified according to a protocol modified from Sprecher et al. [49] as described in Dayan et al. [50]. After 24 h incubation, the barley leaf samples were homogenized (Polytron, PT 3300) in 3 ml of 6% KOH in methanol (w/v), stored for 15 min at room temperature (RT) and then centrifuged for 5 min at 1,300 *g* (Sorvall Swinging Bucket SH-3000 rotor). The supernatant was transferred into new tubes and mixed with 3 ml of petroleum ether (Acros, Fair Lawn, NJ, boiling range 80–110°C). Saturated NaCl solution was added (1.5 ml), mixed and centrifuged for 10 min at 1,300 *g*. A clear partition is formed by centrifugation and an aliquot (1.250 ml) of the epiphase was transferred to disposable cuvettes (methacrylate) followed by UV-spectrophotometric (Shimadzu, UV-3101PC) measurements at 287 nm. All manipulations were performed in a dark room under green light. The phytoene content was calculated by its extinction coefficient (ε) of 1108 mM cm^{-1} and expressed as μg g^{-1} fresh weight (FW) according to equation 1:

$$\mu g\, g^{-1}\, FW = \frac{(A/\varepsilon) \times 0.03}{g\,FW} 10^6 \tag{1}$$

Statistical Analysis

Phytoene accumulation was plotted against inhibitor concentrations to generate dose-response curves. Data were analyzed by a four-parameters log-logistic model [51] using R software (version 2.15.2, R Foundation for Statistical Computing, Vienna, Austria) with the drc module [52]. Means and standard deviations were obtained using the raw data and the half-maximal inhibitory response (I_{50}) was defined as the concentration at which this accumulation was inhibited by 50% compared with controls. I_{50} values were obtained from the parameters in the regression curves. Graphs were generated with Sigma Plot (version 11, Systat Software Inc., San Jose, CA, USA). Means were separated with the Duncan multiple range test at $P = 0.05$ using the Agricolae module [53].

The quality of the assay was determined by calculating the Z′ factor using equation 2 according to Zhang et al. [54]

$$Z' = 1\frac{(3\sigma_{c+} + 3\sigma_{c-})}{|\mu_{c+} - \mu_{c-}|} \tag{2}$$

where $3\sigma_{c+} = 3$ standard deviations of positive control, $3\sigma_{c-} = 3$ standard deviations of negative control, μ_{c+} = mean of positive control and μ_{c-} = mean of negative control. The utilization of the 3 standard deviations ensures 99.73% confidence limit. The Z′ of the phytoene accumulation assay is based on the measurements of 33 positive controls and 33 negative controls.

Results and Discussion

Accumulation of phytoene over time

This simple assay relies on the premise that the accumulation of phytoene resulting from the inhibition of phytoene desaturase by norflurazon is a reflection of the carbon flow through the MEP pathway. Incubation of barley leaves floating on a medium supplemented with 200 μM norflurazon for 24 h caused a time-dependent linear accumulation of phytoene (Figure 3). Therefore, any compound reducing the level of phytoene accumulation during this assay is likely to inhibit one of the many enzymatic steps leading to phytoene synthesis (Figure 1). It has been suggested that inhibiting the MEP pathway of plants could be useful in the search of novel herbicides [55] and the bioassay described herein may be a useful new tool to discover such compounds.

Barley was selected because its small seeds germinate quickly in the dark and contain highly active MEP and carotenoid pathways during its light-induced thylakoid formation, in the transition of etioplasts to chloroplasts during the greening process [56]. Additionally, it is highly sensitive to clomazone [57], which is important because some plants do not metabolize clomazone to ketoclomazone (the putative active form) very rapidly, and their inhibitory effects may not be detected during the time-span of this experiment.

Effect of clomazone, ketoclomazone, and 5-ethoxyclomazone

Clomazone is the only commercial herbicide known to inhibit carotenoid synthesis upstream from phytoene desaturase. Actually, clomazone is inactive, but its metabolite ketoclomazone inhibits DXS [34,56], the thiamine diphosphate-dependent enzyme that catalyzes the first step in the MEP pathway [58].

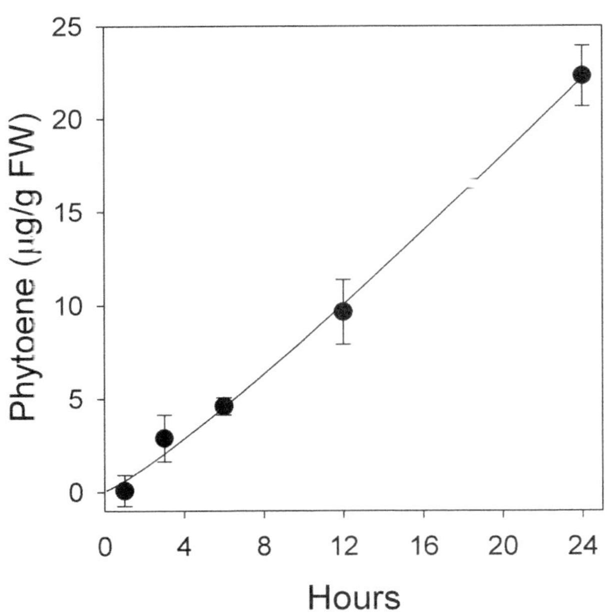

Figure 3. Time-dependent phytoene accumulation in barley (*Hordeum vulgare* L.) exposed to 200 μM norflurazon. Data represent means of three replications with standard deviation.

In our simple barley leaf cutting assay, clomazone inhibited phytoene accumulation in a dose-dependent manner that illustrates the inhibition of carbon flow into the MEP pathway in both green (Figure 4A) and greening etiolated tissues (Figure 4B). Clomazone had an I_{50} for inhibition of phytoene accumulation of 0.6±0.16 and 0.33±0.05 μM in green and greening tissues, respectively. Sandmann and Böger [30] reported an I_{50} value for clomazone of less than 15 μM for inhibition of phytoene and phytol biosynthesis in spinach extracts, suggesting that our *in vivo* assay may be more sensitive as it allows for the metabolic activation of clomazone [36,37].

The requirement for metabolic activation of clomazone was confirmed by repeating the same dose-response curves in the presence of 50 μM phorate. Phorate is an organophosphate insecticide that inhibits cytochrome P450 monooxygenases in plants [59]. In some species, clomazone is rapidly metabolized by P450s [60–62] and phorate can protect plants against the

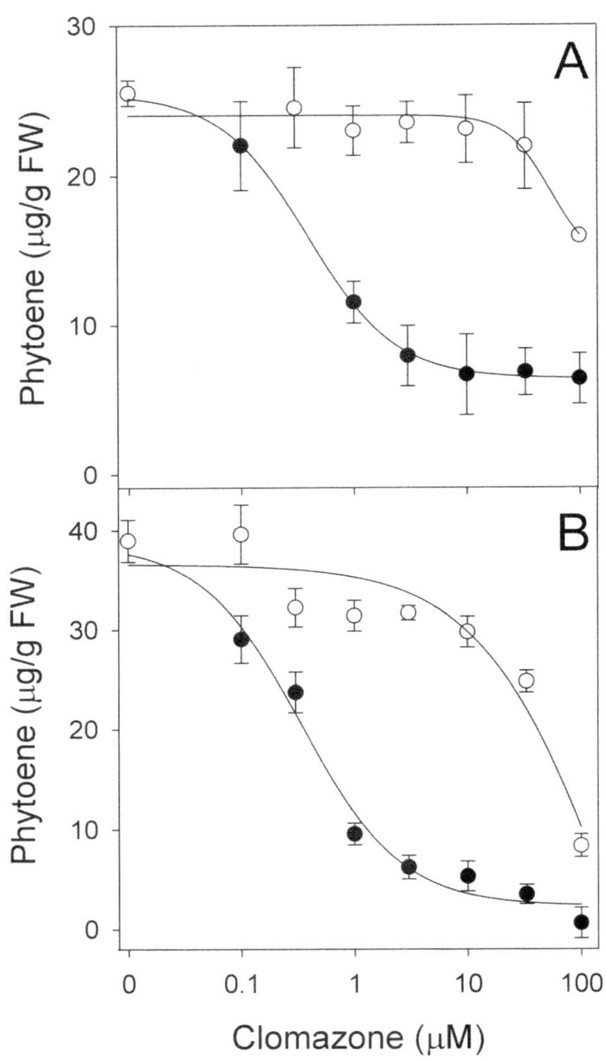

Figure 4. Dose-response curves showing the effect of the herbicide clomazone with (○) and without (●) phorate on phytoene accumulation induced by 200 μM norflurazon. (A) Green and (B) greening etiolated young barley leaves. Data represent means of three replications with standard deviation.

phytotoxic effect of clomazone by preventing this metabolic activation [36,63].

The addition of 50 μM phorate to the solution prevented clomazone from inhibiting phytoene accumulation in both green and greening etiolated, with $I_{50}>100$ μM (Figures 4 A and B). These results are consistent with previous studies indicating that clomazone must be metabolically activated by P450s and that phorate can abolish its herbicidal activity [36,63].

In plants, metabolism of clomazone follows a couple of pathways. One route involves a *N*-dealkylation step yielding 2-chlorobenzyl alcohol. The other route follows the progressive oxidation of carbon 5 to yield 5-hydroxyclomazone and ultimately ketoclomazone [60,62]. In both of these studies, 5-hydroxyclomazone and ketoclomazone were present in equivalent amounts, suggesting that the conversion of clomazone to ketoclomazone is not instantaneous. While ketoclomazone has been proposed as the active form responsible for the herbicidal activity of clomazone [34,56], early work on the development of clomazone by FMC (Philadelphia, PA USA) also reported that 5-hydroxyclomazone and several 5-alkoxy derivatives were also potent herbicides [64]. It is postulated that all these 5-hydroxy derivatives are further metabolized into ketoclomazone.

When tested in our bioassay, 5-ethoxyclomazone was as active as clomazone, and phorate did not negatively affect that activity (Figure 5A). Although still very active, ketoclomazone was 16 times less active in greening tissues than clomazone, with I_{50} values of 5 μM (Figure 5B). This was unexpected since clomazone is a proherbicide and appears to be bioactivated by P450s (Figure 4). Phorate did not affect the activity of ketoclomazone, suggesting that it is not subject to further oxidation by P450s (Figure 5B). Similar results were obtained in green tissues (Figure S1).

From a metabolic perspective, the pattern observed with clomazone, 5-ethoxyclomazone and ketoclomazone, and the effect of phorate on the activity of these compounds is particularly informative. Clomazone is a potent *in vivo* inhibitor of phytoene accumulation but phorate greatly reduces this activity, confirming that clomazone is not herbicidal and that one of its oxidized metabolites must be very active, causing inhibition of the carbon flow at submicromolar concentrations. 5-Ethoxyclomazone is as active as the commercial herbicide. On the other hand, the activity of ketoclomazone was lower than that of clomazone, and phorate had no effect on the herbicidal activity of ketoclomazone, suggesting that either the uptake of ketoclomazone was limiting, or that another intermediate is the active form of the herbicide.

Clomazone, 5-hydroxyclomazone and a series of acylated 5-hydroxy derivatives do not inhibit purified plant DXS, whereas ketoclomazone is highly active, with an I_{50} value of 80 nM [41]. Additionally, the *seco*-ketoclomazone analog with a 3-[methyl(hydroxy)amino]-2,2-dimethyl-3-oxopropanoic acid side chain, which is a known plant metabolite of clomazone [60,62], was recently reported as a potent inhibitor of *Haemophilus influenza* DXS [65]. Analysis of the structure-activity relationship of the compounds tested in that study highlighted the requirement for the hydroxamate, the 2,2-dimethylmethylidene, and the carboxylic functional groups. While this *seco*-ketoclomazone analog has structural features reminiscent of a putative transition state intermediate analog, neither the exact form of the clomazone pharmacophore responsible for inhibition of DXS nor its binding to the enzyme is known.

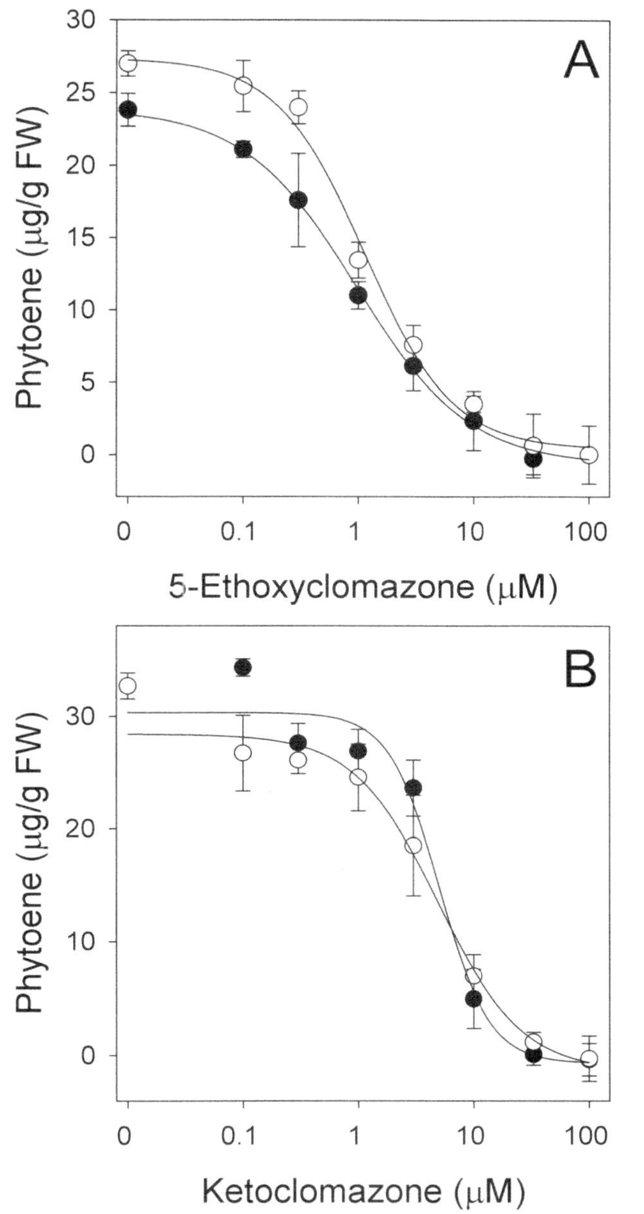

Figure 5. Dose-response curves of (A) 5-ethoxyclomazone and (B) ketoclomazone with (○) and without (●) phorate on greening etiolated barley leaves. Phytoene was caused to accumulate by the presence of 200 μM norflurazon. Data represent means of three replications with standard deviation.

Validity of the bioassay with inhibitors of early steps of the MEP pathway

Keto analogs **1** and **2** are structurally related to ketoclomazone with simple changes in the halogen substitutions on the phenyl ring (Figure 2). In this study, keto analog **1** was a weak inhibitor of phytoene accumulation, with an I_{50} value of 22 μM on greening etiolated tissue (Figure S2A). Its activity on green tissue was even lower, with an I_{50} value of 52 μM (Figure S2B). However, the structural similarity between ketoclomazone and keto analog **1** suggests that this new compound may inhibit DXS. Keto analog **2** did not affect phytoene accumulation (data not shown), suggesting that the position of the halogen on the phenyl ring has a significant effect on the activity of ketoclomazone.

The first compound described to inhibit the MEP pathway was fosmidomycin, also known as FR-31564 and 3-(*N*-formyl-*N*-hydroxyamino) propylphosphonic acid [66,67]. This compound prevents the conversion of labeled DOXP into carotenoid in the *Capsicum* chromoplast system [68] by inhibiting DXR (Figure 1) [66]. Since then, a considerable number of fosmidomycin derivatives have been synthesized to identify new DXR inhibitors [69,70].

In our study, fosmidomycin and its structural analog FR-900098 inhibited phytoene accumulation in a dose-dependent manner with I_{50} values of 5 and 5.5 μM for greening etiolated, respectively (Figure 6). Similar activity was observed on green barley leaves, with I_{50} values of 11 and 17.5 μM for fosmidomycin and FR-900098, respectively (Figure S3).

Interest in inhibitors of the MEP pathway has led a research group at BASF-SE (Ludwigshafen, Germany) to screen for new herbicides targeting this pathway in target-based high throughput *in vitro* assays [41]. Eleven of these experimental compounds were tested in our *in vivo* bioassay.

None of these compounds were very active in our *in vivo* bioassay (Table 1). The most active cluster were **BASF 11** and **BASF 12**, the azolopyrimidine inhibitors of 2*C*-methyl-D-erythritol 4-phosphate cytidyltransferase (IspD), the third step on the MEP pathway (Figure 1), with 63.6% and 61.3% inhibition of phytoene accumulation, respectively (Table 1). Interestingly, these compounds were the most potent MEP inhibitors generated by BASF with *in vitro* I_{50} values against IspD activity at 140 nM and 35 nM, respectively [41]. IspD activity has been validated by antisense experiments to be essential to plant survival and inhibitors of this enzyme were anticipated to be potentially herbicidal [71,72]. All of the other BASF compounds, including the putative inhibitors of 4-diphosphocytidyl-2*C*-methyl-D-erythritol kinase (ispE), had *in vitro* activity in the micromolar range [41] and were not very active in the barley bioassay (Table 1). The difference between the activity reported on the target sites and in the barley leaf bioassay may be due to a number of factors. There is often no correlation between the *in vitro* activity of experimental compounds and their performance as herbicides *in vivo* because their physicochemical properties may be less than ideal for uptake

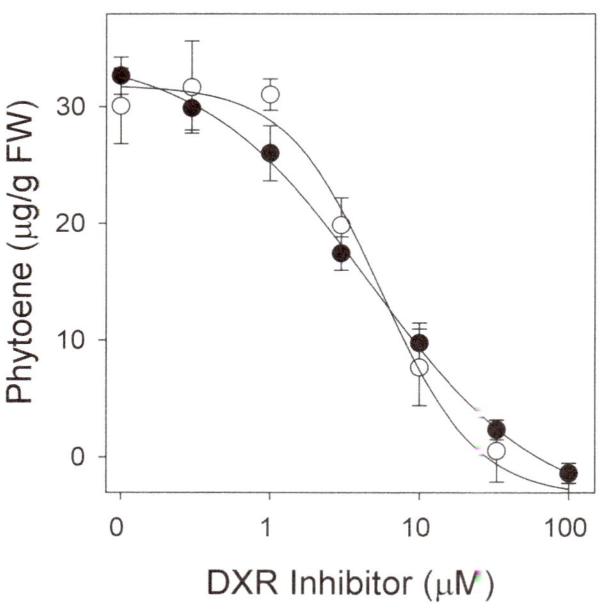

Figure 6. Dose-response curves of the DXR inhibitors fosmidomycin (●)and FR-900098 (○) on greening etiolated barley leaves. Phytoene accumulation was induced in the presence of 200 μM norflurazon. Data represent means of three replications with standard deviation.

and translocation [33]. Metabolic degradation of the compounds may also play a role.

Selectivity of the bioassay

Herbicides have unique affinities for their respective molecular target sites within important plant biochemical pathways and/or physiological processes [35,73]. To test the selectivity of the bioassay to identify inhibitors of the early steps of carotenoid synthesis, herbicides representative of all known modes of action were selected to survey their effects on phytoene accumulation (Table 2). All of the compounds were tested at 100 μM to determine whether they would be detected as false positives in a screening program.

Most of the herbicides targeting enzymes in pathways unrelated to carotenoid synthesis had either little or no effect on the carbon flow into phytoene (−/+). The slight inhibition caused by glyphosate may be a reflection of the secondary effect of this herbicide on the carotenoid pathway. Studies to understand the mechanism of vacuolar sequestration involved in resistance to glyphosate revealed that 2-*C*-methyl-*D*-erythritol-2,4-cyclopyrophosphate, the last intermediate in the MEP pathway, accumulated in the presence of this herbicide [74,75].

Activities of a large number of herbicide families are directly or indirectly influenced by light. Herbicide mechanisms that are light-dependent or enhanced by light fall into several different categories [76]. Herbicides with the strongest indirect inhibitory effects (++) on phytoene accumulation were those generating reactive oxygen species (ROS) that lead to lipid peroxidation via light-dependent processes (e.g., inhibitors of PPO, electron diverters from photosystem I) (Table 2). This effect is most likely due to membrane degradation affecting all the biochemical process in the damaged tissues. In most cases, this interference could be alleviated by performing the assays in the dark.

To explore the impact of oxidative stress on the expression of MEP-pathway enzymes, Chang [12] exposed *Catharanthus roseus* leaf discs to a 0.5 μM paraquat solution. This treatment resulted in the bleaching within 10 h of exposure, whereas control treatment did not show any bleaching over a 30 h period. The paraquat treatment also caused a strong induction of *DXS* transcripts, suggesting deregulation of the MEP pathway.

The precise relationship between the effect of the herbicide dinoterb, a synthetic phenol that is no longer used as a herbicide, on phytoene accumulation and its mechanism of action is not well understood but this compound is a strong generator of ROS that destabilizes membranes under both light and dark conditions. Dinoterb also uncouples oxidative phosphorylation, which may reduce the endogenous levels of ATP, thereby inhibiting some of the steps in the MEP pathway (Figure 1). Endothall, an inhibitor of serine/threonine protein phosphatases, also caused a reduction of phytoene accumulation under both light and dark conditions, but the biochemical basis for this effect is unknown.

Conclusions

The simple *in vivo* bioassay developed in this study proved to be an efficient and inexpensive screening method for putative novel inhibitors of the MEP and terpenoid pathways preceding carotenoid synthesis. The method permitted determination of carbon flow through this pathway with accuracy, reproducibility, and minimal sample consumption. The assay appears to be robust in terms of detecting the inhibitory activity of compounds that target the early steps of carotenoid synthesis. The quality of an assay is most commonly assessed by calculating the Z' factor [77]. It is generally accepted that assays with $0.5<Z'<1$ have good separation of the

Table 1. Effect of BASF experimental compounds on phytoene accumulation in the presence of 200 μM norflurazon.

MEP target site	BASF #[a]	Inhibition[b] (%)	
DXS			
1-Deoxyxylulose-5-phosphate synthase			
	3	11.7	BC
	4	15.6	BC
	5	27.2	CD
	6	44.4	DE
DXR			
1-Deoxyxylulose-5-phosphate reductoisomerase			
	7	(117.7)[c]	A
	8	40.3	DE
IspD			
2C-Methyl-D-erythritol 4-phosphate cytidyltransferase			
	9	20.6	BCD
	10	26.0	CD
	11	63.6	E
	12	61.3	E
IspE			
4-Diphosphocytidyl-2C-methyl-D-erythritol kinase			
	13	28.8	CD

Inhibition of phytoene accumulation by compound treatment was expressed in percentage of inhibition related to maximum accumulation in control assays. All compounds were tested at 100 μM.

[a]Numbering corresponds to structures in Witschel et al. (2013) [41].

[b]Means values followed by the same letter do not differ significantly at the 5% level by Duncan's multip e range test.

[c]Caused greater phytoene accumulation.

Table 2. Effect of herbicides with different modes of action on phytoene accumulation in the presence of 200 μM norflurazon.

WSSA Class	Compound[a]	MOA	Inhibition[b] Light	Dark
A	Diclofop	Acetyl-CoA carboxylase	+	–
B	Imazethapyr	Acetolactate synthase	–	–
C1	Atrazine	Photosystem II	–	–
D	Paraquat	Photosystem I electron diverter	++	–
E	Sulfentrazone	Protoporphyrinogen oxidase	++	+
F2	Sulcotrione	*p*-Hydroxyphenylpyruvate dioxygenase	+	–
F4	Clomazone	Deoxyxylulose-5-phosphate synthase	++	++
G	Glyphosate	Enolpyruvylshikimate synthase	+	–
H	Glufosinate	Glutamine synthetase	–	–
I	Asulam	Dihydropteroate synthase	+	–
K1	Oryzalin	Tubulin	–	–
K3	Alachlor	Very long chain fatty acid elongases	–	–
L	Dichlobenil	Cellulose synthase	–	–
M	Dinoterb	Oxidative phosphorylation uncoupler	++	++
NC	Endothall	Serine/threonine protein phosphatase	++	+
O	Quinclorac	Synthetic auxin	–	–

The bioassay was performed in three replications of each treatment.

[a]All compounds were tested at 100 μM to determine whether they would be detected as false positives in a high throughput screening program.

[b]no inhibition (0–10%) (–), slight inhibition (10 to 50%) (+) and strong inhibition (50 to 100%) (++).

distributions and are excellent assays. The assay developed in this study has a Z′ factor of 0.584 (Figure S4), which indicates that is likely to successfully identify active compounds [5-]. Additionally, it would easily complement enzyme-based high-throughput screenings and evaluate the *in vivo* activity potential performance of inhibitors identified in the *in vitro* assay, without requiring greenhouse tests. The sensitivity of the bioassay to compounds producing reactive oxygen species may be a limitation.

Supporting Information

Figure S1 Dose-response curves of ketoclomazone (a clomazone metabolite) with (○) and without (●) phorate in green barley leaves. Phytoene was caused to accumulate by the presence of 200 μM norflurazon. Data represent means of three replications with standard deviation.

Figure S2 Dose-response to the keto analog 1 on greening etiolated (A) and green (B) young barley leaves (●). Data represent means of three replications with standard deviation.

Figure S3 Dose-response curves of the DXR inhibitors fosmidomycin (○) and FR-900098 (●) on green young barley leaves. Phytoene accumulation was induced in the presence of 200 μM norflurazon. Data represent means of three replications with standard deviation.

Figure S4 Calculation of the Z′ factor of the bioassay.

Acknowledgments

We are grateful to J'Lynn Howell, Susan B. Watson and Robert Johnson for their excellent technical assistance. We also thank BASF-SE (Ludwigshafen, Germany) for providing their experimental compounds.

Author Contributions

Conceived and designed the experiments: FED NC. Performed the experiments: NC FMLS NPDN FED. Analyzed the data: NC MW EDV FED. Contributed reagents/materials/analysis tools: MW FED. Contributed to the writing of the manuscript: NC MW FED.

References

1. Vranová E, Coman D, Gruissem W (2013) Network analysis of the MVA and MEP pathways for isoprenoid synthesis. Ann Rev Plant Biol 64: 665–700.
2. Croteau RB, Kutchan TM, Lewis NG (2000) Natural products (secondary metabolites). In: Buchanan BB, Gruissem W, Jones RL, editors. Biochemistry and Molecular Biology of Plants. Rockville, MD: American Society of Plant Biologists. pp. 1250–1268.
3. Cordoba E, Porta H, Arroyo A, San Román C, Medina L, et al. (2011) Functional characterization of the three genes encoding 1-deoxy-D-xylulose 5-phosphate synthase in maize. J Exp Bot 62: 2023–2038.
4. Bartley GE, Scolnik PA (1995) Plant carotenoids: Pigments for photoprotection, visual attraction, and human health. Plant Cell 7: 1027–1038.
5. DellaPenna D, Pogson BJ (2006) Vitamin synthesis in plants: Tocopherols and carotenoids. Ann Rev Plant Biol 57: 711–738.
6. Botella-Pavia P, Besumbes Ó, Phillips MA, Carretero-Paulet L, Boronat A, et al. (2004) Regulation of carotenoid biosynthesis in plants: evidence for a key role of hydroxymethylbutenyl diphosphate reductase in controlling the supply of plastidial isoprenoid precursors. Plant J 40: 188–199.
7. Malkin R, Niyogi K (2000) Photosynthesis. In: Buchanan BB, Gruissem W, Jones RL, editors. Biochemistry and Molecular Biology of Plants. Rockville, MD: American Society of Plant Biologists. pp. 568–628.
8. Lichtenthaler HK (2010) The non-mevalonate DOXP/MEP (deoxyxylulose 5-phosphate/methylerythritol 4-phosphate) pathway of chloroplast isoprenoid and pigment biosynthesis. In: Rebeiz CA, Benning C, Bohnert HJ, Daniell H, Hoober JK et al., editors. The Chloroplast: Basics and Applications: Springer Science+Business Media B.V. pp. 95–118.
9. Lichtenthaler HK, Rohmer M, Schwender J (1997) Two independent biochemical pathways for isopentenyl diphosphate and isoprenoid biosynthesis in higher plants. Physiol Plant 101: 643–652.
10. Lichtenthaler HK (2000) Sterols and isoprenoids: Non-mevalonate isoprenoid biosynthesis: enzymes, genes and inhibitors. Biochem Soc Trans 28: 785–789.
11. Cunningham FX Jr., Gantt E (1998) Genes and enzymes of carotenoid biosynthesis in plants. Ann Rev Plant Physiol Plant Mol Biol 49: 557–583.
12. Chang W-C, Song H, Liu H-W, Liu P (2013) Current development in isoprenoid precursor biosynthesis and regulation. Curr Opin Chem Biol 17: 571–579.
13. Weimer MR, Balke NE, Buhler DD (1992) Herbicide clomazone does not inhibit *in vitro* geranylgeranyl synthesis from mevalonate. Plant Physiol 1992: 427–432.
14. Duke SO, Paul RN, Becerril JM, Schmidt JH (1991) Clomazone causes accumulation of sesquiterpenoids in cotton (*Gossypium hirsutum* L.). Weed Sci 39: 339–346.
15. Croteau RB (1992) Clomazone does not inhibit the conversion of isopentenyl pyrophosphate to geranyl, farnesyl, or geranylgeranyl pyrophosphate *in vitro*. Plant Physiol 98: 1515–1517.
16. Rohmer M (1999) The discovery of a mevalonate-independent pathway for isoprenoid biosynthesis in bacteria, algae and higher plants. Nat Prod Rep 16: 565–574.
17. Eisenreich W, Bacher A, Arigoni D, Rohdich F (2004) Biosynthesis of isoprenoids via the non-mevalonate pathway. Cell Mol Life Sci 61: 1401–1426.
18. Lichtenthaler HK, Schwender J, Disch A, Rohmer M (1997) Biosynthesis of isoprenoids in higher plant chloroplasts proceeds via a mevalonate-independent pathway. FEBS Lett 400: 271–274.
19. Schwender J, Zeidler J, Groner R, Müller C, Focke M, et al. (1997) Incorporation of 1-deoxy-D-xylulose into isoprene and phytol by higher plants and algae. FEBS Lett 414: 129–134.
20. Schuhr C, Radykewicz T, Sagner S, Latzel C, Zenk M, et al. (2003) Quantitative assessment of crosstalk between the two isoprenoid biosynthesis pathways in plants by NMR spectroscopy. Phytochem Rev 2: 3–16.
21. Laule O, Fürholz A, Chang H-S, Zhu T, Wang X, et al. (2003) Crosstalk between cytosolic and plastidial pathways of isoprenoid biosynthesis in *Arabidopsis thaliana*. Proc Natl Acad Sci USA 100: 6866–6871.
22. Lichtenthaler HK (1999) The 1-deoxy-D-xylulose-5-phosphate pathway of isoprenoid biosynthesis in plants. Ann Rev Plant Physiol Plant Mol Biol 50: 47–65.
23. Gong YF, Liao ZH, Guo BH, Sun XF, Tang KX (2006) Molecular cloning and expression profile analysis of *Ginkgo biloba* DXS gene encoding 1-deoxy-D-xylulose 5-phosphate synthase, the first committed enzyme of the 2-*C*-methyl-D-erythritol 4-phosphate pathway. Planta Med 72: 329–355.
24. Cordoba E, Salmi M, León P (2009) Unravelling the regulatory mechanisms that modulate the MEP pathway in higher plants. J Exp Bot 60: 2933–2943.
25. Koyama T, Ogura K (1999) Isopentenyl diphosphate isomerase and prenyltransferases. In: Cane DE, editor. Comprehensive Natural Products Chemistry: Isoprenoids Including Carotenoids and Steroids. Oxford: Pergamon Press. pp. 69–96.
26. Norris SR, Barrette TR, DellaPenna D (1995) Genetic dissection of carotenoid synthesis in *Arabidopsis* defines plastoquinone as an essential component of phytoene desaturation. Plant Cell 7: 2139–2149.
27. Carol P, Stevenson D, Bisanz C, Breitenbach J, Sandmann G, et al. (1999) Mutations in the Arabidopsis gene IMMUTANS cause a variegated phenotype by inactivating a chloroplast terminal oxidase associated with phytoene desaturation. Plant Cell 11: 57–68.
28. Duke SO, Kenyon WH, Paul RN (1985) FMC 57020 effects on chloroplast development in pitted morningglory (*Ipomoea lacunosa*) cotyledons. Weed Sci 33: 786–794.
29. Dayan FE, Duke SO (2003) Herbicides: Carotenoid biosynthesis inhibitors. In: Plimmer JR, Gammon DW, Ragsdale NN, editors. Encyclopedia of Agrochemicals. New York, NY: John Wiley & Sons. pp. 744–749.
30. Sandmann G, Böger P (1989) Inhibition of carotenoid biosynthesis by herbicides. In: Sandmann G, Böger P, editors. Target Sites of Herbicide Action. Boca Raton, FL: CRC Press. pp. 25–44.
31. Weinberg T, Lalazar A, Rubin B (2003) Effects of bleaching herbicides on field dodder (*Cuscuta campestris*). Weed Sci 51: 663–670.
32. Lee DL, Prisbylla MP, Cromartie TH, Dagarin DP, Howard SW, et al. (1997) The discovery and structural requirements of inhibitors of *p*-hydroxyphenylpyruvate dioxygenase. Weed Sci 45: 601–609.
33. Duke SO (2012) Why have no new herbicide modes of action appeared in recent years? Pest Manag Sci 68: 505–512.
34. Müller C, Schwender J, Zeidler J, Lichtenthaler HK (2000) Properties and inhibition of the first two enzymes of the non-mevalonate pathway of isoprenoid bioynthesis. Biochem Soc Trans 28: 792–793.
35. Dayan FE, Duke SO, Grossmann K (2010) Herbicides as probes in plant biology. Weed Sci 58: 340–350.

36. Ferhatoglu Y, Avdiushko S, Barrett M (2005) The basis for the safening of clomazone by phorate insecticide in cotton and inhibitors of cytochrome P450s. Pestic Biochem Physiol 81: 59–70.
37. Ferhatoglu Y, Barrett M (2006) Studies of clomazone mode of action. Pestic Biochem Physiol 85: 7–14.
38. Rohmer M (1998) Isoprenoid biosynthesis via the mevalonate-independent route, a novel target for antibacterial drugs? Progr Drug Res 50: 135–154.
39. Rohmer M, Grosdemange-Billiard C, Seemann M, Tritsch D (2004) Isoprenoid biosynthesis as a novel target for antibacterial and antiparasitic drugs. Curr Opin Investig Drugs 5: 154–162.
40. Withers S, Keasling J (2007) Biosynthesis and engineering of isoprenoid small molecules. Appl Microbiol Biotechnol 73: 980–990.
41. Witschel M, Röhl F, Niggeweg R, Newton T (2013) In search of new herbicidal inhibitors of the non-mevalonate pathway. Pest Manag Sci 69: 559–563.
42. Service RF (2007) A growing threat down on the farm. Science 316: 1114–1117.
43. Service RF (2013) What happens when weed killers stop killing? Science 341: 1329.
44. Hale I, O'Neill PM, Berry NG, Odom A, Sharma R (2012) The MEP pathway and the development of inhibitors as potential anti-infective agents. Med Chem Comm 3: 418–433.
45. Singh N, Cheve G, Avery MA, McCurdy CR (2007) Targeting the methyl erythritol phosphate (MEP) pathway for novel antimalarial, antibacterial and herbicidal drug discovery: Inhibition of 1-deoxy-D-xylulose-5-phosphate reductoisomerase (DXR) enzyme. Curr Pharm Des 13: 1161–1177.
46. Zhao L, Chang W-c, Xiao Y, Liu H-w, Liu P (2013) Methylerythritol phosphate pathway of isoprenoid biosynthesis. Ann Rev Biochem 82: 497–530.
47. Zamponi GW, Stotz SC, Staples RJ, Andro TM, Nelson JK, et al. (2002) Unique structure−activity relationship for 4-isoxazolyl-1,4-dihydropyridines. J Med Chem 46: 87–96.
48. Deno NC (1947) Diethyl acetals of α-formyl esters. J Am Chem Soc 69: 2233–2234.
49. Sprecher SL, Netherland MD, Stewart AB (1998) Phytoene and carotene response of aquatic plants to fluridone under laboratory conditions. J Aquat Plant Manage 36: 111–120.
50. Dayan FE, Owens DK, Corniani N, Silva FML, Watson SB, et al. (2014) Biochemical markers and enzyme assays for herbicide mode of action and resistance studies. Weed Sci DOI:10.1614/WS-D-1613-00063.00061.
51. Seefeldt SS, Jensen JE, Fuerst EP (1995) Log-logistic analysis of herbicide dose-response relationships. Weed Technol 9: 218–227.
52. Ritz C, Streibig JC (2005) Bioassay analysis using R. J Statist Soft 12: 1–22.
53. de Mendiburu F (2014) Agricolae Version 1.1–4. Practical Manual: 1–60.
54. Zhang J-H, Chung TDY, Oldenburg KR (1999) A simple statistical parameter for use in evaluation and validation of high throughput screening assays. J Biomol Screen 4: 67–73.
55. Lichtenthaler HK, Zeidler J, Schwender J, Müller C (2000) The non-mevalonate isoprenoid biosynthesis of plants as a test system for new herbicides and drugs against pathogenic bacteria and the malaria parasite. Z Naturforsch 55c: 305–313.
56. Zeidler J, Schwender J, Mueller C, Lichtenthaler HK (2000) The non-mevalonate isoprenoid biosynthesis of plants as a test system for drugs against malaria and pathogenic bacteria. Biochem Soc Trans 28: 796–798.
57. Anderson RL (1990) Tolerance of safflower (*Carthamus tinctorius*), corn (*Zea mays*), and proso millet (*Panicum miliaceum*) to clomazone. Weed Technol 4: 606–611.
58. Lois LM, Campos N, Putra SR, Danielsen K, Rohmer M, et al. (1998) Cloning and characterization of a gene from *Escherichia coli* encoding a transketolase-like enzyme that catalyzes the synthesis of D-1-deoxyxylulose 5-phosphate, a common precursor for isoprenoid, thiamin, and pyridoxol biosynthesis. Proc Natl Acad Sci USA 95: 2105–2110.
59. Baerg RJ, Barrett M, Polge ND (1996) Insecticide and insecticide metabolite interactions with cytochrome P450 mediated activities in maize. Pestic Biochem Physiol 55: 10–20.
60. Yasuor H, Zou W Tolstikov VV, Tjeerdema RS, Fischer AJ (2010) Differential oxidative metabolism and 5-ketoclomazone accumulation are involved in *Echinochloa phyllopogon* resistance to clomazone. Plant Physiol 153: 319–326.
61. Weimer MR, Buhler DD, Balke NE (1991) Clomazone selectivity: Absence of differential uptake translocation, or detoxication. Weed Sci 39: 529–534.
62. ElNaggar SF, Creekmore RW, Shcocken MJ, Rosen RT, Robinson RA (1992) Metabolism of clomazone herbicide in soybean. J Agric Food Chem 40: 880–883.
63. Culpepper AS, York AC, Marth JL, Corbin FT (2001) Effect of insecticides on clomazone absorption, translocation, and metabolism in cotton. Weed Sci 49: 613–616.
64. Chang JH, Konz MJ, Aly EA, Sticker RE, Wilson KR, et al. (1987) 3-Isoxazolidinones and related compounds a new class of herbicides. In: Baker DR, Fenyes JG, Moberg WK, Cross B, editors. Synthesis and Chemistry of Agrochemicals. Washington, DC: American Chemical Society. pp. 10–23.
65. Hayashi D, Kuzuyama N, Kato T, Sato Y, Ohkanda J (2013) Antimicrobial *N*-(2-chlorobenzyl)-substituted hydroxamate is an inhibitor of 1-deoxy-D-xylulose 5-phosphate synthase. Chem Comm 49: 5535–5537.
66. Zeidler J, Schwender J, Müller C, Weisner J, Weidemeyer C, et al. (1998) Inhibition of the non-mevalonate 1-deoxy-D-xylulose-5-phosphate pathway of plant isoprenoid biosynthesis by fosmidomycin. Z Naturforsch 54c: 980–986.
67. Rodríguez-Concepción M (2004) The MEP pathway: A new target for the development of herbicides, antibiotics and antimalarial drugs. Curr Pharm Des 10: 2391–2400.
68. Fellermeier M, Kis K, Sagner S, Maier U, Bacher A, et al. (1999) Cell-free conversion of 1-deoxy-D-xylulose 5-phosphate and 2-*C*-methyl-D-erythritol 4-phosphate into β-carotene in higher plants and its inhibition by fosmidomycin. Tetrahed Let: 40: 2743–2746.
69. Ershov YV (2007) 2-*C*-methylerythritol phosphate pathway of isoprenoid biosynthesis as a target in identifying new antibiotics, herbicides, and immunomodulators: A review. Appl Biochem Microbiol 43: 115–138.
70. Jomaa H, Wiesner J, Sanderbrand S, Altincicek B, Weidemeyer C, et al. (1999) Inhibitors of the non-mevalonate pathway of isoprenoid biosynthesis as antimalarial drugs. Science 285: 1573–1576.
71. Fellermeier M, Raschke M, Sagner S, Wungsintaweekul J, Schuhr CA, et al. (2001) Studies on the nonmevalonate pathway of terpene biosynthesis. Eur J Biochem 268: 6302–6310.
72. Rohdich F, Wungsintaweekul J, Eisenreich W, Richter G, Schuhr CA, et al. (2000) Biosynthesis of terpenoids: 4-Diphosphocytidyl-2C-methyl-D-erythritol synthase of *Arabidopsis thaliana*. Proc Natl Acad Sci USA 97: 6451–6456.
73. Duke SO, Dayan FE (2011) Bioactivity of Herbicides. In: Murray M-Y, editor. Comprehensive Biotechnology. 2 ed. Amsterdam: Elsevier. pp. 23–35.
74. Ge X, d'Avignon DA, Ackerman JJH, Duncan B, Spaur MB, et al. (2011) Glyphosate-resistant horseweed made sensitive to glyphosate: low-temperature suppression of glyphosate vacuolar sequestration revealed by ^{31}P NMR. Pest Manag Sci 67: 1215–1221.
75. Ge X, d'Avignon DA, Ackerman JJH, Sammons RD (2012) Observation and identification of 2-*C*-methyl-D-erythritol-2,4-cyclopyrophosphate in horseweed and ryegrass treated with glyphosate. Pestic Biochem Physiol 104: 187–191.
76. Hess FD (2000) Light-dependent herbicides: An overview. Weed Sci 48: 160–170.
77. Sui Y, Wu Z (2007) Alternative statistical parameter for high-throughput screening assay quality assessment. J Biomol Screen 12: 229–234.

8

Taxonomic and Life History Bias in Herbicide Resistant Weeds: Implications for Deployment of Resistant Crops

Jodie S. Holt[1], Shana R. Welles[1]*, Katia Silvera[2], Ian M. Heap[3], Sylvia M. Heredia[1], Alejandra Martinez-Berdeja[1], Kai T. Palenscar[1], Lynn C. Sweet[1], Norman C. Ellstrand[1]

1 Department of Botany and Plant Sciences, University of California Riverside, Riverside, California, United States of America, **2** Smithsonian Tropical Research Institute, Balboa, Ancon, Republic of Panama, **3** International Survey of Herbicide Resistant Weeds, Corvallis, Oregon, United States of America

Abstract

Evolved herbicide resistance (EHR) is an important agronomic problem and consequently a food security problem, as it jeopardizes herbicide effectiveness and increases the difficulty and cost of weed management. EHR in weeds was first reported in 1970 and the number of cases has accelerated dramatically over the last two decades. Despite 40 years of research on EHR, why some weeds evolve resistance and others do not is poorly understood. Here we ask whether weed species that have EHR are different from weeds in general. Comparing taxonomic and life history traits of weeds with EHR to a control group ("the world's worst weeds"), we found weeds with EHR significantly over-represented in certain plant families and having certain life history biases. In particular, resistance is overrepresented in Amaranthaceae, Brassicaceae and Poaceae relative to all weeds, and annuality is ca. 1.5 times as frequent in weeds with EHR as in the control group. Also, for perennial EHR weeds, vegetative reproduction is only 60% as frequent as in the control group. We found the same trends for subsets of weeds with EHR to acetolactate synthase (ALS), photosystem II (PSII), and 5-enolpyruvylshikimate-3-phosphate (EPSP) synthase-inhibitor herbicides and with multiple resistance. As herbicide resistant crops (transgenic or not) are increasingly deployed in developing countries, the problems of EHR could increase in those countries as it has in the USA if the selecting herbicides are heavily applied and appropriate management strategies are not employed. Given our analysis, we make some predictions about additional species that might evolve resistance.

Editor: Boris Alexander Vinatzer, Virginia Tech, United States of America

Funding: Support for NCE was provided by a John Simon Guggenheim Memorial Fellowship and NSF OPUS grant #DEB-1020799. Support for SRW is supported under grant NSFDGE-0813967. The funders had no role in study design, data collection and analysis, decision to publish, or preparation of the manuscript.

Competing Interests: The authors have declared that no competeing interests exist.

* E-mail: shana.welles@email.ucr.edu

Introduction

Evolved herbicide resistance (EHR) has become a threat to agriculture and consequently a food security problem worldwide [1,2]. EHR in weeds was first reported in 1970 [3,4] and widely studied in the 1970s through 1990s [3]. The number of cases has accelerated dramatically over the last two decades [5]. The evolution of resistance to multiple herbicides with different modes of action has also been found within numerous weed species to date [6]. The discovery of resistance to glyphosate (the dominant herbicide worldwide) in the 1990s [7], the introduction of transgenic glyphosate-resistant crops in 1996 [8], and the recent expansion of cases of evolved resistance to glyphosate in weeds, likely due to greater glyphosate usage, have inspired a renewal of interest and resurgence of research into this phenomenon.

Despite 40 years of research on EHR, it is not clear why some weeds evolve resistance faster than others. Baker's list of characteristics that might be expected in the "ideal weed" is well known [9]; one might expect weeds with EHR to possess a subset of these traits. A cursory review of available data in 2001 revealed that some weeds have a greater propensity to evolve resistance than others [10]. This observation was attributed to opportunity, as many resistant weeds are among the world's worst weeds [11,12], are widespread, and occur in many cropping systems [10]. Well before EHR was discovered, heritable variability, breeding system, reproductive capacity, annuality, and population size were predicted to correlate with evolution of herbicide resistance [13]. Other plant factors can affect the evolution of resistance, including mutation frequency, generation time, fitness in absence of the herbicide, plasticity, and soil seed reservoir [14], as well as mode of inheritance of resistance, population size, seed dormancy, and gene flow by pollen and seed [15]. While these factors have been tested in models predicting evolution of resistance [14,16], few have been tested empirically.

Given that taxonomic families are relatively cohesive internally but generally vary from one another in ecological traits, we might expect EHR to be represented nonrandomly among plant families. Similarly, based on observations and reports of the propensity for resistance to evolve within certain genera or species, we would expect weeds with HER to be ecologically and taxonomically different than weeds in general. Here we ask whether weed species that have EHR are different from weeds in general and if evolution of resistance to multiple herbicides follows the same patterns.

Materials and Methods

A list of species with evolved herbicide resistance EHR; (henceforth, the EHR list) was obtained from the database at the website "International Survey of Herbicide Resistant Weeds" http://www.weedscience.org/In.asp created and maintained by Ian Heap with support from the Herbicide Resistance Action Committee, the North American Herbicide Resistance Action Committee, and the Weed Science Society of America [5]. At the time of our study (May 2012), the inventory included 187 species in 31 plant families. For statistical comparison, we chose the two volumes by Holm et al. that inventory the "World's Worst Weeds" and "World Weeds" [11,12] (henceforth the control list) including 201 species in 49 plant families. Entries from different subspecies of the same species were lumped together. The taxonomy for both lists was updated using the USDA ARS Germplasm Resources Information Network website "GRIN Taxonomy for Plants", http://www.ars-grin.gov/cgi-bin/npgs/html/index.pl?language = en [17]. All scientific names were then checked against the "The Plant List" website database, http://www.theplantlist.org/, which was also used to obtain current plant family assignments. The two updated lists are presented in Table S1.

Information regarding plant life history was extracted from the Holm et al. [11,12] publications, if available. For species not present in those publications, the information was obtained from major floras (e.g. [18,19],) or citations for individual species entries in the USDA ARS "GRIN Taxonomy for Plants" [17].

All comparisons between the EHR list and the control list were done using a Parson's chi-square statistic (2-tailed test) with one degree of freedom; $\alpha = 0.05$ was used to determine significance. This test was used because it is appropriate for comparisons between theoretical and experimental populations where a data set is large and observations are independent. Calculations were done using R statistical package [20].

We constructed a phylogenetic tree hypothesis for the relationship among plant families with EHR using Phylomatic 2 and the Angiosperm Phylogeny Group 3 derived megatree (http://www.phylodiversity.net/phylomatic/phylomatic.html). Presence and absence of EHR to acetolactate synthase (ALS)- and photosystem II (PSII)-inhibitors were mapped onto the same tree. Character state reconstruction for each EHR group was performed by maximum likelihood using a marginal probability reconstruction with the Asymmetrical parameter Markov-k model of evolution in Mesquite version 2.74 [21]. The results are displayed as likelihood states reported as proportion of the total likelihood and represented as pie charts in each branch node within the tree.

Results and Discussion

Comparisons between the control and EHR lists showed sixty-two species in 19 families are found in common on both lists, representing 33% of the 187 species on the EHR list and 31% of the 201 species on the control list. When ranked by number of species, the same plant families were dominant on both lists (Table 1). Poaceae, Asteraceae, Amaranthaceae, and Brassicaceae were among the top six families on each list, comprising 75% and 57% of species on the EHR and control lists, respectively. This is not surprising since plant families that are overrepresented in the global weed flora would have a greater probability of herbicide exposure and selection of species with EHR [10].

Despite the taxonomic overlap between the EHR and control lists at the family level, we found significant differences with regard to species abundance within families (Table 2). In particular, EHR was overrepresented relative to the control in three of the four aforementioned families (Poaceae, Amaranthaceae, and Brassicaceae but not Asteraceae) (Table 2). While all but three families on the EHR list were also on the control list, 21 plant families that had one or more species on the control list were lacking altogether on the EHR list.

Table 1. Rank of plant families by number and % of species in control and EHR lists.

Rank	Control list[1] Plant Family	Species (no.)	Species (%)	EHR List Plant Family	Species (no.)	Species (%)
1	Poaceae	47	23	Poaceae	60	32
2	Asteraceae	30	15	Asteraceae	33	18
3	Cyperaceae	12	6	Amaranthaceae	18	10
4	Amaranthaceae	10	5	Brassicaceae	17	9
5	Polygonaceae	8	4	Alismataceae	6	3
6	Brassicaceae	7	4	Polygonaceae	6	3
	Total	114	57	Total	140	75

[1]Control list includes 201 species from "World's Worst Weeds of Holm et al. (1977, 1979); Evolved herbicide resistance (EHR) list includes 187 species from the website http://www.weedscience.org/In.asp.

An early attempt to relate EHR to phylogeny revealed broad trends at the level of the superorder [22]. EHR was not found in woody tropical superorders, rather was found in groups predominantly associated with temperate agriculture, likely reflecting the areas of greatest herbicide use. Perhaps because fewer cases of EHR were known at that time, no phylogenetic trends in EHR were evident, in contrast to our results.

Because the EHR list is heterogeneous with evolved resistance to a variety of herbicide groups, we also asked whether the same taxonomic bias was present for the two most prevalent types of EHR, that is, to ALS- and PSII-inhibitor herbicide groups, as well as for resistance to the world's most important herbicide, glyphosate [8]. For some families overrepresentation of resistance to the three herbicide groups was present. Amaranthaceae were overrepresented in all three herbicide groups, Brassicaceae were overrepresented in EHR to ALS-inhibitors, and Poaceae were overrepresented in EHR to glyphosate, when the EHR list for each of these herbicide groups was compared to the control list (Table 2). Characteristics of these herbicides could partially account for results in some instances. For example, many grasses are naturally resistant to ALS-inhibitors and therefore are less likely to be treated with herbicides from this group. so it is not surprising that Poaceae is not overrepresented in this herbicide group. While natural variation in resistance to herbicides could account for lack of representation by some taxa, it cannot account for overrepresentation of EHR in these families relative to weeds in general. These results suggest that there is a strong phylogenetic signal in these families for frequent selection of resistance to specific herbicide groups [14,15].

We examined the phylogenetic relationship between EHR and each of these herbicide groups and found different trends (Figure 1). The breadth of phylogenetic distribution appears to correlate with the ease of selection for resistance to distinct herbicide groups. EHR to ALS-inhibitors is spread widely over plant families (Figure 1). This trend can be explained by the strong selection pressure exerted by ALS inhibitors as well as the relative ease with which plants evolve resistance to ALS inhibitors. Several point mutations can confer EHR to ALS-inhibitors, which is

Table 2. Comparison of plant family and life history traits.

		All Resistance				
		Control list (%)[1]	**Complete EHR list (%)**[1]	**χ^2 statistic**	***p*-value**	**Difference from control list**[2]
Family	Amaranthaceae	5	10	8.55	0.004	+
	Brassicaceae	4	9	17.55	2.87×10^{-5}	+
	Poaceae	23	32	7.91	0.0049	+
Life-history traits	Annuality[3]	53	86	64.26	1.08×10^{-15}	+
	Perenniality	43	11	64.32	1.06×10^{-15}	−
	Vegetative Reproduction[4]	68	41	18.11	2.08×10^{-5}	−
		ALS Resistance				
		Control list (%)	**EHR to ALS-inhibitors (%)**	**χ^2 statistic**	***p*-value**	**Difference from control list**
Family	Amaranthaceae	5	11	7.88	0.005	+
	Brassicaceae	4	15	40.1	2.4×10^{-8}	+
	Poaceae	23	23	0.013	NS	0
Life-history traits	Annuality	53	87	35.18	3.08×10^{-9}	+
	Perenniality	43	9	39	4.23×10^{-10}	−
	Vegetative Reproduction	68	48	4.91	0.03	−
		PSII Resistance				
		Control list (%)	**EHR to PSII-inhibitors (%)**	**χ^2 statistic**	***p*-value**	**Difference from control list**
Family	Amaranthaceae	5	24	51.8	6.06×10^{-13}	+
	Brassicaceae	4	6	1.3	NS	0
	Poaceae	23	30	2.21	NS	0
Life-history traits	Annuality	53	97	35.41	2.7×10^{-9}	+
	Perenniality	43	2	43.34	4.59×10^{-10}	−
	Vegetative Reproduction	68	25	6.6	0.01	−
		Glycine Resistance				
		Control list (%)	**EHR to glycine herbicides (%)**	**χ^2 statistic**	***p*-value**	**Difference from control list**
Family	Amarthaceae	5	14	3.85	0.049	+
	Brassicaceae	4	0			None with glycine resistance
	Poaceae	23	48	6.88	0.0087	+
Life-history traits	Annuality	53	176	3.73	0.05	+
	Perenniality	43	17	4.95	0.026	−
	Vegetative Reproduction	68	43			Not enough data
		Multiple Resistance				
		Control list (%)	**Multiple resistance list (%)**	**χ^2 statistic**	***p*-value**	**Difference from control list**
Family	Amaranthaceae	5	13	4.79	0.0286	+
	Brassicaceae	4	8	1.92	0.1658	+
	Poaceae	23	48	12.99	0.000314	+
Life-history traits	Annuality	53	94	20.38	6.35×10^{-6}	+
	Perenniality	43	3	19.66	9.26×10^{-6}	−

Table 2. Cont.

	Multiple Resistance				
	Control list (%)	**Multiple resistance list (%)**	**χ^2 statistic**	***p*-value**	**Difference from control list**
Vegetative Reproduction	68	67			Not enough data

Plant family representation and life history traits in complete EHR list and in lists of weeds with EHR to ALS-inhibitors, PSII-inhibitors, glycine herbicides, and with multiple herbicide resistance. Pearson's chi-square statistic (2-tailed test) with 1 degree of freedom was used to assess the difference between EHR lists and the control list. An $\alpha = .05$ was used to determine significance. Calculations were done using R statistical package [20]).

[1]Control list, n = 201; complete EHR list, n = 187; EHR to ALS inhibitors, n = 99; EHR to PSII inhibitors, n = 66; EHR to glycine herbicides, n = 21; multiple resistance, n = 40.

[2]+, EHR is over-represented in the family or life history trait; −, EHR is under-represented in the family or life history trait; 0, no difference between control and EHR list.

[3]Calculation of % annuality excluded species that could also be biennial or perennial; calculation of % perenniality excluded species that could also be annual or biennial.

[4]Vegetative reproduction is a % of non-annuals.

followed by rapid evolution because the encoded gene is nuclear inherited and easily transmitted by seed and pollen [23].

In contrast, although PSII inhibitors have long soil residual activity, selection of EHR to the PSII-inhibitor herbicide group is more difficult because in most cases reported it requires a specific point mutation in the psbA gene, which is generally maternally inherited [24]. Accordingly, our analysis reveals that evolved PSII-inhibitor resistance is much more phylogenetically restricted than that to ALS-inhibitors. Poaceae is the only monocot family that has EHR to PSII-inhibitors; in eudicot families, there is a notable concentration in the Caryophyllales and Eurosid II clades with the other cases present in only four weakly related eudicot families (Figure 1).

EHR to glyphosate is more complex, involving target and non-target site resistance mechanisms, changes in the translocation of glyphosate to meristematic regions [20], and virtually no soil residual activity. Although there is abundant intraspecific variability in glyphosate susceptibility in weeds, EHR to this herbicide was not discovered for two decades following its introduction [8]. The relatively slow appearance of EHR to glyphosate compared to other herbicide groups was initially attributed to genetic and biochemical constraints, restricted use, and lack of soil residual activity resulting in low selection pressure for resistance [8]. Indeed, EHR to glyphosate currently shows a very restrictive phylogenetic distribution with almost half of species found in a single family, Poaceae, and six of the eight genera with EHR in that family are in two closely related subfamilies, Panicoideae and Chloridoideae. Dramatic increases in glyphosate use where transgenic glyphosate-resistant crops are planted have intensified selection pressure for EHR, which might alter these phylogenetic patterns in the future.

Cumulatively, our results for the three herbicide groups tested suggest that as selection of resistance to an herbicide group becomes increasingly difficult, the taxonomic distribution of EHR will be subject to increasing phylogenetic constraint. Genetic variation for resistance appears not to be uncommon in weed populations, however, little is known about actual rates of mutation to EHR for any weed species [15]. More information about genetics of resistance to a particular herbicide (frequency, number, dominance, and fitness cost of resistance genes) as well as about natural mutation frequencies, would improve our ability to understand underlying causes of the phylogenetic patterns detected here.

We also compared the full EHR list and the control list with respect to plant habit (annuality, bienniality, perenniality). We found that annual weeds are 1.5 times as frequent, whereas perennial weeds are only 0.27 times as frequent in the EHR list as in the control list (Table 2). We attribute this difference to the fact that short-lived species are recognized to have a more rapid response to directional selection than long-lived species, leading to greater opportunity for evolution of EHR [14,15]. If length of life cycle is inversely correlated with EHR, then we would predict that because vegetative reproduction effectively extends generation time, perennials with vegetative reproduction should be highly underrepresented among weeds with EHR. As expected, we found the proportion of vegetative reproduction in perennials with EHR to be only 0.6 times that in the control list. The same trends were found in the three herbicide groups evaluated with respect to plant habit and vegetative reproduction (Table 2).

While many traits are predicted to correlate with EHR [13,14,15], some (e.g., fitness, plasticity, soil seed reserve, population size, and gene flow) are strongly influenced by environment and difficult to include in a phylogenetic analysis. Other life history traits, e.g., outcrossing rate, spontaneous hybridization, fecundity, and seed size, might help explain phylogenetic patterns in EHR. Unfortunately, available data are insufficient to allow further quantitative assessment or meta-analysis of these traits for weeds with EHR compared to weeds in general.

A total of 139 species across 44 families on the control list had no reports of EHR. All of these species are found in countries for which reports of EHR occur [5]; thus, the absence of EHR cannot be explained by lack of herbicide exposure or EHR reporting. We note that aquatic and wetland species are significantly more numerous on the control list than the EHR list ($\chi^2 = 11.69$, df = 2, p = 0.00063) [9,10]. Despite the fact that aquatic weeds are widespread, restricted herbicide use in wetland habitats (with the exception of rice production) apparently precludes selection for EHR [25]. For other species on the control list where no obvious factors account for the lack of EHR, it is likely that habitat, cropping system, or other agronomic factors influence the probability of exposure to herbicides, and thus, selection for EHR.

We also examined patterns between individuals with resistance to multiple herbicides in different functional groups (herbicides with different modes of action). We found that there is a significant increase in multiple resistance in the Amaranthaceae and Poaceae compared with the control list. The abundance of multiple resistance in these families is not surprising given that they are also a subset of the families that have high levels EHR. We also analyzed multiple resistance from the perspective of the herbicides involved. We found that the majority of herbicides have a similar proportion of species with resistance to that herbicide and species with multiple resistance. However, we found that 62% of species with glyphosate resistance have multiple resistance compared to

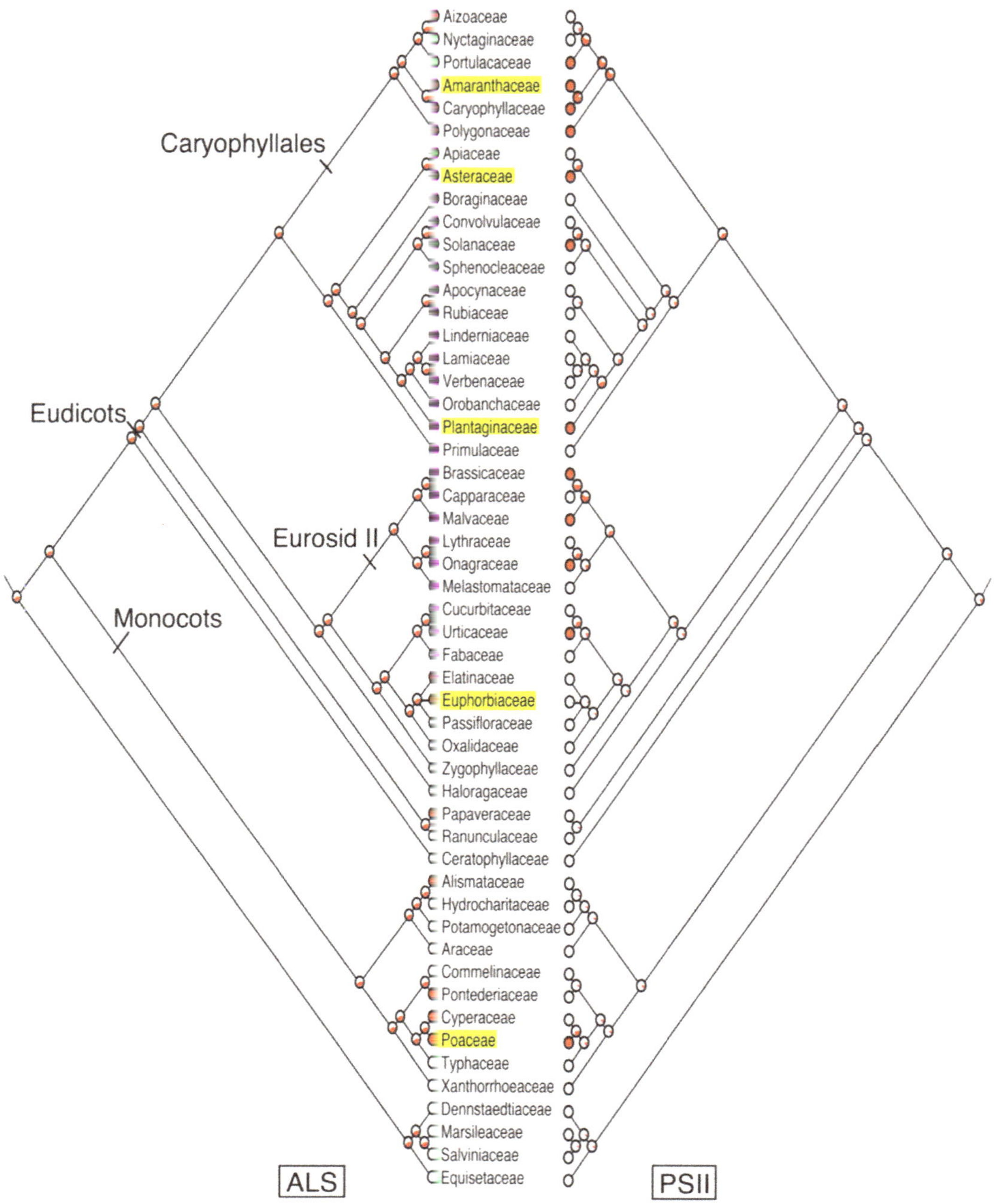

Figure 1. Phylogenetic tree showing the relationship among 52 plant families. EHR to ALS-inhibitors (left panel) and PS-II inhibitors (right panel) were mapped onto the tree using Mesquite version 2.74. Families with EHR to glyphosate are highlighted in yellow. Relevant plant lineages are labeled within the tree nodes. Lineages that show EHR are depicted in red. The red area within each pie chart indicates the relative support for different ancestor states.

the approximately 30% among most other herbicide families. It is possible that the cause of increased multiple resistances that include glyphosate is due to the fact that glyphosate is a newer herbicide and is often applied to areas that already have resistance to other herbicides. Therefore, there are increased opportunities for evolution of resistance to glyphosate on a background of resistance to other herbicides.

A major mechanism for multiple herbicide resistance in plants is metabolism by cytochrome P450 monooxygenases [6,26,27]. As these enzymes can metabolize herbicides with different modes of action, their existence in weeds with EHR could strongly influence phylogenetic analysis. Unfortunately, while much is known about the role of these enzymes in insecticide resistance [28], little is known about their existence and role in weeds [6,27]. To date most of the reports of cytochrome P450 monooxygenases have been for grass weeds [6], but the limited number of species for which P450-based EHR is known precludes using this information in our analysis.

We expect that herbicide resistance will continue to evolve in regions with the strongest herbicide selection pressure, in particular, in major agricultural countries. For example, considering the case of glyphosate in 2010, the United States, Brazil, and Argentina were #1, #2, and #3 with respect to acreage of crops engineered for glyphosate resistance, collectively comprising more than 80% of the world's acreage, which comprised 134 million ha in 2009 [29,30]. Presently, there are 112 known instances of EHR to glyphosate [5]; 86 of these were found in the United States, Brazil, and Argentina (more than 75% of the total). While we acknowledge that glyphosate is used for more than weed control in transgenic crops, the strong relationship between their acreage and EHR cannot be ignored.

Our analysis has immediate implications for developing countries that are accelerating or about to accelerate their use of herbicides. As reported by the National Research Council [1], without careful attention to herbicide resistance management (e.g., appropriate herbicide rotations in combination with non-chemical control methods), increased use of herbicides could bring about new instances of EHR [31]. Our results show that new instances of EHR are more likely to be found for annual weeds in the families that we identified as EHR prone. To take a specific example, the recent historic information on glyphosate use is easily available. In 2010, in terms of area planted to glyphosate resistant crops, Paraguay ranked fifth in the world with 2.6 million hectares of glyphosate resistant soybean [29]. That crop was first introduced to Paraguay in 2004 [29]. With six years of intense use, there are now two reports of glyphosate resistance in Paraguay, both for *Digitaria insularis* (sourgrass, once in 2006 and later in 2008). The situation is not trivial; the latter report indicates an increasing infestation on the order of hundreds of acres at over a dozen locations. We note that this species follows the phylogenetic trends our analysis has identified. Sourgrass is a member of the Poaceae, a family overrepresented for both EHR in general and glyphosate resistance in particular. With continued glyphosate use in Paraguay we would predict some of the following species to rapidly evolve resistance: *Echinochloa colona* and *Eleusine indica*, reported to have EHR to other herbicide groups in South America, and even more worrisome, *Lolium* spp. and *Sorghum halepense*, reported to have glyphosate resistance elsewhere in South America. As herbicide resistant crops (transgenic or not) are increasingly deployed in developing countries, the problems of EHR will likely spread to those countries [32].

Pesticide resistance was first discovered in insects, then fungi, then plants [33], yet this type of analysis of phylogenetic patterns has never been done except at the gene level (e.g. [34,35], but see [22]). We have shown that weeds with EHR are far from a random sample of weedy plant species. Our analysis indicates that certain species in certain families, particularly those species with shorter life cycles, will rapidly evolve herbicide resistance under high selection pressure. Accordingly, local weed managers and scientists can anticipate this evolution and employ recommended herbicide resistance management techniques to delay evolution toward herbicide resistance and preserve the efficacy of a given herbicide [31]. Further analysis of other traits of EHR will refine such strategies.

Author Contributions

Conceived and designed the experiments: NCE JSH. Performed the experiments: SRW SMH AMB KTP LCS. Analyzed the data: SRW KS. Contributed reagents/materials/analysis tools: IH. Wrote the paper: JSH SRW NCE.

References

1. National Research Council Committee on the Impact of Biotechnology on Farm-Level Economics and Sustainability (2010) The Impact of Genetically Engineered Crops on Farm Sustainability in the United States: The National Academies Press, Washington, DC.
2. Yuan JS, Tranel PJ, Stewart CN Jr (2007) Non-target site herbicide resistance: A family business. Trends in Plant Science 12: 6–13.
3. Holt JS (1992) History of identification of herbicide resistant weeds. Weed Technology 6: 615–620.
4. Ryan GF (1970) Resistance of common groundsel to simazine and atrazine. Weed Science 18: 614–616.
5. Heap I (2013) The International Survey of Herbicide Resistant Weeds. Online Internet. Available: www.weedscience.org. Accessed 2012 May 31.
6. Powles SB, Yu Q (2010) Evolution in action: Plants resistant to herbicides. Annual Review of Plant Biology 61: 317–347.
7. Powles SB, Lorraine-Colwill DF, Dellow JJ, Preston C (1998) Evolved resistance to glyphosate in rigid ryegrass (*Lolium rigidum*) in Australia. Weed Science 46: 604–607.
8. Duke SO, Powles SB (2008) Mini-review. Glyphosate: A once-in-a-century herbicide. Pest Management Science 64: 319–325.
9. Baker HG (1965) Characteristics and modes of origin of weeds. In The Genetics of Colonizing Species: Baker HG, Stebbins GL, Eds, Academic Press, New York, pp 147–172.
10. Heap I, LeBaron H (2001) Introduction and overview of resistance. In Herbicide Resistance and World Grains: Powles SB, Shaner DL, Eds, CRC Press, Boca Raton, FL, pp 1–22.
11. Holm LG, Doll J, Holm E, Pancho JV, Herberger JP (1997) World Weeds: Natural Histories and Distribution: John Wiley and Sons, Inc., New York, NY.
12. Holm LG, Plucknett DL, Pancho JV, Herberger JP (1977) The World's Worst Weeds: Distribution and Biology: East-West Center, University Press of Hawaii, Honolulu, HI.
13. Harper JL (1956) The evolution of weeds in relation to resistance to herbicides. 3rd British Weed Control Conference Proceedings, pp 179–188.
14. Gressel J, Segel LA (1978) The paucity of plants evolving genetic resistance to herbicides: Possible reasons and implications. Journal of Theoretical Biology 75: 349–371.
15. Jasieniuk MJ, Brûlé-Babel AL, Morrison IN (1996) The evolution and genetics of herbicide resistance in weeds. Weed Science 44: 176–193.
16. Maxwell BD, Roush ML, Radosevich SR (1990) Predicting the evolution and dynamics of herbicide resistance in weed populations. Weed Technoogy 4: 2–13.
17. USDA ARS National Genetic Resources Program (2011) Germplasm Resources Information Network - (GRIN) Online Database: National Germplasm Resources Laboratory, Beltsville, Maryland. Available: http://www.ars-grin.gov/cgi-bin/npgs/html/index.pl. Accessed 2011 Mar 15.
18. Hickman JC (1993) The Jepson Manual: Higher Plants of California: University of California Press, Berkeley, CA.
19. Stace C (2010) New Flora of the British Isles, 3rd Edition: Cambridge University Press, Cambridge, UK.
20. R Core Team (2012) R: A Language and Environment for Statistical Computing, R Foundation for Statistical Computing, Vienna, Austria. ISBN 3-900051-07-0. Available: http://R-project.org/.
21. Maddison WP, Maddison DR (2010) Mesquite: A modular system for evolutionary analysis. Version 2.74. Available: http://mesquiteproject.org.
22. Hill RJ (1982) Taxonomy and biological considerations of herbicide-resistance and herbicide tolerant biotypes. In Herbicide Resistance in Plants: LeBaron HM, Gressel J, Eds, John Wiley & Sons, New York, pp 81–98.
23. Tranel PJ, Wright TR (2002) Resistance of weeds to ALS-inhibiting herbicides: What have we learned? Weed Science 50: 700–712.
24. Hirschberg J, Bleeker A, Kyle DJ, McIntosh L, Arntzen CJ (1984) The molecular basis of triazine-resistance in higher plant chloroplasts. Z Naturforsch 39C: 412–419.
25. Radosevich SR, Holt JS, Ghersa CM (2007) Ecology of Weeds and Invasive Plants. Relationship to Agriculture and Natural Resource Management, 3rd Edition: John Wiley and Sons, Inc., New York, NY.
26. Hall LM, Holtum JAM, Powles SB (1994) Mechanisms responsible for cross resistance and multiple resistance. In Herbicide Resistance in Plants: Biology and Biochemistry: Powles SB, Holtum JAM, Eds, Lewis Publishers, Boca Raton, FL, pp 243–261.
27. Yuan JS, Tranel PJ, Stewart CN Jr (2007) Non-target-site herbicide resistance: a family business. Trends in Plant Science 12: 6–13.

28 Li X, Schuler MA, Berenbaum MR (2007) Molecular mechanisms of metabolic resistance to synthetic and natural xenobiotics. Annual Review of Entomology 52: 231–253.
29. James C (2005) Global Status of Commercialized Biotech/GM Crops: ISAAA Brief No. 34. ISAAA: Ithaca, NY.
30. James C (2010) Global Status of Commercialized Biotech/GM Crops: ISAAA Brief No. 42. ISAAA: Ithaca, NY.
31. Mortensen DA, Egan JF, Maxwell BD, Ryan MR, Smith RG (2012) Navigating a critical juncture for sustainable weed management. BioScience 62: 75–84.
32. Shaner DL (2000) The impact of glyphosate-tolerant crops on the use of other herbicides and on resistance management. Pest Management Science 56: 320–326.
33. Georghiou GP (1986) The magnitude of the resistance problem. In Pesticide Resistance: Strategies and Tactics for Management: Committee on Strategies for the Management of Pesticide Resistant Pest Populations, Board on Agriculture, National Research Council, National Academy Press, Washington, DC, pp 14–43.
34. Kanazin V, Marek LF, Shoemaker RC (1996) Resistance gene analogs are conserved and clustered in soybean. Proceedings of the National Academy of Sciences USA 93: 11746–11750.
35. Yang Z, Chen J, Chen Y, Jiang S (2010) Molecular cloning and characterization of an acetylcholinesterase cDNA in the brown planthopper, *Nilaparvata lugens*. Journal of Insect Science 10: 102.

9

High-Content Screening in Zebrafish Embryos Identifies Butafenacil as a Potent Inducer of Anemia

Jessica K. Leet, Casey D. Lindberg, Luke A. Bassett, Gregory M. Isales, Krystle L. Yozzo, Tara D. Raftery, David C. Volz*

Department of Environmental Health Sciences, Arnold School of Public Health, University of South Carolina, Columbia, South Carolina, United States of America

Abstract

Using transgenic zebrafish (*fli1:egfp*) that stably express enhanced green fluorescent protein (eGFP) within vascular endothelial cells, we recently developed and optimized a 384-well high-content screening (HCS) assay that enables us to screen and identify chemicals affecting cardiovascular development and function at non-teratogenic concentrations. Within this assay, automated image acquisition procedures and custom image analysis protocols are used to quantify body length, heart rate, circulation, pericardial area, and intersegmental vessel area within individual live embryos exposed from 5 to 72 hours post-fertilization. After ranking developmental toxicity data generated from the U.S. Environmental Protection Agency's (EPA's) zebrafish teratogenesis assay, we screened 26 of the most acutely toxic chemicals within EPA's ToxCast Phase-I library in concentration-response format (0.05–50 μM) using this HCS assay. Based on this screen, we identified butafenacil as a potent inducer of anemia, as exposure from 0.39 to 3.125 μM butafenacil completely abolished arterial circulation in the absence of effects on all other endpoints evaluated. Butafenacil is an herbicide that inhibits protoporphyrinogen oxidase (PPO) – an enzyme necessary for heme production in vertebrates. Using *o*-dianisidine staining, we then revealed that severe butafenacil-induced anemia in zebrafish was due to a complete loss of hemoglobin following exposure during early development. Therefore, six additional PPO inhibitors within the ToxCast Phase-I library were screened to determine whether anemia represents a common adverse outcome for these herbicides. Embryonic exposure to only one of these PPO inhibitors – flumioxazin – resulted in a similar phenotype as butafenacil, albeit not as severe as butafenacil. Overall, this study highlights the potential utility of this assay for (1) screening chemicals for cardiovascular toxicity and (2) prioritizing chemicals for future hypothesis-driven and mechanism-focused investigations within zebrafish and mammalian models.

Editor: Zhiyuan Gong, National University of Singapore, Singapore

Funding: Funding was provided by a U.S. Environmental Protection Agency (EPA) Science to Achieve Results (STAR) Grant No. R835169 to DCV. The contents of this manuscript are solely the responsibility of DCV and do not necessarily represent the official views of the U.S. EPA. Further, the U.S. EPA does not endorse the purchase of any commercial products or services mentioned in the publication. The funders had no role in study design, data collection and analysis, decision to publish, or preparation of the manuscript.

Competing Interests: The authors have declared that no competing interests exist.

* Email: volz@mailbox.sc.edu

Introduction

Baseline toxicity and chemical mode-of-action (MOA) data are lacking for the majority of chemicals currently in commerce within the United States and around the world. In most cases, this uncertainty reflects the fact that conventional, animal-based toxicity tests are resource-intensive, require high animal numbers, and, due to a principal focus on apical endpoints, provide little to no information about chemical MOA. Based on these limitations, increased public concern, and an interest in reducing animal usage, organizations around the world have advocated for the development and application of rapid and cost-effective assays that support MOA- or pathway-based chemical screening and prioritization for further hazard characterization [1,2,3].

Zebrafish are well studied and established as a model species for developmental and environmental toxicology, and have become a desirable alternative to mammalian animal models for toxicity testing – especially since embryos and eleutheroembryos are considered non-protected life stages within the European Union and United States [4,5,6,7]. Due to transparency, small size, and rapid development *ex utero*, zebrafish are highly amenable to whole-organism imaging during embryogenesis and, as a result, high-content screening (HCS) assays for identifying and prioritizing hazardous compounds for further testing. To this end, using transgenic zebrafish (*fli1:egfp*) that stably express enhanced green fluorescent protein (eGFP) within vascular endothelial cells, we recently developed and optimized a 384-well HCS assay that enables us to screen and identify chemicals affecting cardiovascular development and function at non-teratogenic concentrations [8]. Within this assay, automated image acquisition procedures and custom image analysis protocols are used to quantify body length, heart rate, circulation, pericardial area, and intersegmental vessel area within individual live embryos exposed from 5 to 72 hours post-fertilization (hpf). Although other HCS zebrafish assays have been developed for evaluating cardiac function or cardiovascular morphology [9,10,11], our assay incorporates both functional and morphological endpoints, provides increased initial sample sizes (32 embryos per treatment), and allows us to simultaneously test a wide range of concentrations (12 treatment groups, including one vehicle control).

The objective of this study was to evaluate the potential utility of this HCS assay for screening and prioritizing chemicals for cardiovascular toxicity testing. As this was a pilot study, we restricted our screen to the most acutely toxic chemicals from the U.S. Environmental Protection Agency's (EPA's) ToxCast Phase-I chemical library in order to minimize potential false negative results due to limited chemical transfer across the embryonic chorion. Therefore, chemicals within the EPA's ToxCast Phase-I chemical library were ranked from most to least acutely toxic based on available data for zebrafish survival and gross malformations present at 6 days post fertilization (dpf) following an 8-hpf to 5-dpf exposure [12]. Using protocols described in Yozzo et al. [8], 26 of the most acutely toxic chemicals within this library (Table S1) were first screened in concentration-response format (0.05–50 μM) to evaluate the ability of this assay to detect cardiovascular toxicity at non-teratogenic concentrations. Based on results from this pilot screen, we then conducted initial hit validation studies to (1) determine whether exposure to chemicals with a common site of action resulted in similar effects on cardiovascular development and (2) begin characterizing the potential mechanism leading to cardiovascular toxicity.

Materials and Methods

Animals

For all assays described below, we relied on a robust line of transgenic zebrafish (*fli1:egfp*) that stably express eGFP within vascular endothelial cells [13]. Adult *fli1:egfp* zebrafish were maintained on a 14-h:10-h light:dark cycle within a zebrafish stand-alone system (Aquatic Habitats, Inc., Apopka, FL) containing photoperiod light-cycle enclosures and recirculating conditioned reverse osmosis (RO) water (~27–28°C). Adult females and males were bred directly on-system using in-tank breeding traps suspended within 3-L tanks, directly within a Mini Mass Embryo Production System (Aquatic Habitats), or bred off-system within a light- and temperature-controlled incubator using breeding traps suspended within 1-L tanks. For all experiments described below, newly fertilized eggs were staged according to previously described methods [14]. All fish were handled and treated in accordance with approved Institutional Animal Care and Use Committee protocols at the University of South Carolina – Columbia (protocol # 2021-100372-072511). To minimize discomfort and distress, zebrafish husbandry and euthanasia methods were consistent with the National Research Council's *2011 Guide for the Care and Use of Laboratory Animals* and American Veterinary Medical Association's *2007 Guidelines on Euthanasia*, respectively.

Chemicals

Chemicals were purchased from ChemService, Inc. (West Chester, PA) and Sigma Aldrich (St. Louis, MO) (Table S1). Stock solutions of each chemical were prepared by dissolving chemicals in high performance liquid chromatography (HPLC)-grade dimethyl sulfoxide (DMSO) (50 mM), and then performing two-fold serial dilutions into DMSO to create stock solutions for each working solution. All stock solutions were stored at room temperature within 2-mL amber glass vials containing polytetrafluoroethylene-lined caps. For each individual plate, working solutions of tricaine methanesulfonate (MS-222) (Western Chemical, Inc., Ferndale, WA) were freshly prepared by dissolving MS-222 into embryo media (EM) (5 mM NaCl, 0.17 mM KCl, 0.33 mM $CaCl_2$, 0.33 mM $MgSO_4$), and working solutions of all treatments were freshly prepared by spiking stock solutions into EM, resulting in 0.1% DMSO within all vehicle control and treatment groups.

High-content screening (HCS) assay

Exposure setup. Black 384-well microplates containing 0.17-mm glass-bottom wells (Matrical Bioscience, Spokane, WA) were used for all HCS assays. Immediately following spawning, newly fertilized eggs were collected and placed in groups of approximately 50 per glass petri dish within a light- and temperature-controlled incubator until 5 hpf. For each HCS assay, 384 viable *fli1:egfp* embryos were arrayed at 5 hpf into a 384-well plate (one embryo per well) containing 50 μL per well of vehicle (0.1% DMSO) or treatment solution, and then incubated at 28°C under a 14-h:10-h light:dark cycle and static conditions until 72 hpf.

Image acquisition. At 72 hpf, the plate was removed from the incubator, and zebrafish embryos were anesthetized with 100 mg/L MS-222 by adding 25 μL of 300 mg/L MS-222 to 50 μL of vehicle or treatment solution. The plate was then centrifuged at 200 rpm for 2 minutes to help orient hatched embryos into right or left lateral recumbency. Using automated image acquisition protocols and parameters previously optimized [8] for our ImageXpress Micro (IXM) Widefield High-Content Screening System (Molecular Devices, Sunnyvale, CA), each embryo was imaged to analyze the following endpoints: heart rate, arterial circulation, pericardial area, body length, and intersegmental vessel area. During the entire image acquisition period, internal temperature within the IXM system was maintained between 25–27°C by removing panels on both sides of the IXM system and blowing air from left to right through the IXM with a portable fan; internal temperature was monitored and recorded at initiation and termination of each imaging protocol using a digital thermometer. In accordance with National Institutes of Health (NIH) guidelines [15], 72-hpf embryos were then euthanized by placing the plate at 4°C for 30 minutes.

Data extraction. Within MetaXpress 4.0.0.24 software (Molecular Devices, Sunnyvale, CA), custom journal scripts for extraction of heart rate, arterial circulation, pericardial area, body length, and intersegmental vessel area data were used as previously described [8]. Prior to data extraction, stream acquisitions within each well were inspected within MetaXpress to assess embryo orientation and survival. Coagulated embryos, unhatched embryos, grossly malformed embryos, or developed embryos lacking a heartbeat were considered dead. Using these survival criteria, only hatched and live embryos positioned in right or left lateral recumbency were analyzed. Interactive semi-automated journal scripts were used to isolate regions of interest and quantify heart rate, arterial circulation, pericardial area, and intersegmental vessel area, whereas a fully automated journal script was used to quantify body length. Examples of raw and analyzed images for each endpoint as well as additional details of the data extraction and analysis process are described within Yozzo et al. [8].

Hemoglobin staining

Based on protocols previously described by Paffett-Lugassy and Zon [16], *o*-dianisidine staining was used to detect the presence of hemoglobin within intact 30-, 48-, and 72-hpf zebrafish embryos exposed to vehicle (0.1% DMSO) or butafenacil (0.195 and 0.78 μM). Dechorionated or hatched embryos were stained in the dark for 30 min at room temperature within a solution containing *o*-dianisidine (0.6 mg/ml), 0.01 M sodium acetate (pH 4.5), 0.65% H_2O_2, and 40% (vol/vol) ethanol. Once stained, embryos were

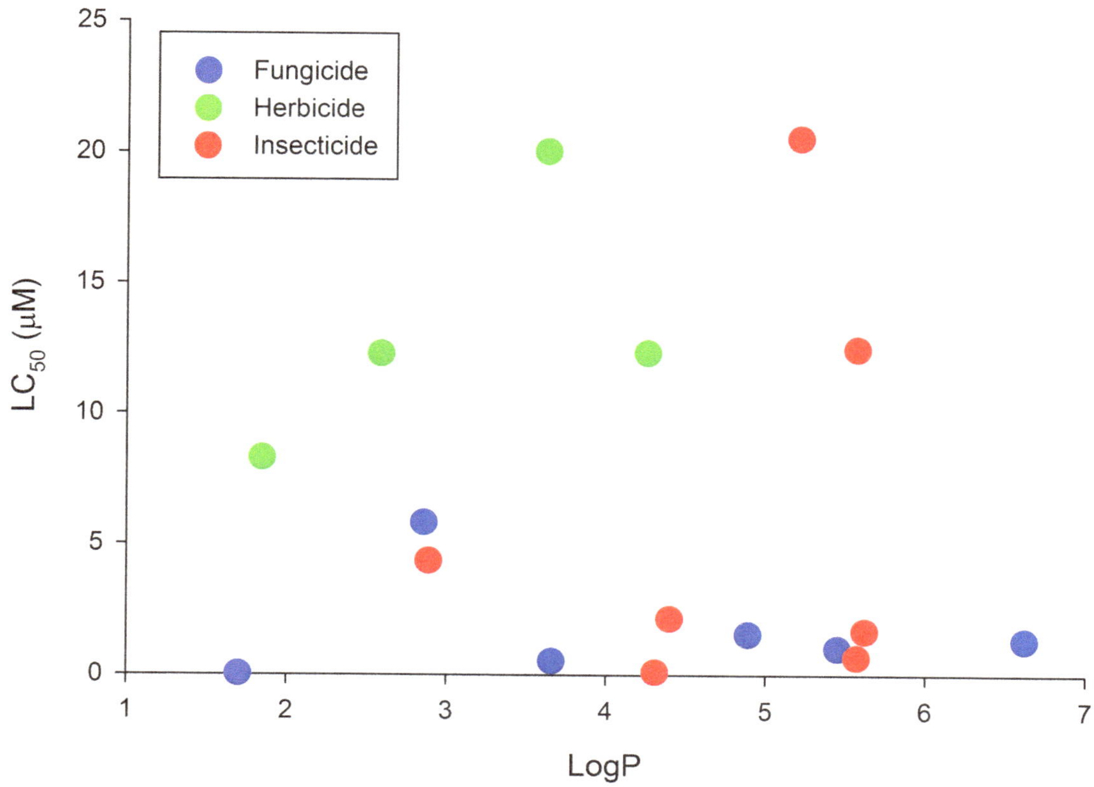

Figure 1. Lethal concentration (LC_{50}) vs. octanol:water coefficient (LogP) for 17 chemicals screened from the EPA's ToxCast Phase-I library.

washed with RO water and then fixed in 4% paraformaldehyde for at least 1 h at room temperature. Pigments were removed from fixed embryos by incubating in a solution of 0.8% KOH, 0.9% H_2O_2, and 0.1% Tween-20 for 30 min at room temperature. Embryos were then washed with phosphate-buffered saline (PBS) containing 0.1% Tween-20, and then fixed again in 4% paraformaldehyde for at least 3 h before storage in PBS at 4°C. All embryos were oriented in dorsal recumbency and imaged using an Olympus BX51 Research System Microscope equipped with an Olympus DP72 Digital Color Camera and Olympus DP2-BSW Imaging Software (Olympus Corp., Center Valley, PA). Images were analyzed using a color range tool within Adobe Photoshop CS4 (San Jose, CA). Sample colors were selected within vehicle control images to represent the *o*-dianisidine stain, and then control color samples were used to measure the stained area on the ventral side of the yolk sac and pericardial region of embryos across all treatments.

Statistical analyses

All statistical procedures were performed using SPSS Statistics 19.0 (Chicago, IL). A general linear model (GLM) analysis of variance (ANOVA) ($\alpha = 0.05$) was used for all data, as these data did not meet the equal variance assumption for non-GLM ANOVAs. Pair-wise Tukey-based multiple comparisons of least square means were performed to identify significant treatment-related effects. To estimate the median lethal concentration (LC_{50}) after static exposure to each chemical from 5 to 72 hpf, a four-parameter concentration-response curve was fit to percent mortality data using log-transformed chemical concentrations within Prism 6.0 (GraphPad Software Inc., La Jolla, CA).

Results

Chemical screening

For this study, we first identified 31 of the most acutely toxic chemicals from the EPA's ToxCast Phase-I chemical library (Table S1). Two chemicals (pyridaben and cypermethrin) were previously screened using this HCS assay [8]. In addition, two chemicals (tefluthrin and milbemectin) were not commercially available (as of August 2013), and fentin was insoluble in DMSO at concentrations as low as 2.5 mM. Therefore, these five chemicals were not included in the final pilot screen for this study, resulting in a total of 26 chemicals screened in concentration-response format. Microsoft Excel spreadsheets containing raw data for all HCS assays are provided within Dataset S1. To determine targeted effects on the cardiovascular system, heart rate, arterial circulation, pericardial area, and intersegmental vessel area were quantified within treatment groups with >70% survival and/or <10% decrease in body length relative to vehicle controls [8]. All chemicals in this initial screen had at least one concentration that resulted in an effect on either body length or survival (0.05–50 μM) (Figures S1, S2, S3, S4, S5, S6, S7, S8, S9, S10, S11, S12, S13, S14, S15, S16, S17, S18, S19, S20, S21, S22, S23, S24, S25, S26). Of the chemicals tested, thiram was the most acutely toxic based on survival, as exposure to greater than 0.02 μM – the highest concentration analyzed – resulted in <70% survival (Figure S2). Based on survival data, we calculated an LC_{50} for approximately 65% of the chemicals screened (17 chemicals) and, for these chemicals, found no relationship between the chemical-specific LC_{50} and octanol:water partition coefficient (logP) (Figure 1). We then compared the lowest observable effect concentrations (LOECs) for functional endpoints (heart rate and circulation) to identify the ability of this assay to discriminate chemicals with specific effects on cardiac vs. vascular function. Based on this comparison, butafenacil was identified as the only chemical that significantly impacted circulation in the absence of effects on heart rate (Figure 2).

Butafenacil is an herbicide that inhibits protoporphyrinogen oxidase (PPO) [20] – an enzyme necessary for heme production in vertebrates – and was the only chemical to induce a significant concentration-dependent effect on any of the endpoints measured within our HCS assay. Survival was <70% for butafenacil concentrations higher than 12.5 µM (Figures S3). Therefore, 12.5 µM was the highest concentration analyzed for all endpoints. At 0.39 µM and higher, butafenacil exposure resulted in a concentration-dependent decrease in arterial circulation in the absence of effects on heart rate and intersegmental vessel area (Figure 3). Although butafenacil exposure increased pericardial area, this effect was only significant at the two highest concentrations analyzed (6.25 and 12.5 µM). As indicated by our quantitative data, exposure of embryos to butafenacil at 0.39 µM and higher resulted in visibly decreased circulation (Video S1) compared to embryos exposed to vehicle alone (Video S2).

To determine whether other commercially available PPO inhibitors mimicked butafenacil-induced anemia, we relied on classifications defined by the Herbicide Resistance Action Committee (http://www.hracglobal.com/Education/ClassificationofHerbicideSiteofAction.aspx), Insecticide Resistance Action Committee (http://www.irac-online.org/documents/moa-classification/?ext=pdf), and Fungicide Resistance Action Committee (http://www.frac.info/publication/anhang/FRAC%20Code%20List%202013-update%20April-2013.pdf) to first assign MOAs to all pesticides within the ToxCast Phase-I chemical library and, if necessary, then test additional PPO inhibitors not included within our original pilot screen. In addition to butafenacil, five PPO inhibitors (fluthiacet-methyl, pyraflufen-ethyl, lactofen, flufenpyr-ethyl, and carfentrazone-ethyl) were included within our initial pilot screen; six additional (but less acutely toxic) PPO inhibitors (flumiclorac-pentyl, oxadiazon, flumioxazin, oxyfluorfen, sulfentrazone, and acifluorfen) not included within our initial pilot screen were also tested using our HCS assay (Figures S27, S28, S29, S30, S31, S32). Out of 11 PPO inhibitors screened (excluding butafenacil), flumioxazin was the only chemical to induce anemia and significantly decrease arterial circulation at more than three concentrations (Figure S29), albeit the phenotype was not as severe compared to butafenacil (Figure 3 and Figure S3).

Hemoglobin staining

The presence of hemoglobin was assessed in embryos exposed to vehicle (0.1% DMSO), 0.195, or 0.78 µM butafenacil from 5 hpf to either 30, 48, or 72 hpf. Hemoglobin-positive cells appeared within vehicle-exposed embryos shortly after circulation commenced at 30 hpf, but these cells were absent following exposure to 0.195 or 0.78 µM butafenacil (Figure 4). At 48 and 72 hpf, embryos exposed to 0.195 µM butafenacil resulted in significantly decreased hemoglobin compared to vehicle controls (Figure 4), whereas embryos exposed to 0.78 µM butafenacil showed negligible or no *o*-dianisidine staining, indicating complete abolishment of hemoglobin due to butafenacil exposure (Figure 4).

Discussion

In this study, we leveraged our recently developed and optimized 384-well HCS assay to screen and identify chemicals

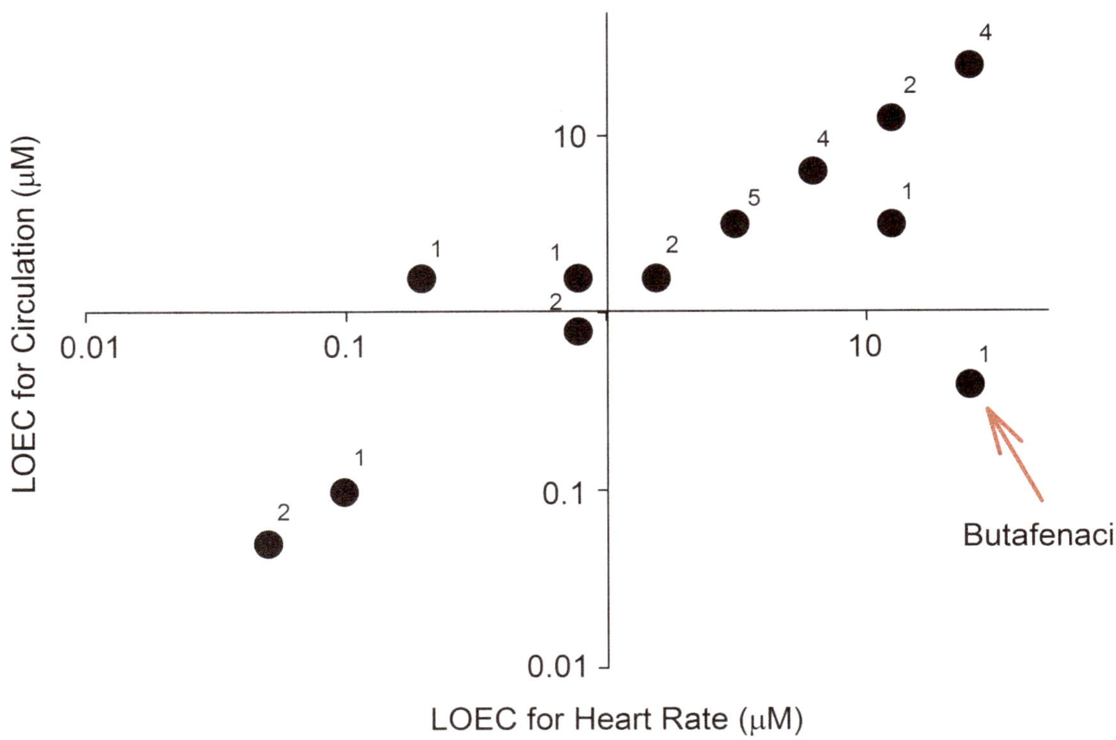

Figure 2. Comparison of lowest observable effect concentrations (LOECs) for heart rate and circulation for 26 of the most acutely toxic chemicals from the EPA's ToxCast Phase-I library. Butafenacil was the only chemical with a concentration-dependent effect on circulation in the absence of effects on heart rate. Numbers above the points denote the number of chemicals each point represents.

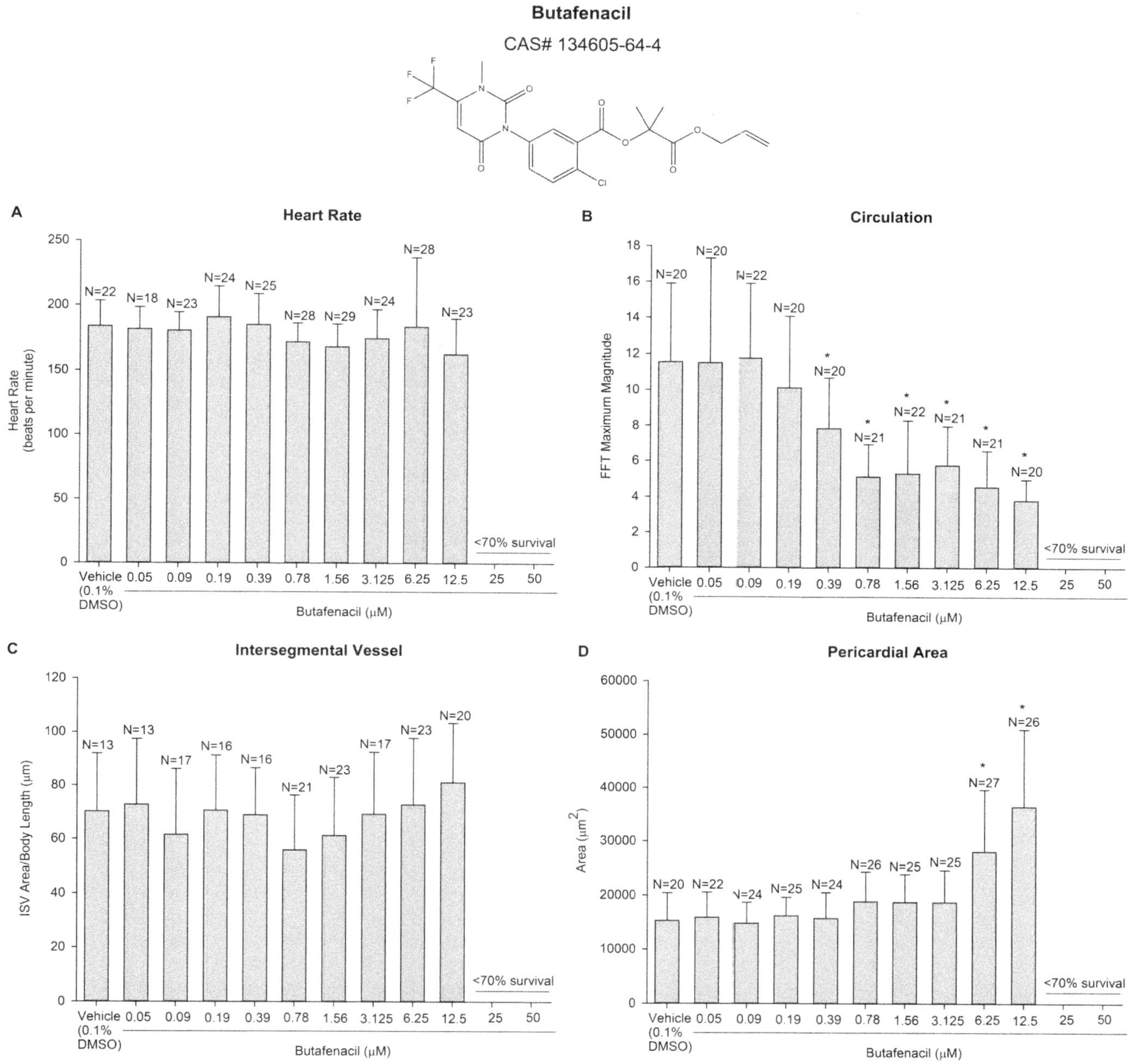

Figure 3. Butafenacil exposure resulted in a concentration-dependent effect on circulation. Based on decision criteria defined by Yozzo et al. [8], concentrations with <70% survival were not analyzed for potential effects on circulation, pericardial area, heart rate, or intersegmental vessel area. An asterisk denotes a significant difference from vehicle controls ($p<0.05$). N = final number of embryos analyzed per treatment.

affecting cardiovascular development and function at non-teratogenic concentrations. After ranking acute toxicity data generated from the U.S. EPA's zebrafish teratogenesis assay, we first screened 26 of the most acutely toxic chemicals within EPA's ToxCast Phase-I chemical library in concentration-response format (0.05–50 μM). Based on this screen, we identified butafenacil as a potent inducer of anemia, as exposure to 0.39–3.125 μM butafenacil completely abolished arterial circulation in the absence of effects on all other endpoints evaluated. Using *o*-dianisidine staining, we then revealed that severe butafenacil-induced anemia in zebrafish was due to a complete loss of hemoglobin following exposure during early development.

All ToxCast Phase-I chemicals in our initial pilot screen resulted in significant effects on body length or survival at 50 μM or lower. Therefore, an absence of targeted effects on the cardiovascular system was not likely due chemical sorption to well walls or limited uptake across the embryonic chorion. Although previous research has suggested that acute toxicity (LC_{50}) and hydrophobicity (logP) are positively correlated [12], findings from this study and others [17,18] suggest that there is no relationship between these two parameters (Figure 1). However, these discrepancies may be due to differences in experimental design such as the use of glass inserts within 96-well plates vs. adding treatment solutions directly to wells of 96- and 384-well plates. Interestingly, we were unable to

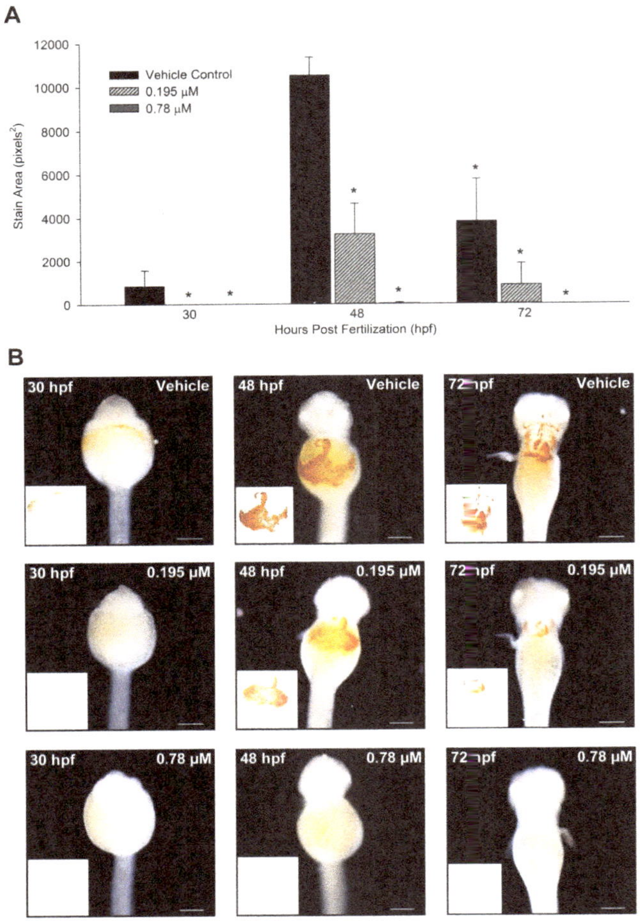

Figure 4. *o*-Dianisidine staining in embryos exposed to vehicle (0.1% DMSO) or butafenacil (A), and representative images at each time point for vehicle, 0.195, and 0.78 μM butafenacil (B). All embryos were oriented in dorsal recumbency for imaging. An asterisk denotes a significant difference from vehicle controls ($p<0.05$). White boxes represent stained areas within representative images. Scale bar = 200 μm.

determine an LC_{50} for six chemicals screened, five of which have a logP>5.5 and one of which has a logP<1, suggesting that there may be an optimal logP range for chemicals tested within our assay.

Based on our screen, butafenacil was the only chemical to induce a significant concentration-dependent effect on any of the endpoints evaluated. Butafenacil is an herbicide that inhibits PPO, an enzyme that catalyzes oxidation of protoporphyrinogen IX to protoprophyrin IX – the penultimate step of chlorophyll and heme biosynthesis in plants and animals, respectively [19]. As a result, this post-emergent herbicide is phytotoxic on contact and provides rapid knockdown of broadleaf and grass weeds [20]. While there are currently no registered butafenacil-containing products within the United States, butafenacil is registered for use within Australia, Argentina, Brazil, Japan, Switzerland, and Thailand [20]. Registered uses include cotton and citrus, as well as uses on cereal grains (except rice) up to 100 to 200 g ai/ha [20].

Given that butafenacil is a PPO inhibitor and completely eliminated hemoglobin production and circulation, butafenacil likely blocked heme biosynthesis and, as a result, hemoglobin and red blood cell production within zebrafish embryos. Similarly, butafenacil exposure in rodent models (rats and mice) resulted in decreased hemoglobin, hematocrit, and mean corpuscular volume [21]. While ruptured vessels and extravasation of blood cells may also lead to abolishment of circulation, butafenacil-induced anemia was likely not mediated by this MOA, as hemorrhaging was absent based on images generated from hemoglobin staining and our HCS assay. Although circulation was abolished in embryos exposed to 0.39 to 3.125 μM butafenacil, these embryos developed in the absence of gross malformations – a finding that is consistent with zebrafish mutants that develop little to no blood cells. These mutants appear normal for the first five days during development, but fail to inflate a swim bladder and die within two weeks [22]. The lack of developmental malformations is because oxygen demands of developing zebrafish are met by bulk diffusion until about two weeks post-fertilization [23]. By 12 to 14 dpf, functional hemoglobin is needed for proper swim bladder inflation and the transition to gill uptake and convective oxygen transport [24]; therefore, these zebrafish mutants do not survive beyond approximately 14 dpf. Similarly, if irreversible, butafenacil-exposed embryos would likely not survive after two weeks following the transition from passive oxygen diffusion to convective oxygen transport.

Out of 11 PPO inhibitors screened (excluding butafenacil), flumioxazin was the only PPO inhibitor to induce an anemic phenotype similar to butafenacil, albeit the flumioxazin-induced phenotype was not as severe as butafenacil. With the possible exception of sulfentrazone and acifluorfen (the least acutely toxic chemicals), it is unlikely that differential uptake across the chorion accounts for an absence of anemia within embryos exposed to the remaining eight PPO inhibitors screened, as significant effects on one or more endpoints were observed within at least one concentration tested within our assay. Rather, potency differences may be attributable to relative differences in chemical affinity to zebrafish PPO, but this was not directly tested within our study or, to date, within any other laboratories. While there are over 700 assays within the ToxCast program, to our knowledge none of these assays directly address chemical-mediated PPO inhibition nor impacts on hemoglobin production, suggesting that our targeted, whole-organism HCS assay has the potential to identify a unique mechanism (PPO inhibition) and phenotype (anemia) not currently covered by the large battery of existing ToxCast assays.

In summary, results from this study suggest that our HCS assay not only has potential for screening potential chemicals for cardiovascular toxicity, but also has potential for identifying chemicals impacting hematopoiesis. Adverse outcomes due to inhibition of PPO are detectable by this assay, and PPO is functionally conserved between zebrafish and humans. Moreover, PPO is established as an important target for certain human blood disorders, such as variegate porphyria for which preventative therapies are still needed [25]. While certain zebrafish mutant strains with disrupted blood cell development or differentiation are available, these strains have mainly been used to investigate the molecular mechanisms of hematopoiesis rather than HCS applications [26,27]. Therefore, using *fli1:egfp* zebrafish and butafenacil as a positive control, this HCS assay may provide a robust whole-organism-based platform for pre-clinical toxicology and/or drug discovery efforts focused on rapid identification of pro- or anti-anemic chemicals.

Supporting Information

Figure S1 Rotenone did not have a concentration dependent effect on any endpoints. Based on decision criteria defined by Yozzo et al. [8], hashed bars represent concentrations that were not analyzed for potential effects on circulation, pericardial area, heart rate, or intersegmental vessel area. An asterisk denotes a significant difference from vehicle controls ($p<0.05$). N = final number of embryos analyzed per treatment.

Figure S2 Thiram did not have a concentration dependent effect on any endpoints. Based on decision criteria defined by Yozzo et al. [8], hashed bars represent concentrations that were not analyzed for potential effects on circulation, pericardial area, heart rate, or intersegmental vessel area. An asterisk denotes a significant difference from vehicle controls ($p< 0.05$). N = final number of embryos analyzed per treatment.

Figure S3 Butafenacil significantly decreased circulation without having a concentration dependent effect on any other endpoint. Based on decision criteria defined by Yozzo et al. [8], hashed bars represent concentrations that were not analyzed for potential effects on circulation, pericardial area, heart rate, or intersegmental vessel area. An asterisk denotes a significant difference from vehicle controls ($p<0.05$). N = final number of embryos analyzed per treatment.

Figure S4 Flumetralin did not have a concentration dependent effect on any endpoints. Based on decision criteria defined by Yozzo et al. [8], hashed bars represent concentrations that were not analyzed for potential effects on circulation, pericardial area, heart rate, or intersegmental vessel area. An asterisk denotes a significant difference from vehicle controls ($p<0.05$). N = final number of embryos analyzed per treatment.

Figure S5 Fluthiacet-methyl did not have a concentration dependent effect on any endpoints. Based on decision criteria defined by Yozzo et al. [8], hashed bars represent concentrations that were not analyzed for potential effects on circulation, pericardial area, heart rate, or intersegmental vessel area. An asterisk denotes a significant difference from vehicle controls ($p<0.05$). N = final number of embryos analyzed per treatment.

Figure S6 Abamectin did not have a concentration dependent effect on any endpoints. Based on decision criteria defined by Yozzo et al. [8], hashed bars represent concentrations that were not analyzed for potential effects on circulation, pericardial area, heart rate, or intersegmental vessel area. An asterisk denotes a significant difference from vehicle controls ($p<0.05$). N = final number of embryos analyzed per treatment.

Figure S7 Propargite did not have a concentration dependent effect on any endpoints. Based on decision criteria defined by Yozzo et al. [8], hashed bars represent concentrations that were not analyzed for potential effects on circulation, pericardial area, heart rate, or intersegmental vessel area. An asterisk denotes a significant difference from vehicle controls ($p<0.05$). N = final number of embryos analyzed per treatment.

Figure S8 Pyraclostrobin did not have a concentration dependent effect on any endpoints. Based on decision criteria defined by Yozzo et al. [8], hashed bars represent concentrations that were not analyzed for potential effects on circulation, pericardial area, heart rate, or intersegmental vessel area. An asterisk denotes a significant difference from vehicle controls ($p<0.05$). N = final number of embryos analyzed per treatment.

Figure S9 (Z, E)-Fenpyroximate did not have a concentration dependent effect on any endpoints. Based on decision criteria defined by Yozzo et al. [8], hashed bars represent concentrations that were not analyzed for potential effects on circulation, pericardial area, heart rate, or intersegmental vessel area. An asterisk denotes a significant difference from vehicle controls ($p<0.05$). N = final number of embryos analyzed per treatment.

Figure S10 Tribufos did not have a concentration dependent effect on any endpoints. Based on decision criteria defined by Yozzo et al. [8], hashed bars represent concentrations that were not analyzed for potential effects on circulation, pericardial area, heart rate, or intersegmental vessel area. An asterisk denotes a significant difference from vehicle controls ($p<0.05$). N = final number of embryos analyzed per treatment.

Figure S11 Famoxadone did not have a concentration dependent effect on any endpoints. Based on decision criteria defined by Yozzo et al. [8], hashed bars represent concentrations that were not analyzed for potential effects on circulation, pericardial area, heart rate, or intersegmental vessel area. An asterisk denotes a significant difference from vehicle controls ($p<0.05$). N = final number of embryos analyzed per treatment.

Figure S12 Fluoxastrobin did not have a concentration dependent effect on any endpoints. Based on decision criteria defined by Yozzo et al. [8], hashed bars represent concentrations that were not analyzed for potential effects on circulation, pericardial area, heart rate, or intersegmental vessel area. An asterisk denotes a significant difference from vehicle controls ($p<0.05$). N = final number of embryos analyzed per treatment.

Figure S13 Pyraflufen-ethyl did not have a concentration dependent effect on any endpoints. Based on decision criteria defined by Yozzo et al. [8], hashed bars represent concentrations that were not analyzed for potential effects on circulation, pericardial area, heart rate, or intersegmental vessel area. An asterisk denotes a significant difference from vehicle controls ($p<0.05$). N = final number of embryos analyzed per treatment.

Figure S14 Trifloxystrobin did not have a concentration dependent effect on any endpoints. Based on decision criteria defined by Yozzo et al. [8], hashed bars represent concentrations that were not analyzed for potential effects on circulation, pericardial area, heart rate, or intersegmental vessel area. An asterisk denotes a significant difference from vehicle controls ($p<0.05$). N = final number of embryos analyzed per treatment.

Figure S15 Dazomet did not have a concentration dependent effect on any endpoints. Based on decision criteria defined by Yozzo et al. [8], hashed bars represent concentrations that were not analyzed for potential effects on circulation, pericardial area, heart rate, or intersegmental vessel area. An asterisk denotes a significant difference from vehicle controls ($p<0.05$). N = final number of embryos analyzed per treatment.

Figure S16 Esfenvalerate did not have a concentration dependent effect on any endpoints. Based on decision criteria defined by Yozzo et al. [8], hashed bars represent concentrations that were not analyzed for potential effects on circulation, pericardial area, heart rate, or intersegmental vessel area. An asterisk denotes a significant difference from vehicle controls ($p<0.05$). N = final number of embryos analyzed per treatment.

Figure S17 Chlorothalonil did not have a concentration dependent effect on any endpoints. Based on decision criteria defined by Yozzo et al. [8], hashed bars represent concentrations that were not analyzed for potential effects on circulation, pericardial area, heart rate, or intersegmental vessel area. An asterisk denotes a significant difference from vehicle controls ($p<0.05$). N = final number of embryos analyzed per treatment.

Figure S18 Lactofen did not have a concentration dependent effect on any endpoints. Based on decision criteria defined by Yozzo et al. [8], hashed bars represent concentrations that were not analyzed for potential effects on circulation, pericardial area, heart rate, or intersegmental vessel area. An asterisk denotes a significant difference from vehicle controls ($p<0.05$). N = final number of embryos analyzed per treatment.

Figure S19 Clodinafop-propargyl did not have a concentration dependent effect on any endpoints. Based on decision criteria defined by Yozzo et al. [8], hashed bars represent concentrations that were not analyzed for potential effects on circulation, pericardial area, heart rate, or intersegmental vessel area. An asterisk denotes a significant difference from vehicle controls ($p<0.05$). N = final number of embryos analyzed per treatment.

Figure S20 Tebufenpyrad did not have a concentration dependent effect on any endpoints. Based on decision criteria defined by Yozzo et al. [8], hashed bars represent concentrations that were not analyzed for potential effects on circulation, pericardial area, heart rate, or intersegmental vessel area. An asterisk denotes a significant difference from vehicle controls ($p<0.05$). N = final number of embryos analyzed per treatment.

Figure S21 Fenpropathrin did not have a concentration dependent effect on any endpoints. Based on decision criteria defined by Yozzo et al. [8], hashed bars represent concentrations that were not analyzed for potential effects on circulation, pericardial area, heart rate, or intersegmental vessel area. An asterisk denotes a significant difference from vehicle controls ($p<0.05$). N = final number of embryos analyzed per treatment.

Figure S22 Cyfluthrin did not have a concentration dependent effect on any endpoints. Based on decision criteria defined by Yozzo et al. [8], hashed bars represent concentrations that were not analyzed for potential effects on circulation, pericardial area, heart rate, or intersegmental vessel area. An asterisk denotes a significant difference from vehicle controls ($p<0.05$). N = final number of embryos analyzed per treatment.

Figure S23 Indoxacarb did not have a concentration dependent effect on any endpoints. Based on decision criteria defined by Yozzo et al. [8], hashed bars represent concentrations that were not analyzed for potential effects on circulation, pericardial area, heart rate, or intersegmental vessel area. An asterisk denotes a significant difference from vehicle controls ($p<0.05$). N = final number of embryos analyzed per treatment.

Figure S24 Chlorpyrifos (ethyl) oxon did not have a concentration dependent effect on any endpoints. Based on decision criteria defined by Yozzo et al. [8], hashed bars represent concentrations that were not analyzed for potential effects on circulation, pericardial area, heart rate, or intersegmental vessel area. An asterisk denotes a significant difference from vehicle controls ($p<0.05$). N = final number of embryos analyzed per treatment.

Figure S25 Flufenpyr-ethyl did not have a concentration dependent effect on any endpoints. Based on decision criteria defined by Yozzo et al. [8], hashed bars represent concentrations that were not analyzed for potential effects on circulation, pericardial area, heart rate, or intersegmental vessel area. An asterisk denotes a significant difference from vehicle controls ($p<0.05$). N = final number of embryos analyzed per treatment.

Figure S26 Carfentrazone-ethyl did not have a concentration dependent effect on any endpoints. Based on decision criteria defined by Yozzo et al. [8], hashed bars represent concentrations that were not analyzed for potential effects on circulation, pericardial area, heart rate, or intersegmental vessel area. An asterisk denotes a significant difference from vehicle controls ($p<0.05$). N = final number of embryos analyzed per treatment.

Figure S27 Flumiclorac-pentyl did not have a concentration dependent effect on any endpoints. Based on decision criteria defined by Yozzo et al. [8], hashed bars represent concentrations that were not analyzed for potential effects on circulation, pericardial area, heart rate, or intersegmental vessel area. An asterisk denotes a significant difference from vehicle controls ($p<0.05$). N = final number of embryos analyzed per treatment.

Figure S28 Oxadiazon did not have a concentration dependent effect on any endpoints. Based on decision criteria defined by Yozzo et al. [8], hashed bars represent concentrations that were not analyzed for potential effects on circulation, pericardial area, heart rate, or intersegmental vessel area. An asterisk denotes a significant difference from vehicle controls ($p<0.05$). N = final number of embryos analyzed per treatment.

Figure S29 Flumioxazin did not have a concentration dependent effect on any endpoints. Based on decision criteria defined by Yozzo et al. [8], hashed bars represent concentrations that were not analyzed for potential effects on circulation, pericardial area, heart rate, or intersegmental vessel area. An asterisk denotes a significant difference from vehicle controls ($p<0.05$). N = final number of embryos analyzed per treatment.

Figure S30 Oxyfluorfen did not have a concentration dependent effect on any endpoints. Based on decision criteria defined by Yozzo et al. [8], hashed bars represent concentrations that were not analyzed for potential effects on circulation, pericardial area, heart rate, or intersegmental vessel area. An asterisk denotes a significant difference from vehicle controls ($p<0.05$). N = final number of embryos analyzed per treatment.

Figure S31 Sulfentrazone did not have a concentration dependent effect on any endpoints. Based on decision criteria defined by Yozzo et al. [8], hashed bars represent concentrations that were not analyzed for potential effects on circulation, pericardial area, heart rate, or intersegmental vessel area. An asterisk denotes a significant difference from vehicle controls ($p<0.05$). N = final number of embryos analyzed per treatment.

Figure S32 Acifluorfen did not have a concentration dependent effect on any endpoints. Based on decision criteria defined by Yozzo et al. [8], hashed bars represent concentrations that were not analyzed for potential effects on circulation, pericardial area, heart rate, or intersegmental vessel area. An asterisk denotes a significant difference from vehicle controls ($p<0.05$). N = final number of embryos analyzed per treatment.

Table S1 Chemical name, chemical formula, CAS registry number, vendor, and purity of all chemicals screened.

Video S1 Circulation of blood through a 72-hpf zebrafish embryo exposed to 3.125 µM butafenacil.

Video S2 Circulation of blood through a 72-hpf zebrafish embryo exposed to vehicle control (0.1% DMSO).

Dataset S1 Microsoft Excel spreadsheets containing raw data for all chemical screening assays.

Acknowledgments

We gratefully thank Dr. Robert Tanguay (Oregon State University) for providing founder fish to establish our *fli1:egfp* zebrafish colony, Sylvia de Bruin (Molecular Devices) for providing technical assistance during custom journal script development, and David Catron for providing custom Microsoft Excel macros for FFT analysis.

Author Contributions

Conceived and designed the experiments: JKL DCV. Performed the experiments: JKL CDL LAB GMI KLY TDR. Analyzed the data: JKL CDL LAB KLY. Contributed to the writing of the manuscript: JKL DCV.

References

1. Collins FS, Gray GM, Bucher JR (2008) Toxicology. Transforming environmental health protection. Science 319: 906–907.
2. Dix DJ, Houck KA, Martin MT, Richard AM, Setzer RW, et al. (2007) The ToxCast program for prioritizing toxicity testing of environmental chemicals. Toxicol Sci 95: 5–12.
3. NRC, editor (2007) Toxicity Testing in the 21st Century: A Vision and a Strategy. Washington, DC: National Academies Press. 196 p.
4. Belanger SE, Balon EK, Rawlings JM (2010) Saltatory ontogeny of fishes and sensitive early life stages for ecotoxicology tests. Aquatic Toxicology 97: 88–95.
5. Embry MR, Belanger SE, Braunbeck TA, Galay-Burgos M, Halder M, et al. (2010) The fish embryo toxicity test as an animal alternative method in hazard and risk assessment and scientific research. Aquatic Toxicology 97: 79–87.
6. Halder M, Leonard M, Iguchi T, Oris JT, Ryder K, et al. (2010) Regulatory aspects on the use of fish embryos in environmental toxicology. Integr Environ Assess Manag 6: 484–491.
7. Scholz S, Fischer S, Gundel U, Kuster E, Luckenbach T, et al. (2008) The zebrafish embryo model in environmental risk assessment–applications beyond acute toxicity testing. Environ Sci Pollut Res Int 15: 394–404.
8. Yozzo KL, Isales GM, Raftery TD, Volz DC (2013) High-content screening assay for identification of chemicals impacting cardiovascular function in zebrafish embryos. Environ Sci Technol 47: 11302–11310.
9. Letamendia A, Quevedo C, Ibarbia I, Virto JM, Holgado O, et al. (2012) Development and validation of an automated high-throughput system for zebrafish in vivo screenings. PLoS One 7: e36690.
10. Tran TC, Sneed B, Haider J, Blavo D, White A, et al. (2007) Automated, quantitative screening assay for antiangiogenic compounds using transgenic zebrafish. Cancer Res 67: 11386–11392.
11. Vogt A, Cholewinski A, Shen X, Nelson SG, Lazo JS, et al. (2009) Automated image-based phenotypic analysis in zebrafish embryos. Developmental Dynamics 238: 656–663.
12. Padilla S, Corum D, Padnos B, Hunter DL, Beam A, et al. (2012) Zebrafish developmental screening of the ToxCast Phase I chemical library. Reproductive Toxicology 33: 174–187.
13. Lawson ND, Weinstein BM (2002) In vivo imaging of embryonic vascular development using transgenic zebrafish. Developmental Biology 248: 307–318.
14. Kimmel CB, Ballard WW, Kimmel SR, Ullmann B, Schilling TF (1995) Stages of Embryonic-Development of the Zebrafish. Developmental Dynamics 203: 253–310.
15. NIH (2007) Guidelines for use of zebrafish in the NIH Intramural Research Program.
16. Paffett-Lugassy NN, Zon LI (2005) Analysis of hematopoietic development in the zebrafish. Methods Mol Med 105: 171–198.
17. Truong L, Reif DM, St Mary L, Geier MC, Truong HD, et al. (2014) Multidimensional in vivo hazard assessment using zebrafish. Toxicol Sci 137: 212–233.
18. Van den Bulck K, Hill A, Mesens N, Diekman H, De Schaepdrijver L, et al. (2011) Zebrafish developmental toxicity assay: A fishy solution to reproductive toxicity screening, or just a red herring? Reproductive Toxicology 32: 213–219.
19. Kawamura S, Kato T, Matsuo M, Katsuda Y, Yasuda M (1996) Species Difference in Protoporphyrin IX Accumulation Produced by anN-Phenylimide Herbicide in Embryos between Rats and Rabbits. Toxicology and Applied Pharmacology 141: 520–525.
20. NRA (2002) Evaluation of the new active butafenacil in the products Logran B-Power Herbicide & Touchdown B-Power Herbicide. Canberra: National Registration Authority for Agricultural and Veterinary Chemicals.
21. EPA (2003) Butafenacil; Pesticide Tolerance. In: Programs OoP, editor. Washington. pp. 54818–54827.
22. Weinstein BM, Schier AF, Abdelilah S, Malicki J, Solnica-Krezel L, et al. (1996) Hematopoietic mutations in the zebrafish. Development 123: 303–309.
23. Jacob E, Drexel M, Schwerte T, Pelster B (2002) Influence of hypoxia and of hypoxemia on the development of cardiac activity in zebrafish larvae. Am J Physiol Regul Integr Comp Physiol 283: R911–917.
24. Pelster B, Burggren WW (1996) Disruption of hemoglobin oxygen transport does not impact oxygen-dependent physiological processes in developing embryos of zebra fish (Danio rerio). Circ Res 79: 358–362.

25. Dooley KA, Fraenkel PG, Langer NB, Schmid B, Davidson AJ, et al. (2008) montalcino, A zebrafish model for variegate porphyria. Experimental Hematology 36: 1132–1142.
26. Hanaoka R, Dawid IB, Kawahara A (2007) Cloning and expression of zebrafish genes encoding the heme synthesis enzymes uroporphyrinogen III synthase (UROS) and protoporphyrinogen oxidase (PPO). DNA Seq 18: 54–60.
27. Ransom DG, Haffter P, Odenthal J, Brownlie A, Vogelsang E, et al. (1996) Characterization of zebrafish mutants with defects in embryonic hematopoiesis. Development 123: 311–319.

Size Changes in Honey Bee Larvae Oenocytes Induced by Exposure to Paraquat at Very Low Concentrations

Marianne Cousin[1], Elaine Silva-Zacarin[2], André Kretzschmar[3], Mohamed El Maataoui[4], Jean-Luc Brunet[1], Luc P. Belzunces[1]*

1 INRA, UR 406 Abeilles & Environnement, Laboratoire de Toxicologie Environnementale, Avignon, France, **2** Universidade Federal de São Carlos, Laboratory of Structural and Functional Biology, Sorocaba, São Paulo State, Brazil, **3** INRA, UR Biostatistiques et Processus Spatiaux, Avignon, France, **4** Université d'Avignon, Pôle Agrosciences, Avignon, France

Abstract

The effects of the herbicide Paraquat were investigated in honey bee larvae with attention focused on oenocytes. Honey bee larvae were exposed to Paraquat at different concentrations in the food: 0, 0.001, 0.01, 0.1 and 1 µg/kg. In controls, between 24 h and 48 h, oenocytes grew from 630.1 to 1643.8 μm^2 while nuclei changed in size from 124.9 to 245.6 μm^2. At 24 h, Paraquat induced a slight decrease in the size of oenocytes and nuclei. N-acetylcysteine (NAC), an antioxidant substance, slightly lowered the effects of Paraquat. At 48 h, Paraquat elicited a strong concentration-dependent decrease in the size of oenocytes, even at the lowest concentration. NAC reversed the effect of Paraquat at a concentration of ≥0.01 µg/kg. This reversion suggested different modes of action of Paraquat, with an oxidant action prevalent at concentrations ≥0.01 µg/kg. This study is the first which reports an effect of a pesticide at the very low concentration of 1 ng/kg, a concentration below the detection limits of the most efficient analytic methods. It shows that chemicals, including pesticides, are likely to have a potential impact at such exposure levels. We also suggest that Paraquat could be used as a suitable tool for investigating the functions of oenocytes.

Editor: Nicolas Desneux, French National Institute for Agricultural Research (INRA), France

Funding: This work was funded by INRA and Avignon University. The funders had no role in study design, data collection and analysis, decision to publish, or preparation of the manuscript.

Competing Interests: The authors have read the journal's policy and have the following conflicts: In 2011, LPB has reviewed a project, not related to this work, for Union National d'Apiculture Française (UNAF).

* E-mail: luc.belzunces@avignon.inra.fr

Introduction

The honey bee *Apis mellifera* is a pollinator insect of agro-environmental, scientific and economic importance [1]. In order to gather food (pollen and nectar) and plant substances to make the antiseptic propolis, honey bees visit plants in a radius up to12 km around the hive, thus being exposed to a large variety of environmental stressors [2]. The honey bee is regarded as involved in 80% of global agricultural pollination and contributes to the plant productions by increasing the qualitative and quantitative yields of fruits and vegetables [3]. In 2009, the economic value of insect pollination for agriculture was estimated at close to 20 billion € per year in Europe and approximately 153 billion € worldwide, which represented 9.5% of the value of the world agricultural production used for human food in 2005 [4]. In the European Community, the direct value of honey produced is estimated at about € 140 million, and the total added value of the crop pollination service has recently been estimated at € 14.2 billion [5]. On a scientific level, the honey bee is considered as one of the best models for investigating cognitive functions such as vision, orientation, and memory [6,7].

The constant decline of honey bee populations has been described and studied for the past 15 years [8]. Several factors have been identified to explain this decline, including pathogens and parasites, decrease of food resources and pesticides [9]. Environmental stressors, such as pesticides, may induce effects that impair not only the longevity but also the vitality of bees [5]. At very low doses, they may act at lethal and sublethal levels and elicit physiological, neural, metabolic and behavioral effects [10]. Sublethal effects can be very detrimental to the fate of the colony by altering the physiology and the functioning of individuals and by causing rapid depopulation.

Embryonic and post-embryonic developments are crucial phases during which environmental physiological disruptors may cause irreversible damages to the individuals. In the very early stages of honey bee larval development, trophocytes and oenocytes represent two main groups of cells that are easily observed in the larvae fat body [11]. Oenocytes exhibit a well-developed smooth endoplasmic reticulum and contribute to numerous physiological processes [12]. They are not only involved in the production of lipids and lipoproteins [13], but they also appear to be the source of precursors for cuticulin formation and play a role in the constitution of external cuticle in both larvae and adults [14]. In addition, they are involved in intermediary metabolism [15] and synthesize hydrocarbons to waterproof cuticle or to make beeswax [16,17]. Furthermore, oenocytes secrete hormones, especially those involved in larval and adult development, such as the ecdysteroids that induce metamorphosis and trigger the remodeling of larval tissues [18]. They are also described as the major cells expressing NADPH-cytochrome P450 reductase (CPR) and are involved in detoxification of xenobiotics and in metabolic resistance to insecticides [19]. Some pesticides have been proved to modulate the number, size, structure and activity of oenocytes. Sublethal doses of the organophosphates Parathion and Demethon decrease the number and increase the volume of oenocytes in nymphs and adults of the aphid *Pemphigus bursarius* [20]. These changes are linked to the loosening of chromatin in the nuclei and the amitotic division of oenocytes. Dimilin, a chitin inhibitor, reduces the production of polysaccharides and consequently disturbs the formation of cuticle [21].

A few studies have focused on the toxicity of pesticides to brood and the effects have mainly been investigated on adult bees, with special attention to insecticides and fungicides. The recent release of a new method to analyze the toxicity to the honey bee brood enables more accurate researches on the effects of pesticides on embryonic and post-embryonic development [22,23]. In the present study, the effects of a model substance, the herbicide Paraquat, on honey bee development have been examined by exposing the brood to very low concentrations of Paraquat in food. Although Paraquat was banned from the European Union in 2007 and in other countries, it is still used, at field rate of 0.3–2 kg/ha, on more than 100 crops in about 100 countries and particularly in developing countries [24]. Paraquat was shown to be non-toxic to slightly toxic to adult bees [25,26]. Thus, the aim of this study was to see whether substances different from insecticides and generally considered as non-toxic to adult bees, such as herbicides, fungicides and insect growth regulators (IGR), can induce adverse effects on brood. The effects of Paraquat were also investigated in the presence of N-acetylcysteine (NAC), an aminothiol precursor of intracellular cysteine, which is a radical scavenger that exhibits antioxidant properties and attenuates the detrimental effects of Paraquat [27]. Paraquat has been chosen because it can elicit negative effects not related to its herbicide action in vertebrates and invertebrates. Paraquat, a quaternary ammonium herbicide of the bispyridinium family [28], is widely used in agriculture and can affect different non-target organisms such as fishes [29], Collembola [30], birds [31], Gastropoda [32] and mice [33] and is a non-specific inducer of apoptosis [34]. In human beings, Paraquat can trigger lethal lung fibrosis [35]. It is also strongly implicated in the development of Parkinson's disease in farmers [36] and, acting as a pro-oxidant, can induce the formation of reactive oxygen species that elicit the destruction of dopaminergic neurons in the striatum [37].

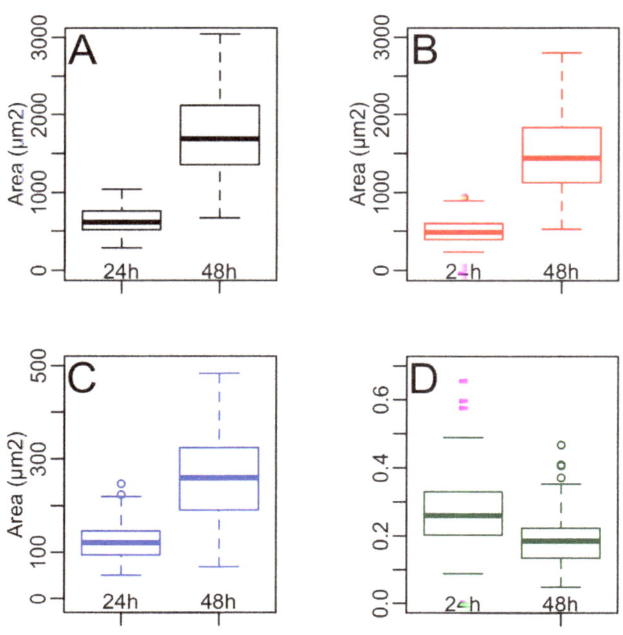

Figure 1. Evolution of oenocytes between 24 and 48 h. Changes of oenocyte biometrical parameters were noticed at 24 h (n = 138) and 48 h (n = 169) and were expressed as areas (µm^2). At the given time, larvae were picked out, immediately fixed in formaldehyde and processed for histochemical analysis. Oenocytes were detected and total cell and nuclei areas were measured on each oenocytes. Three compartments were analyzed : whole cell (A), cytoplasm (B, by subtracting nucleus area from total cell area) and nucleus (C), and the nucleo-cytoplasmic ratio NCR (D) was calculated. Surface values distribution at 24 and 48 h were compared with pair T-tests which assess significant differences for all parameters studied ($p<1.10^{-5}$).

Materials and Methods

Materials

Larvae of the honey bee *Apis mellifera* were randomly collected from three queen-right colonies, from the INRA apiary (Avignon, France), containing 45,000 to 50,000 bees and 6–7 brood frames. The honey bee colonies were carefully monitored to check their health status. The day before the experiment, empty combs were inserted in hives. Twenty four hours after, cells containing eggs were identified by means of a transparent grids and the combs were left in the hives for 3 days. In the early morning of day four, the combs were transported into the laboratory and newly hatched larvae, within previously identified cells, were exposed to Paraquat for two days in laboratory. The effects of Paraquat on oenocytes were observed after 24 and 48 h of exposure, which means that exposure to Paraquat through the food was perfectly controlled during the first two days of observation. Two brood combs per hive were collected to facilitate the gathering of 200 larvae for each experimental group, and were placed in rearing boxes with controlled temperature (33±1°C) and relative humidity (90%). Paraquat (98% pure; 1,1′-Diméthyl-4,4′-bipyridinium; CAS number: 1910-42-5) was purchased from Cluzeau-Info Labo, France. The concentrations of the stock solutions of Paraquat were checked by high-performance liquid chromatography with a method optimized for Paraquat solutions that enables reaching a limit of quantification of 8 ng/L and a limit of detection of 2.4 ng/L in simple matrices [38,39]. The difference between nominal and measured concentrations did not exceed 4%. Yeast extract (Y1000) and N-acetylcysteine were obtained from Sigma-Aldrich (France).

Experimental procedures

The marked larvae were randomly divided into five groups of 200 larvae (3 colonies, 65–67 larvae per colony and per Paraquat concentration), each of which were exposed to a specific concentration of Paraquat, namely: 0 (Control), 0.001, 0.01, 0.1 and 1 µg/kg of food, during larval development, in the presence (NAC) or in the absence (NoNAC) of N-acetyl-cysteine (NAC) at a concentration of 50 µg/kg. Thus, for each treatment condition (Paraquat at 5 concentrations, plus or minus NAC), 32–34 bees were individually analyzed per colony. The larvae from each colony were clearly identified for statistical analysis. The larvae were fed three times, at 0, 24 and 48 h, with a basic diet containing 50% (v/v) of royal jelly and 50% (v/v) of an aqueous (distilled water) solution containing yeast extracts (1%, w/v), D-Glucose (6%, w/v) and D-fructose (6%, w/v) [40]. Paraquat was added directly, from stock solutions in water, into the distilled water used for food preparation. For each marked larva, food was delivered

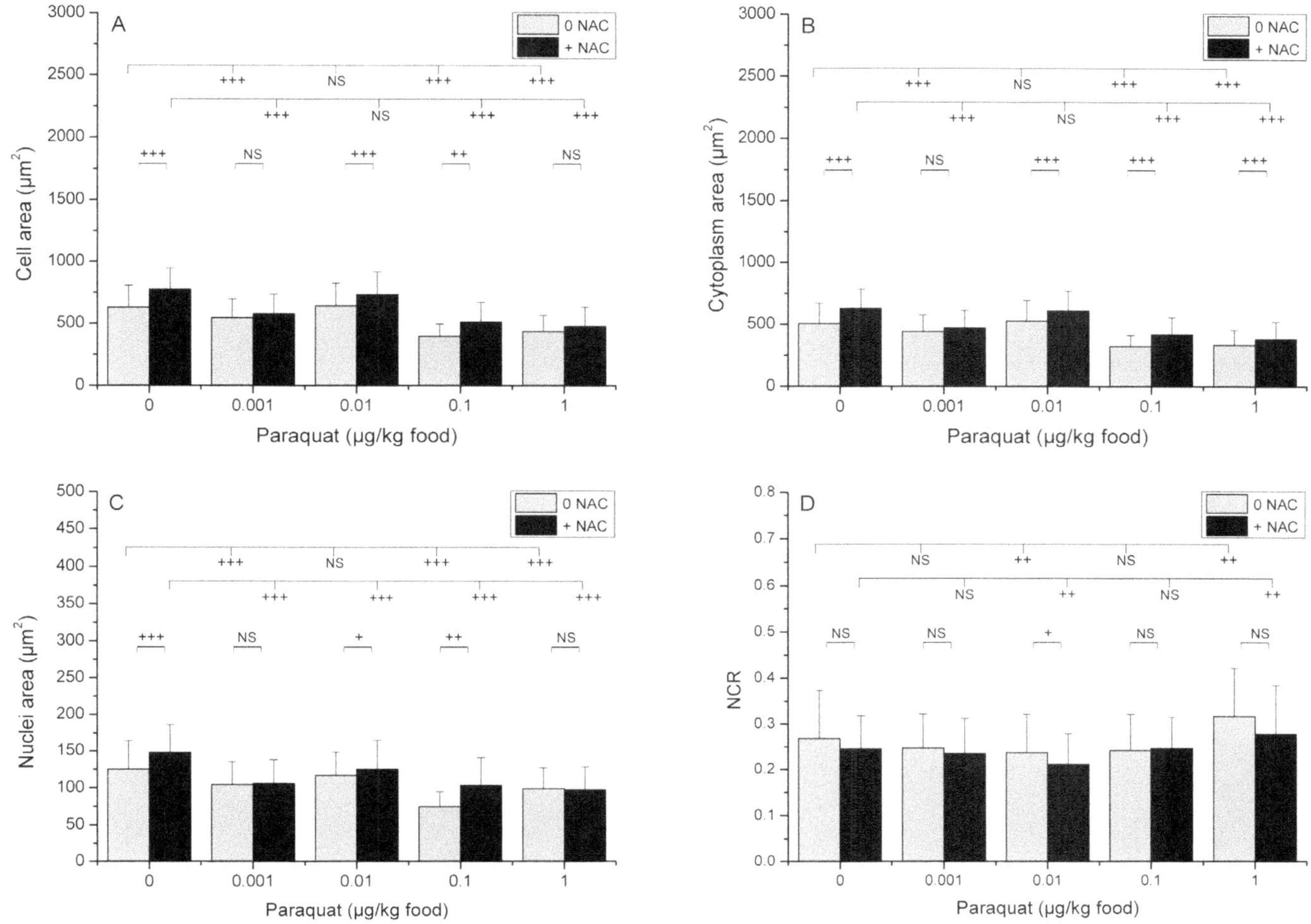

Figure 2. Effects of Paraquat at 24 h on the size of three oenocyte compartments. Larvae were exposed at different concentrations of Paraquat (0 (control), 0.001, 0.01, 0.1 and 1 µg/kg of food) in the absence (NoNAC) or in the presence of NAC. They were then picked out, immediately fixed in formaldehyde and processed for histochemical analysis. Statistically significant effects were observed after 24 h of exposure to Paraquat. Three compartments were analyzed: whole cell (A), cytoplasm (B, by subtracting nucleus area from total cell area) and nucleus (C), and the nucleo-cytoplasmic ratio (NCR) was calculated (D). (+), $p<0.01$; (++), $p<0.001$; (+++), $p<0.0001$; NS, not significant ($p>0.05$).

by means of a micropipette with a volume of 5 µL at 0 h, 8 µL at 24 h and 10 µL at 48 h of larval development.

Histochemistry

For each experimental group, larvae were randomly sampled at 24 and 48 h (they were fed once or twice respectively) and separately prepared for light microscopy. For morphological analysis, the larvae were fixed in 4% (v/v) formaldehyde in 0.1 M sodium phosphate buffer pH 7.2 for 48 h at 4°C. Larvae were then rinsed in a phosphate buffer, sequentially dehydrated in ethanol [from 70% to 100% (v/v)] and embedded in resin (Kit Technovit 7100; Heraeus-Kulzer GmbH, Wehrheim, Germany) according to the instructions of the manufacturer. Tissue sections (3 µm thick) were obtained using a retraction microtome (Reichert-Young Supercut 2065, Vienna, Austria), collected on microscope slides, dried and stored in dust-proof containers until staining. During the next stages, the slides were hydrated in distilled water before staining in order to visualize polysaccharides and proteins using periodic acid-Schiff's reagent (PAS) and naphtol blue black procedures, respectively [41]. After staining, the slides were dried and cover slipped in Surgipath Micromount medium (Surgipath, Richmond, Canada). Sections were photographed using a light microscope VWR (TR500 High End Tri.) at 100X magnification.

Oenocyte analysis

Images were processed with a Motic Images Plus camera 2300. For each group of treatment, all the oenocytes of each section from the different larvae were analyzed using ImageJ software 1.43u (US NIH) to measure the cell (Ac) and nucleus (An) areas. The mean number of oenocytes analyzed per larvae was 81±30. The nucleo-cytoplasmic ratio NCR was estimated by the following formula:

$$NCR = An/(Ac - An)$$

For the global statistical analysis, data were processed by a generalized linear model (GLM) with a log link function for analyzing the effect of the different experimental factors (i.e., larvae stage, Paraquat concentration and NAC). In order to assess the effects of the factor levels, statistical differences between treatment groups were analyzed with a multi-T-test, after normality had been checked to fit the distribution of observed measurements at each factor level. For non-independent tests, a Bonferroni correction was applied. The level of significance was determined when the p value was greater than 0.001. The concentration-effect relationship of Paraquat after 48 h of exposure was explored by log-log linear regression analysis.

Results

NCR and cell size in control oenocytes

At concentrations used in this study, Paraquat did not induce significant mortality in honey bee larvae (T test: p value $=0.471$). Significant growth of oenocytes was observed between 24 and 48 h with respective sizes of 630.12±175.96 and 1643.77±479.02 μm^2 (T-test: p value $=2.2.10^{-16}$) (Fig. 1). In the same period, the size of the nuclei increased significantly from 24.89±38.95 to 245.55±89.74 μm^2 (T-test: p value $=2.2.10^{-16}$). As a consequence, the variation of NCR between 24 and 48 h decreased significantly from 0.269±0.106 to 0.187 ±0.07[illegible] (T-test: p value $=5.982.10^{-13}$).

Effects of Paraquat on oenocytes

The global analysis with GLM showed significant effects of the experimental factors on the different oenocyte compartments. In the case of entire cells, larvae stage and Paraquat concentration had a *p* value $<2.10^{-16}$ while the *p* value for NAC was 0.0287. Similar results were obtained for the cytoplasm compartment whereas, for nuclei, there was no significant effect of NAC. Additionally, no effect of larvae stage was observed for NCR whereas the other factors were highly significant (*p* value $<2.10^{-16}$ and $1.39.10^{-8}$ for Paraquat concentration and NAC, respectively).

The effects of Paraquat on biometrical parameters of oenocytes were studied in the absence (noNAC) and in the presence (NAC) of N-Acetylcysteine, an antioxidant substance that prevents the cells from the oxidative effects of Paraquat (Fig. 2, 3 and 4).

At 24 hours, in the absence of NAC, Paraquat induced a slight decrease in the size of oenocytes (except at the concentration of 0.01 µg/kg) and nuclei (Fig. 2). Consequently, a decrease in the cytoplasm area was observed only at the two highest concentrations (0.1 and 1 µg/kg) (Fig. 2, and 4A, 4B, for 1 µg/kg), which therefore resulted in an increase of NCR at the highest concentration. NAC only slightly modified the effects induced by Paraquat at 24 h and, like in the absence of NAC, a decrease of cell and nuclei sizes was observed in the presence of NAC (Fig. 2). However, at 24 h, NAC slightly attenuated the reduction of size of

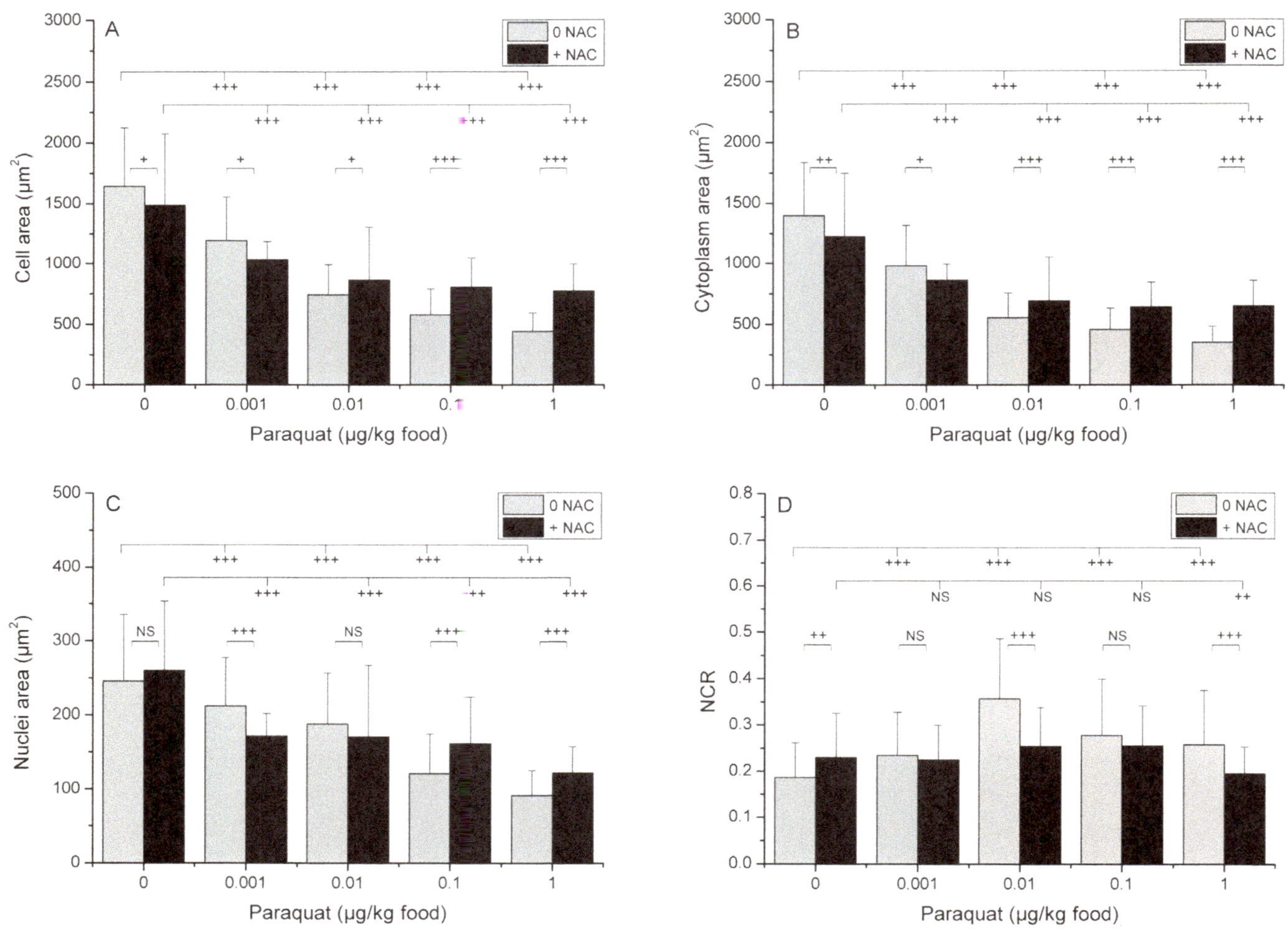

Figure 3. Effects of Paraquat at 48 h on the size of three oenocyte compartments. Larvae were exposed at different concentrations of Paraquat (0 (control), 0.001, 0.01, 0.1 and 1 µg/kg of food) in the absence (NoNAC) or in the presence of NAC. Then, they were picked out, immediately fixed in formaldehyde and processed for histochemical analysis. Biometrical effects were observed after 48 h of exposure to Paraquat. Three compartments were analyzed, whole cell (A), cytoplasm (B, by subtracting nucleus area from total cell area) and nucleus (C), and the nucleo-cytoplasmic ratio (NCR) was calculated (D). (+), $p<0.01$; (++), $p<0.001$; (+++), $p<0.0001$; NS, not significant ($p>0.05$).

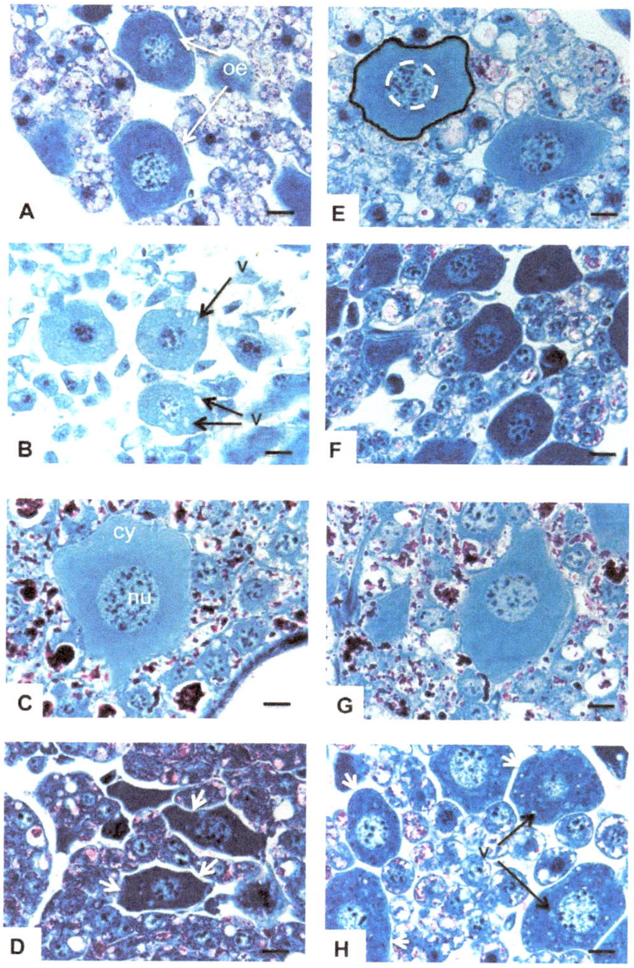

Figure 4. Effects of Paraquat on oenocyte morphology. Larvae were subjected to exposure to Paraquat at different concentrations in the food in the absence (A, B, C, D) or in the presence (E, F, G, H) of NAC for 24 and 48 h. Larvae were then fixed in formaldehyde 4% (v/v), dehydrated in ethanol and embedded in methacrylate resin. Sections (3 µm) were stained with the PAS-NBB procedure and observed under light microscopy. (A and E), controls at 24 h. (A), identification of oenocytes. (E), method of area measurement: black line, total area, and white dotted line, nuclei area. (B and F), oenocytes at 24 h in larvae exposed to Paraquat at 1 µg/kg. Notice the smaller nuclei and cytoplasm areas by comparison with controls (A and E) and also small vacuoles in the cytoplasm (black arrow). (C and G), controls at 48 h. Observe the increase in cell and nuclei areas between 24 h (A and E) and 48 h. (D and H), oenocytes at 48 h in larvae exposed to Paraquat at 1 µg/kg. Note the smaller nuclei and cytoplasm areas by comparison with controls (C and G) and also small vacuoles in the cytoplasm (black arrow). The cytoplasm of oenocytes from larvae exposed to Paraquat was more contracted in the absence of NAC (D, white arrow heads) than in the presence of NAC (H, white arrow heads). cy = cytoplasm, n = nuclei, oe = oenocyte and v = vacuole. Scale bar = 10 µm; magnification: 600.

cells and nuclei induced by Paraquat (Fig. 2 and 4A, B, E, F). Thus, as a result, NCR remained nearly stable.

At 48 h, in the absence of NAC, Paraquat had a major significant concentration-dependent effect upon oenocytes by decreasing the size of the cell, cytoplasm and nucleus (Fig. 3). Concentration-effect relationship was explored with log-log linear regression analysis (adjusted R-square 0.9785) and enabled to determine an effective concentration 50% of 16.00±2.24 ng/kg. It is noteworthy that a significant decrease was already observed at the lowest concentration (0.001 µg/kg). The size of entire cells was reduced by 50% at the concentration of 0.01 µg/kg, compared to the control. The intensity of the size decrease at the two highest concentrations was less marked because the decrease of the cell volume is limited by the nucleus (Fig. 4D). The concentration-dependent decrease of nuclei area presented a more regular pattern and, consequently, NCR increased with the concentration. The decrease elicited at the highest concentration was important enough to almost abolish the cell growth between 24 and 48 h: 630.12 µm^2 and 1643.77 µm^2 for controls at 24 and 48 h respectively and, 434.48 µm^2 and 443.55 µm^2 for 1 µg/kg food at 24 and 48 h, respectively. At 48 h, the average decrease in size due to the highest concentration compared with control (73.0% and 62.9% for total cell and nuclei areas, respectively) was more than twice the decrease at 24 h (31.05% and 20.83% for total cell and nuclei areas, respectively) (Fig. 2 and 4C, D). When the effect of Paraquat was observed in the presence of NAC at 48 h, NAC in turn had a significant effect (Fig. 2 and. 4C, D, G, H). NAC attenuated the Paraquat-induced decrease of cell and cytoplasm areas at the three highest concentrations and the nuclei area at the two highest concentrations. For the control and for the lowest Paraquat concentration (0.001 µg/kg), NAC induced a slight reduction in the size of the cell but not of the nuclei. As a consequence, cytoplasm and total cell areas presented identical concentration-response patterns. It is notable that oenocytes from exposed larvae, but not from controls, exhibited vacuoles in the cytoplasm (Fig. 4).

Discussion

In the honey bee, oenocytes have been described early at different developmental stages [42]. The size of these cells increases gradually during the larval development, as in other species such as *Drosophila* [43]. In the present study, we show that Paraquat induces a concentration-dependent decrease of total cell and nuclei areas, to such a degree that the growth between 24 and 48 h (usually about tripling in size) does not take place. The effect of Paraquat is also observed at the lowest concentration (0.001 µg Paraquat/kg food), which demonstrates the high sensitivity of oenocytes. This concentration is indeed very low and much lower than the limit of detection, in complex matrices, of very efficient analytical methods involving mass spectrometry [44]. In other words, it will be impossible to detect Paraquat in larvae food even at higher concentrations close to 100 ng/kg, for which a significant effect is observed. Thus, this is the first time that biological effects of pesticides can be observed at such very low exposure levels.

The effects of Paraquat demonstrated here raise the question of the impacts of quaternary ammonium herbicides, and especially pyridinium active substances, on the development of the honey bee. The herbicide Diquat also triggers an oxidative stress [45] and can accumulate in the neuromelanin of frogs and mice [46]. MPP+ (1-methyl-4-phenyl pyridinium), a neurotoxic metabolite of the drug MPTP (1-methyl-4-phenyl-1,2,3,6-tetrahydropyridine), has been registered as the herbicide Cyperquat [47]. Like Paraquat and rotenone, it induces Parkinson's disease by destroying the dopaminergic neurons in the *substantia nigra* but acts through a mechanism close to that of rotenone and different from that of Paraquat [48]. The honey bee brain is devoid of structure like *substantia nigra* [49], and a different mechanism of action could be expected for Paraquat in the honey bee larvae. However, recent results suggest that Paraquat may exhibit common action pathways in the honey bee and mammals. Paraquat induces the condensation of cytoplasm and irregular cell shapes not only in bees, but also in mouse hippocampus cells [50]. These changes are associated with a decay of learning and memory performances and the induction of an oxidative stress in mouse. Thus, it would be worth investigating the effects of Paraquat on larvae cells, and especially neural cell precursors, and their impacts on neural functions in adult bees.

The accentuated presence of vacuoles in the cytoplasm of oenocytes from exposed larvae, but not from the controls, could be interpreted as a primary sign of cell death in insect tissue [51] that could impair hormone synthesis. In addition, oenocytes are involved in the synthesis of ecdysteroid hormones that regulate different genes [*AmEcR-A, AmE74, AmE75, AmBR-C, AmHR-38, Mblk-1 (AmE93)* and *AmUsp (Ultraspiracle)*] preferentially and differentially expressed in the mushroom bodies of the queen and the worker brain, particularly in the Kenyon interneurons [52,53]. Since Kenyon cells are regulated by ecdysteroids and involved in cognitive functions and brain plasticity [54,55], it would be interesting to further explore the cognitive effects of Paraquat in adult bees exposed at larval stages. However, the influence of Paraquat should not be restricted to neural functions and the fertility of the queen should also be taken into consideration because *AmEcR-A* and *AmE74* genes are strongly expressed in the ovary of the queen [52,56].

NAC is known as an antioxidant substance and a free radical scavenger [57]. In the honey bee larvae, the protective action of NAC against the Paraquat-induced effects in oenocytes is more pronounced after 48 h of exposure than after 24 h. At 48 h, NAC moderates the Paraquat-induced reduction of the size of cells and cytoplasm at Paraquat concentrations ≥0.01 μg/kg, and of nuclei at Paraquat concentrations ≥0.1 μg/kg. In this study, NAC shows a protective effect at a concentration of 50 μg/kg, which is 100 to 1000 times lower than the concentrations used in studies on Paraquat in vertebrates [58]. Thus, at this concentration, NAC also seems to present antioxidant properties to counteract the oxidative damages induced by Paraquat in honey bee larvae. Yet, the effects of Paraquat are moderated but not completely abolished by NAC, whatever the Paraquat concentration. This reveals that Paraquat has a strong effect on larvae cells, even at very low concentrations. NAC displays a differential action at low and high Paraquat concentrations. From 0.01 to 1 μg/kg Paraquat, NAC reduces the effect of Paraquat whereas in controls and at 0.001 μg/kg, it decreases the cell and cytoplasm areas. This suggests different modes of action of Paraquat, depending on the concentration, with an oxidant action of Paraquat prevalent at concentrations above 0.01 μg/kg.

In this study, we have demonstrated the high sensitivity of oenocytes to Paraquat at very low concentrations. However, considering that the Paraquat-induced effects in bees are also observed in mammalian brain cells [50], it is legitimate to think that they could also occur in other cell types and at different developmental stages. Hence, changes in oenocyte morphology could be regarded as a high sensitivity biomarker of exposure to pesticides and as a biological model to study the effects of these latter. Very few cells have been described as markers of intoxication in honey bees, especially, midgut epithelial cells and Malpighian tubule cells [59]. Oenocytes are large and independent cells embedded in fat body and readily accessible [16] and thus represent a good study model [42]. The sensitivity of oenocytes, and the very low concentrations at which Paraquat induces effects, indicate that these cells could respond not only to direct exposure but also to pollen and wax contamination [60,61].

Conclusion

In this study, we have established that Paraquat, a model pesticide, can disrupt the larval development of oenocytes at very low concentrations. The most important point is that concentrations around ng/kg are able to induce damages in living organisms and especially in honey bees. Since a model molecule has proved to be efficient at this contamination level, it seems legitimate to think that the same phenomenon is likely to take place with other pesticides. Regarding the sensitivity of oenocytes to low concentrations of Paraquat, we advance that the morphology of oenocytes could be used in developing biomarkers of exposure to pesticides. The development of this biomarker will require further in-depth studies with other chemicals, at other developmental stages and in relatively large scales involving spatial and temporal variations. The impairment of oenocyte development at very low exposure levels makes Paraquat a valuable tool for the investigation of the functions of oenocytes.

Acknowledgments

We heartily thank Jean-Paul Vermandère and Jean Aptel, the beekeepers of the UR 406 *Abeilles & Environnement* Research Unit, for their skilled bee-rearing and for providing brood combs.

Author Contributions

Conceived and designed the experiments: LPB MEM. Performed the experiments: MC. Analyzed the data: LPB MC AK JLB ESZ. Wrote the paper: LPB MC ESZ.

References

1. Klein A-M, Vaissiere BE, Cane JH, Steffan-Dewenter I, Cunningham SA, et al. (2007) Importance of pollinators in changing landscapes for world crops. P Royal Soc B - Biol Sci 274: 303–313.
2. Beekman M, Ratnieks FLW (2000) Long-range foraging by the honey-bee, *Apis mellifera* L. Functional Ecology 14: 490–496.
3. Breeze TD, Bailey AP, Balcombe KG, Potts SG (2011) Pollination services in the UK: How important are honeybees? Agr Ecosyst Enviro 142: 137–143.
4. Gallai N, Salles J-M, Settele J, Vaissiere BE (2009) Economic valuation of the vulnerability of world agriculture confronted with pollinator decline. Ecol Econ 68: 810–821.
5. Moritz RFA, de Miranda J, Fries I, Le Conte Y, Neumann P, et al. (2010) Research strategies to improve honeybee health in Europe. Apidologie 41: 227–242.
6. Gauthier M (2010) State of the art on insect nicotinic acetylcholine receptor function in learning and memory. In: Thany SH, editor. Insect Nicotinic Acetylcholine Receptors. Berlin: Springer-Verlag. pp. 97–115.
7. Srinivasan MV (2011) Honeybees as a model for the study of visually guided flight, navigation, and biologically inspired robotics. Physiol Rev 91: 413–460.
8. Potts SG, Biesmeijer JC, Kremen C, Neumann P, Schweiger O, et al. (2010) Global pollinator declines: Trends, impacts and drivers. Trends Ecol Evol 25: 345–353.
9. Ratnieks FLW, Carreck NL (2010) Clarity on Honey Bee Collapse? Science 327: 152–153.
10. Desneux N, Decourtye A, Delpuech J-M (2007) The sublethal effects of pesticides on beneficial arthropods. Annu Rev Entomol. pp. 81–106.
11. Paes de Oliviera VT, da Cruz-Landim C (2003) Morphology and function of insect fat body cells: Review. Biocencias 11: 195–205.
12. Martins GF, Ramalho-Ortigao JM (2012) Oenocytes in insects. Inver Surv J 9: 139–152.
13. Cruz Landim CD (1985) Oenocytes of honey bee workers - Structural modifications during their adult life. Rivista Di Biologia-Biology Forum 78: 107–122.
14. Wigglesworth VB (1970) Structural lipids in the insect cuticle and the function of the oenocytes. Tissue Cell 2: 155–179.
15. Miller A (1950) Chapter 6. The internal anatomy and histology of the imago of *Drosophila melanogaster*. In: Demerec M, editor. Biology of Drosophila. New York: John Wiley and Sons. pp. 420–534.

16. Fan YL, Zurek L, Dykstra MJ, Schal C (2003) Hydrocarbon synthesis by enzymatically dissociated oenocytes of the abdominal integument of the German Cockroach, *Blaftella germanica*. Naturwissenschaften 90: 121–126.
17. Hepburn HR, Bernard RTF, Davidson BC, Muller WJ, Lloyd P, et al. (1991) Synthesis and secretion of beeswax in honeybees. Apidologie 22: 21–36.
18. Rachinsky A, Strambi C, Strambi A, Hartfelder K (1990) Caste and metamorphosis - Hemolymph titers of juvenile-hormone and ecdysteroids in last instar honeybee larvae. Gen Comp Endocrinol 79: 31–38.
19. Lycett GJ, McLaughlin LA, Ranson H, Hemingway J, Kafatos FC, et al. (2006) *Anopheles gambiae* P450 reductase is highly expressed in oenocytes and in vivo knockdown increases permethrin susceptibility. Insect Mol Biol 15: 321–327.
20. Amiressami M (1973) The effect of insecticides on the oenocytes of the different generations of *Pemphigus bursarius* L. (Aphidina). Z Angew Zool 60: 321–330.
21. Denneulin JC, Lamy M (1977) Effects of dimilin, a chitin inhibitor 1 (4 chlorophenyl) 3 (2-6 difluorobenzoyl) urea on oenocytes and molting in processionary caterpillar (*Thaumetopoea pityocampa* Schiff) (Lepidoptera). Annal Endocrinol 38: 405–406.
22. Aupinel P, Fortini D, Michaud B, Marolleau F, Tasei J-N, et al. (2007) Toxicity of dimethoate and fenoxycarb to honey bee brood (*Apis mellifera*), using a new in vitro standardized feeding method. Pest Manage Sci 63: 1090–1094.
23. Hendriksma HP, Haertel S, Steffan-Dewenter I (2011) Honey bee risk assessment: new approaches for in vitro larvae rearing and data analyses. Method Ecol Evol 2: 509–517.
24. Cocenza DS, de Moraes MA, Beppu MM, Fraceto LF (2012) Use of biopolymeric membranes for adsorption of Paraquat herbicide from water. Water Air Soil Poll 223: 3093–3104.
25. FAO (2003) FAO Specification and evaluations for agricultural pesticides. Paraquat dichloride, 1,1'-dimethyl-4,4'-bispyridinium dichloride. FAO. 21 p.
26. US-EPA (1997) Paraquat dichloride. Registration Eligibility Decision (RED), EPA 738-F-96-018. EPA. 267 p.
27. Hoffer E, Shenker L, Baum Y, Tabak A (1997) Paraquat-induced formation of leukotriene B-4 in rat lungs: Modulation by N-acetylcysteine. Free Rad Biol Med 22: 567–572.
28. Lacoume S, Bressac C, Chevrier C (2009) Male hypofertility induced by Paraquat consumption in the non-target parasitoid *Anisopteromalus calandrae* (Hymenoptera: Pteromalidae). Biol Control 49: 214–218.
29. Eisler R (1990) Paraquat hazards to fish, wildlife, and invertebrates. A synoptic review. In: Eisler R, editor. Biological Report 85(122), Contaminant Hazard Reviews. Washington, DC: Fish and Wildlife Service, U.S. Dept. of the Interior. pp. 1–38.
30. Choi WI, Neher DA, Ryoo MI (2008) Life-history trade-offs of *Paronychiurus kimi* (Lee) (Collembola: Onychiuridae) populations exposed to Paraquat. Ecotoxicol Environ Saf 69: 227–232.
31. Bauer CA (1985) Effects of Paraquat on reproduction and growth in northern bobwhite. J Wildlife Manage 49: 1066–1073.
32. Bacchetta R, Mantecca P, Vailati G (2002) Oocyte degeneration and altered ovipository activity induced by Paraquat in the freshwater snail *Physa fontinalis* (Gastropoda : Pulmonata). J Mollusc Studies 68: 181–186.
33. Dial CAB, Dial NA (1987) Effects of Paraquat on reproduction and mortality in 2 generations of mice. Arch Environ Contam Toxicol 16: 759–764.
34. Yang W, Tiffany-Castiglioni E, Koh HC, Son I-H (2009) Paraquat activates the IRE1/ASK1/JNK cascade associated with apoptosis in human neuroblastoma SH-SY5Y cells. Toxicol Lett 191: 203–210.
35. Zhi Q, Sun H, Qian X, Yang L (2011) Edaravone, a novel antidote against lung injury and pulmonary fibrosis induced by Paraquat? Int Immunopharmacol 11: 96–102.
36. Tanner CM, Kamel F, Ross GW, Hoppin JA, Goldman SM, et al. (2011) Rotenone, Paraquat, and Parkinson's disease. Environ Health Perspect 119: 866–872.
37. Patel S, Singh V, Kumar A, Gupta YK, Singh MP (2006) Status of antioxidant defense system and expression of toxicant responsive genes in striatum of maneb- and Paraquat-induced Parkinson's disease phenotype in mouse: Mechanism of neurodegeneration. Brain Res 1081: 9–18.
38. Wang A, Zhai Z (2010) Determination of Paraquat in vegetables by high performance liquid chromatography. J Environ Health 27: 261–262.
39. Restek (2006) Simple, sensitive HPLC/UV analysis for Paraquat and Diquat, using high-recovery solid phase extraction and an ultra Quat HPLC column. Restek Corporation, Applications Note 580006. 4 p.
40. Vandenberg JD, Shimanuki H (1986) Two commercial preparations of the beta-exotoxin of *Bacillus thuringiensis* influence the mortality of caged adult honey bees, *Apis-mellifera* (hymenoptera, Apidae). Environ Entomol 15: 166–169.
41. Cousin M, El Maataoui M (1998) Female reproductive organs in self-compatible almond (*Prunus dulcis* (Mill.) D.A. Webb) Lauranne and fertilization patterns. Sci Hort 72: 287–297.
42. Snodgrass RE (1956) Anatomy of the honey bee. New York: Cornell University Press. 334 p.
43. Johnson MB, Butterworth FM (1985) Maturation and aging of adult fat-body and oenocytes in *Drosophila* as revealed by light microscopic morphometry. J Morphol 184: 51–59.
44. Kolberg DIS, Mack D, Anastassiades M, Hetmanski MT, Fussell RJ, et al. (2012) Development and independent laboratory validation of a simple method for the determination of Paraquat and diquat in potato, cereals and pulses. Anal Bioanal Chem 404: 2465–2474.
45. Salama AK, Osman KA, Aly NM (2001) Protection against oxidative stress induced by Paraquat and diquat in female rats. Toxicology 164: 193–194.
46. Lindquist NG, Larsson BS, Lydén-Sokolowski A (1988) Autoradiography of [^{14}C]Paraquat or [^{14}C]diquat in frogs and mice: Accumulation in neuromelanin. Neurosci Lett 93: 1–6.
47. Thakar JH, Hassan MN (1988) Effects of 1-methyl-4-phenyl-1,2,3,6-tetrahydropyridine (MPTP), cyperquat (MPP^+) and Paraquat on isolated-mitochondria from rat striatum, cortex and liver. Life Sci 43: 143–149.
48. Richardson JR, Quan Y, Sherer TB, Greenamyre JT, Miller GW (2005) Paraquat neurotoxicity is distinct from that of MPTP and rotenone. Toxicol Sci 88: 193–201.
49. Brandt R, Rohlfing T, Rybak J, Krofczik S, Maye A, et al. (2005) Three-dimensional average-shape atlas of the honeybee brain and its applications. J Comp Neurol 492: 1–19.
50. Chen Q, Niu Y, Zhang R, Guo H, Gao Y, et al. (2010) The toxic influence of Paraquat on hippocampus of mice: Involvement of oxidative stress. NeuroToxicology 31: 310–316.
51. Silva-Zacarin ECM, Tomaino GA, Brocheto-Braga MR, Taboga SR, de Moraes RLMS (2007) Programmed cell death in the larval salivary glands of *Apis mellifera* (Hymenoptera, Apidae). J Biosci 32: 309–328.
52. Paul RK, Takeuchi H, Matsuo Y, Kubo T (2005) Gene expression of ecdysteroid-regulated gene E74 of the honeybee in ovary and brain. Insect Mol Biol 14: 9–15.
53. Yamazaki Y, Shirai K, Paul RK, Fujiyuki T, Wakamoto A, et al. (2006) Differential expression of HR38 in the mushroom bodies of the honeybee brain depends on the caste and division of labor. FEBS Lett 580: 2667–2670.
54. Maronde U (1991) Common projection areas of antennal and visual pathways in the honeybee brain, *Apis mellifera*. J Comp Neurol 309: 328–340.
55. Velarde RA, Robinson GE, Fahrbach SE (2009) Coordinated responses to developmental hormones in the Kenyon cells of the adult worker honey bee brain (*Apis mellifera* L.). J Insect Physiol 55: 59–69.
56. Takeuchi H, Paul RK, Matsuzaka E, Kubo T (2007) EcR-A expression in the brain and ovary of the honeybee (*Apis mellifera* L.). Zool Sci 24: 596–603.
57. Price TO, Uras F, Banks WA, Ercal N (2006) A novel antioxidant N-acetylcysteine amide prevents gp120- and Tat-induced oxidative stress in brain endothelial cells. Exp Neurol 201: 193–202.
58. Yeh ST-Y, Guo H-R, Su Y-S, Lin H-J, Hou C-C, et al. (2006) Protective effects of N-acetylcysteine treatment post acute Paraquat intoxication in rats and in human lung epithelial cells. Toxicology 223: 181–190.
59. Malaspina O, Silva-Zacarin ECMd, da Silva-Zacarin ECM (2006) Cell markers for ecotoxicological studies in target organs of bees. Braz J Morphol Sci 23: 303–309.
60. Chauzat MP, Faucon JP, Martel AC, Lachaize J, Cougoule N, et al. (2006) A survey of pesticide residues in pollen loads collected by honey bees in France. J Econ Entomol 99: 253–262.
61. Chauzat M-P, Faucon J-P (2007) Pesticide residues in beeswax samples collected from honey bee colonies (*Apis mellifera* L.) in France. Pest Manage Sci 63: 1100–1106.

11

Using Bioassays and Species Sensitivity Distributions to Assess Herbicide Toxicity towards Benthic Diatoms

Floriane Larras*, Agnès Bouchez, Frédéric Rimet, Bernard Montuelle

Institut National de la Recherche Agronomique, UMR 0042 Carrtel, Thonon, France

Abstract

Although benthic diatoms are widely used in ecological studies of aquatic systems, there is still a dearth of data concerning species sensitivities towards several contaminants. Within the same community, different species may respond differently depending on their physiological and ecological characteristics. This lack of knowledge makes specific appropriate risk assessment impossible. To find out whether species sensitivity distribution (SSD) could be used to estimate the risk of herbicide toxicity for diatoms, we need to know whether their sensitivity depends on their physiological and ecological characteristics. We carried out single-species bioassays on 11 diatom species exposed to 8 herbicides. Dose-responses relationships were used to extrapolate the Effective Concentration 5 (EC_5) and the Effective Concentration 50 (EC_{50}) for each exposure. These data were used to fit a SSD curve for each herbicide, and to determine the Hazardous concentration 5 (HC_5) and 50 (HC_{50}). Our results revealed a high level of variability of the sensitivity in the set of species tested. For photosystem-II inhibitor (PSII) herbicides, diatoms species displayed a typical grouping of sensitivity levels consistent with their trophic mode and their ecological guild. N-heterotroph and "motile" guild species were more tolerant of PSII inhibitors, while N-autotroph and "low profile" guild species were more sensitive. Comprehensive SSD curves were obtained for 5 herbicides, but not for sulfonylurea herbicides or for dimetachlor, which had toxicity levels that were below the range of concentration tested. The SSD curves provided the following ranking of toxicity: diuron> terbutryn> isoproturon> atrazine> metolachlor. The HC that affected 5% of the species revealed that, even at the usual environmental concentrations of herbicides, diatom assemblages could be affected, especially by isoproturon, terbutryn, and diuron.

Editor: Sandra Maria Feliciano de Oliveira Azevedo, Federal University of Rio de Janeiro, Brazil

Funding: This work was financially supported by the Programme Assessing and reducing environmental risks from plant protection products (PESTICIDES) and funded by the French Ministry in charge of Ecology and Sustainable Development (project IMPALAC), grant CV2100204294. The funders had no role in study design, data collection and analysis, decision to publish, or preparation of the manuscript.

Competing Interests: The authors have declared that no competiting interests exist.

* E-mail: floriane.larras@thonon.inra.fr

Introduction

Rivers and lakes are often compromised by contaminants such as pesticides derived from watershed runoff or urban discharges [1–3]. In particular, the coastal zones of lakes are exposed to higher concentrations of toxicants than the pelagic zone [4], leading to a higher exposure of organisms in these zones. Among the various herbicides found in water bodies, atrazine (s-triazines), diuron and isoproturon (phenylureas) are some of the main substances monitored in France [5]. Some of these herbicides are banned or restricted, but they are still found in European inland water ecosystems, and sometimes reach concentrations above the Annual Average- Environmental Quality Standard (AA-EQS) defined in the Water Framework Directive [2,5–7]. In small rivers, ponds, and coastal zones of lakes, microalgae and especially diatoms are crucial primary producers and account for a considerable proportion of the fixed biomass as well as carrying out some of the main biochemical processes. Benthic diatoms are known to be good indicators of ecosystem water quality, especially of the trophic level and physical disturbances [8,9].

The impact of herbicides, such as triazines and phenylureas, on phototrophic organisms is well known. They inhibit photosynthesis by disrupting the electron transmission chain in the thylakoid membrane, leading to the dissipation of luminous energy [10]. As a consequence, benthic diatoms are potential targets due to the phytotoxic characteristics of these herbicides [10,11]. Other herbicides belonging chloroacetamide and sulfonylurea families are also found in water. For example, metolachlor was found in 37% of stations analyzed station in the Adour-Garonne basin in France [7]. They act on cells by inhibiting the biosynthesis of very long chain fatty acids or the biosynthesis of amino-acids [11,12]. Some authors have already demonstrated their phytotoxicity on microalgae [13–15].

Single-species laboratory bioassays for microalgae and herbicides are mostly carried on phytoplanktonic species, such as *Scenedesmus* spp. or *Chlorella* spp., which are easy to grow under controlled laboratory conditions. A few studies have investigated the sensitivity of benthic diatoms under single-species conditions. These studies showed that a) the sensitivity towards the same herbicide is very variables in different species [16–18] and b) pre-exposure of diatoms can increase their herbicide tolerance [18]. Laboratory single-species bioassays provide usefull data for assessing individual sensitivities, however in the natural environment, herbicides could disrupt communities in which many species are present and which may interact and respond differently, depending on their physiological and chemical parameters [19]. The variation of species sensitivity in a community or an assemblage towards one or several toxicants can be expressed in

Table 1. Characteristics of the 11 diatom species: taxonomy, references, and ecological guild.

Order	Species	Code[a]	TCC[b]	Ecological guild[c]
Naviculales	*Craticula accomoda*, (Hustedt) D.G. Mann	CRAC	107	Motile
	Eolimna minima, (Grunow) Lange-Bertalot	EOMI	524	Motile
	Mayamaea fossalis, (Krasske) Lange-Bertalot	MAFO	366	Motile
Cymbellales	*Encyonema silesiacum*, (Bleisch) D.G. Mann	ESLE	678	Low profile
	Gomphonema parvulum, (Kützing) Kützing	GPAR	653	High profile
Fragilariales	*Fragilaria capucina* var. *vaucheriae*, (Kützing) Lange-Bertalot	FCVA	752	High profile
	Fragilaria ulna, (Nitzsch) Lange-Bertalot	FULN	365	High profile
	Fragilaria rumpens (Kützing) G.W.F. Carlson	FRUM	666	High profile
Bacillariales	*Nitzschia palea*, (Kützing) W. Smith	NPAL	139–2	Motile
Achnantes	*Achnanthidium minutissimum*, (Kützing) Czarnecki	ADMI	746	Low profile
Thalassiosirales	*Cyclotella meneghiniana*, Kützing	CMEN	755	Low profile

[a]Species code in the OMNIDIA database [48]. [b]Strain reference in the INRA Thonon Culture Collection. [c]Ecological guild from Passy [42] and refined by Berthon [46].

terms of a cumulative distribution known as the Species Sensitivity Distribution (SSD) [20]. Such distributions are based on selected benchmarks obtained from single-species bioassays. SSDs are used in risk assessment, as predictive models, and to derive thresholds in order to protect environmental biodiversity. The threshold concentration is often defined as the Hazardous Concentration for 5% of the species (HC_5), which is expected to spare 95% of the species in the assemblage [20,21].

Benthic diatoms are known to dominate biofilms of benthic littoral zones in terms of abundance [22,23]. These algae present a great diversity, which enabled some authors to develop various biotic indices for nutrient and organic matter assessment in rivers and lakes [24,25]. Within this great diversity, species exhibit various sensitivity to herbicides [14,16,26]. Even if others sorts of algae, as chlorophytes and cyanobacteria are more sensitive [27], they are not as abundant as diatoms which appeared as relevant and representative organisms for the benthic zone. With the view to assess herbicides risk assessment of benthic littoral zone, benthic diatoms appeared then as ecologically relevant tools.

The two aims of this study were a) to determine the variation of the sensitivity of benthic diatom species towards herbicides in an artificial assemblage and b) to find out whether SSD models based on benthic diatoms sensitivities can be used to find out whether herbicides constitute a risk at concentrations found in the environment. In this study we develop a data base of EC_5 and EC_{50} thresholds based on single-species bioassays for 11 benthic diatoms exposed to 8 agricultural herbicides. These two thresholds were chosen because of their environmental relevance and because they are widely used in the literature, which made it easier to compare our results with those of other studies. These EC_5 and EC_{50} values were then used to construct SSD models in order to derive HC_5 values for a diatom assemblage. Experiments were carried out as single-species bioassays, using diatom strains derived from benthic biofilms (taken from the Geneva Lake coastal zone) or from the INRA Thonon Culture Collection. The selected endpoint was the growth rate calculated from the chlorophyll a fluorescence.

Methods

Benthic diatoms cultures

Eleven species of diatom were selected from the species found in biofilms from the benthic coastal zone of Lake Geneva and according to the literature (Table 1). Strain collection in Lake Geneva was made on public area, not requiring any specific permission, and did not involve protected species. In order to ensure that we were working on a diatom diversity representative of freshwater benthic ecosystems, we selected diatoms with a variety of life forms and taxonomic diversity. In natural biofilms, pennate diatoms outnumber centric diatoms. Consequently, most of the species tested were pennate and only one was centric (*Cyclotella meneghiniana*, CMEN). Most of them are classified as benthic, and in fact all had been isolated from benthic biofilms and came from the Thonon Culture collection (Thonon-Les-Bains, France, http://www.inra.fr/carrtel-collection). The diatoms selected for this study were: *Fragilaria capucina* var. *vaucheriae* (FCVA), *Fragilaria rumpens* (FRUM), *Fragilaria ulna* (FULN), *Craticula accomoda* (CRAC), *Mayamaea fossalis* (MAFO), *Eolimna minima* (EOMI), *Nitzschia palea* (NPAL), *Achnanthidium minutissimum* (ADMI), *Cyclotella meneghiniana* (CMEN), *Encyonema silesiacum* (ESLE) and *Gomphonema parvulum* (GPAR). Cultures were maintained in DV culture media (http://www6.inra.fr/carrtel-collection_eng/Culture-media/Composition-of-the-culture-media) filtered at 0.22 μm (Millipore). They were grown in a 300 mL Erlenmeyer flask at 21±2°C, and with a 16:8h light:dark cycle at 66 μmol. m^{-2}. sec^{-1}.

Herbicides

The herbicides used in this study have all been detected in the coastal zone of Lake Geneva, and were selected on the basis of toxicity and mode of action (Table 2). They had all been detected regularly in the Lake since 2004 [4].We used atrazine (purity 99.9%), terbutryn (purity 99.3%), diuron (purity 99.5%), isoproturon (purity 99.9%), metolachlor (purity 98%), dimetachlor (purity 99.9%), amidosulfuron (purity 99.9%) and foramsulfuron (purity 97.5%). These substances were obtained from Sigma-Aldrich (St Louis, MO 63103, USA). They are representative of 4 different modes of action and have different Kow values (Table 2).

Bioassays

Herbicide stock solutions were prepared by dissolving herbicides in DV growth media before filtering on a 0.22 μm filter unit (Millipore) and stored at 4°C in the dark. Due to their low solubility, atrazine and diuron were first dissolved in 0.05% Dimethyl sulfoxide (DMSO), and then sonicated for 30 minutes.

Table 2. Chemical properties and mode of action of the 8 herbicides.

Family	Herbicide	Mode of action	Log Kow	Solubility (mg/l)
Phenylureas	Diuron	Photosystem 2 inhibition (site B)	2.87	35.6
	Isoproturon		2.5	70.2
Triazines	Atrazine	Photosystem 2 inhibition (site A)	2.7	35
	Terbutryn		3.65	25
Chloroacetamides	Metolachlor	Inhibition of the synthesis of very long chain fatty acid	3.4	530
	Dimetachlor		2.17	2300
Sulfonylureas	Foramsulfuron	Growth regulators (higher plants)	−1.56	3293
	Amidosulfuron		−0.78	3070

The non-toxicity of 0.05% DMSO dissolved in DV media was confirmed on diatoms before the bioassays (data not shown).

The 11 diatom species were exposed to the 8 herbicides in single-species laboratory bioassays (11diatoms*8 herbicides, i.e. 88 bioassays), and their growth rates were measured as an endpoint. Bioassays were carried out in triplicate in 30 mL glass tubes, with a cap that permits gaseous exchange. Each tube contained 1 mL of a contaminated solution and 1 mL of an 8 to 10 day old strain culture in the exponential growth phase. The range of concentrations to be used in bioassays for each species and each herbicide had previously been determined by means of preliminary tests in 96 well microplates. At the start of the experiment cell density was approximately 10000±2000 cells/mL according to the Standard Guide for Conducting Static Toxicity Tests with Microalgae [28]. Seven to 10 pesticide concentrations were tested, and each of the diatoms were exposed for 96 h under the same experimental conditions as used for the cultures. At the end of the experiment, each tube was shaken (MS2 Minishaker, IKA) for a few seconds to resuspend the algae in the media, and to homogenize the solution. The remaining fixed algae were carefully scraped off with a rod. From each tube, 250 µL of solution was transferred into a 96 well black microplate in quadruplicate. The growth rate was determined by measuring the chlorophyll a fluorescence and was quantified in Reference Fluorescence Units (RFUs) using a Fluoroskan (Fluoroskan Ascent,Thermo-Scientific Finland) with a 430 nm excitation filter and a 680 nm emission filter after incubating for 20 minutes in the dark. To assess the reference growth of diatoms during assay, the fluorescence was checked at the beginning and at the end of the experiment in the control. The linear correlation between chlorophyll a fluorescence (measured using the Fluoroskan) and the cell (optically counted) had previously been checked on the basis of 5 measurements for each species. As chl a fluorescence is correlated with cell density, the growth rate was then calculated as described by Debenest et al.[29] for cell density:

$$\mu = \frac{\beta t2/\beta t1}{t}$$

Where $\beta t2$ and $\beta t1$ represent the fluorescence at the end and at the beginning of the experiment respectively. t is the exposure time, in days.

Data analysis

Dose-response modeling. From each single-species bioassay, the dose-response curve was fitted with R software and the "drc" package [30]. Two non-linear regression models were used depending on the features of the dose-response curve. In the case of a monotically-decreasing function, a log-logistic model (equation 1) was applied where d was the upper limit of the curve, c was the lower limit, and b was the relative slope around the EC_{50} known as e [30].

Equation 1: $f(x,(b,c,d,e)) = c + \dfrac{d-c}{1+\exp(b(\log(x)-\log(e)))}$

Hormesis, or the "greening-effect", leads to a biphasic dose-response in which curves have a typical inverted U-shape or J-shape [31]. In the context of hormesis, the Cedergreen-Ritz-Streibig model (equation 2) was used to fit the data better. This model considered the stimulation phase as the f parameter [32]. The higher f, the greater the hormesis effect.

Equation 2:

$$f(x) = c + \frac{d-c+f\exp(-1/x^{\alpha})}{1+\exp(b(\log(x)-\log(e)))}$$

The Effective Concentration (EC) affecting the growth of 5% (EC_5) or 50% (EC_{50}) of the population was numerically determined from each dose-response curve. For low effect levels, many ECs are used (e.g.:NOEC, LOEC, EC_{10}, EC_{20}). In this study, we preferred to work with the EC_5, because in most of the tests this threshold fell between two concentrations tested. This meant that EC_5 was a reliable value which could be determined from the regression curve. For each herbicide, 11 EC_5 and 11 EC_{50} values were expected. Their 95% confidence intervals were calculated using bootstrap methods.

Species Sensitivity Distribution modeling. SSD curves were fitted with R software using the EC values derived from dose-response curves. For each herbicide, two kinds of SSD curves were fitted: the SSD-EC_5 and the SSD-EC_{50}, which were based on the 11 EC_5 and the 11 EC_{50} values, respectively. From each SSD-ECx curve, the slope and 2 Hazardous Concentrations (HC_5, HC_{50}) were numerically derived. They correspond to the herbicide concentrations affecting 5% and 50% of the species in an assemblage, respectively.

For each herbicide i, the SSD-EC_{50i} curve was fitted using log-logistic regression (equation 3).Several distributions are proposed to fit the SSD model, but in this study the criterion used was the best fit of the data. For a small dataset there is no significant difference between log normal and log-logistic regression [33] and so, as some studies have already shown that the log logistic regression fits the data best [34,35] we chose to use this. According to Chèvre et al [36], $HC50_{EC50,i}$ was the hazardous concentration affecting 50% of species, and $EC50_{i,s}$ was the EC_{50} of herbicide i for species s.

Equation 3:

$$Fractionaffected_{EC50i,s} = \frac{100}{1 + 10^{(\log HC50_{EC50,i} - \log EC50_{i,s}) * slope}}$$

Results

Individual dose-response relationship

In this study, 88 bioassays were carried out and the results are presented on Table 3. Thirty-five of these tests failed to provide reliable EC estimations for various reasons. According to this endpoint, the highest concentrations tested did not produce enough effect to allowed us to fit reliable dose-response curves for the sulfonylurea family. For the chloroacetamide family, only 3 and 7 species, respectively, had an EC_{50} below this concentration for dimethachlor (data not shown) and metolachlor respectively. For atrazine, no value was available for ESLE because the culture failed. The exposure of each species to a wide range of concentrations of each herbicide did not lead to 100% inhibition of growth in all cases. The inhibition was usually between 80 and 100%.

The 11 species of diatoms tested displayed very variable sensitivity for the same herbicide at the EC_5 and EC_{50} levels (Table 3). Phenylurea EC_{50} values varied from 4.03 µg/l (FCVA) to 4236 µg/l (EOMI) for diuron and from 44 µg/l (ESLE) to 1664 µg/l (MAFO) for isoproturon. Triazines EC_{50} ranged from 306 µg/l (FULN) to 8297 µg/l (MAFO) for atrazine and from 5.15 µg/l (ESLE) to 3133 µg/l (EOMI) for terbutryn. Finally,

Table 3. EC_{50} and EC_5 (µg/l) values extrapolated from dose-response curves for 5 herbicides and 11 diatom species. 95% confidence intervals are shown in square brackets. A dash indicates a failed culture.

Species	ECx	Herbicides Diuron	Atrazine	Isoproturon	Terbutryn	Metolachlor
Craticula accomoda	EC_{50}	1734 [1578;1985]	919 [865;1011]	853. [808; 901]	814 [751; 877]	30147 [17134; 44657]
	EC_5	261 [210; 292]	524 [514;534]	277 [251; 294]	92[72; 113]	2575 [1729; 2999]
Eolimna minima	EC_{50}	4236 [3905; 4529]	2510 [2313;2748]	1566 [1536; 1594]	3133 [2728; 3512]	>50000
	EC_5	3007 [2809; 3173]	1443 [1397;1499]	747 [611; 1139]	1450 [1211; 1790]	
Mayamaea fossalis	EC_{50}	463 [311; 658]	8297 [7893;8900]	1664 [1625;1713]	63 [61; 67]	10313 [8510;12020]
	EC_5	74 [62;89]	4766 [3861; 5389]	559 [546; 576]	13 [10;18]	3393 [1791; 4389]
Encyonema silesiacum	EC_{50}	8.79 [7.51;9.88]	-	44 [42;47]	5.15 [4.26; 6.30]	6399 [5946; 6522]
	EC_5	3.11 [2.10;4.21]	-	12.51 [11.25; 13.47]	0.55 [0.40;0.78]	54 [52; 56]
Gomphonema parvulum	EC_{50}	2255 [1920; 2518]	907 [837; 995]	1014 [962; 1065]	464 [409; 564]	4054 [3397;4384]
	EC_5	904 [739; 1144]	588 [554; 633]	652 [505; 892]	22 [20;26]	299 [160; 391]
Fragilaria capucina var *vaucheriae*	EC_{50}	4.03 [3.93; 4.16]	801[689; 966]	117 [98; 131]	60 [59; 62]	>50000
	EC_5	0.069 [0.062; 0.073]	205 [125;287]	35 [31;38]	31 [28;34]	
Fragilaria ulna	EC_{50}	51 [48; 54]	306 [289; 318]	74.38 [65.91;85.24]	56 [45; 60]	3314 [2609; 3570]
	EC_5	12.6 [11.37; 13.0]	191 [173; 207]	30 [27;33]	0.85 [0.73; 0.95]	60 [52; 68]
Fragilaria rumpens	EC_{50}	122 [117; 127]	629 [596; 756]	357 [319; 450]	477 [457; 497]	>50000
	EC_5	18 [13;20]	210 [164;304]	22 [18;30]	0.70 [0.51; 0.90]	
Nitzschia palea	EC_{50}	1539 [1336; 1697]	3988 [3519; 4467]	1577 [1364; 1881]	1414 [1186; 1663]	>50000
	EC_5	106 [77; 129]	596 [411;798]	222 [189; 271]	81 [41; 108]	
Achnanthidium minutissimum	EC_{50}	108 [89; 137]	748 [562;888]	173 [150; 205]	411 [372; 447]	8551 [7067; 9853]
	EC_5	3.15 [2.61; 3.87]	129 [109;143]	24.78 [16.83; 24.90]	63 [58; 69]	5957 [5573; 6236]
Cyclotella meneghiniana	EC_{50}	23 [21;25]	812 [726;908]	46 [43;48]	9.62 [9.39; 9.97]	3476 [2582; 4042]
	EC_5	1.59 [0.90; 2.05]	58 [45;67]	3.16 [2.57; 3.64]	5.38 [5.21; 5.48]	1282 [1084; 1443]

metolachlor EC_{50} varied from 3314 (FULN) to over 50000 µg/l (NPAL, EOMI, FRUM). EC_5 values derived from dose-response curves for FCVA exposed to diuron and FULN exposed to terbutryn were below the lower concentration tested.

Species sensitivity patterns

EC_5 and EC_{50} values obtained from dose-response curves were used to plot SSD-EC_5 and SSD-EC_{50} curves (Fig. 1). HC_5, HC_{50} and slope values were obtained from SSD-EC_5 (SSD-$EC_{5\ HC5}$ or SSD-$EC_{5\ HC50}$) and SSD-EC_{50} (SSD-$EC_{50\ EC5}$ or SSD-$EC_{50\ HC50}$) (Table 4). From SSD-EC_5 to SSD-EC_{50}, the curves shifted toward higher concentrations, due to the higher values of EC_{50}. On SSD-EC_5 and SSD-EC_{50} curves of atrazine terbutryn, diuron and isoproturon, the least sensitive species were always EOMI, GPAR, NPAL, CRAC and MAFO (except for terbutryn). For SSD-EC_{50} curves of all PSII inhibitors, FCVA, FULN, CMEN and ELSE were always the most sensitive species among species of the assemblage. For metolachlor, CRAC, MAFO, ADMI and GPAR were the least sensitive species, while CMEN, FULN and ELSE were more sensitive.

Order of toxicity between herbicides

According to the HC_5 values extracted from SSD-EC_{50}, the toxicity ranking of the herbicides was: diuron (1.43 µg/l) > terbutryn (3.33 µg/l) > isoproturon (14 µg/l) > atrazine (203 µg/l) > metolachlor (1784 µg/l) (Table 4, Fig. 2). We found the same ranking for other HC_x values of SSD-ECx, with the exception of SSD-$EC_{5\ HC50}$, where terbutryn was slightly more toxic than diuron.

Discussion

Ecotoxic responses of benthic strains

In this study, diatoms of benthic origin were cultured as plankton, which mean that the biofilm structure and its possible consequences for the responses of the microalgae were not taken into account. It has in fact been shown that the biofilm can have a protective effect, or increase the exposure of the diatoms [37]. Consequently, the sensitivity level of the diatoms calculated here usually reflects the specific physiological characteristics of the strains tested rather than the effect of the biofilm on exposure to the toxic substance. Depending on its mechanism of action, each herbicide produces a specific toxic effect on these species, and the toxicity rankings found in this study are similar to those already observed for microalgae and/or macrophytes [27,36,38]. PSII inhibitors seem to inhibit diatom growth more effectively than herbicides with other modes of action, but a considerable difference in toxicity was found between atrazine and diuron. Their mechanism of action involves competing with plastoquinone so as to block the flux of electrons in the thylakoid membrane, thus leading to the dissipation of light energy [10]. As photosynthesis is the key functional process in primary producers, inhibiting this process more readily affects the growth of algae than inhibiting other functions. Certain herbicides in the chloroacetamide family, such as metolachlor, block the germination of seeds by inhibiting the synthesis of the very long chain fatty acids. In microalgae, this inhibition can impair the stiffness and permeability of the plasma membrane, resulting in inhibition of cell division [11,15,39]. Some studies have also demonstrated the teratological potential of metolachlor for diatoms, and its ability to increase the volume of the cells [14,15]. The herbicides in the sulfonylurea family are used in agriculture to inhibit the growth of higher plants. They act by eliminating the biosynthesis of amino acids via inhibition of the enzyme acetolactate synthase, which implies that they have true phytotoxic potential [12]. Some authors have worked on the toxicity of various different herbicides belonging to this group (e.g.: metsulfuron methyl, nicosulfuron, chlorsulfuron) and have demonstrated considerable variation in the toxicities of these substances towards phytoplanktonic species [12,40].

Overall, the EC_{50} values obtained in this study were very high compared to the concentrations of herbicides found in the environment, and it is therefore unlikely that the *in-situ* exposure levels would produce drastic effects on the diatom species tested. However, the EC_5 values are sometimes fairly close to the concentrations encountered in the environment (e.g.: terbutryn [<1 µg/l], diuron [<3 µg/l]) for some strains, such as FCVA, CMEN, ESLE, FRUM, and FULN. These species are associated with high values of the sensitivity index in the context of the generic index for diatoms [8]. In contrast, Rimet and Bouchez [41] have shown ESLE to be relatively resistant during its benthic phase, and they suggest that the fact that this species has a mucous tube might protect the cells when they develop in a biofilm. However, no structure corresponding to a mucous tube was observed in this strain in our study.

The EC_{50} and EC_5 values calculated from the dose-response curves have demonstrated considerable variability in the sensitivity of the strain, regardless of the herbicide being tested. However, the general pattern of the sensitivity of the species, revealed by the SSD-EC_{50} curves, does seem to be similar for all four PSII-inhibitor herbicides. The biological and ecological characteristics of the diatoms are beginning to be well attested [8,9,42], but little work has been done to relate the ecological characteristics of the diatoms to their sensitivity towards toxic agents [14,40]. For the PSII inhibitors, the 4 least sensitive species all share the same trophic mode: they are all heterotrophs, either obligatory (CRAC, NPAL) or optional (mixotrophs, such as EOMI, GPAR) depending on their Van Dam trophic index [9]. These strains are likely to be less affected by PSII inhibitors than autotrophic species, because they are less dependent on photosynthesis products for their growth. The other more sensitive species (with an EC_{50} below the median) are all autotrophs, with the exception of CMEN. Facultative heterotroph diatoms can change their trophic mode according to changes in environmental conditions, which permits better adaptation [43]. Some authors have noted the presence or even dominance of heterotrophic species in the context of PSII inhibitor exposure as a result of adaptation and selection processes. Debenest et al. [29] showed an increase in the abundance of facultative heterotrophic diatoms after exposure to 30 µg/l of isoproturon. Pérès et al.[44], observed that some of the tolerant species were heterotrophic, such as *GPAR* and *Sellaphora seminulum*. An increase in the relative abundance of heterotrophic and tolerant autotrophic species was observed by Ricart et al. [45] following diuron exposure.

It looks as though the tolerance to PSII inhibitors of these species may also be correlated with the 3 ecological guilds defined by Passy [42], the taxonomic list of which has been enhanced by Berthon et al. [46]. The "high profile" guild (in our tests GPAR, and *Fragilaria spp.*) includes large-size species that are advantaged by weak flow velocity, in contrast to the "low profile" guild (in our tests: *ADMI, CMEN,* and *ESLE*), which consists of small-sized species able to withstand flow velocity as a result of their position in the lower part of the biofilm and encompass slow-moving species. The "motile" guild (in our tests *CRAC, NPAL, EOMI* and *MAFO*) which includes species capable of moving around within the biofilm display the greatest tolerance in our tests, regardless of the pesticide tested. All of *Fragilaria* species (FCVA, FULN, FRUM) in this study show high sensitivity to PSII inhibitors that seem to be correlated to theirs common guild. Fragilariales order is

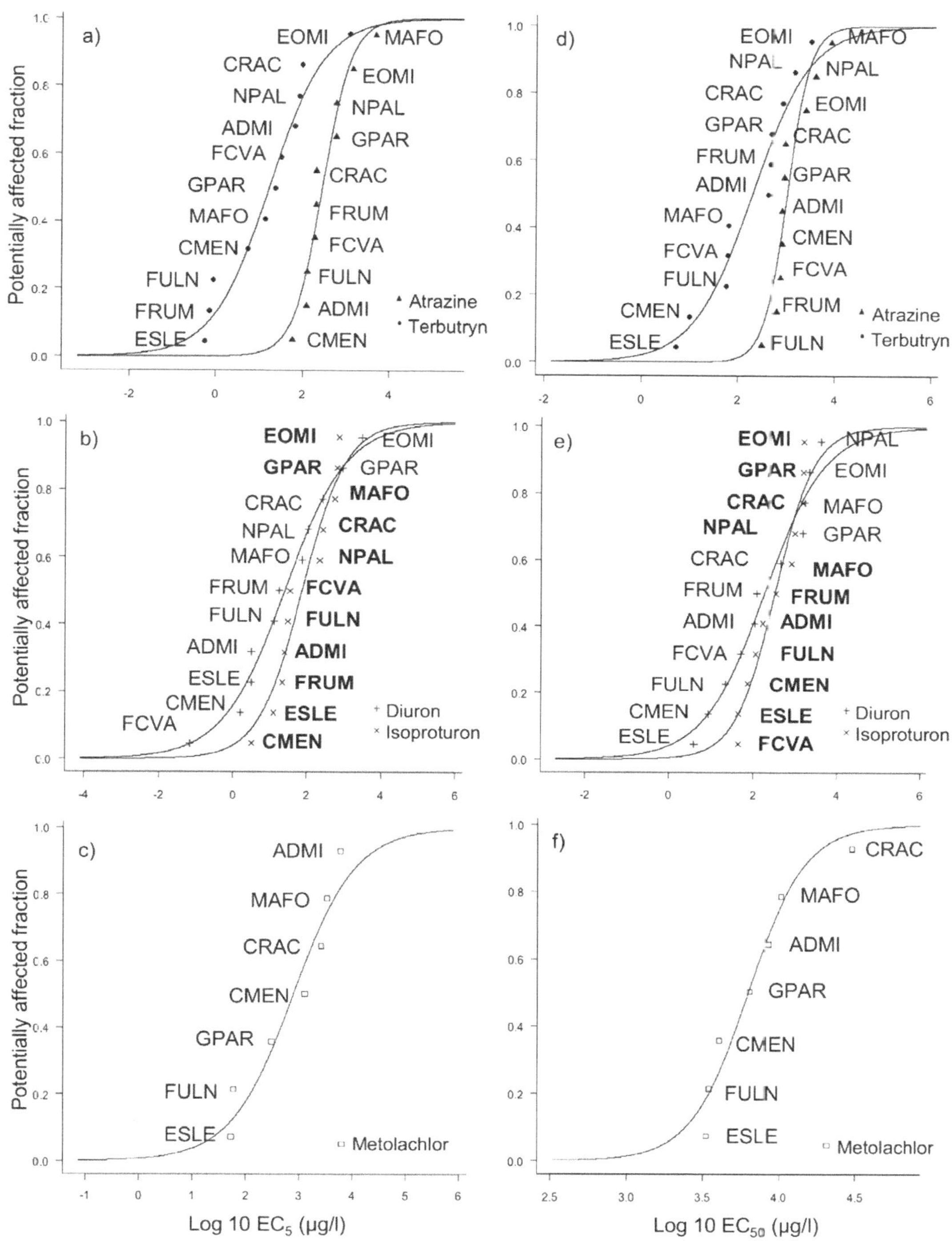

Figure 1. SSD curves of the 5 herbicides and details of species sensitivity ranking. Curves are based on EC_5 values a), b), c), and EC_{50} values d), e), f) for five herbicides grouped according to their mode of action. In b) and e) the code name in bold type corresponds to the isoproturon curve.Species references: FCVA: *Fragilaria capucina* var. *vaucheriae*; FRUM: *Fragilaria rumpens*; FULN: *Fragilaria ulna*; CRAC: *Craticula accomoda*; MAFO: *Mayamaea fossalis*; EOMI: *Eolimna minima*; NPAL: *Nitzschia palea*; ADMI: *Achnanthidium minutissimum*; CMEN: *Cyclotella meneghiniana*; ELSE: *Encyonema silesiacum*; GPAR: *Gomphonema parvulum*.

one of the most diverse for benthic diatoms in terms of life forms [47]. It encompasses many species with different morphological and physiological characteristics. For this reason, Berthon et al. [46] have shared this order between high and low profile guild. In our study, the three different species of *Fragilaria* we worked on all belong to the "high profile" guild. Two of them, FCVA and FRUM, have the same shape, length, biovolume, and ecological preferences, while FULN have cells which are about ten times longer and present a biovolume more than 100 times larger [48]. It is important to highlight that GPAR, which is one of the most

Table 4. Hazardous concentration (HC_5 and HC_{50}) values extrapolated from SSD curves based on EC_5 or EC_{50} of diatoms for 5 herbicides.

	SSD-EC_5		SSD-EC_{50}	
Herbicide	**HC_5 (µg/l)**	**HC_{50} (µg/l)**	**HC_5 (µg/l)**	**HC_{50} (µg/l)**
Diuron	0.09	24.83	1.43	188.21
Isoproturon	1.38	73.46	13.90	332.24
Atrazine	29.90	283.52	202.91	1020.88
Terbutryn	0.21	16.48	3.33	195.2
Metolachlor	14.93	787.69	1784.36	6312.69

resistant species of this study belong to this "high profile" guild too. Contrary to *Fragilaria* species, GPAR have heterotroph capabilities which may confer it a higher resistance to herbicide contamination.

Rimet and Bouchez [41] observed a reduction in the abundance of the species of the "high profile" guild, which is the opposite of the situation for those in the "motile" guild, and they attribute this to the fact that the "high profile" species may face greater exposure to dissolved substances. Our findings therefore establish a link between the trophic mode and the response of the strains to the PSII inhibitors. However, this hypothesis is not applicable to the herbicides in this study that display some other mode of action. In the case of metolachlor, Roubeix et al. [14] have observed a reduction in the abundance of ADMI, NPAL and *Mayamaea permitis* at 30 µg/l, whereas the values of the LOEC and EC_{50} for ADMI obtained with their single-species bioassay are much higher (180 and 1880 µg/l respectively). Debenest et al. [29] also demonstrated a reduction in the growth rate via chlorophyll c and the cell density at 30 µg/l of metolachlor in the periphyton. Other more pertinent and sensitive endpoints than the growth or the biomass, such as the level of very long-chain fatty acids, would provide complementary information for the assessment of the toxicity of the chloroacetamides. Finally, too little information is available about the toxicity of amidosulfuron and of foramsulfuron to make it possible to determine the effects of these herbicides on diatoms. What is certain, is that the cellular mechanisms targeted by the sulfonylureas and the chloroacetamides have less effect on the growth of benthic diatoms than the photosynthesis targeted by PSII inhibitors.

Risk evaluation and SSD models

Analysis of the EC_{50} has revealed that there is considerable variability for a given substance depending on the strain considered. The basic assumptions underlying SSD is that different species can display differing sensitivities towards the same toxicant in the environment. The SSD model allowed us to demonstrate this variation of sensitivity among an assemblage using data obtained by means of single-species bioassays [20] and, *in fine*, to extrapolate the HC for a natural community, which is more realistic from an environmental standpoint. Our SSD curves were based on sensitivity data for 11 diatom species obtained using the same protocol. Wheeler et al [34] recommended using more than 10 data in order to obtain a more reliable HC. The HC_5 values obtained in our study and determined from the SSD-EC_{50} and SSD-EC_5 curves are very variable, and tend to be higher in the case of metolachlor and atrazine. However, the fact that this modeling was however carried out for single-species, implies that the distribution ignores any interaction between the species [20,49]. In this study, the equations used to establish the SSDs do indeed model the differences in sensitivity of the diatoms within the reconstituted assembly, and the level of toxicity of each herbicide. To ensure environmental pertinence, Forbes and Calow [50] recommend carrying out SSDs from species that are representative of the environment being studied. The diatoms selected for this study are not all strictly associated with a benthic life form, but they are representative and cover the coastal zones as well as providing wide diversity in terms of taxonomy, of sensitivity and ecological traits. The HC_5 values that we determined from our SSD-EC_{50} and SSD-EC_5 curves vary considerably for different herbicides, and tend to be higher in the case of metolachlor and atrazine.

Some studies have used scientific literature and toxicity databases to construct the SSD curves [1,27,36], but the amount of data available varies considerably depending on the substances. In general, the paucity of toxicity data hinders the construction of

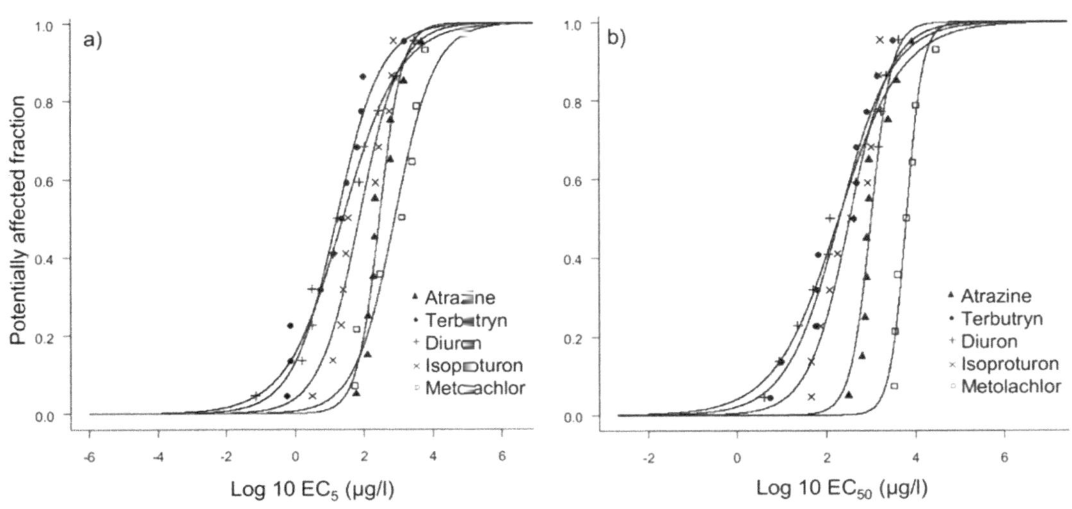

Figure 2. SSD curves of each herbicide based on benthic diatom sensitivities. Curves are based on a) EC_5 and b) EC_{50} values. Each symbol represents one species.

specific curves for a functional group and for a given substance, resulting in curves that are not very robust, based on too few data or on pooled groups and differing endpoints. Some authors have constructed SSD-EC_{50} curves for atrazine, isoproturon or diuron based on the sensitivity of primary producers including macrophytes and microalgae [1,36]. The various HC values extrapolated from these curves were much lower than those we found in our study. The SSD-EC_{50} curves based on the sensitivity of several primary producers, not only benthic organisms, plotted by Schmitt-Jansen and Altenburger [27] revealed greater sensitivity of the assemblages towards atrazine and isoproturon than in our study. This means, that the HC values obtained from SSD curves based on various endpoints and using phototrophic organisms are lower than those specifically involving benthic diatoms. These differences could be explained by the differences in the sensitivities of the algal classes towards the PSII inhibitors or to compounds with other mechanisms of action and by the species composition. Guasch et al [43] observed that periphyton communities in which diatoms dominated were more resistant to atrazine than those dominated by green algae. Specific SSD curves based on chlorophyte and diatom sensitivities to atrazine are different in shape [27], and the median EC_{50} found for diatoms was more than twice as high as those found for chlorophytes. The fact that benthic diatoms seemed more resistant to PSII inhibitors than chlorophytes led to a threshold not enough protective for chlorophytes wich make up small proportions of biofilms.

For diuron and terbutryn, these HC_5 values correspond to concentrations sometimes found in water column. In France, peaks of about 3 µg/l of diuron were found in 2009 in superficial water of some stations of Adour-Garonne catchment, [7].In other European river, peak level of isoproturon and diuron of 1,96 and 0.86 µg/l, respectively, have been reported [2]. HC5 obtained from SSD-EC5 for these two herbicides are lower than these environmental concentrations found in water. Finally, the AA-EQS of diuron defined as 0.2 µg/l, exceed the SSD-EC5 HC5 value. In view of our results, this implies that the concentrations of diuron and isoproturon found in the environment carrya potential risk of reducing diversity in communities of benthic diatoms. However terbutryn is not a prioritary substance, and we can say that in the light of its low HC5, very low concentrations of this substance can affect benthic diatom communities. Finally, as SSDs based on EC_5 values gave HC5 values below the environmental concentration, curves based on these thresholds are probably more appropriate to ensure that benthic diatom communities are protected.

Single-species bioassays of these 11 species revealed the huge variation of sensitivities present in a given algal group for the same herbicide. The main finding to emerge is mainly that the trophic mode plays an important role in the resistance of benthic diatoms towards PSII inhibitors. For risk assessment, SSD curves based on log-logistic regression fitted well. If the panel of algal strains and the endpoint selected to some extent determine the final HC_5, this study has demonstrated that diuron, isoproturon and terbutryn, even at low levels, constitute a risk for natural communities of benthic diatoms. In view of these results, risk assessments of herbicides for the total biofilm algal community should be done using specific SSD curves for each of the algal groups present in biofilm or in the phytoplankton communities.

In addition to performing studies, improving risk assessment involves taking the real *in-situ* exposure levels facing these benthic diatoms into account, i.e. taking into account the mixtures of substances. The trials are currently under way to include these factors in the evaluation of the risks linked to the pesticides in aquatic ecosystems.

Acknowledgments

The authors would like to thank N. Chèvre and V. Grégorio (EPF Lausanne, CH) for their scientific assistance.

Author Contributions

Conceived and designed the experiments: FL AB FR BM. Performed the experiments: FL. Analyzed the data: FL BM AB. Contributed reagents/materials/analysis tools: FL FR. Wrote the paper: FL AB BM.

References

1. Knauert S, Escher B, Singer H, Hollender J, Knauer K (2008) Mixture toxicity of three photosystem II inhibitors (atrazine, isoproturon, and diuron) toward photosynthesis of freshwater phytoplankton studied in outdoor mesocosms. Environ Sci Technol 42: 6424–6430.
2. Loos R, Gawlik BM, Locoro G, Rimaviciute E, Contini S, et al. (2009) EU-wide survey of polar organic persistent pollutants in European river waters. Environ Pollut157: 561-568.
3. Montuelle B, Dorigo U, Bérard A, Volat B, Bouchez A, et al. (2010) The periphyton as a multimetric bioindicator for assessing the impact of land use on rivers: an overview on the Ardières-Morcille experimental watershed (France). Hydrobiologia 657: 123-141.
4. Ortelli D, Edder P, Rapin F, Ramseier S (2009) Metals and organic micropollutants in Geneva Lake waters. Commission Internationale pour la Protection des Eaux du Léman contre la pollution, Campagne 2008.
5. Commissariat General au Développement Durable (2011) Bilan de présence des micropolluants dans les milieux aquatiques continentaux. Période 2007-2009. 59-71 pp.
6. Lepper P (2005) Manual on the Methodological Framework to Derive Environmental Quality Standards for Priority Substances in Accordance with Article 16 of the Water Framework Directive (2000/60/EC). Schmallenberg (DE): Fraunhofer-Institute Molecular Biology and Applied Ecology.
7. Agence de l'eau Adour-Garonne (2009) Qualité des eaux et produits phytosanitaires sur le bassin Adour-Garonne.
8. Rumeau A, Coste M (1988) Introduction into the systematic of freshwater diatoms, for a useful generic diatomic index. Bull Fr Peche Piscicult 309: 1-69.
9. Van Dam H, Mertens A, Sinkeldam J (1994) A coded checklist and ecological indicator values of freshwater diatoms from the Netherlands. Neth J Aquatic Ecol 28: 117-133.
10. Bérard A, Pelte T. (1999) The impact of Photosystem II (PSII) inhibitors on algae communities and dynamics. Rev Sci Eau 12: 333-361.
11. Böger P (2003) Mode of action for chloroacetamides and functionally related compounds. J Pestic Sci 28: 324-329.
12. Nyström B, Blanck H (1998) Effects of the sulfonylurea herbicide metsulfuron methyl on growth and macromolecular synthesis in the green alga Selenastrum capricornutum. Aquat Toxicol 43: 25–39.
13. Seguin F, Druart JC, Le Cohu R (2001) Effects of atrazine and nicosulfuron on periphytic diatom communities in freshwater outdoor lentic mesocosms. Ann Limnol- Int J Lim 37: 3-8.
14. Roubeix V, Mazzella N, Mechin B, Coste M, Delmas F (2011) Impact of the herbicide metolachlor on river periphytic diatoms: experimental comparison of descriptors at different biological organization levels. Ann Limnol. – Int J lim 47: 1–11.
15. Vallotton N, Moser D, Eggen RIL, Junghans M, Chèvre N (2008) S-metolachlor pulse exposure on the alga *Scenedesmus vacuolatus*: Effects during exposure and the subsequent recovery. Chemosphere 73: 395–400.
16. Roubeix V, Mazzella N, Schouler L, Fauvelle V, Morin S et al. (2011) Variations of periphytic diatom sensitivity to the herbicide diuron and relation to species distribution in a contamination gradient: implications for biomonitoring. J Environ Monitor 13: 1768–1774.
17. Magnusson M, Heimann K, Negri AP (2008) Comparative effects of herbicides on photosynthesis and growth of tropical estuarine microalgae. Mar Pollut Bull 56: 1545–1552.
18. Nelson KJ, Hoagland KD, Siegfried B (1999) Chronic effects of atrazine on tolerance of a benthic diatom. Environ Toxicol Chem 18: 1038–1045.
19. Pesce S, Bouchez A, Montuelle B (2011) Effects of organic herbicides on phototrophic microbial communities in freshwater ecosystems. Rev Environ Contam T 214: 87-124.
20. Posthuma L, Traas TP, Sutter II GW (2002) General introduction to species sensitivity distributions. In Posthuma L, Traas TP, Sutter II GW, eds, Species Sensitivity Distributions in Ecotoxicology, Lewis, Boca Raton, FL, USA. pp. 3–10.
21. Van Straalen NM (2002) Threshold models for species sensitivity distributions applied to aquatic risk assessment for zinc. Environ Toxicol Phar 11: 167–172.

22. Hoagland K D, Roemer S C, Rosowski J R (1982) Colonization and community structure of two periphyton assemblages, with emphasis on the diatoms (Bacillariophyceae). American Journal of Botany 69: 18 –213.
23. Guerrero M, Rodriguez L (1991) Species composition of phytobenthos in the littoral of lake xolotlan (Managua). Hydrobiol. Bull. 25: 117-120 .
24. Schaumburg J, Schranz C, Hofmann G, Stelzer D, Schneider S, et al. (2004) Macrophytes and phytobenthos as indicators of ecological status in German lakes – a contribution to the implementation of the Water Framework Directive. Limnologica 34: 302–314.
25. Lenoir A, Coste M (1996) Development of a practical diatom index of overall water quality applicable to the french national water board network. In Whitton BARott E, eds, Use of Algae for Monitoring Rivers II, Innsbruck, Austria (09, 17-19,1995). pp 29-43.
26. Tang J X, Hoagland K D, Siegfried B D (1997) Differential toxicity of atrazine to selected freshwater algae. Bull. Environ. Contam. Toxicol. 59:631-637.
27. Schmitt-Jansen M, Altenburger R (2005) Predicting and observing responses of algal communities to photosystem II-herbicide exposure using pollution-induced community tolerance and species-sensitivity distributions. Environ Toxicol Chem 24: 304–312.
28. ASTM International (2007) Standard Guide for Conducting Static Toxicity Tests with Microalgae. Designation: E1218-04. 14 p.
29. Debenest T, Pinelli E, Coste M, Silvestre J, Mazzella N, et al. (2009) Sensitivity of freshwater periphytic diatoms to agricultural herbicides. Aquat Toxicol 93: 11-17.
30. Ritz C, Streibig JC (2005) Bioassay Analysis using R. J Stat Softw 12: 1-22.
31. Calabrese EJ, Baldwin LA (2002) Defining hormesis. Hum exp toxicol 21: 91-97.
32. Cedergreen N, Ritz C, Streibig JC (2005) Improved empirical models describing hormesis. Environ Toxicol Chem 24: 3166-3172.
33. Aldenberg T, Jaworska JS (2000) Uncertainty of the hazardous concentration and fraction affected for normal species sensitivity distributions. Ecotox Environ Safe 46: 1-18.
34. Wheeler JR, Grist EPM, Leung KMY, Morritt D, Crane M (2002) Species sensitivity distributions: data and model choice. Mar. Pollut. Bull. 45: 192-202.
35. De Zwart D (2002) Observed regularities in species sensitivity distributions for aquatic species. In Posthuma L, Suter GWI, Traas TP, eds, Species-Sensitivity Distributions in Ecotoxicology. Lewis,Boca Raton, FL, USA. pp 133–15.
36. Chèvre N, Loepfe C, Singer H, Stamm C, Fenner K, et al. (2006) Including mixtures in the determination of water quality criteria for herbicides in surface water. Environ Sci Technol 40: 426-435.
37. Dorigo U, Bérard A, Bouchez A, Rimet F, Montuelle B (2010) Transplantation of Microbenthic Algal Assemblages to Assess Structural and Functional Recovery After Diuron Exposure. Arch Environ Contam Toxicol 59: 555–563.
38. Kotrikla A, Gatidou G, Lekkas TD (1999) Toxic effects of atrazine, deethyl-atrazine, deisopropyl-atrazine and metolachlor on chlorella fusca var fusca. Global Nest: the Int J 1: 39-45.
39. Rivard L (2003) Environmental Fate of Metolachlor, Environmental Monitoring Branch, Department of Pesticide Regulation, Sacramento, CA 95812.
40. Nyström B, Björnsäter B, Blanck H (1999) Effects of sulfonylurea herbicides on non-target aquatic micro-organisms growth inhibition of micro-algae and short-term inhibition of adenine and thymidine incorporation in periphyton communities. Aquat Toxicol 47: 9–22.
41. Rimet F, Bouchez A (2011) Use of diatom life-forms and ecological guilds to assess pesticide contamination in rivers: Lotic mesocosm approaches. Ecol Indic11: 489–499.
42. Passy SI (2007) Diatom ecological guilds display distinct and predictable behavior along nutrient and disturbance gradients in running waters. Aquat Bot 86: 171–178.
43. Guasch H, Ivorra N, Lehmann V, Paulsson M, Real M et al. (1998) Community composition and sensitivity of periphyton to atrazine inflowing waters: the role of environmental factors. J App Phycol 10: 203–213.
44. Pérès F, Florin D, Grollier T, Feurtet-Mazel A, Coste M, et al. (1996) Effects of the phenylurea herbicide isoproturon on periphytic diatom communities in freshwater indoor microcosms. Environ Pollut 94: 141-152.
45. Ricart M, Barceló D, Geiszinge A, Guasch H, López de Alda M, et al. (2009) Effects of low concentrations of the phenylurea herbicide diuron on biofilm algae and bacteria. Chemosphere76: 1392–1401.
46. Berthon V, Bouchez A, Rimet F (2011) Using diatom life-forms and ecological guilds to assess organic pollution and trophic level in rivers: a case study of rivers in south-eastern France. Hydrobiologia 673: 259–271.
47. Rimet F, Bouchez A (2012) Life-forms, cell-sizes and ecological guilds of diatoms in European rivers. Knowledge and management of Aquatic Ecosystems 406, 1.
48. Lecointe C, Coste M, Prygiel J (1993) "Omnidia": software for taxonomy, calculation of diatom indices and inventories management. Hydrobiologia 269/270, 509-513.
49. Maltby L, Blake N, Brock TCM, Van Den Brink PJ (2005) Insecticide species sensitivity distributions: importance of tests species selection and relevance to aquatic ecosystems. Environ Toxicol Chem 24: 379–388.
50. Forbes VE, Calow P (2002) Species sensitivity distributions revisited: A critical appraisal. Hum Ecol Risk Assess 8: 473–492.

12

Photosynthetic Physiological Response of Radix Isatidis (*Isatis indigotica* Fort.) Seedlings to Nicosulfuron

Xiangyang Yuan[1☯¶], Liguang Zhang[1,2☯¶], Na Ning[1], Yinyuan Wen[1], Shuqi Dong[1], Meiqiang Yin[1], Meijun Guo[1], Binqiang Wang[1], Lei Feng[1], Pingyi Guo[1*]

1 Laboratory of Crop Chemical Regulation and Chemical Weed Control, Agronomy College, Shanxi Agricultural University, Taigu, the People's Republic of China, **2** Institute of Crop Sciences, Shanxi Academy of Agricultural Sciences, Taiyuan, the People's Republic of China

Abstract

Radix Isatidis (Isatis indigotica Fort.) is one of the most important traditional Chinese medicine plants. However, there is no suitable herbicide used for weed control in Radix Isatidis field during postemergence stage. To explore the safety of sulfonylurea herbicide nicosulfuron on Radix Isatidis (Isatis indigotica Fort.) seedlings and the photosynthetic physiological response of the plant to the herbicide, biological mass, leaf area, photosynthetic pigment content, photosynthetic rate, chlorophyll fluorescence characteristics, and P_{700} parameters of Radix Isatidis seedlings were analyzed 10 d after nicosulfuron treatment at 5th leaf stage in this greenhouse research. The results showed that biological mass, total chlorophyll, chlorophyll a, and carotenoids content, photosynthetic rate, stomatal conductance, PS II maximum quantum yield, PS II effective quantum yield, PS II electron transport rate, photochemical quenching, maximal P_{700} change, photochemical quantum yield of PS I, and PS I electron transport rate decreased with increasing herbicide concentrations, whereas initial fluorescence, quantum yield of non-regulated energy dissipation in PS II and quantum yield of non-photochemical energy dissipation due to acceptor side limitation in PS I increased. It suggests that nicosulfuron ≥1 mg L−1 causes the damage of chloroplast, PS II and PS I structure. Electron transport limitations in PS I receptor side, and blocked dark reaction process may be the main cause of the significantly inhibited growth and decreased photosynthetic rate of Radix Isatidis seedlings.

Editor: Manuel Reigosa, University of Vigo, Spain

Funding: This paper was supported by "Scientific and Technological Project in Shanxi Province, China" (20100321103), "Key Research Foundation for Returness in Shanxi Province, China" (201004), and "Shanxi Agricultural University Technology Innovation Fund" (201023). The funders had no role in study design, data collection and analysis, decision to publish, or preparation of the manuscript.

Competing Interests: The authors have declared that no competing interests exist.

* Email: yuanxiangyang200@126.com

☯ These authors contributed equally to this work.

¶ XY and LZ are co-first authors on this work.

Introduction

As one of the most important traditional Chinese medicine plants, Radix Isatidis (*Isatis indigotica* Fort.), one kind of cruciferous plants, is extensively cultivated in China. However, weeds seriously limit its yield and quality. Compared with conventional manual control of weeds, chemical control is more effective, yet there is no suitable herbicide used for weed control in Radix Isatidis field during postemergence stage [1]. Nicosulfuron (2-((4,6-dimethoxypyrimidin-2-ylcarbamoyl)sulfamoyl)-N, N-dimethylnicotinamide) belongs to the acetyl lactic acid synthase (ALS) inhibitor, and can effectively control many perennial and annual grasses as well as certain broadleaf weeds [2,3]. Is nicosulfuron safe to Radix Isatidis seedlings or not?

It is reported that nicosulfuron can effectively control large crabgrass (*Digitaria sanguinalis*) cornfield, goosegrass (*Eleusine indica*), spiny amaranth (*Amaranthus spinosus* L), amaranthus blitum (*Amaranthus ascendens* Loisel.), speargrass (*Imperata cylindrica*), and purslane (*Portulaca oleracea*) in maize field [4]. However, there are significant differences during the sensitivity of maize varieties to nicosulfuron [5], and sweet corn is much more sensitive to nicosulfuron than the others [6]. The leaves of sensitive maize variety may show symptoms of chlorosis and shrinking 7 d after applying nicosulfuron [7]. The weeding efficiency of nicosulfuron herbicide is related to weed species, leaf age and how long it is applied after rain. [8]. Meanwhile, the resistance of plant to nicosulfuron depends on its metabolic rate, sensitivity to ALS, and nicosulfuron dosage [9–11].

However, there is few paper reporting the application of post-emergence herbicide controlling many grasses and broadleaf weeds in Radix Isatidis fields [1]. It is shown that rational application of trifluralin, pendimethalin and glyphosate at pre-emergence stage can control weeds effectively in Radix Isatidis fields [12,13]. It is also reported that clethodim and quizalofop-p-ethyl at post-emergence stage can be used for grassy weed control in Radix Isatidis fields [12,14].

Primary mode of action of the ALS-inhibiting herbicides that interfere with the activity of ALS enzyme seems no longer in doubt [15]. But, secondary effects of ALS inhibition, such as decreased photosynthesis, disturbed respiration, and synthesis of branched chain amino acids, etc., need to be investigated, which have also been implicated in the mechanism of plant death. Detection of chlorophyll fluorescence dynamics is a rapid and non-invasive probe of researching plant photosynthetic functions, which has been widely applied to study the effects of herbicides in plants [1,16–19]. Therefore, the objectives of this research were to (1) assess the possibility of application of nicosulfuron in Radix Isatidis field, and (2) understand the related photosynthetic physiological mechanism.

Materials and Methods

Materials and experiment design

Radix Isatidis (*Isatis indigotica* Fort.) seeds were supplied by Anguo Lixin Medicinal Materials Co., Ltd., Hebei province, China. Nicosulfuron (40%, OF) was provided by Xianda Chemical Co., Ltd., Shandong province, China.

This study was conducted in a greenhouse at Shanxi Agricultural University, China. The experiment was designed as a randomized complete block design with three replications and each replicate containing three pots. Fifteen Radix Isatidis seeds were grown equidistantly in 23-cm diam containers filled with a 1:2 mixture of sand and loam soil with 57.2 g kg^{-1} of organic matter, 0.92 g kg^{-1} of total nitrogen, 14.37 mg kg^{-1} of available phosphorus, and 114.3 mg kg^{-1} of rapidly-available potassium. The seeds were covered with 1 cm of 1:2 sand/soil mixtures and each pot was carefully watered. Seedlings were grown under greenhouse conditions of 24/16±3°C day/night temperatures and were thinned to three plants per container at three-leaf stage.

Radix Isatidis seedlings were treated at five-leaf stage. These plants were treated with 0, 0.5, 1, 2, and 4 times the labeled use concentrations of nicosulfuron in corn. The herbicide product contained 40% nicosulfuron, large amounts of water and a small amount of oil adjuvants, and the recommended effective concentrations were 1 mg L^{-1} nicosulfuron. Herbicides were applied with a laboratory pot-sprayer equipped with a nozzle, calibrated to deliver 450 L ha^{-1}. Agronomic characters and photosynthetic physiology parameters of Radix Isatidis seedling were determined every 10 d after herbicide treatment. Except the control, seedlings treated with nicosulfuron wilted or died 20 d after herbicide treatment, so the data 10 d after herbicide treatment was determined.

Measurements

The third fully expanded leaf of Radix Isatidis seedlings were sampled for the following tests. Leaf area was determined using a laser leaf area meter CI-203 and the CI-203CA conveyor attachment (United States CID Inc.).

Photosynthetic gas exchange was analyzed with a GFS-3000 optical instrument (Germany WALZ company) which can control photosynthesis by means of light intensity, leaf temperature, air flow rate and CO_2 concentration in the cuvette. Photosynthetic rate (P_n), transpiration rate (T_r), stomatal conductance (G_s) and intercellular CO_2 concentrations (C_i) were measured simultaneously with the light intensity at (800±0.4 μmol·m^{-2}·s^{-1}) and CO_2 concentration (379±0.4 μmol·mol^{-1}). Air flow rate was set at 750 μmol s^{-1} and air temperature (20.9±0.4°C) was also recorded automatically by the instrument. Stomatal limitation value $(L_s) = 1-C_i/C_a$ (C_a is the atmospheric CO_2 concentration). Non-stomatal limitation value was calculated by C_i/G_s [20]. For the measurement of photosynthetic pigments, leaves were extracted from leaf discs with 80% (v/v) acetone and assayed spectrophotometrically using extinction coefficients according to Porra et al. [21].

Chlorophyll fluorescence and P_{700} parameter were measured simultaneously by Dual-PAM-100 measurement system (Germany WALZ company), using the automated "Induction Curve" routine provided by the Dual PAM software [22]. Prior to measurements, treated plants were placed in darkroom for 30 min, and fluorescence induced curve (Slow Kinetics) was determined in "Fluo+P_{700} mode". Then, the kinetics of chlorophyll fluorescence induction and P_{700} oxidation were recorded simultaneously by the instrument. Firstly, the initial fluorescence (F_o) was established and subsequently the maximum fluorescence (F_m) was determined by the "Saturation Pulse" method. Secondly, the maximal P_{700} change (P_m) was determined by application of a saturation pulse (SP) after far-red pre-illumination. Thirdly, actinic illumination was started and SP was given every 20 s, with the same pulses serving for fluorescence and P_{700} analysis.

PS II maximum quantum yield (F_v/F_m) was evaluated as $F_v/F_m = (F_m-F_o)/F_m$. Other PS II energy dissipation parameters were estimated by the Dual PAM software. $q_P = (F_m'-F)/(F_m'-F_o')$ was used as indicator to reflect a ratio of light energy absorbed in PS II being used to photochemical electron transport. Apparent electron transfer efficiency in PS II in light was calculated according to ETR(II) = PAR×0.84×0.5×Y(II), and was used to measure electron transfer of carbon fixation resulted from photochemical reactions. Three complementary quantum yields of energy conversion in PS II were calculated: PS II effective quantum yield (Y(II)) was evaluated as $(Y(II)) = (F_m'-F)/F_m'$, the yield of non-photochemical losses via non-regulated pathways of PS II as $Y(NO) = 1/(NPQ+1+q_L\cdot(F_m/F_o-1))$, $NPQ = F_m/F_m'-1$, $q_L = q_P\cdot F_o'/F$, where quantum yield of regulated energy dissipation in PS II as Y(NPQ) = 1–Y(II)–Y(NO) [23].

P_{700} oxidation was monitored by absorbance changes in the near-infrared (830–875 nm) [24]. The maximal P_{700} signal observed upon full oxidation was denoted by Pm. Y(NA), the quantum yield of non-photochemical energy dissipation due to acceptor-side limitation, was calculated according to: $Y(NA) = (P_m'-P_m)/P_m'$. Photochemical quantum yield of PSI as Y(I) was estimated according to $Y(I) = (P_m'-P)/P_m$; Quantum yield of non-photochemical energy dissipation due to donor side limitation in PS I as Y(ND) was calculated by $Y(ND) = (P-P_o)/P_m$. The total value of three quantum yields was one: Y(I)+Y(ND)+Y(NA) = 1. The electron transfer efficiency of PS I as ETR (I) was provided by the Dual PAM software.

Statistical analysis

Microsoft Office Excel 2003 and Statistics Analysis System 8.0 were used in statistical analysis of the data. Mean values were compared by a one-way analysis of variance (ANOVA), and Duncan's test was used to determine the significant differences among the treatments. We used $P = 0.05$ as the statistical significance threshold.

Results

Effect of nicosulfuron on agronomic characteristics of Radix Isatidis seedlings

The effects of nicosulfuron on agronomic characteristics of Radix Isatidis seedlings are shown in Table 1. Fresh weight of Radix Isatidis seedlings decreased with the increasing concentrations of nicosulfuron. Leaf area was significantly decreased by nicosulfuron at 4 mg L^{-1}, whereas this was not affected by other treatment. The whole plant fresh weight and shoot fresh weight showed the similar trend, and differences between the treatment of nicosulfuron at 0.5 mg L^{-1} and the control were significant. Shoot fresh weight declined by 41.64%, 52.90%, 53.58% and 59.04%, respectively, from 0.5 to 4 mg L^{-1} of nicosulfuron. The reduction in root fresh weight was not significant by nicosulfuron, suggesting

Table 1. Effect of different concentrations of nicosulfuron on agronomic traits of radix isatidis seedlings.

Nicosulfuron (mg L^{-1})	*LA* (cm^2)	*WPFW* (g)	*SFW* (g)	*RFW* (g)
0	10.54 ± 0.24^a	3.18 ± 0.04^a	2.93 ± 0.05^a	0.25 ± 0.05^a
0.5	9.93 ± 0.75^a	1.94 ± 0.18^b	1.71 ± 0.16^b	0.23 ± 0.03^a
1	11.11 ± 0.37^a	1.58 ± 0.10^{bc}	1.38 ± 0.04^{bc}	0.20 ± 0.04^a
2	9.54 ± 0.24^a	1.52 ± 0.14^{bc}	1.36 ± 0.27^{bc}	0.17 ± 0.02^a
4	7.37 ± 0.71^b	1.36 ± 0.29^c	1.20 ± 0.18^c	0.16 ± 0.01^a

Note: The data in the table is mean ± SD. The different letters in the same column indicate significantly different at $P<0.05$ level by Duncan's new multiple range test. *LA*, *WPFW*, *SFW* and *RFW* represent leaf area, the whole plant fresh weight, shoot fresh weight, and root fresh weight, respectively.

that the suppression of nicosulfuron on the aboveground parts of Radix Isatidis was greater than root.

Effect of nicosulfuron on photosynthetic pigment contents in leaves of Radix Isatidis seedlings

As shown in Table 2, each nicosulfuron treatment caused different degrees of decline in photosynthetic pigment contents in leaves of Radix Isatidis seedlings. It seems that 1 mg L^{-1} of nicosulfuron inhibited *Chl* by 33.57%, *Chl* a by 37.04% and *Car* by 47.37, and the differences between the treatment and the control were significant. However, *Chl* b was not significantly decreased until the herbicide concentrations reached up to 2 mg L^{-1}, and it was inhibited by 37.14%. Although nicosulfuron declined chlorophyll a/b, there was no significant effect.

Effect of nicosulfuron on photosynthetic characteristics in leaves of Radix Isatidis seedlings

Nicosulfuron decreased P_n and G_s in leaves of Radix Isatidis seedling significantly. As concentration of nicosulfuron increases, inhibition of P_n changed by 58.21%, 70.02%, 70.69%, and 78.89%, respectively (Table 3). G_s inhibition also increases as concentration of the herbicide increases by 86.03%, 91.10%, 91.21%, and 96.49%, respectively when nicosulfuron varies from 0.5 to 4 mg L^{-1} (Table 3). However, the results of P_n and G_s revealed no significant differences ($P>0.05$) from 1 to 4 mg L^{-1} of nicosulfuron. C_i increased first then declined with increasing of nicosulfuron concentrations, peaked at 0.5 mg L^{-1}, and was the lowest at 4 mg L^{-1}. Compared to C_i, the change trend of L_s was just the opposite. L_s was the least at 4 mg L^{-1}, and the largest at 1 mg L^{-1}. For C_i/G_s, the value for treated plants was higher than the control in different degrees.

Effect of nicosulfuron on chlorophyll fluorescence parameters in leaves of Radix Isatidis seedlings

As shown in Fig. 1-A, F_o increased with the increase in nicosulfuron concentration. F_o was a bit higher than the control for 2 mg L^{-1} treatment, but significantly higher for 4 mg L^{-1} treatment. F_v/F_m was slightly higher than the control for 0.5 mg L^{-1} concentration, and then decreased with the increase concentrations of nicosulfuron.

Changes of Y(II), ETR and q_P were consistent, and declined in a concentration-dependent manner. Compared with the control, Y(II) was reduced by 7.01%, 36.07%, 70.57% and 81.95%, ETR declined by 7.10%, 36.14%, 70.44% and 81.91%, and q_P decreased by 8.64%, 29.54%, 70.19% and 82.51%, respectively (Fig. 1-B, 1-D).

Y(NO) showed a fluctuant increase as nicosulfuron concentration increases, slightly higher than the control for 0.5 mg L^{-1} and 1 mg L^{-1} concentration, and significantly (49.70%) higher for 2 mg L^{-1}. Compared to Y(NO), the trend of Y(NPQ) was the opposite (Fig. 1-C). It showed that the degree of the relative excess light damage induced by nicosulfuron in Radix Isatidis leaf was more severe with increasing nicosulfuron concentration.

Effect of nicosulfuron on P_{700} parameters in leaves of Radix Isatidis seedlings

Increasing nicosulfuron concentration induced a decline in P_m, ETR (I), and PS I. Nicosulfuron treatment at 1 mg L^{-1} had dramatically influence on P_m, while the differences between 1 mg L^{-1} and 2 mg L^{-1} were not significant (Fig. 2-A). As shown in Fig. 2-B, the trend of ETR (I) and PS I was consistent, and the values of them at 2 mg L^{-1} were much smaller than 1 mg L^{-1} (Fig. 2-B). In Fig. 2-C, nicosulfuron induced an increase of Y(NA), while Y(ND) increased first and then decreased with the increasing of nicosulfuron concentration, and the peak was at 1 mg L^{-1} (Fig. 2-C).

Table 2. Effect of different concentrations of nicosulfuron on photosynthetic pigment content in leaves of Radix isatidis seedling.

Nicosulfuron (mg L^{-1})	*Chl* (mg g^{-1})	*Chl* a (mg g^{-1})	*Chl* b (mg g^{-1})	*Car* (mg g^{-1})	*Chl* a/b
0	1.43 ± 0.10^a	1.08 ± 0.08^a	0.35 ± 0.01^a	0.19 ± 0.01^a	3.12 ± 0.11^a
0.5	1.00 ± 0.03^b	0.75 ± 0.02^b	0.25 ± 0.05^{ab}	0.13 ± 0.02^b	2.98 ± 0.68^a
1	0.95 ± 0.05^{bc}	0.68 ± 0.001^{bc}	0.27 ± 0.04^{abc}	0.10 ± 0.01^b	2.49 ± 0.35^a
2	0.82 ± 0.07^c	0.60 ± 0.09^c	0.22 ± 0.02^{bc}	0.12 ± 0.01^{bc}	2.76 ± 0.61^a
4	0.57 ± 0.11^d	0.40 ± 0.09^d	0.17 ± 0.02^d	0.08 ± 0.02^c	2.34 ± 0.30^a

Note: The data in the table is mean ± SD. The different letters in the same column indicate significantly different at $P<0.05$ level by Duncan's new multiple range test. *Chl*, *Chl* a, *Chl* b, *Car* and *Chl* a/b represent chlorophyll, chlorophyll a, chlorophyll b, carotenoid, and chlorophyll a/b, respectively.

Table 3. Effect of different concentrations of nicosulfuron on photosynthetic characteristics in leaves of Radix isatidis seedling.

Nicosulfuron (mg L^{-1})	P_n (μmol $CO_2 \cdot m^{-2} \cdot s^{-1}$)	C_i (μmol·mol^{-1})	G_s (mmol$H_2O \cdot m^{-2} \cdot s^{-1}$)	L_s	C_i/G_s
0	11.94 ± 1.857^a	296.[illegible]$\pm 10.62^a$	264.57 ± 76.31^a	0.21 ± 0.03^a	1.16 ± 0.29^c
0.5	4.99 ± 0.23^b	336.[illegible]$\pm 13.18^{ab}$	36.96 ± 0.24^b	0.11 ± 0.04^{ab}	9.10 ± 0.30^{bc}
1	3.58 ± 1.26^b	309.[illegible]$\pm 70.20^{ab}$	23.54 ± 13.04^b	0.05 ± 0.01^b	19.98 ± 7.05^{ab}
2	3.50 ± 0.60^b	235.[illegible]$\pm 63.10^b$	23.25 ± 14.89^b	0.38 ± 0.17^b	13.85 ± 1.58^{ab}
4	2.52 ± 1.57^b	115.[illegible]$\pm 10.38^c$	9.29 ± 17.92^b	0.70 ± 0.01^b	21.60 ± 2.32^a

Note: The data in the table is mean±SD. The different letters in the same column indicate significantly different at $P<0.05$ level by Duncan's new multiple range test. P_n, C_i, G_s, L_s and C_i/G_s represent photosynthetic rate, intercellular CO_2 concentrations, stomatal conductance, stomatal limitation value, and non-stomatal limitation value, respectively.

Discussion

The safety of herbicides on plants may be represented through agronomic traits and physiological indexes [1,25–26]. In this study, nicosulfuron at the recommended usage (1 mg L^{-1}) was not safe to Radix Isatidis seedlings, which was reflected by reduced biomass. It was coincident with Yuan et al. [1], mesosulfuron-iodosulfuron was unsafe to Radix Isatidis seedlings and decreased the leaf area

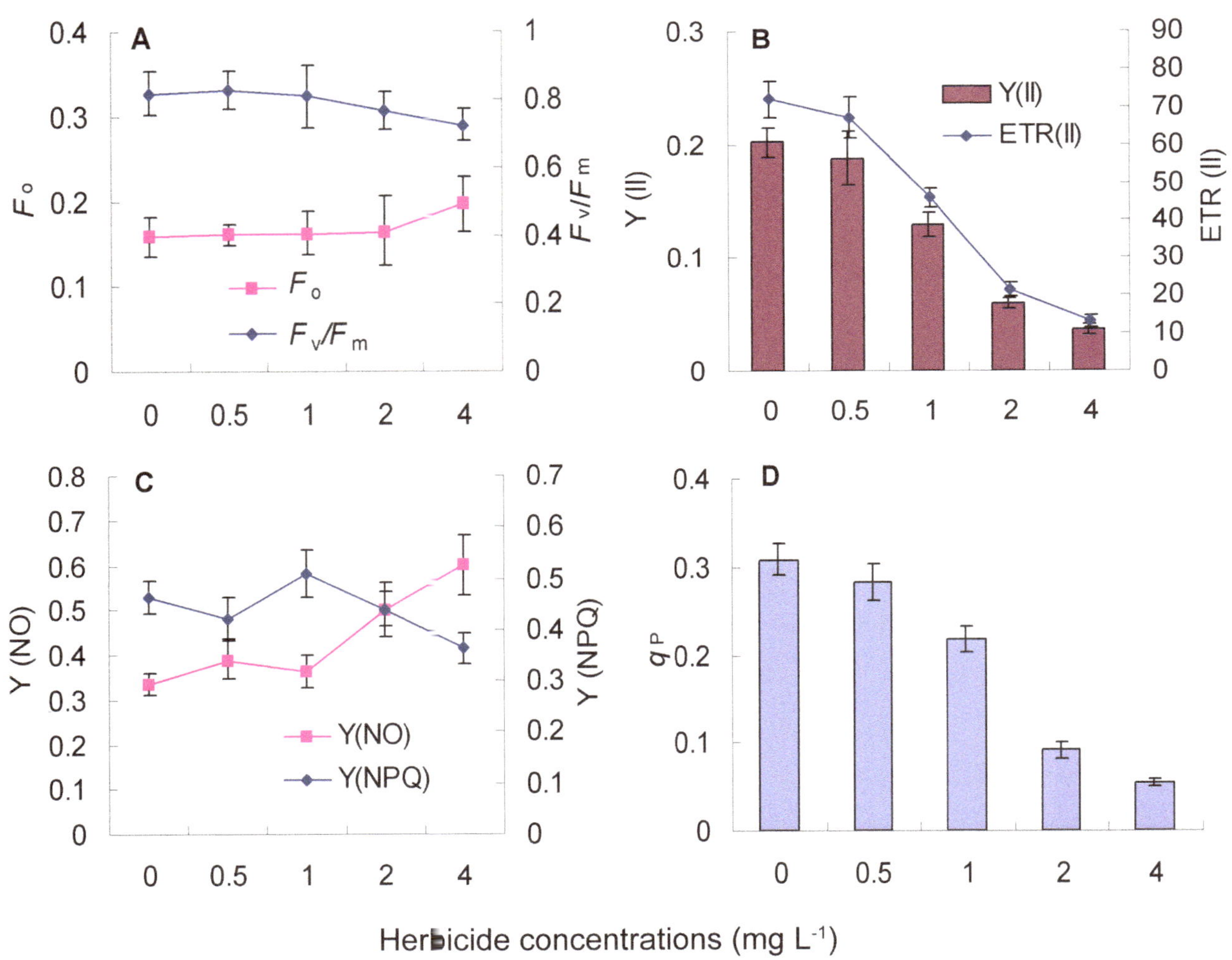

Figure 1. Effect of nicosulfuron on chlorophyll fluorescence parameters in leaves of Radix Isatidis seedlings. Values represent the means and vertical bars indicate the standard deviation of three separate experiments. F_o, F_v/F_m, Y(II), ETR (II), Y(NO), Y(NPQ) and q_P represent initial fluorescence, PS II maximum quantum yield, PS II electron transport rate, PS II effective quantum yield, quantum yield of non-regulated energy dissipation in PS II, quantum yield of regulated energy dissipation in PS II, and photochemical quenching, respectively. The abscissa in the figure represents the concentration of nicosulfuron and the unit is "mg L^{-1}".

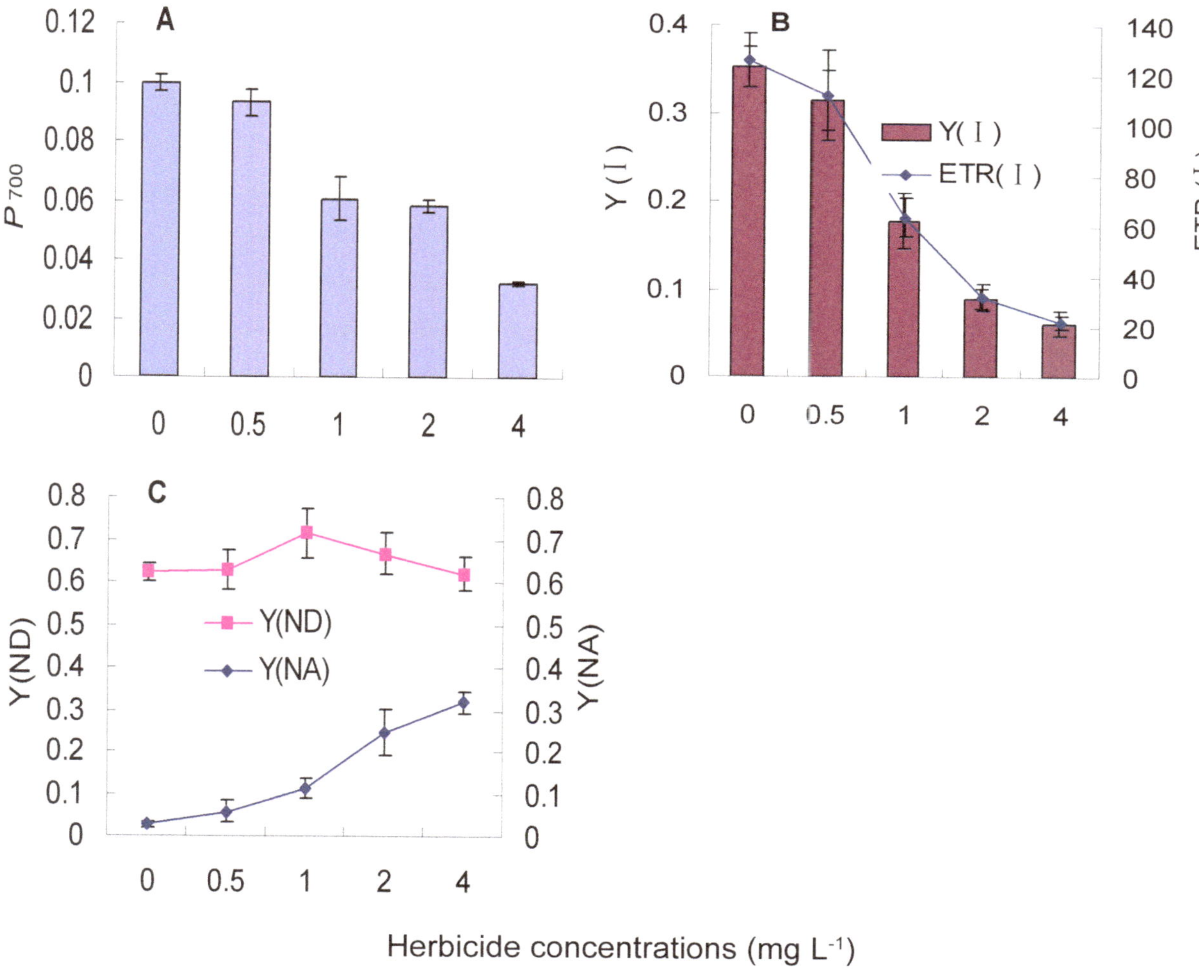

Figure 2. Effect of nicosulfuron on P700 parameters in leaves of Radix Isatidis seedlings. Values represent the means and vertical bars indicate the standard deviation of three separate experiments. P_m, Y(I), ETR (I), Y(ND) and Y(NA) represent maximal P_{700} change, photochemical quantum yield of PS I, PS I electron transport rate, quantum yield of non-photochemical energy dissipation due to donor side limitation in PS I, and quantum yield of non-photochemical energy dissipation due to acceptor side limitation in PS I. The abscissa in the figure represents the concentration of nicosulfuron and the unit is "mg L^{-1}".

and fresh weight significantly. Meanwhile, nicosulfuron was unsafe to Radix Isatidis seedlings being reflected by reduced chlorophyll content and P_n.

Is the decline in P_n resulted by stomatal factors or non-stomatal factors? In our current research, P_n and G_s at 0.5 mg L^{-1} were significant lower than the control, however, C_i and C_i/G_s were a little higher than the control. It may suggest that the decline in P_n is mainly caused by stomatal limitation. P_n, G_s and C_i treated by nicosulfuron ≥2 mg L^{-1} were significantly lower than the control, and L_s and C_i/G_s were significant higher than the control, showing that both stomatal and non-stomatal factors may limit the photosynthesis. Nicosulfuron at 0.5 mg L^{-1} decreased *Chl* a and *Car* content significantly. It may suggest that nicosulfuron destructs the chloroplast structure of Radix Isatidis leaf, reduces the thylakoid stacking level [27], increases the risk of photo-oxidation damage, reduces the light absorption, transmission, distribution between PS II and PSI [28], and affects the synthesis of ATP and NADPH.

Chloroplasts of photosynthetic apparatus, PS II and PS I in thylakoid membranes are the most sensitive parts to environmental changes. Reversible inactivation or damage of PS II reaction center can cause increase in F_o [29]. Paraquat and norflurazon at 100 μg L^{-1} significantly reduced F_v/F_m, Y(II) and q_P in leaves of Lemna minor, but enhanced NPQ markedly [30]. Acetochlor and fluoroglycofen decreased the photochemical efficiency of photosystem II (Y(II)) in the light and increased non-photochemical quenching (NPQ) [17]. Previous studies [19] have shown that Sigma Broad causes damage to PS II complex, block photosynthetic electron transfer, reduce F_v/F_m, Y(II) and q_P significantly, and lead to increase in initial fluorescence, quantum yield of non-regulated energy dissipation in PS II in Radix Isatidis seedlings. Similar to the previous, F_o and Y(NO) increased, whereas F_v/F_m, Y(II), ETR(II), q_P and Y(NPQ) decreased in leaves treated by nicosulfuron in this study. It may suggest that nicosulfuron causes excess excitation energy accumulation in PS II reaction center, the higher reduction state of Q_A, net loss of D_1 protein, reversible deactivation or destruction in PS II reaction centers [31], opening percentage to decrease, harmful effect on photosynthetic oxygen-evolving complex, electron transport efficiency to decline, ATP and NADPH to reduce in Radix Isatidis leaves.

Typical feature of PS I photoinhibition is the decline in maximum oxidation-reduction ability of PS I [32]. In this research, P_m, Y(I) and ETR (I) declined with the increase of nicosulfuron usage, and which at 1 mg L^{-1} were significantly lower than the control. It may suggest that herbicide suppressed the activity of PS I in Radix Isatidis leaves, and electron transfer was blocked at its receptor side. Y(ND) reflects the state of electron donor in PS I, and it is affected by the transmembrane proton gradient and PS II damage degree. Y(NA) reflects the state of electron acceptor in PS I, and it is affected by dark adaptation and CO_2 fixation damage level. Y(NA) >0.5 mg L^{-1} increased significantly, showing that nicosulfuron aggravated the injury of PS II in Radix Isatidis leaves, declined transmembrane proton gradient, blocked dark reaction process, and reduced the fixed amount of CO_2.

In conclusion, recommended usage of nicosulfuron for maize is not safe to Radix Isatidis seedlings. It causes the damage of chloroplast, PS II and PS I structure. Electron transport limitations in PS I receptor side, blocked dark reaction process may be the main cause of the significantly inhibited growth and decreased photosynthetic rate of Radix Isatidis seedlings. Effect of nicosulfuron on the activities of key enzymes in the Calvin cycle of Radix Isatidis seedlings will be researched in the later experiment.

Acknowledgments

We thank Professor Yuguo Wang and Hongfu Wang of Shanxi Agricultural University for suggestions on the manuscript. Thanks are also due to two anonymous reviewers for their thoughtful and valuable comments and suggestions, which helped in improving the manuscript.

Author Contributions

Conceived and designed the experiments: XY PG LZ. Performed the experiments: XY LZ NN MY. Analyzed the data: XY LZ SD. Contributed reagents/materials/analysis tools: BW LF YW MG. Wrote the paper: XY LZ.

References

1. Yuan X, Guo P, Qi X, Ning N, Wang H, et al. (2013) Safety of herbicide Sigma Broad on Radix Isatidis (*Isatis indigotica Fort.*) seedlings and their photosynthetic physiological responses. Pestic Biochem Physiol 10[illegible] 45–50.
2. Lum AF, Chikoye D, Adesiyan SO (2005) Control of *Imperata cylindrica* (L.) Raeuschel (speargrass) with nicosulfuron and its effects on the growth, grain yield and food components of maize. Crop Prot 24 (1): 41–47.
3. Hennigh DS, Al-Khatib K, Tuinstra MR (2010) Response of Acetolactate synthase-resistant grain sorghum to nicosulfuron plus rimsulfuron. Weed Technol 24 (4): 411–415.
4. Zhang L, Hu Y, Wang W, Guo L, Wang Y (2012) Controlling effect of nicosulfuron 6% SC on weeds in cron fields. Weed Sci 30 ([illegible]): 58–60. (in Chinese with English abstract).
5. Dong X, Wang J, Bi J, Liu Y, Zhang X (2007) The sensitivity of different maize varieties to the nicosulfuron. Acta Phytophy Sin 34 (2): [illegible]2–186. (in Chinese with English abstract).
6. Jonathan NN, Williams MM (2008) Acommon genetic basis in sweet corn inbred Cr1 for cross sensitivity to multiple cytochrome P450-metabolized herbicides. Weed Sci 56: 376–382.
7. Môro FV, Damiao Filho CF (1999) Morphological and anatomical alterations in corn leaves submitted to nicosulfuron. Planta Daninha, 1[illegible] (3): 331–337.
8. Wu J, Mathiassen S, Kudsk P (2003) The influence of weed species, leaf stage and rainfastness on the performance of nicosulfuron. Chin J Pestic Sci 5 (1): 77–81. (in Chinese with English abstract).
9. Sullivan JO', Bouw WJ (1998) Sensitivity of processing sweet corn (*Zea mays*) cultivars to nicosulfuron/rimsulfuron. Can J Plant Sci 78 ([illegible]): 151–154.
10. Gallaher K, Mueller TC, Hayes RM, Schwartz O, Barrett M (1999) Absorption, translocation, and metabolism of primisulfuron and nicosulfuron in broadleaf signalgrass (*Brachiaria platyphylla*) and corn. Weed Sci 1: 3–12.
11. Hennigh DS, Al-Khatib K (2010) Response of Barnyardgrass (*Echinochloa crus-galli*), Green Foxtail (*Setaria virdis*), Longspine Sandbur (*Cenchrus longispinus*), and Large Crabgrass (*Digitaria sanguinalis*) to Nicosulfuron and Rimsulfuron. Weed Sci 3: 189–194.
12. Yuan S, Guo F, Guo Q, Zhang Y, Wang X (2003) Demonstration on chemical weed control in *Isatis tinctoria* L field, J Shanxi Agric Sci 3[illegible] 78–80. (in Chinese with English abstract).
13. Li J, Lv P, Zou Y (2011) The chemical control of Indigowoad root field broad leaves, Value Eng 30: 328. (in Chinese with English abstract).
14. Wang X, Rong E, Zhang J, Yuan X, Wen Y, et al. (2012) Effect of quizalofop on protective enzymes and photosynthesis in Radix Isatidis, J Med Plants Res 6: 1770–1776.
15. Zhou Q, Liu W, Zhang Y, Liu K (2007) Action mechanisms of acetolactate synthase-inhibiting herbicides. Pestic Biochem Physiol 89: [illegible]9–96.
16. Kocurek V, Smutny V, Filova J (2009) Chlorophyll fluorescence as an instrument for the assessment of herbicide efficacy. Cereal Res Commun 37: 239–292.
17. Tan W, Li Q, Zhai H (2012) Photosynthesis and growth responses of grapevine to acetochlor and fluoroglycofen. Pestic Biochem Physiol 1[illegible]3: 210–218.
18. Maibrift H, Laurence M, Bruno B, Claude A., Sylvie R, et al. (2006) An easy and rapid method using microscopy to determine herbicide effects in Poaceae weed species. Pest Manag Sci 62: 515–521.
19. Wang Z, Zhou L, Guo W, Zhu X, Li C, et al. (2011) Effects of herbicides on photosynthesis and chlorophyll fluorescence parameters in wheat leaves. J Agro-Environ Sci 30: 1037–1043. (in Chinese with English abstract).
20. Ramanjulu S, Sreenivasulu N, Sudhakar C (1998) Effect of water stress on photosynthesis in two mulberry genotypes with different drought tolerance. Photosynthetica 35: 279–283.
21. Porra RJ, Thompson WA., Kriedmann PE (1989) Determination of accurate extinction coefficients and simultaneous equations for assaying chlorophylls a and b extracted with four different solvents: verification of the concentration of chlorophyll standards by atomic absorption spectroscopy. Biochem Biophys Acta 975: 384–394.
22. Pfündel E, Klughammer C, Schreiber U (2008) Monitoring the effects of reduced PS II antenna size on quantum yields of photosystems I and II using the Dual-PAM-100 measuring system. PAM Application Notes 1: 21–24.
23. Kramer DM, Johnson G, Kiirats O, Edwards GE (2004) New fluorescence parameters for the determination of QA redox state and excitation energy fluxes. Photosynth Res 79: 209–218.
24. Klughammer C, Schreiber U (2008) Saturation Pulse method for assessment of energy conversion in PS I. PAM Application Notes 1: 11–14.
25. Zhao R, Guo P, Yuan X, Wang J, Han M (2010) Effect of paraquat on the antioxidative enzyme activities and lipid peroxidation in opium poppy (*Papaver somniferum* L), J Plant Dis Protect 2: 55–59.
26. Qian H, Lu T, Peng X, Han X, Fu Z, et al. (2011) Enantioselective phytotoxicity of the herbicide Imazethapyr on the response of the antioxidant system and starch metabolism in Arabidopsis thaliana. PLoS ONE 6 (5): e19451. doi:10.1371/journal.pone.0019451.
27. Liu Z, Shi L, Bai L, Zhao K (2007) Effects of salt stress on the contents of chlorophyll and organic solutes in Aeluropus littoralis var. sinensis Debeaux. J Plant Physiol Mol Biol 33: 165–172.
28. Murata N, Takahashi S, Nishiyama Y, Allakhverdiev SI (2007) Photoinhibition of photosystem IIunder environmental stress. Biochim Biophys Acta 1767: 414–421.
29. Demmig B, Winter K, Kruger A, Czygan FC (1987) Photoinhibition and zeaxanthin formation in intact leaves: A possible role of the xanthophyll cycle in the dissipation of excess light energy. J Plant Physiol 84: 218–224.
30. Frankart C, Eullaffroy P, Vernet G (2003) Comparative effects of four herbicides on non-photochemical fluorescence quenching in Lemna minor. Environ Exp Bot 49 (2): 159–168.
31. Schrader SM, Wise RR, Waholtz WF, Ort DR, Sharkey TD (2004) Thylakoid membrane responses to moderately high leaf temperature in pima cotton. Plant Cell Environ 27 (6): 725–735.
32. Scheller HV, Haldrup A (2005) Photoinhibition of photosystem I. Planta 221: 5–8.

De novo Assembly and Characterization of the Barnyardgrass (*Echinochloa crus-galli*) Transcriptome Using Next-Generation Pyrosequencing

Xia Yang, Xin-Yan Yu, Yong-Feng Li*

Institute of Plant Protection, Jiangsu Academy of Agricultural Sciences, Nanjing, China

Abstract

Background: Barnyardgrass (*Echinochloa crus-galli*) is an important weed that is a menace to rice cultivation and production. Rapid evolution of herbicide resistance in this weed makes it one of the most difficult to manage using herbicides. Since genome-wide sequence data for barnyardgrass is limited, we sequenced the transcriptomes of susceptible and resistant barnyardgrass biotypes using the 454 GS-FLX platform.

Results: 454 pyrosequencing generated 371,281 raw reads with an average length of 341.8 bp, which made a total length of 126.89 Mb (SRX160526). *De novo* assembly produced 10,142 contigs (~5.92 Mb) with an average length of 583 bp and 68,940 singletons (~22.13 Mb) with an average length of 321 bp. About 244,653 GO term assignments to the biological process, cellular component and molecular function categories were obtained. A total of 6,092 contigs and singletons with 2,515 enzyme commission numbers were assigned to 151 predicted KEGG metabolic pathways. Digital abundance analysis using Illumina sequencing identified 78,124 transcripts among susceptible, resistant, herbicide-treated susceptible and herbicide-treated resistant barnyardgrass biotypes. From these analyses, eight herbicide target-site gene groups and four non-target-site gene groups were identified in the resistant biotype. These could be potential candidate genes involved in the herbicide resistance of barnyardgrass and could be used for further functional genomics research. C_4 photosynthesis genes including *RbcS*, *RbcL*, *NADP-me* and *MDH* with complete CDS were identified using PCR and RACE technology.

Conclusions: This is the first large-scale transcriptome sequencing of *E. crus-galli* performed using the 454 GS-FLX platform. Potential candidate genes involved in the evolution of herbicide resistance were identified from the assembled sequences. This transcriptome data may serve as a reference for further gene expression and functional genomics studies, and will facilitate the study of herbicide resistance at the molecular level in this species as well as other weeds.

Editor: Nicholas James Provart, University of Toronto, Canada

Funding: This work was supported by National Natural Science Foundation of China (No. 30971926), Special Fund for Agro-scientific Research in the Public Interest (No. 201303031) and Jiangsu Agriculture Science and Technology Innovation Fund (No. CX(12)1003). The funders had no role in study design, data collection and analysis, decision to publish, or preparation of the manuscript.

Competing Interests: The authors have declared that no competing interests exist.

* E-mail: liyongfeng_2010@hotmail.com

Introduction

Barnyardgrass (*Echinochloa crus-galli* [L.] Beauv.) is one of the main problematic grass weeds that grows along with important staple crops such as rice [1]. During cultivation even when the ratio of rice plants to barnyardgrass is 10:1 rice biomass is reduced by 75% and yield by about 50% [2]. Many herbicides are being used to destroy barnyardgrass which in turn would improve rice production. However, persistent use of herbicides results in rapid development of herbicide resistance [3–4]. In the last two decades, it has been reported that *E. crus-galli* worldwide has developed resistance to nine herbicide groups: ALS inhibitors (e.g. penoxsulam, bispyribac-sodium), ACCase inhibitors (e.g. cyhalofop-butyl), synthetic auxin (e.g. quinclorac), photosystem II (e.g. atrazine), ureas and amides (inhibition of photosynthesis at photosystem II, e.g. propanil), dinitroanilines (microtubule assembly inhibition, e.g. pendimethalin), thiocarbamates (inhibition of lipid synthesis, e.g. thiobencarb), chloroacetamides (inhibition of cell division, e.g. butachlor) and isoxazolidinoes (inhibition of carotenoid biosynthesis, e.g. clomazone) [4–8]. The increasing resistance of *E. crus-galli* to herbicides has drastically threatened rice production and alternate weed management strategies should be considered during the cultivation of direct-seeded rice. Therefore, an understanding of the fundamental molecular mechanisms behind development of herbicide resistance is necessary to minimize and manage resistance development, and increase crop yield [3–4].

Recent advances in genomic sequencing technologies are radically changing biological research, and will also have a major impact on crop improvement [9–11]. However, genomics and bioinformatics studies on the rapidly evolving weeds of modern agriculture are limited [12]. Therefore, developing new genomics resources is necessary to study the troublesome weeds in crop fields.

In recent years, the next generation sequencing (NGS) technologies have enabled inexpensive and quick generation of large-scale sequence data when compared to conventional Sanger sequencing [9,13]. The 454 pyrosequencing technology has been used to analyze the transcriptome of grass species such as *Brachypodium distachyon*, waterhemp (*A. tuberculatus*), horseweed (*C. canadenisis*), grain amaranth (*A. hypochondriacus*), 11 composite weeds and switchgrass (*Panicum virgatum*) [14–19]. Besides obtaining the transcriptomic data from these species, some candidate genes potentially involved in herbicide resistance have also been analyzed. The preliminary sequence data for all of the major target-site genes such as 4-hydroxyphenylpyruvate dioxygenase and glutamine synthetase were obtained in *A. tuberculatus* [15]. The functions of ABC transporters involved in non-target glyphosate resistance in *C. canadenisis* were demonstrated by real-time PCR experiments [16]. These data indicate that 454 GS-FLX pyrosequencing is a powerful tool for the development of genomic resources for functional genomics on grasses. They would also potentially benefit future research efforts in weed science.

To increase the genomic resources for weeds, we sequenced the trancriptomes of the susceptible and resistant biotypes of barnyardgrass (*E. crus-galli*) using the 454 GS-FLX platform. The data were assembled *de novo* followed by sequence annotation and clustering into putative functional categories using the Gene Ontology (GO) framework and grouping into pathways using the Kyoto Encyclopedia of Genes and Genomes (KEGG) database. Moreover, transcript abundance analysis was performed using Illumina sequencing and the observed differences between the resistant and susceptible *E. crus-galli* biotypes are presented here. Then, potential candidate sequences involved in the development of herbicide resistance were also analyzed. Finally, C_4 photosynthesis genes including *RbcS*, *RbcL*, *NADP-me* and *MDH* with complete CDS were identified using PCR and RACE technology.

De novo sequence assembly was then performed, which involved comparison of sequences, finding overlapping fragment pairs and merging as many fragments as possible to create a consensus sequence [27]. This assembly resulted in 10,142 contigs (~5.92 Mb) with an average length of 583 bp and 68,940 singletons (~22.13 Mb) with an average length of 321 bp (Table 1). The length distribution of all assembled contigs and singletons is shown in Fig. 3. About 9,432 contigs and 64,272 singletons were >100 bp, 5,103 (50.32%) contigs were longer than 500 bp and 2,935 (28.94%) were longer than 700 bp. The longest contig and singleton was 6,708 and 878 bp respectively. The N50 value for the contigs was 900 bp and represents the contig length of half of the total contigs.

Materials and Methods

We promise that no specific permissions were required for the barnyardgrass species in the described locations in this manuscript, and the field studies did not involve endangered or protected species.

Plant Materials and Growth

Barnyardgrass (*E. crus-galli*) seeds from susceptible and resistant biotypes were collected from the rice field at Lu-jiang county in An-hui province, China. Before the experiments, this resistant biotype was first confirmed to be resistant to quinclorac (inhibitor of synthetic auxin and also inhibitor of cell wall), penoxsulam (ALS inhibitor) and bispyribac-sodium (ALS inhibitor) group of herbicides (Fig. 1). For this, the seeds were sterilized in 3% NaClO for 10 min, washed three times with tap water and allowed to germinate at 30°C. These germinating seeds were then planted in plastic pots, filled with a 2:1:1 mixture of soil: peat: sand in a climate chamber at 28°C/25°C (day/night) with a 16 h photoperiod. Three weeks later, leaves from the susceptible and resistant biotype seedlings were treated with 1.143 mg/ml quinclorac (BASF AG, German), 4.114 mg/ml penoxsulam (Dow AgroSciences, USA) and 3.429 mg/ml bispyribac-sodium (Jiangsu Institute of Ecomones Co., LTD, China). After 24 h of spraying, both treated and untreated seedlings were harvested and immediately frozen in liquid nitrogen, and stored at −80°C until RNA extraction.

RNA Isolation and cDNA Library Construction

Total RNA from the seedlings was extracted using Trizol reagent (Invitrogen, USA) according to the manufacturer's protocol. RNA quantity and quality were analyzed using gel

Figure 1. The growth of resistant (R) and susceptible (S) barnyardgrass biotypes at various concentrations of the herbicides quinclorac (A), bispyribac-sodium (B) and penoxsulam (C). The dose of each herbicide is in µmol.

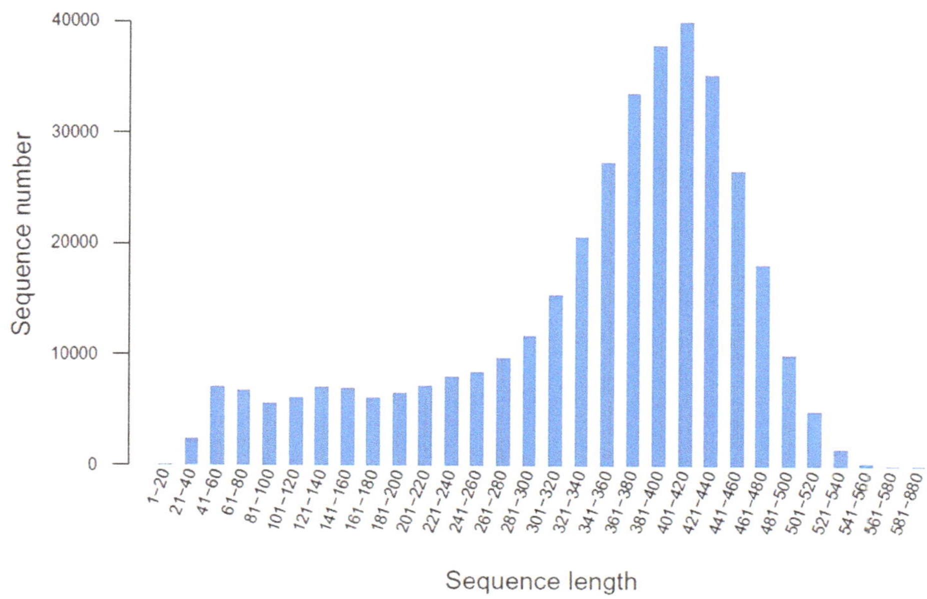

Figure 2. Distribution of the raw read sequence lengths from the barnyardgrass 454 transcriptome.

electrophoresis and Nanodrop 2000 spectrophotometer (Nanodrop, USA). cDNA library construction was performed at Shanghai Majorbio Bio-pharm Biotechnology Co., Ltd. (Shanghai, China). Briefly, Oligo-dT beads were used to isolate poly(A_+) mRNA from a total RNA pool consisting of equal quantities of total RNA from three sample types: untreated susceptible barnyardgrass, untreated resistant barnyardgrass and herbicide-treated resistant barnyardgrass. Then, mRNA was converted into cDNA using the SMART cDNA synthesis kit (Clontech, USA) with a modified 3′ primer containing a *Bsg*I site (5′-ATTCTAGAGGCCGAGGCGTGCAG d(T)18 VN-3′). The newly synthesized cDNA were treated with *Bsg*I enzyme to remove the poly (A) tails, fragmented, blunt ended and ligated with the customized 454 adaptors (A and B) to form the single-stranded cDNA library for emulsion PCR amplification and Roche GS-FLX 454 sequencing.

454 Sequencing, Assembly and Annotation

The cDNA pool representing untreated susceptible barnyardgrass, untreated resistant barnyardgrass and herbicide-treated resistant barnyardgrass was used for high throughput sequencing using the 454 GS-FLX platform (Roche, USA). A half-plate run was performed at the Shanghai Majorbio Bio-pharm Biotechnology Co., Ltd. (Shanghai, China) following the manufacturer's protocol. GS *de novo* assembler (Newbler v2.3, Roche) was used for 454 sequencing assembly. For similarity searches, all assembled contigs and singletons were compared with the proteins in the NCBI non-redundant (nr) protein database using BLASTx [20]. The BLAST E-value threshold was set at 1×10^{-5}, with a minimum alignment length of 60 bp. To determine gene functions of the sequences, these results were imported to Blast2GO, a software that retrieves gene ontology (GO) terms [21]. GO terms were determined with respect to Cellular Component, Biological Process and Molecular Function. Kyoto Encyclopaedia of Genes and Genomes (KEGG) pathways were assigned to the assembled sequences and Enzyme Commission (EC) numbers were used to automatically color and retrieve KEGG pathway maps [22]. The assembled sequences were mapped to the KEGG biochemical pathways according to the enzyme distribution in the pathway database. InterPro annotations were performed using InterProScan (http://www.ebi.ac.uk/Tools/InterProScan/).

Illumina Sequencing and Digital Expression Analysis

Oligo-dT beads were used to yield poly(A_+) mRNA from a total RNA pool consisting of equal quantities of total RNA from four sample types: untreated susceptible barnyardgrass, herbicide-treated susceptible barnyardgrass, untreated resistant barnyard-

Table 1. Summary of barnyardgrass 454 transcriptome sequencing, assembly and annotation.

	Reads (n)	Bases (Mb)	Average length (bp)
454 raw reads	371,281	126.89	341
Assembled contigs	10,142	5.92	583
Singletons	68,940	22.13	321
Annotated sequences	45,438	18.92	416
Sequences assigned with GO terms	40,399	17.03	421
Sequences assigned with EC numbers	12,846	5.72	445

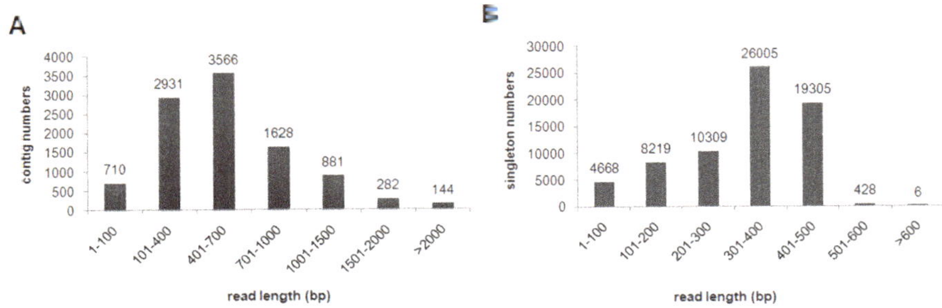

Figure 3. Sequence length distribution of assembled contigs (A) and singletons (B) from the barnyardgrass 454 transcriptome.

grass and herbicide-treated resistant barnyardgrass. Then, mRNA bound to the beads was converted into first-strand cDNA using oligo-dT and second-strand cDNA using random primers. Paired-ended (PE) library with approximate 200 bp fragment was synthesized using the Genomic DNA Sample Prep Kit (Illumina, USA) according to the manufacuter's instructions. PE library products were sequenced on the Illumina Genome Analyzer II platform equipped with a paired-end module according to the manufacturer's instruction. Then the PE reads were assembled using ABySS assembler tool [23]. The sequences were aligned to the assembled contigs and singletons using the bowtie software [24]. The expressions of each reads between sample pairs (untreated susceptible barnyardgrass VS herbicide-treated susceptible barnyardgrass, untreated resistant barnyardgrass VS herbicide-treated resistant barnyardgrass) were calculated using the numbers of reads with a specific match. Among the four samples, a minimum of a two-fold difference in log2 expression were considered as expression differences. Genes with different expression were identified using R package DEGseq with MARS (MA-plot-based method with random sampling Model) [25].

Quantitative Real-time PCR Analysis

Quantitative real-time PCR analysis of two candidate genes, *ALS* and *ACCase*, was performed using gene specific primers. *ALS* primers were 5′-GCCTTCCAGGAGACGCCAATCG-3′ and 5′-GGTCATCTCCACAAAGCGGCACA-3′. *ACCase* primers were 5′-GCTGCGTGGAGGAGCTTGGG-3′ and 5′-CACCTTTAGCTGCCATTCTGAGGGA-3′. The *β-actin* gene

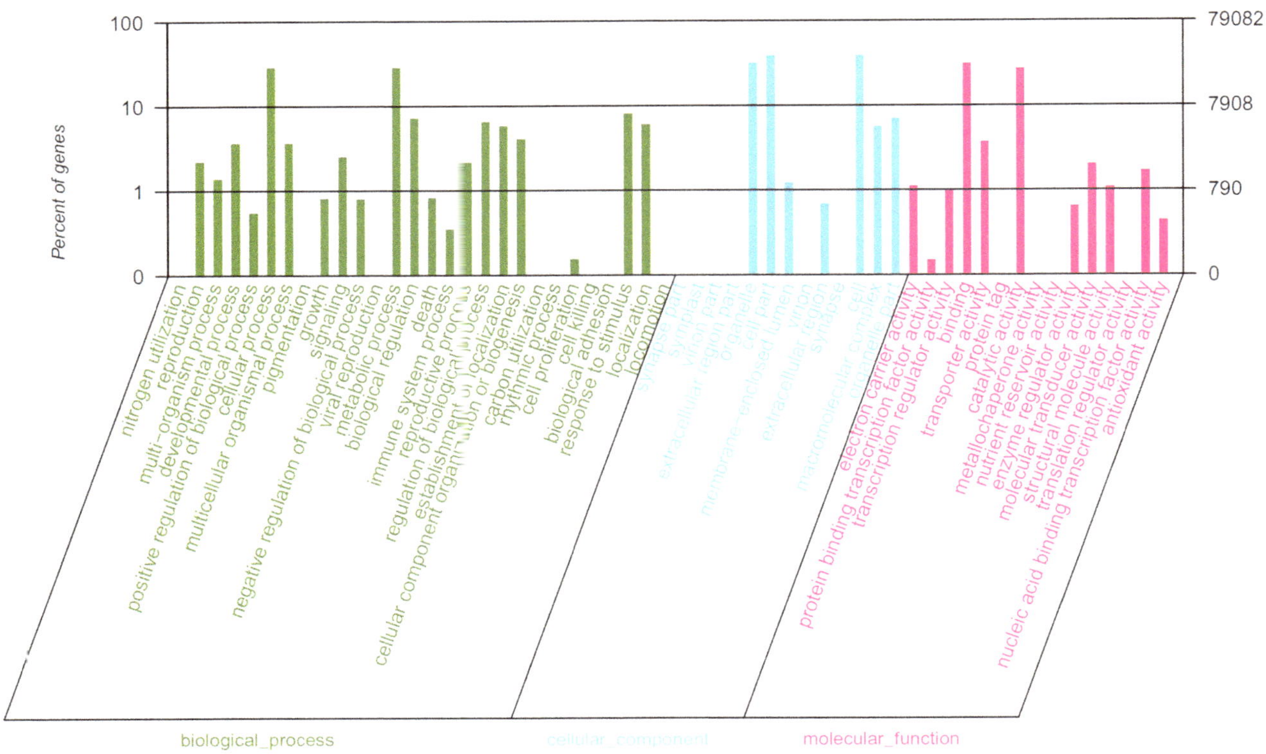

Figure 4. Functional annotation of assembled sequences based on gene ontology (GO) terms. GO analysis was performed at level 2 for the three main categories (biological process, cellular component and molecular function).

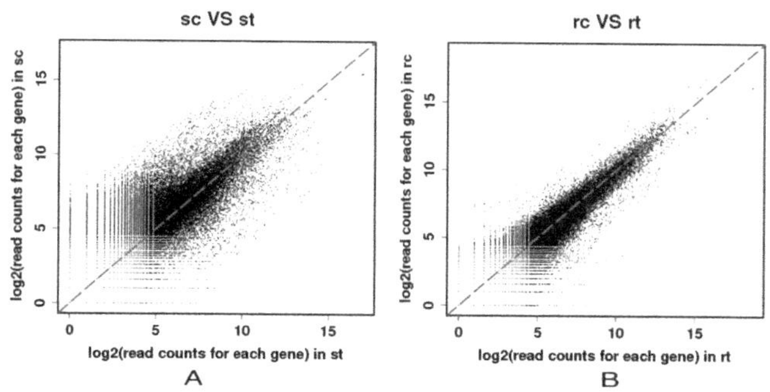

Figure 5. Scatter plot analysis of two sample pairs (sc VS st, rc VS rt) from barnyardgrass. sc: untreated susceptible biotype; st: herbicide-treated susceptible biotype; rc: untreated resistant biotype; rt: herbicide-treated resistant biotype. Log2 transformed read counts of the mapped contigs and singletons were used for comparison.

was used as an internal control. The primers used to amplify *β-actin* were 5′-CACACTGGTGTTATGGTAGG-3′ and 5′-AGAAGGTGTGATGCCAAAT-3′. Real-time PCR was performed using SYBR® *Premix Ex Taq*™ kit (TaKaRa, Japan) on a ABI-7500 Fast Real-Time PCR System (ABI, USA). About 2μl cDNA from each of the four samples (untreated and herbicide-treated susceptible barnyardgrass, untreated and herbicide-treated resistant barnyardgrass) was mixed with 0.2 μM primers, 1×ROX Reference Dye II and SYBR® *Premix Ex Taq*™ (including TaKaRa Ex Taq HS, dNTP mixture, Mg^{2+} and SYBR Green I) in a final volume of 20 μl. The amplification program was as follows: 30 s at 95°C for denaturation followed by 40 cycles of 3 s at 95°C and 30 s at 60°C. Each experiment had three independent repetitions. Values were presented as mean ± SD of three independent RT-PCR analysis. Error bars were calculated based on values from three replicates. Relative expression of the target genes to the *β-actin* control was calculated using the $2^{-\Delta\Delta CT}$ method [26].

Results and Discussion

454 Sequencing and *de novo* Assembly

We performed 454 pyrosequencing of the barnyardgrass (*E. crus-galli*) cDNA to identify target genes for functional genome studies. Our choice to sequence the transcriptome was based on the difficulty in sequencing the polyploid barnyardgrass genome by whole genome sequencing. A pooled cDNA library from three samples (untreated susceptible biotype, untreated resistant biotype and herbicide-treated resistant biotype) was run on a half-plate using the 454 GS-FLX Titanium Chemistry to obtain sequences of low- and high- abundance transcripts. This biotype-treatment combo could produce the transcripts related to herbicide resistance, herbicide-treated susceptible biotype was therefore absent in this 454 pyrosequencing,. This generated 371,281 raw reads with an average length of 341.8 bp and a total length of 126.89 Mb (Table 1). Sequence length distribution of raw reads is shown in Fig. 2.

Sequence Annotation

In order to identify new *E. crus-galli* specific sequences, all 79,082 assembled sequences (10,142 contigs and 68,940 singletons) were used as queries to search the NCBI nr protein database using BLASTx followed by analysis using Blast2GO. In total 45,438 assembled sequences including 7,407 (73.03%) contigs and 38,031 (55.17%) singletons were annotated. The remaining 44.83% singletons and 26.97% contigs were not annotated presumably due to their short average lengths. Average length of the annotated contigs was 690 bp, while that of non-annotated contigs was only 260 bp. The average length of annotated singletons was 360 bp, while that of non-annotated singletons was 272 bp.

In these annotations, only 24 sequences were hits to proteins from the *Echinochloa* species while 23,657 (52.06%) were hits to proteins from *Sorghum bicolor*, 9,584 (21.09%) to *Zea mays*; 7,718 (16.99%) to *Oryza sativa* and 1,764 (3.88%) to *Hordeum vulgare*. Such low numbers of proteins from barnyardgrass is most likely due to the absence of a well annotated reference genome for the species. Almost half of the sequences (42.54%) assembled in this study could not be annotated by searching the NCBI nr protein database. Absence of sufficient sequences from species phylogenetically close to barnyardgrass in the public databases could have also contributed to the high percentage of non-annotated sequences. These non-annotated sequences could be resourceful for pursuing taxon-guided searches of specific genes of interest [15].

GO terms were assigned to the annotated EST sequences using Blast2GO. One or more GO terms were assigned to 40,399

Table 2. The top 15 genes with highest expression levels in the transcriptome data.

No.	Sequence ID	Annotation	Digital expression			
			SC	ST	RC	RT
1	contig04556	asparagines synthetase 2	8,433	6,834	108,302	50,444
2	contig04555	asparagines synthetase 2	5,984	4,222	74,988	39,573
3	contig06106	glycine-rich RNA-binding protein 2	7,229	4,267	20,464	18,167
4	contig07670	metallothionein-like protein	9,599	498	10,924	16,650
5	contig03094	light-induced protein 1-like	15,289	4,400	14,736	14,398
6	contig04500	sucrose synthase 1	7,879	13,895	8,938	12,881
7	contig00684	pyruvate orthophosphate dikinase	22,739	7,861	5,464	12,814
8	contig00581	elongation factor	15,557	6,714	11,676	12,365
9	contig04470	acetyl-coenzyme A carboxylase	16,114	2,472	12,989	12,088
10	contig05353	light-induced protein 1-like	11,779	3,203	13,083	11,876

Table 3. Comparison of the herbicide target gene families and non-target gene families from transcriptome of barnyardgrass, waterhemp, horseweed and grain amaranth

	Barnyardgrass	Waterhemp	Horseweed	Grain amaranth
Herbicide target gene family				
tubulin	25	33	29	34
glutamine synthetase	16	7	9	11
acetyl-CoA carboxylase	8	8	6	13
phytoene desaturase	7	1	5	2
1-deoxy-D-xylulose-5-phosphate synthase	3	1	6	3
5-enolpyruvylshikimate-3-phosphate synthase	2	3	2	1
acetolactate synthase	2	2	6	5
d hydropteroate synthase	1	2	1	1
4-hydroxyphenylpyruvate dioxygenase	0	2	1	3
protoporphyrinogen oxidase	0	8	2	5
D1 protein (plastidic gene)	0	2	4	2
Non-target gene family				
cytochrome P450 monooxygenase	36	191	125	-
glycosyltransferase	31	84	76	-
ABC transporter genes	64	192	151	-
glutathione S-transferase	19	22	7	-

(51.09%) sequences with 244,653 GO assignments in total for the biological process, cellular component and molecular function categories. These three groups were not mutually exclusive because many contigs were assigned to more than one type of GO term. The largest proportion was represented by the cellular process (GO: 0009987; 25.16%) and metabolic process (GO: 0008152; 25.05%) under biological process, cell (GO: 0005623; 31.40%), cell part (GO: 0044464; 31.40% and organelle (GO:0043226; 25.40%) under cellular component, and binding (GO: 0005488; 44.36%) and catalytic activity (GO: 0003824; 38.66%) under molecular function (Fig. 4). Representation of 56 GO terms in the transcriptome provided an indication of the diversity of genes expressed in *E. crus-galli*. Similar results were reported for the biological process category in *Conyza canadensis*, while the predominant subgroups in molecular function (enzyme

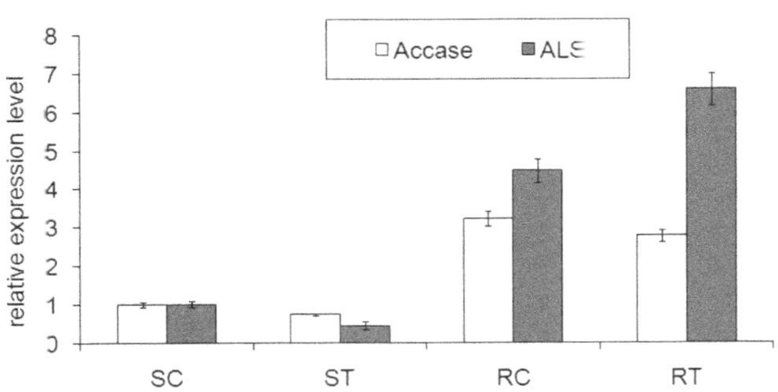

Figure 6. Real-time PCR analysis of *ALS* and *ACCase* genes in the seedlings of barnyardgrass. ALS: Acetolactate synthase; ACCase: Acetyl-CoA carboxylase; SC: untreated susceptible biotype; ST: herbicide-treated susceptible biotype; RC: untreated resistant biotype; RT: herbicide-treated resistant biotype. The *β-actin* gene was used as an internal control. Values represent mean ± SD. Error bars were calculated based on values from three replicates.

activity and transferase activity) and cellular component (intracellular components and cytoplasmic components) are different from our data [16]. To our interests were the GO terms related to herbicidal mechanisms: response to auxin stimulus (152), photosystem II (114), cellulose biosynthetic process (106), lipid catabolic process synthesis (95), acetyl-CoA biosynthetic process (65), carotenoid biosynthetic process (37) and acetolactate synthase activity (6). Quinclorac used in our study is an inhibitor of synthetic auxin (acting like indolylacetic acid) and also an inhibitor of cell wall (cellulose) biosynthesis [4]. About 51 sequences under auxin response factor and 28 sequences under cellulose synthase were obtained. These sequences could be further investigated to determine their specific function in the evolution of herbicide resistance in *E. crus-galli*.

Functional Classification by KEGG

In addition, KEGG pathways were assigned to all 79,082 assembled sequences by comparing to the KEGG database using BLASTx with an E-value cutoff of $<1\times10^{-5}$. A total of 6,092 contigs and singletons with 2,515 enzyme commission (EC) numbers were assigned to 151 predicted KEGG metabolic pathways and the numbers of barnyardgrass sequences in the different pathways ranged from 1 to 4,424. This analysis showed that 4,424 sequences were present in the major metabolic pathways and 1,865 in the biosynthesis of secondary metabolites. The sequences represented in the metabolic pathways of major bio-molecules included purine, starch and sucrose, terpenoids and steroids, methane, pyrimidine, phenylalanine, amino sugar and nucleotide sugar, glycerophospholipid, glycerolipid, etc. About 894 sequences with 109 EC numbers were mapped to the biosynthesis of plant hormones pathway. In the metabolism pathway, we found seven sequences assigned to acetolactate synthase (ALS, EC 2.2.1.6), which is the target for herbicides, penoxsulam and bispyribac-sodium. When these sequences were compared in the NCBI nr database using BLASTx, only two (contig05544 and

Table 4. Sequence ID, annotation, hit size and digital expression profile of tubulin genes in barnyardgrass (*E. crus-galli*).

No.	Sequence ID	Annotation	Hit size (bp)	Digital expression			
				SC	ST	RC	RT
1	contig02007	α-tubulin	447	4,561	429	5,173	5,029
2	contig02008	α-tubulin	935	10,817	1,185	8,054	7,396
3	contig02010	α-tubulin	935	9,359	1,043	7,354	7,069
4	contig02031	α-tubulin	911	6,782	890	5,483	6,575
5	contig02033	α-tubulin	309	1,276	190	1,147	1,266
6	contig02034	α-tubulin	330	1,252	221	1,253	1,502
7	contig04855	α-tubulin	1,461	4,939	1,449	2,076	3,832
8	contig04932	α-tubulin	1,364	3,400	1,509	5,048	3,828
9	G81ESPA01D3IDY	α-tubulin	439	1,203	209	921	1,141
10	G81ESPA01BB9ZQ	α-tubulin	364	1,304	167	839	814
11	G81ESPA01AXQ2P	α-tubulin	129	1,489	232	1,288	1,270
12	G81ESPA01BU5X0	α-tubulin	305	1,084	191	888	1,097
13	G81ESPA01DWSAD	α-tubulin	387	1,206	224	517	560
14	G81ESPA01D0ZJ4	α-tubulin	461	1,609	228	1,042	1,034
15	G81ESPA01D6KVH	α-tubulin	366	258	217	358	275
16	G81ESPA01AG82W	α-tubulin	181	1,350	220	1,181	1,138
17	G81ESPA01BDRNJ	α-tubulin	349	1	2	6	3
18	contig00720	β-tubulin	415	508	185	849	491
19	contig00721	β-tubulin	603	2,198	270	1,597	1,517
20	contig00723	β-tubulin	681	1,514	123	837	1,126
21	contig00724	β-tubulin	939	2,136	591	2,308	1,695
22	contig00725	β- tubulin	789	1,608	603	1,347	1,127
23	G81ESPA01CF9IL	β- tubulin	291	112	18	42	53
24	G81ESPA01BZVDB	β- tubulin	283	99	14	31	42
25	G81ESPA01BQRNR	β- tubulin	161	67	12	23	27

SC: untreated susceptible biotype; ST: herbicide-treated susceptible biotype; RC: untreated resistant biotype; RT: herbicide-treated resistant biotype.

G81ESPA01CWAJO) sequences were found to be highly similar (99% and 99%) to ALS genes from *Echinochloa phyllopogon* [28], while the remaining five (contig03089, contig06883, contig08547, G81ESPA01BLGZC and G81ESPA01ECFXF) were annotated to hypothetical proteins of unknown function.

Illumina Sequencing and Digital Expression Analysis

To identify large scale differences in gene expression of susceptible and resistant barnyardgrass, quantitative transcriptome analysis using Illumina Genome Analyzer II was performed. It produced quantitative expression scores, which should be accurate and highly consistent with other quantitative methods such as quantitative RT-PCR [29]. A mixed cDNA pool from four samples (untreated and herbicide-treated susceptible barnyardgrass, untreated and herbicide-treated resistant barnyardgrass) was sequenced using Illumina. This yielded 2.8 GB data with 14,769,229, 14,610,811, 12,341,146 and 15,099,517 assembled paired-ended reads respectively. These transcripts were mapped to the assembled sequences from 454 pyrosequencing and those that were mapped successfully were further used to determine and

Table 5. Sequence ID, annotation, hit size and digital expression profile of acetyl-CoA carboxylase.

No.	Sequence ID	Hit size (bp)	Digital expression			
			SC	ST	RC	RT
1	contig04470	6,708	16,114	2,472	12,989	12,088
2	contig04540	2,225	2,903	1,495	1,379	1,515
3	G81ESPA01C7K8F	295	74	28	59	88
4	G81ESPA01EALO0	355	103	37	78	113
5	G81ESPA01DANWF	346	410	113	283	276
6	G81ESPA01D05V2	221	135	22	145	121
7	G81ESPA01CVZK6	301	350	50	263	242
8	G81ESPA01CW4DK	438	479	78	513	478

compare the expression profiles of the transcripts in the four samples. Digital expression analysis identified a total of 78,124 sequences from 454 pyrosequencing including 9,737 contigs and 68,387 singletons showing significant abundance differences between sample pairs (untreated susceptible barnyardgrass VS herbicide-treated susceptible barnyardgrass, untreated resistant barnyardgrass VS herbicide-treated resistant barnyardgrass) (Fig. 5). The expression of many non-annotated genes with unknown function was revealed in this digital gene expression analysis. Thus this transcriptome could be a potentially rich source of genetic material for the systematic discovery of genes involved in the mechanisms of herbicide resistance [17].

In the digital gene expression analysis, we found that asparagines synthetase 2 had the highest expression in the herbicide-treated resistant barnyardgrass. The top 10 annotated sequences with highest expression level in the barnyardgrass transcriptome are listed in Table 2. Except sucrose synthase, the other nine genes had higher expression in herbicide-treated resistant barnyardgrass than in herbicide-treated susceptible barnyardgrass biotype. Metallothionein-like protein, elongation factor, acetyl-coenzyme A carboxylase, light-induced protein 1 and pyruvate orthophosphate dikinase had higher expression levels in the herbicide-treated resistant barnyardgrass than the untreated resistant barnyardgrass biotype. In addition, the expression levels of these genes were lower in the herbicide-treated susceptible barnyardgrass and lowest in the untreated susceptible barnyardgrass biotype. Functional analysis of these genes will be conducted in future to determine the relationship between these genes and herbicide resistance.

Table 6. Sequence ID, annotation, hit size and digital expression profile of acetolactate synthase.

No.	Sequence ID	Annotation	Hit size (bp)	Digital expression			
				SC	ST	RC	RT
1	contig05544	acetolactate synthase	991	482	143	557	385
2	G81ESPA01CWAJ0	acetolactate synthase	409	233	28	271	235

Table 7. A list of *Echinochloa* genes with homology to barnyardgrass sequences.

Echinochloa	Gene	NCBI accession number
E. crus-galli	ALS	JQ319[illegible]6, JX415268~JX415271 [34]
	ACCase	AY763[illegible]5, AH014492, AJ966453, AJ966423, AY762963, HQ395758, HQ395759
	NADP-me	FJ6033[illegible]5, FJ644948
	PPDK	DQ083[illegible]57
	Rps3	FJ7662[illegible]8, FJ766300
	matK	HM850[illegible]73, HQ593274, HQ599953, HQ599954, HQ599962
	PEPC	AY251[illegible]32, AY995212
	rbcL	HQ590[illegible]71
E. phyllopogon	ALS	AB636[illegible]0, AB636581 [28]
	ACCase	AB636[illegible]2~AB636588 [28]
E. frumentacea	PPDK	AY792[illegible]9, AB289641 [35]
E. colona	matK	HM850[illegible]72 [36]
E. esculenta	matK	AF164[illegible]2 [37]
	PEPC	AM690[illegible]15 [38]
	rcbL	EF1251[illegible]8 [39]
E. muricata	matK	HQ593[illegible]5, HQ596677
	rcbL	HQ590[illegible]2
E. oryzicola	Rps3	FJ7662[illegible]

Comparison of the Sequences Involved in Herbicide Resistance

The candidate genes involved in herbicide resistance were obtained by comparing our transcriptome (454) with the trascriptomes of waterhemp [15], horseweed [16] and grain amaranth [17]. The comparison of eleven herbicide target gene families and four non-target gene families is presented in Table 3. More glutamine synthetase (16) and phytoene desaturase (7) transcripts from the herbicide target gene families were identified in barnyardgrass than in the other three species. Notable absence in the barnyardgrass transcriptome were 4-hydroxyphenylpyruvate dioxygenase, protophenylpyruvate dioxygenase and D1 protein. The reason of such absence could be due to similar expression patterns between the susceptible and resistant biotypes. From our analysis we could conclude that 215 assembled sequences could be involved in the evolution of herbicide resistance. Since there is no reference genome to assemble these short reads into scaffolds and these 215 sequences could be redundant, the actual number of genes could be less than 215. In future work, we will focus on the functional analysis of these candidate genes.

Microtubules are polymers of α-tubulin and β-tubulin dimers and play an important role in many essential cellular processes including mitosis, cytokinesis and vesicular transport [3]. The digital expression of 24 tubulin genes including α-tubulin and β-tubulin was much higher in the herbicide-treated resistant biotype (RT) than the herbicide-treated susceptible biotype (ST) of *E. crus-galli*, except one α-tubulin gene (G81ESPA01BDENJ) and three β-tubulin genes (G81ESPA01CF9IL, G81ESPA01BZVDB and G81ESPA01BQRNR) with lower expression Table 4). After treatment with herbicides, the digital expression levels of tubulin genes in the susceptible biotype (ST) declined when compared with untreated susceptible biotype (SC). However, the digital expressions of tubulin genes in the resistant biotype remained the same or declined below the susceptible biotype. The microtubule growth was more disrupted in the susceptible biotype than in the resistant biotype after the herbicide treatment. Tubulin genes play an important role in herbicide induced growth inhibition. For instance, in goosegrass (*Eleusine indica*), nucleotide differences were observed in the three α-tubulin genes between resistant (R) and susceptible (S) biotypes, although missense mutations were not present in any of the four β-tubulin genes in the two R-biotype lines [30,31]. This indicated the possible association between these missense mutations in the α-tubulin genes and herbicide resistance in goosegrass. Similar results were found between the resistant and sensitive biotypes of green foxtail (*Setaria viridis* L. Beauv.) [32]. Therefore, the isolation and functional analysis of the full-length α-tubulin and β-tubulin genes from barnyardgrass is necessary to understand the mechanisms of herbicide resistance. Acetyl-CoA carboxylase (ACCase, EC 6.4.1.2) is a key enzyme in lipid biosynthesis that catalyzes the formation of malonyl-CoA from acetyl-CoA [3]. Similar to tubulin genes, the digital expression of eight ACCase genes also showed higher expression in the resistant biotype than in the susceptible biotype after herbicide treatment (Table 5). Two ACCase belonging to *E. crus-galli* biotype Geqiushan-S (HQ395758) and Geqiushan-R (HQ395759) were found in NCBI. These genes shared 96% identity with contig04470 and no significant similarity with contig04540. Further studies on the ACCase genes in barnyardgrass are needed to compare the differences among these genes, and determine their role in herbicide resistance. Penoxsulam and bispyribac-sodium herbicides, which were used in our study, were ALS-inhibiting herbicides [33]. In the barnyardgrass transcriptome and digital expression analysis, higher levels of ALS sequences (contig05544 and G81ESPA01CWAJO) were present in the resistant biotype than in the susceptible biotype after treatment with herbicides (Table 6). Based on real-time PCR experiments, the relative expression of *ACCase* and *ALS* genes were analyzed in each untreated and herbicide-treated biotype. Compared to herbicide treated susceptible plants, *ALS* and *ACCase* were highly expressed in herbicide treated resistant plants. The expressions of *ALS* and *ACCase* genes decreased when the susceptible biotype was treated with herbicides. However, the expression of *ACCase* declined slightly and *ALS* increased slightly in the herbicide-treated resistant biotype (Fig. 6). This trend was consistent with the expression profiles of ALS and ACCase sequences in the transcriptome. These genes are therefore regarded as good preliminary targets for further functional studies of herbicide resistance.

Comparison of the Genes in *Echinochloa*

About 26 assembled sequences had best hits to *Echinochloa* genes in the NCBI nr database and included *ALS* [34], *ACCase* [28], NADP-dependent malic enzyme (*NADP-me*), pyruvate orthophosphate dikinase (*PPDK*) [35], *Rps3*, maturase K (*matK*) [36,37], phosphoenolpyruvate carboxylase (*PEPC*) [38] and ribulose-1,5-bisphosphate carboxylase/oxygenase large subunit (*rbcL*) [39]. These seven genes were obtained from different varieties of *Echinochloa* including *E. phyllopogon*, *E. frumentacea*, *E. colona*, *E. esculenta*, *E. muricata* and *E. oryzicola*, in addition to *E. crus-galli* (Table 7). According to our barnyardgrass transcriptome data, C_4 photosynthesis genes with complete coding sequence (CDS) including ribulose-1,5-bisphosphate carboxylase/oxygenase small subunit (*rbcS*) (KC478601), *rbcL* (KC478602), *NADP-me* (KC478603) and malate dehydrogenase (*MDH*) (JX683695) were identified from *E. crus-galli* using PCR and RACE technology. These genes in general will be benefit future genomic and phylogenetic studies of grasses. Functional analysis of these genes is necessary to understand the evolution of C_4 photosynthesis during angiosperm diversification and phylogenetic patterns in grasses [38,40].

Conclusions

The susceptible and resistant biotypes of *E. crus-galli*, with and without treatment with three herbicides, were used to generate the first large-scale transcriptome sequencing data using the 454 GS-FLX platform. *De novo* assembly and functional annotation of the assembled contigs and singletons were performed using Newbler. Interestingly, eight herbicide target-site gene groups and four non-target-site gene groups were identified in the resistant biotype of *E. crus-galli* that would enable isolation of the full-length genes, which could be involved in herbicide resistance. These genes could also be used for further functional genomics research. Compared to the previously *Echinochloa* reported genes, four additional genes including *RbcS*, *RbcL*, *NADP-me* and *MDH* with complete CDS were identified using PCR and RACE technology. This study provides a valuable genetic resource for grasses in general and should be useful for weed control which should result in an improvement in agriculture.

Author Contributions

Conceived and designed the experiments: YFL. Performed the experiments: XY XYY. Analyzed the data: XY. Contributed reagents/materials/analysis tools: XYY YFL. Wrote the paper: XY.

References

1. Chauhan BS, Abugho SB (2012) Effect of growth stage on the efficacy of postemergence herbicides on four weed species of direct-seeded rice. ScientificWorldJournal doi:10.1100/2012/123071.
2. He HB, Wang HB, Fang CX, Lin ZH, Yu ZM et al. (2012) Separation of allelopathy from resource competition using rice/barnyardgrass mixed-cultures. PLos One 7: e37201.
3. Powles SB, Yu Q (2010) Evolution in action: plant resistant to herbicides. Annu Rev Plant Biol 61: 317–347.
4. Heap I [April 29, 2013] International Survey of Herbicide Resistant Weeds [Online]. Available: http://www.weedscience.org.
5. Malik MS, Burgos NR, Talbert RE (2010) Confirmation and control of propanil-resistant and quinclorac-resistant barnyardgrass (*Echinochloa crus-galli*) in rice. Weed Tech 24: 226–233.
6. Yu X, Ge L, Liu L, Li Y (2010) Resistance of barnyardgrass (*Echinochloa crus-galli*) to multiple herbicides in direct-seeded rice. J Agri Sci 26: 1438–1440.
7. Rahman MM, Bin SI, Juraimi AS (2010) Study on resistant biotypes of *Echinochloa crus-galli* in Malaysia. Aust J Crop Sci 4: 107–115.
8. Juliano LM, Casimero MC, Llewellyn R (2010) Multiple herbicide resistance in barnyardgrass (*Echinochloa crus-galli*) in direct-seeded rice in the Philippines. Int J Pest Manage 56: 299–307.
9. Varshney RK, Nayak SN, May GD, Jackson SA (2009) Next-generation sequencing technologies and their implications for crop genetics and breeding. Trends Biotechnol 27: 522–530.
10. Mochida K, Shinozaki K (2010) Genomics and bioinformatics resources for crop improvement. Plant Cell Physiol 51: 497–523.
11. Edwards D, Batley J (2010) Plant genome sequencing: applications for crop improvement. Plant Biotechnol J 8: 2–9.
12. Duke SO, Gressel J (2010) Weed genomics advance: a commentary. Pest Manag Sci 66: 1041.
13. Ozsolak F, Milos PM (2011) RNA sequencing: advances, challenges and opportunities. Nat Rev Genet 12: 87–98.
14. International Brachypodium Initiative (2010) Genome sequencing and analysis of the model grass *Brachypodium distachyon*. Nature 463: 763–768.
15. Riggins CW, Peng Y, Stewart CN, Tranel PJ (2010) Characterization of *de novo* transcriptome for waterhemp (*Amaranthus tuberculatus*) using GS-FLX 454 pyrosequencing and its application for studies of herbicide target-site genes. Pest Manag Sci 66: 1042–1052.
16. Peng Y, Abercrombie LL, Yuan JS, Riggins CW, Sammons RD et al. (2010) Characterization of the horseweed (*Conyza canadensis*) transcriptome using GS-FLX 454 pyrosequencing and its application for expression analysis of candidate non-target herbicide resistance genes. Pest Manag Sci 66: 1053–1062.
17. Délano-Frier JP, Avilés-Arnaut H, Casarrubias-Castillo K, Casique-Arroyo G, Castrillón-Arbeláez PA et al. (2011) Transcriptomic analysis of grain amaranth (*Amaranthus hypochondriacus*) using 454 pyrosequencing: comparison with *A. tuberculatus*, expression profiling in stems and in response to biotic and abiotic stress. BMC Genomics 12: 363–380.
18. Lai Z, Kane NC, Kozik A, Hodgins KA, Dlugosch KM et al. (2012) Genomics of Compositae weeds: EST libraries, microarrays, and evidence of introgression. Am J Bot 99: 209–218.
19. Wang Y, Zeng X, Iyer NJ, Bryant DW, Mockler TC et al. (2012) Exploring the Switchgrass Transcriptome Using Second-Generation Sequencing Technology. PloS One 7: e34225.
20. Johnson M, Zaretskaya I, Raytselis Y, Merezhuk Y, McGinnis S et al. (2008) NCBI BLAST: a better web interface. Nucleic Acids Res 36: W5–W9.
21. Götz S, Garcia-Gómez JM, Terol J, Williams TD, Nagaraj SH et al. (2008) High-throughput functional annotation and data mining with the Blast2GO suite. Nucleic Acids Res 36: 3420–3435.
22. Moriya Y, Itoh M, Okuda S, Yoshizawa A, Kanehisa M (2007) KAAS: an automatic genome annotation and pathway reconstruction server. Nucleic Acids Res 35: W182–W185.
23. Birol I, Jackman SD, Nielsen CB, Qian JQ, Varhol R et al. (2009) *De novo* transcriptome assembly with ABySS. Bioinformatics 25: 2872–2877.
24. Langmead B, Trapnell C, Pop M, Salzberg SL (2009) Ultrafast and memory-efficient alignment of short DNA sequences to the human genome. Genome Biology 10: R25.
25. Wang L, Feng Z, Wang X, Zhang X (2010) DEGseq: an R package for identifying differentially expressed genes from RNA-seq data. Bioinformatics 26: 136–138.
26. Livak KJ, Schmittgen TD. (2001) Analysis of relative gene expression data using real-time quantitative PCR and the 2 method. Methods 25: 402–408.
27. Fredrik S, Joakim L (2000) Sequence analysis of genes and genomes. J Biotechnol 76: 1–31.
28. Iwakami S, Uchino A, Watanabe H, Yamasue Y, Inamura T (2012) Isolation and expression of genes for acetolactate synthase and acetyl-CoA carboxylase in *Echinochloa phyllopogon*, a polyploidy weed species. Pest Manag Sci 68: 1098–1106.
29. Wilhelm BT, Landry JR (2009) RNA-seq-quantitative measurement of expression through massively parallel RNA-sequencing. Methods 48: 249–257.
30. Yammoto E, Zeng L, Baird WV (1998) α-tubulin missense mutations correlate with antimicrotubule drug resistance in *Eleusine indica*. Plant Cell 10: 29–308.
31. Yamamoto E, Baird WV (1999) Molecular characterization of four β-tubulin genes from dinitroaniline susceptible and resistant biotypes of *Eleusine indica*. Plant Mol Bio 39: 45–61.
32. Délye C, Menchari Y, Michel S, Darmency H (2004) Molecular bases for sensitivity to tubulin-binding herbicides in green foxtail. Plant Physiol 136: 3920–3932.
33. Pátek M (2005) Branched-chain amino acids. Microbiol Monogr 5: 129–162.
34. Riar DS, Norsworthy JK, Srivastava V, Nandula V, Bond JA et al. (2013) Physiological and molecular basis of acetolactate synthase-inhibiting herbicide

resistance in barnyardgrass (*Echinochloa crus-galli*). J Agric Food Chem 61: 278–289.
35. Wang J, Li R (2008) Intergration of C4-specific PPDK gene of *Echinochloa* to C2 upland rice and its photosynthesis characteristics analysis. Afr J Biotechnol 7: 783–787.
36. Schaefer H, Hardy OJ, Silva L, Barraclough TG, Savolainen V (2011) Testing Darwin's naturalization hypothesis in the Azores. Ecol Lett 14: 389–396.
37. Hilu KW, Alice AL, Liang H (1999) Phylogeny of Poaceae inferred from matK sequences. Ann Mo Bot Gard 86: 835–851.
38. Christin PA, Salamin N, Savolainen V, Duvall MR, Besnard G (2007) C_4 photosynthesis evolved in grasses via parallel adaptive genetic changes. Curr Biol 17: 1241–1247.
39. Bouchenak-Khelladi Y, Verboom GA, Hodkinson TR, Salamin N, Francois O et al. (2009) The origins and diversification of C_4 grasses and savanna-adapted ungulates. Glob Chang Biol 15: 2397–2417.
40. Christin PA, Samaritani E, Petitpierre B, Salamin N, Besnard G (2009) Evolutionary insights on C_4 photosynthetic subtypes in grasses from genomics and phylogenetics. Genome Biol Evol 1: 221–230.

14

Increasing Cropping System Diversity Balances Productivity, Profitability and Environmental Health

Adam S. Davis[1]*, Jason D. Hill[2], Craig A. Chase[3], Ann M. Johanns[4], Matt Liebman[5]

1 United States Department of Agriculture/Agricultural Research Service, Global Change and Photosynthesis Research Unit, Urbana, Illinois, United States of America, **2** Department of Bioproducts and Biosystems Engineering, University of Minnesota, St. Paul, Minnesota, United States of America, **3** Leopold Center for Sustainable Agriculture, Iowa State University, Ames, Iowa, United States of America, **4** Department of Economics, Iowa State University Extension and Outreach, Osage, Iowa, United States of America, **5** Department of Agronomy, Iowa State University, Ames, Iowa, United States of America

Abstract

Balancing productivity, profitability, and environmental health is a key challenge for agricultural sustainability. Most crop production systems in the United States are characterized by low species and management diversity, high use of fossil energy and agrichemicals, and large negative impacts on the environment. We hypothesized that cropping system diversification would promote ecosystem services that would supplement, and eventually displace, synthetic external inputs used to maintain crop productivity. To test this, we conducted a field study from 2003–2011 in Iowa that included three contrasting systems varying in length of crop sequence and inputs. We compared a conventionally managed 2-yr rotation (maize-soybean) that received fertilizers and herbicides at rates comparable to those used on nearby farms with two more diverse cropping systems: a 3-yr rotation (maize-soybean-small grain + red clover) and a 4-yr rotation (maize-soybean-small grain + alfalfa-alfalfa) managed with lower synthetic N fertilizer and herbicide inputs and periodic applications of cattle manure. Grain yields, mass of harvested products, and profit in the more diverse systems were similar to, or greater than, those in the conventional system, despite reductions of agrichemical inputs. Weeds were suppressed effectively in all systems, but freshwater toxicity of the more diverse systems was two orders of magnitude lower than in the conventional system. Results of our study indicate that more diverse cropping systems can use small amounts of synthetic agrichemical inputs as powerful tools with which to tune, rather than drive, agroecosystem performance, while meeting or exceeding the performance of less diverse systems.

Editor: John P. Hart, New York State Museum, United States of America

Funding: Funding for the study was provided by the US Department of Agriculture National Research Initiative (Projects 2002-35320-12175 and 2006-35320-16548), the Leopold Center for Sustainable Agriculture (Projects 2004-E06, 2007-E09, and 2010-E02), the Iowa Soybean Association, and the Organic Center. The funders had no role in study design, data collection and analysis, decision to publish, or preparation of the manuscript.

Competing Interests: The authors have declared that no competing interests exist.

* E-mail: adam.davis@ars.usda.gov

Introduction

One of the key challenges of the 21st century is developing ways of producing sufficient amounts of food while protecting both environmental quality and the economic well-being of rural communities [1,2]. Over the last half century, conventional approaches to crop production have relied heavily on manufactured fertilizers and pesticides to increase yields, but they have also degraded water quality and posed threats to human health and wildlife [3–6]. Consequently, attention is now being directed toward the development of crop production systems with improved resource use efficiencies and more benign effects on the environment [1,7]. Less attention has been paid to developing better methods of pest management, especially for weeds. Here we explore the potential benefits of diversifying cropping systems as a means of controlling weed population dynamics while simultaneously enhancing other desirable agroecosystem processes [8]. We focus on crop rotation, an approach to cropping system diversification whereby different species are placed in the same field at different times.

Rotation systems have been used for millennia to maintain soil fertility and productivity and to suppress pests, and can increase yields even in situations where substantial amounts of fertilizers and pesticides are applied [9,10]. Rotation systems also foster spatial diversity, since different crops within the rotation sequence are typically grown in different fields on a farm in the same year. Diversification through crop rotation can be an especially useful strategy in farming systems that integrate crop and livestock production. The addition of forage crops, including turnips and clovers, to cereal-based systems in northwestern Europe and England in the 1600s and 1700s enhanced nitrogen supply through fixation by legumes, and increased nutrient cycling due to greater livestock density and manure production. These changes allowed the intensification of both crop and livestock production and increased yields substantially [11,12]. Integrated crop–livestock systems remained widespread in northern Europe, England, and much of the humid, temperate regions of North America until the 1950s and 1960s, when increased availability of relatively low-cost synthetic fertilizers made mixed farming and nutrient recycling biologically unnecessary and specialized crop and livestock production more economically attractive. In recent years, there has been interest in reintegrating crop and livestock systems as a strategy for reducing reliance on fossil fuels, minimizing the use of increasingly expensive fertilizers, and limiting water pollution by nutrients, pathogens and antibiotics [13,14].

Weeds are a ubiquitous and recurrent problem in essentially all crop production systems, and chemicals applied for weed control dominate the world market for pesticides [15]. With the introduction of crop genotypes engineered to tolerate herbicides, especially glyphosate, and with the continuing availability of older, relatively inexpensive herbicides, such as atrazine, successful weed management in conventional crop production systems has been largely taken for granted since the mid-1990s. Now, however, with expanded recognition of herbicides as environmental contaminants [4] and the increasing prevalence of herbicide resistant weeds [16], there is an important need to develop weed management strategies that are less reliant on herbicides and that subject weeds to a wide range of stress and mortality factors [17]. We believe that cropping system diversification may play an important role in the development of such strategies.

Here, we report the results of a large-scale, long-term experiment examining the consequences of cropping system diversification on agronomic, economic, and environmental measures of system performance. The experiment was conducted during 2003–2011 in Boone County, Iowa, within the central U.S. maize production region, and comprised three contrasting cropping systems varying in length of crop sequence, levels of chemical inputs, and use of manure. We compared a conventionally managed 2-yr rotation (maize-soybean) that received fertilizers and herbicides at rates comparable to those used on surrounding commercial farms with two more diverse cropping systems: a 3-yr rotation (maize-soybean-small grain + red clover) and a 4-yr rotation (maize-soybean-small grain + alfalfa-alfalfa) managed with reduced N fertilizer and herbicide inputs and periodic applications of composted cattle manure. Triticale was used as the small grain crop in 2003–2005; oat was used in 2006–2011. The 2-yr rotation is typical of cash grain farming systems in the region, whereas the 3-yr and 4-yr rotations are representative of farming systems in the region that include livestock. Details of the experimental site, management practices, sampling procedures, and data analyses are provided in the online SI section (Text S1, Figure S1, Tables S1-S4).

A central hypothesis framing our study was that cropping system diversification would result in the development of ecosystem services over time that would supplement, or eventually displace, the role of synthetic external inputs in maintaining crop productivity and profitability. Based on this hypothesis, we predicted that input requirements of the more diverse systems would initially be similar to that of the less diverse system, but would increasingly diverge from the less diverse system over time as the systems matured. We also predicted that crop yields, weed suppression, and economic performance of the three systems would be similar throughout the study. Finally, we predicted that reduced requirements for external synthetic inputs for pest management would result in a lower toxicological profile of the more diverse systems compared to the less diverse system.

Results

Crop Yields and Net Profitability

Cropping system diversification enhanced yields of maize and soybean grain and system-level harvested crop mass (grain, straw, and hay) while maintaining economic returns. The most parsimonious linear statistical models for each of these measures of system performance contained terms for main effects of *year* and *system*, but no interaction term ($AIC_{with\ interaction} = 319$; $AIC_{no\ interaction} = 315$). Over the 2003 to 2011 period, maize grain yield was on average 4% greater in the 3-yr and 4-yr rotations than in the 2-yr rotation (means for the 2-yr, 3-yr and 4-yr rotations are hereafter referred to as μ_2, μ_3 and μ_4, respectively; $\mu_2 = 12.3 \pm 0.1$ Mg ha^{-1}; $\mu_3 = 12.7 \pm 0.2$ Mg ha^{-1}; $\mu_4 = 12.9 \pm 0.2$ Mg ha^{-1}; pre-planned 1 d.f. contrast of *system*: $F_{1,7} = 8$, $P = 0.03$), and similar in the 3-yr and 4-yr rotations (Fig. 1a). Soybean grain yield during the same period was on average 9% greater in the 3-yr and 4-yr rotations than in the 2-yr rotation ($\mu_2 = 3.4 \pm 0.07$ Mg ha^{-1}; $\mu_3 = 3.8 \pm 0.08$ Mg ha^{-1}; $\mu_4 = 3.8 \pm 0.08$ Mg ha^{-1}; $F_{1,7} = 11.3$, $P = 0.01$) and similar in the 3-yr and 4-yr rotations (Fig. 1b). Harvested crop mass, averaged over the various crop phases comprising each cropping system, followed a similar pattern to maize and soybean grain yields. Mean crop biomass for 2003 to 2011 was 8% greater in the 3-yr and 4-yr rotations than in the 2-yr rotation ($\mu_2 = 7.9 \pm 0.08$ Mg ha^{-1}; $\mu_3 = 8.5 \pm 0.1$ Mg ha^{-1}; $\mu_4 = 8.6 \pm 0.2$ Mg ha^{-1}; *system*: $t_6 = 5.1$, $P = 0.002$), and similar in the 3-yr and 4-yr rotations (Fig. 1c).

We examined system profitability by calculating net returns to land and management, which represent profits to a farm operation without accounting for costs of land (e.g., rent or mortgage payments), management time (e.g., marketing), and federal subsidies. Profitability was analyzed for two temporal periods. From 2003 to 2005, considered the "startup" phase for the study, there were no differences among cropping systems in net profit, either through an analysis of main effects of *system* ($\mu_2 = \$448 \pm 17$ ha^{-1}; $\mu_3 = \$402 \pm 17$ ha^{-1}; $\mu_4 = \$457 \pm 15$ ha^{-1}; $F_{2,6} = 0.12$, $P = 0.89$) or by pre-planned 1-d.f. contrasts (2-yr vs. 3-yr and 4-yr rotations: $F_{1,7} = 0.10$, $P = 0.77$) (Fig. 1d). From 2006 to 2011, the "established" phase of the study, there were again no differences among systems, either through main effects of *system* ($\mu_2 = \$953 \pm 36$ ha^{-1}; $\mu_3 = \$965 \pm 34$ ha^{-1}; $\mu_4 = \$913 \pm 26$ ha^{-1}; $F_{2,6} = 0.62$, $P = 0.57$) or by pre-planned 1-d.f. contrasts (2-yr vs. 3-yr and 4-yr rotations: $F_{1,7} = 0.03$, $P = 0.86$).

Stability of system performance over time, as measured through a comparison of variances for the various products of the system, was similar for maize grain yield ($F_{2,6} = 2.4$, $P = 0.17$), soybean grain yield ($F_{2,6} = 0.95$, $P = 0.44$) and net returns to land and management during the startup phase of the study, 2003 to 2005 ($F_{2,6} = 0.05$, $P = 0.95$). Two system products, harvested crop mass from 2003 to 2011 and profit during the established phase of the study, 2006 to 2011, showed considerable differences in system stability over time, but in contrasting ways. Variance in mean harvested crop mass was greater in the 3-yr and 4-yr rotations than in the 2-yr rotation ($\sigma_2^2 = 0.27$; $\sigma_3^2 = 0.60$; $\sigma_4^2 = 0.95$; $F_{1,7} = 16$, $P = 0.005$). Conversely, cropping system diversification was associated with lower variance in profit during the established phase of the study. Variance in profit from 2006 to 2011 was lower in the 3-yr and 4-yr rotations than in the 2-yr rotation ($\sigma_2^2 = 1.5 \times 10^5$; $\sigma_3^2 = 8.1 \times 10^3$; $\sigma_4^2 = 6.3 \times 10^3$; $F_{1,7} = 16$, $P = 0.005$).

Agrichemical, Labor and Energy Inputs

Application rates of the primary agrichemicals used in this study, manufactured N fertilizer ($F_{2,14} = 117$, $P < 0.0001$) and herbicides ($F_{2,14} = 287$, $P < 0.0001$), both showed strong effects of cropping system. Manufactured N fertilizer applications were higher in the 2-yr rotation than in the 3-yr and 4-yr rotations ($\mu_2 = 80 \pm 3$ kg N ha^{-1}; $\mu_3 = 16 \pm 3$ kg N ha^{-1}; $\mu_4 = 11 \pm 2$ kg N ha^{-1}; $F_{1,17} = 16$, $P = 0.005$), with the difference between systems increasing over the course of the study ($F_{2,14} = 11.6$, $P = 0.001$) (Fig. 1e). Herbicide application rates followed a similar pattern, with greater amounts of herbicide applied in the 2-yr rotation than in the 3-yr and 4-yr rotations ($\mu_2 = 1.9 \pm 0.06$ kg a.i. ha^{-1};

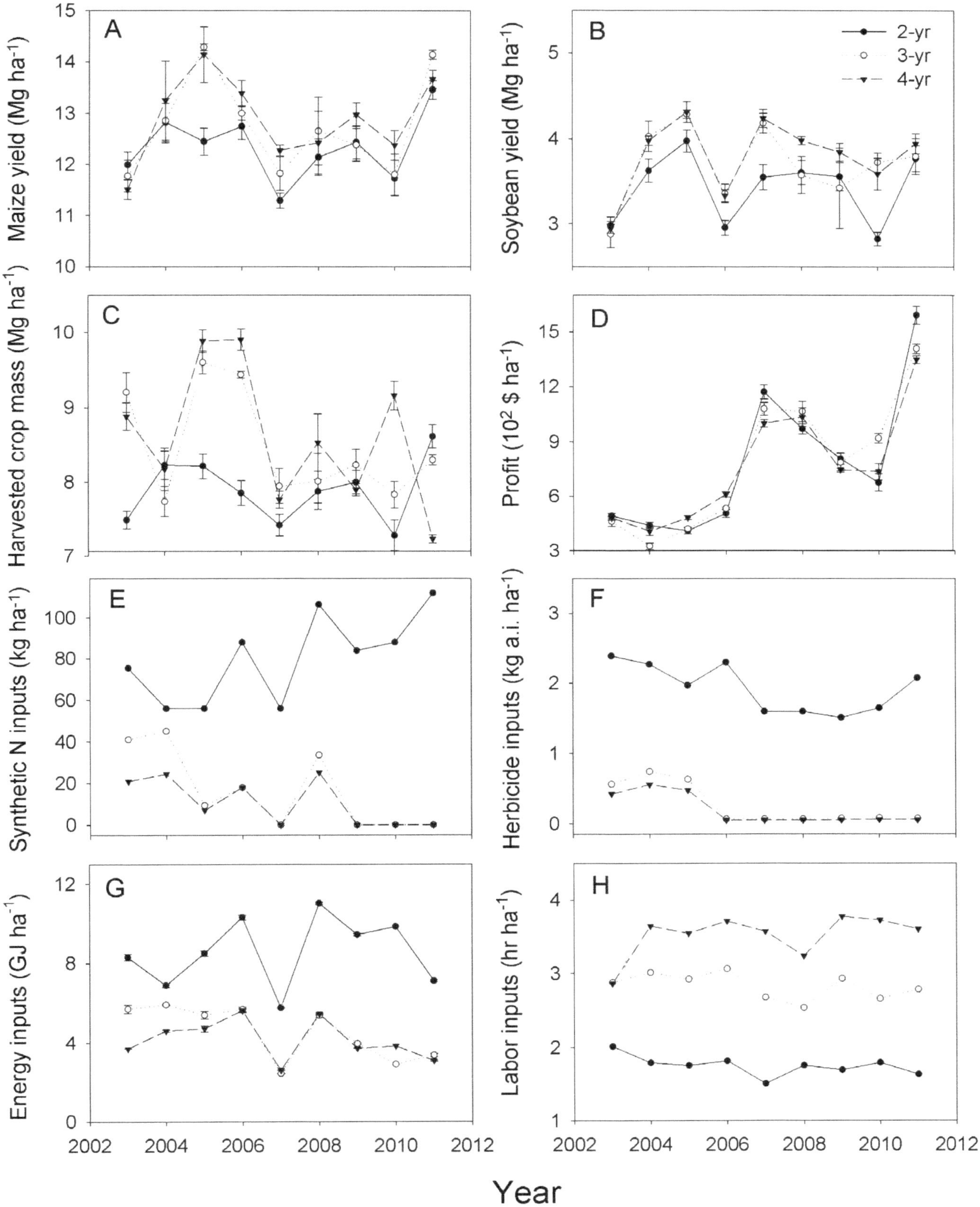

Figure 1. Cropping system performance over time. Annual performance of maize-soybean (2-yr), maize-soybean-small grain/red clover (3-yr), and maize-soybean-small grain/alfalfa-alfalfa (4-yr) cropping systems in Boone, IA, from 2003 to 2011. Performance metrics included: a) maize yield, b) soybean yield, c) rotation-level harvested crop mass, d) net returns to land and management, e) manufactured N fertilizer application rate, f) herbicide

application rate, g) fossil energy use, and h) labor requirements. Symbols represent the mean ± SEM of four replicate experimental blocks (N = 36 per cropping system).

$\mu_3 = 0.26 \pm 0.05$ kg a.i. ha^{-1}; $\mu_4 = 0.20 \pm 0.03$ kg a.i. ha^{-1}; $F_{1,17} = 610$, $P < 0.0001$); differences among systems, however, did not increase over time (Fig. 1f).

Fossil energy use was strongly influenced by cropping system in both the startup ($F_{2,6} = 94$, $P < 0.0001$) and established ($F_{2,6} = 116$, $P < 0.0001$) phases of the study, with no difference in energy use between experimental phases ($F_{1,92} = 0.39$, $P = 0.53$) (Fig. 1g). From 2003 to 2011, inputs of energy were greater in the 2-yr rotation than in the 3-yr and 4-yr rotations ($\mu_2 = 8.6 \pm 0.1$ GJ ha^{-1}; $\mu_3 = 4.5 \pm 0.1$ GJ ha^{-1}; $\mu_4 = 4.2 \pm 0.04$ GJ ha^{-1}; $F_{1,7} = 55$, $P = 0.0001$). The partial correlations between energy use in a given cropping system and energy use in the maize phase of that rotation, taking into account the amount of N fertilizer applied to maize, were 0.94, 0.81 and 0.70 in the 2-yr, 3-yr and 4-yr systems, respectively (SI, Table S5). This indicated that synthetic N fertilizer use in the maize phase of the various cropping systems drove energy use within the maize phase, which in turn drove energy use by a given cropping system.

Demand for labor differed among the three cropping systems in both the startup ($F_{2,4} = 26$, $P = 0.005$) and established ($F_{2,10} = 299$, $P < 0.0001$) study phases, but followed a contrasting pattern to energy requirements (Fig. 1h). Labor inputs were more than 33% lower in the 2-yr rotation than in the 3-yr and 4-yr rotations from 2003 to 2005 ($F_{1,5} = 35$, $P = 0.002$) and from 2006 to 2011 ($F_{1,11} = 59$, $P < 0.0001$). Overall, there was a strong negative correlation ($r = -0.79$, $P < 0.0001$) between fossil energy and labor inputs over time in the three cropping systems.

Divergent Weed Management Systems

Two lines of evidence indicate that weeds were managed effectively in all three cropping systems in both the 'startup' and 'established' phases, in spite of reducing herbicide use by 88% in the 3-yr and 4-yr rotations compared to the 2-yr rotation. First, weed seedbanks declined at an equal rate in all study systems (Fig. 2a). Selection among linear mixed effects regression models incorporating temporal autocorrelation among seedbank measurements over time supported different intercepts (*system*: $F_{2,6} = 16.8$, $P = 0.0035$) but did not support inclusion of a *year* by *system* interaction term ($AIC_s = 182$; $AIC_{s*y} = 185$), thus indicating a common slope ($b_1 = -0.18$). For all three systems, the time to decline to 95% of the weed seedbank levels in 2003 was 16.6 years. Declines in weed seedbanks reflected a focus of management attention on the timing of weed management activities and herbicide choices in all three systems, as well as the increased number and diversity of stress and mortality factors present in the 3-yr and 4-yr rotations [8,21]. Higher densities of weed seeds in the 3-yr and 4-yr rotations, as indicated by their greater intercept values than for the 2-yr rotation (Fig. 2a.), were the result of poorer weed control in the 3-yr and 4-yr rotations during the set-up of the experiment plots in 2002.

The second line of evidence concerns weed biomass, which was very low in all three cropping systems for the duration of the study (Fig. 2b), never exceeding 0.3% of harvested crop mass. Weed biomass was the same within a given crop phase, regardless of the cropping system in which it occurred (main effect of *system*: maize, $F_{2,6} = 1.47$; $P = 0.30$; soybean, $F_{2,6} = 0.88$; $P = 0.46$; small grain, $F_{1,3} = 1.24$; $P = 0.31$). There were differences in mean weed biomass among cropping systems ($\mu_2 = 0.0003 \pm 0.00007$ Mg ha^{-1}; $\mu_3 = 0.0076 \pm 0.0012$ Mg ha^{-1}; $\mu_4 = 0.009 \pm 0.001$ Mg ha^{-1}; $F_{2,6} = 12.7$; $P < 0.007$). These differences arose mainly due to the presence of a small grain phase in the 3-yr and 4-yr rotation crop sequences. Weed biomass did not differ between maize and soybean in any of the cropping systems ($F_{1,202} = 2.1$; $P = 0.15$), however weed biomass in the small grain phase of the 3-yr and 4-yr rotations was greater than weed biomass in the maize and soybean phases ($F_{1,206} = 174$; $P < 0.0001$). In the 4-year system, weed biomass in alfalfa was intermediate between weed biomass levels in the maize/soybean and small grain phases.

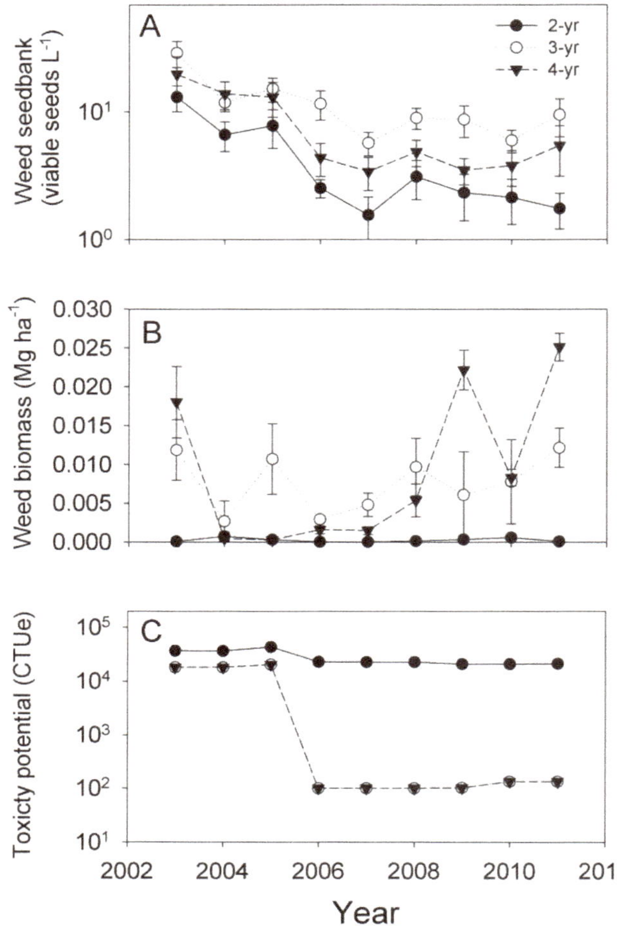

Figure 2. Divergent weed management systems. Weed management characteristics in maize-soybean (2-yr), maize-soybean-small grain/red clover (3-yr), and maize-soybean-small grain/alfalfa-alfalfa (4-yr) cropping systems in Boone, IA, from 2003 to 2011. Performance metrics included a) weed seed density in soil, b) weed aboveground biomass, and c) freshwater toxicity potential expressed in comparative toxic units (CTU_e). Symbols represent the mean ± SEM of four replicate experimental blocks (N = 36 per cropping system).

Environmental toxicity, in relation to ecotoxicological profiles for herbicides used in this study (Fig. 2c), showed a strong effect of *system* ($F_{2,14} = 1673$, $P < 0.0001$), with lower toxicity potential in the 3-yr and 4-yr rotations compared to the 2-yr rotation (*type*: $F_{1,17} = 2691$, $P < 0.0001$). Ecotoxicity in the diversified and conventional systems diverged as the systems matured over time [*type* x *phase*: $F_{1,16} = 7.4$, $P = 0.015$], transitioning from a two-fold difference during 2003 to 2005 to a two hundred-fold difference in toxicity from 2006 to 2011 (Fig. 2c).

Discussion

Our results support the hypothesis that the development of ecosystem services over time in more diverse cropping rotations increasingly displaces the need for external synthetic inputs to maintain crop productivity. From 2003 to 2011, as predicted, the desired products (crop yield, weed suppression, and economic performance) of the more diverse and less diverse cropping rotations were similar, whereas external inputs and environmental impacts differed greatly among the systems (Fig. 3). Comparing these metrics of system performance by experimental phase (initial three years of system establishment versus the following six years) confirmed our prediction that system inputs and environmental impacts would diverge over time, whereas yield and profit would remain similar among more diverse and less diverse rotations. In the more diverse rotations, small amounts of synthetic agrichemical inputs thus served as powerful tools with which to tune, rather than drive, agroecosystem performance.

Grain production in the U.S. is dominated by short rotation systems designed to maximize grain yield and profit. These are important goals but represent only a portion of the many ecosystem services that managed lands may provide [18] and that should be considered when evaluating alternative production systems [1,19]. We believe that these functions are complementary, rather than competing, considerations for agroecosystem design. The results of this study demonstrate that more rotationally diverse cropping systems may be optimized in multiple dimensions, leveraging small agrichemical inputs with biological synergies arising from enhanced diversity of crop species and management tactics.

An example of the synergizing effects of cropping system diversification can be found in weed management in the 3-yr and 4-yr rotations. Weeds were suppressed as effectively in these systems as in the 2-yr rotation, with declining soil seedbanks and negligible weed biomass, yet herbicide inputs in the 3-yr and 4-yr rotation plots were 6 to 10 times lower, and freshwater toxicity 200 times lower, than in the 2-yr rotation. Improved efficiency and environmental sustainability of weed management in the 3-yr and 4-yr rotations resulted from integrating multiple, complementary tactics in an ecological weed management framework [8,20]. Mounting evidence for unintended effects resulting from heavy reliance on herbicides highlights the need to re-think the role of herbicides in weed management. Non-target impacts of herbicides include reproductive abnormalities and mortality in vertebrates [5,21–23] and potential for diminished non-crop nectar resources for key pollinator species [17,24,25]. Herbicide overuse has also resulted in widespread, accelerating evolution of weed genotypes resistant to one or more modes of herbicide action [26,27]. Our

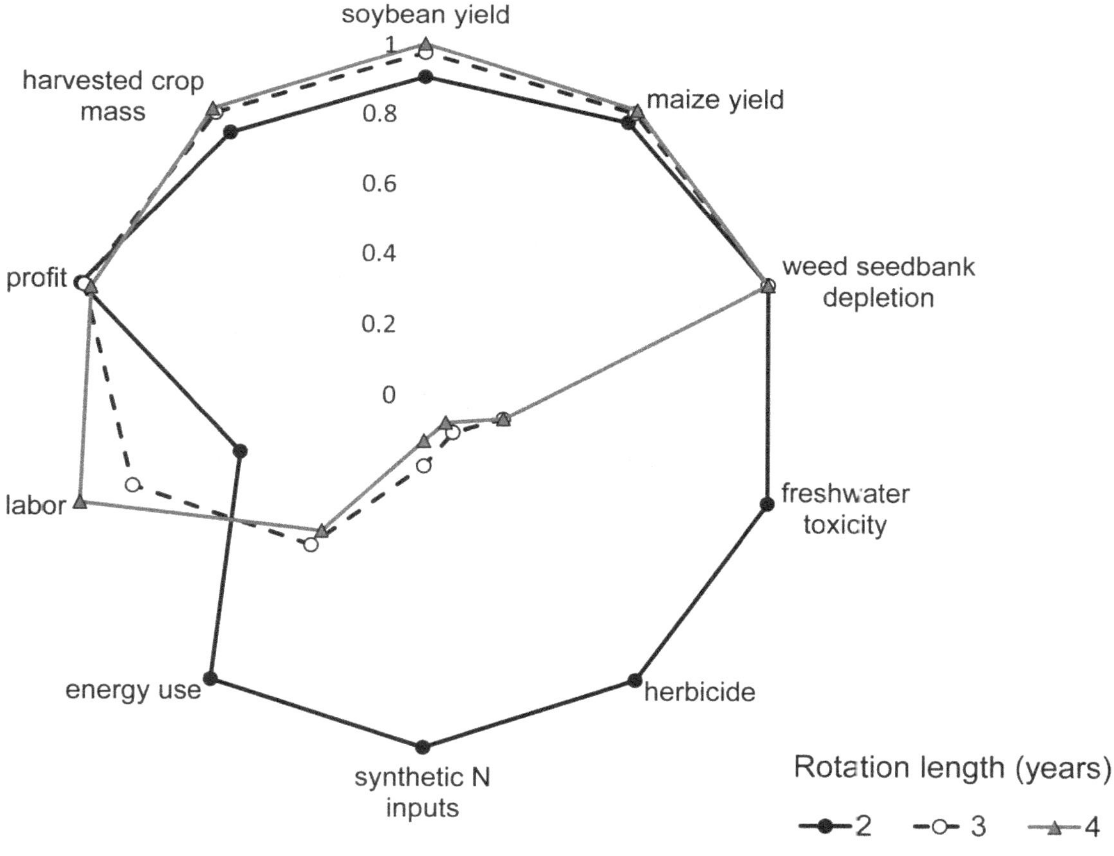

Figure 3. Multiple indicators of cropping system performance. Comparative long-term performance of maize-soybean (2-yr), maize-soybean-small grain/red clover (3-yr), and maize-soybean-small grain/alfalfa-alfalfa (4-yr) cropping systems in Boone, IA, averaged over the 2003–2011 study period. Variable means are normalized on a 0 to 1 scale, with 1 representing the cropping system with the largest absolute value for that variable (N = 36 per cropping system). Performance metrics included: maize and soybean yield, rotation-level harvested crop mass, net returns to land and management, manufactured N fertilizer and herbicide application rate, fossil energy use, labor requirements, freshwater toxicity potential and weed seedbank decline (measured as exponential decay constant).

data indicate that, in the context of a cropping system with weed suppressive characteristics, small herbicide inputs may contribute to a diverse suite of tactics that cumulatively provide effective, reliable, and more durable weed management.

The diversity-productivity-stability relationship has long been a key theme in ecology [28,29]. Recently, it has been applied in the context of bioenergy crop production to describe increases in biomass and ecosystem services, such as C sequestration, associated with increasing species diversity in polycultures of bioenergy feedstock crop species [30]. Our work supports the application of this concept to cropping systems more broadly. Future gains in more diverse systems may depend upon the application of ecological principles surrounding this relationship to cropping system design [31,32]. Cropping system diversification in this study included both crop species and management practices. In contrast to the 2-yr rotation, with two species, both of the 3-yr and 4-yr rotations included four crop species. In the 4-yr rotation, further temporal diversification was achieved by including a perennial-only crop phase (alfalfa hay) for one quarter of the rotation sequence. Our results showed productivity gains associated with greater diversity in system-level harvested crop mass and maize and soybean seed yields. We also observed increased stability of profit, with similar long-term means in the 3-yr and 4-yr rotations compared to the 2-yr rotation.

Similar profits were attained through different pathways in the 3-yr and 4-yr rotations and the 2-yr rotation (Fig. 3). Increased labor, information intensive management and ecosystem services arising from increased biological N fixation (via the clover and alfalfa crops) and contrasting crop phenologies and competitive abilities were substituted in 3-yr and 4-yr rotations for the higher inputs of manufactured N, herbicides and energy from fossil fuels driving the 2-yr rotation. Energy use in maize drove differences among the cropping systems, and manufactured N inputs to maize contributed most strongly to energy balances for this crop. The high sensitivity of agricultural energy use to N fertilizer inputs provides a high-priority target for the redesign of cropping systems for increased sustainability.

Reintegration of crop and livestock production, as represented by the forage legumes and manure applications present in the more diverse systems, is not simply another aspect of cropping system diversification. Instead, it embodies an important principle in sustainable agriculture: system boundaries should be drawn to minimize externalities. Animal manure is produced regardless of whether feed grains are shipped to centralized concentrated animal feeding operations, or produced within integrated crop-livestock farming operations. In the former case the manure may become a waste product and water pollutant if quantities exceed available land area for field application [33], whereas in the latter case, it contributes directly to crop nutrient requirements, improves soil quality, and reduces fossil fuel subsidies associated with grain transport and external N fertilizer inputs [14].

Substantial improvements in the environmental sustainability of agriculture are achievable now, without sacrificing food production or farmer livelihoods. When agrichemical inputs are completely eliminated, yield gaps may exist between conventional and alternative systems [19]. However, such yield gaps may be overcome through the strategic application of very low inputs of agrichemicals in the context of more diverse cropping systems. Although maize is grown less frequently in the 3-yr and 4-yr rotations than in the 2-yr rotation, this will not compromise the ability of such systems to contribute to the global food supply, given the relatively low contribution of maize and soybean production to direct human consumption and the ability of livestock to consume small grains and forages [34]. Through a balanced portfolio approach to agricultural sustainability, cropping system performance can be optimized in multiple dimensions, including food and biomass production, profit, energy use, pest management, and environmental impacts.

Materials and Methods

Site Details and Agronomic Management

To investigate the relative performance of conventional and more diverse cropping systems, we conducted a 9-hectare experiment at the Iowa State University Marsden Farm (Figure S1), in Boone County, IA (42°01′ N; 93°47′ W; 333 m above sea level). The experiment site lies within a region of intensive rain-fed maize and soybean production and is surrounded by farms with high levels of productivity. Soils at the site are deep, fertile Mollisols. The experimental cropping system treatments included a conventionally managed 2-yr rotation (maize/soybean) that received agrichemicals at rates comparable to those used on commercial farms in the region, and more diverse cropping systems – a 3-yr rotation (maize/soybean/small grain + red clover green manure) and a 4-yr rotation (maize/soybean/small grain + alfalfa/alfalfa hay) – managed with reduced N fertilizer and herbicide inputs.

The entire site was planted with oat in 2001 and the cropping systems experiment was established in 2002 using a randomized complete block design with each crop phase of each rotation system present every year in four replicate blocks. Plots were 18 m x 85 m and managed with conventional farm machinery. Spring triticale was used as the small grain in 2003–2005, whereas oat was used in 2006–2010. Synthetic fertilizers were applied in the 2-yr rotation at conventional rates based on soil tests. In the 3-yr and 4-yr rotations, composted cattle manure was applied before maize production at a mean dry matter rate of 8.3 Mg ha^{-1} and substantial amounts of N were added through fixation by red clover and alfalfa [35,36,37]. Manure and legume N-fixation in the 3-yr and 4-yr rotations were supplemented with synthetic fertilizers based on soil tests, including the late-spring soil nitrate test for maize production [38]. Weed management in the 2-yr rotation was based largely on herbicides applied at conventional rates. In the 3-yr and 4-yr rotations, herbicides were applied in 38-cm-wide bands in maize and soybean and inter-row zones were cultivated; no herbicides were applied in small grain and forage legume crops. Choices of post-emergence herbicides used in each of the systems were made based on the identities, densities, and sizes of weed species observed in the plots. Other details of the farming practices used in the different cropping systems are described in Liebman et al. [39] and in the online SI materials (Text S1). Sampling procedures for determining crop yields, weed biomass and weed seed densities in soil are also described in the online SI materials (Text S1).

Energy and Economic Analyses

Energy inputs were divided into five categories: seed, fertilizer, pesticides, fuel for field operations, and propane and electricity used for drying maize grain after harvest. Data were obtained from logs describing all field operations, material inputs, and crop moisture characteristics for the experimental plots during the study period. Economic analyses measured performance characteristics of whole rotation systems under contrasting management strategies. We evaluated net returns to land and management on a unit land area basis, with land units divided in two equal portions for maize and soybean in the 2-yr rotation; three equal portions for maize, soybean, and small grains with red clover in the 3-yr rotation; and four equal portions for maize, soybean, small grains with alfalfa, and alfalfa in the 4-yr rotation. Net returns to land and management represented returns to a farm operation calculated without accounting for costs of land (e.g., rent or mortgage payments), management time (e.g., marketing), or possible federal subsidies. Data sources and assumptions for the energy and economic analyses are shown in the online SI materials.

Ecotoxicological Calculations

Freshwater ecotoxicity of pesticide use was estimated using the USEtox model [40–42]. Characterization factors (CFs) of ecotoxicity potential for active ingredients included transport to freshwater via surface water, soil, and air. CFs were available for eight of ten active ingredients applied in the three rotations. The two active ingredients for which CFs were unavailable are not of particular concern for freshwater ecotoxicity due either to their low toxicity (mesotrione) or low infiltration and persistence in freshwaters (lactofen) [43].

Statistical Analyses

The experiment was arranged in a randomized complete block design, with all entry points of the three crop rotations (i.e. all crops within each of the rotations) represented in four replicate blocks in each year of the study, for a total of 36 plots. Cropping system effects in time series data were analyzed using hierarchical linear mixed effects repeated measures models, modeling temporally correlated errors with an ARMA (auto-regressive moving average) correlation structure in the *nlme* package of R v.2.14.1 [44,45]. Fixed effects included *cropping system* and *experimental phase* (startup = 2003 to 2005; established = 2006 to 2011), and random effects included *replicate block* nested within *cropping system* and *year*. Partial correlations were estimated using the *corpcor* package in R v.2.14.1. In contrast to data for quantitative observations (e.g. crop yield or weed biomass) that varied by replicate block and year, data for input variables, such as synthetic fertilizer or herbicides and associated environmental toxicity metrics, did not vary among blocks for a particular rotation entry point in a given year, but did vary among years. Therefore, site-year was treated as the source of experimental replication for these latter variables in our statistical tests for effects of *cropping system* and *experimental phase*. This led to contrasting degrees of freedom in reported F-tests for these two data types. Finally, for variables with non-constant variance among cropping systems over time (crop biomass and profit), we used the 'varIdent' variance function within the *nlme* package to explicitly model differences in variances among cropping systems for these variables within our mixed effects models.

Supporting Information

Figure S1 Aerial view of Marsden Farm study, Boone IA. Crop abbreviations: m = maize, sb = soybean, g = small grain, a = alfalfa.

Table S1 Mean monthly air temperature and total monthly precipitation during the 2003–2011 growing seasons, and long-term temperature and precipitation averages. Data were collected about 1 km from the experimental site in Boone Co., IA.

Table S2 Crop identities and seeding rates in 2003–2011.

Table S3 Macronutrients applied in manufactured fertilizers, herbicide adjuvants, and manure in 2003–2011. Manufactured N, P, and K fertilizers were applied at rates that varied among years and rotations in response to soil test results. Manure was applied at a rate of 15.7 Mg ha^{-1} in maize phases of the 3-year and 4-year rotation systems, but moisture and nutrient concentrations varied among years, resulting in variable rates of macronutrient additions.

Table S4 Herbicide applications in 2003–2011 to maize and soybean in the three rotation systems. No herbicides were used for triticale, oat, red clover, and alfalfa grown within the 3-yr and 4-yr systems. Reported application rates reflect the effect of banding of herbicides over crop rows in the 3-yr and 4-yr systems.

Table S5 Simple and partial correlations between energy use within a given crop phase and mean rotation energy use and between energy use within a given crop phase and N fertilizer application rates.

Text S1 Detailed description of experimental site, management practices, scientific methods and statistical approach.

Acknowledgments

We are grateful to the many students, postdoctoral research associates and technicians who have made the Marsden Farm study possible.

Author Contributions

Conceived and designed the experiments: ASD ML. Performed the experiments: CAC AMJ ML. Analyzed the data: ASD JDH CAC AMJ ML. Contributed reagents/materials/analysis tools: ML. Wrote the paper: ASD JDH ML.

References

1. Foley JA, Ramankutty N, Brauman KA, Cassidy ES, Gerber JS, et al (2011) Solutions for a cultivated planet. Nature 478: 337–342.
2. Robertson GP, Swinton SM (2005) Reconciling agricultural productivity and environmental integrity: a grand challenge for agriculture. Front Ecol Environ 3: 38–46.
3. Dubrovsky NM, Burow KR, Clark GM, Gronber JM, Hamilton PA, et al. (2010) The quality of our nation's waters: Nutrients in the nation's streams and groundwater, 1992–2004. Circular 1350. Reston, VA: U.S. Geological Survey. http://pubs.usgs.gov/circ/1350/. Accessed 2012 Jul 10.
4. Gilliom RJ, Barbash JE, Crawford CG, Hamilton PA, Martin JD, et al. (2006) The quality of our nation's waters: pesticides in the nation's streams and ground water, 1992–2001. Circular 1291. Reston, VA: U.S. Geological Survey. http://pubs.usgs.gov/circ/2005/1291/. Accessed 2012 Jul 10.
5. Relyea RA (2005) The impact of insecticides and herbicides on the biodiversity and productivity of aquatic communities. Ecol Appl 15: 618–627.
6. Rohr JR, McCoy KA (2010) A qualitative meta-analysis reveals consistent effects of atrazine on freshwater fish and amphibians. Environ Health Persp 118: 20–32.
7. Tilman D, Cassman KG, Matson PA, Naylor R, Polasky S (2002) Agricultural sustainability and intensive production practices. Nature 418: 671–677.
8. Liebman M, Staver CP (2001) Crop diversification for weed management. In: Liebman M, Mohler CL, Staver CP, editors. Ecological management of agricultural weeds. Cambridge: Cambridge University Press. 322–374.
9. Bennett AJ, Bending GD, Chandler D, Hilton S, Mills P (2012) Meeting the demand for crop production: the challenge of yield decline in crops grown in short rotations. Biol Rev 87: 52–71.
10. Karlen DL, Varvel GE, Bullock DG, Cruse RM (1994) Crop rotations for the 21st century. Adv Agron 53: 1–45.
11. Grigg DB (1974) The agricultural systems of the world: An evolutionary approach. Cambridge: Cambridge University Press. 358 pp.

12. Grigg DB (1989) English agriculture: An historical perspective. Oxford: Basil Blackwell. 320 pp.
13. Magdoff F, Lanyon L, Liebhardt B (1997) Nutrient cycling, transformations, and flows: implications for a more sustainable agriculture. Adv Agron 60: 1–73.
14. Naylor R, Steinfeid H, Falcon W, Galloways J, Smil V, et al. (2005) Losing the links between livestock and land. Science 310: 1621–1622.
15. U.S. Environmental Protection Agency (EPA) (2011) Pesticides industry sales and usage: 2006 and 2007 market estimates. Washington, D.C.: U.S. Environmental Protection Agency. www.epa.gov/opp00001/pestsales/07pestsales/market_estimates2007.pdf. Accessed 2012 Jul 10.
16. Heap I (2012) International survey of herbicide resistant weeds. http://www.weedscience.org. Accessed 2012 Jul 10.
17. Mortensen DA, Egan JF, Maxwell BD, Ryan MR, Smith RG (2012) Navigating a critical juncture for sustainable weed management. BioSci 62: 75–84.
18. Jordan N, Warner KD (2010) Enhancing the multifunctionality of US agriculture. BioSci 60: 60–66.
19. Seufert V, Ramankutty N, Foley JA (2012) Comparing the yields of organic and conventional agriculture. Nature 485: 229–232.
20. Liebman M, Davis AS (2009) Managing weeds in organic farming systems: an ecological approach. In: Organic farming: The ecological system. Francis C, editor. Madison: American Society of Agronomy. 173–196.
21. Hayes TB, Collins A, Lee M, Mendoza M, Noriega N, et al. (2002) Hermaphroditic, demasculinized frogs after exposure to the herbicide atrazine at low, ecologically relevant doses. Proc Nat Acad Sci USA 99: 54376–55480.
22. Orton F, Lutz I, Kloas W, Rutledge EJ (2009) Endocrine disrupting effects of herbicides and pentachlorophenol: in vitro and in vivo evidence. Environ Sci Technol 43: 2144–2150.
23. Relyea RA (2005) The lethal impact of Roundup on aquatic and terrestrial amphibians. Ecol Appl 15: 1118–1124.
24. Egan JF, Mortensen DA (2012) Quantifying vapor drift of dicamba herbicides applied to soybean. Envir Toxicol Chem 31: 1023–1031
25. Freemark K, Boutin C (1995) Impacts of agricultural herbicide use on terrestrial wildlife in temperate landscapes - a review with special reference to North America. Agric Ecosyst Environ 52: 67–91.
26. Powles SB, Yu Q (2010) Evolution in action: plants resistant to herbicides. Ann Rev Plant Biol 61: 317–347.
27. Tranel PJ, Riggins CW, Bell MS, Hager AG (2011) Herbicide resistances in *Amaranthus tuberculatus*: A call for new options. J Agric Food Chem 59: 5808–5812.
28. Chase JM, Leibold MA (2002) Spatial scale dictates the productivity-biodiversity relationship. Nature 416: 427–430.
29. Tilman D, Reich PB, Knops J, Wedin D, Mielke T, et al. (2001) Diversity and productivity in a long-term grassland experiment. Science 294: 843–845.
30. Tilman D, Hill J, Lehman C (2006) Carbon-negative biofuels from low-input high-diversity grassland biomass. Science 314: 1598–1600.
31. Smith RG, Gross KL, Robertson GP (2008) Effects of crop diversity on agroecosystem function: Crop yield response. Ecosyst 11: 355–366.
32. Smith RG, Mortensen DA, Ryan MR (2010) A new hypothesis for the functional role of diversity in mediating resource pools and weed-crop competition in agroecosystems. Weed Res 50: 185–185.
33. Jackson LL, Keeney DR, Gilbert EM (2000) Swine manure management plans in north-central Iowa: Nutrient loading and policy implications. J Soil Water Cons 55: 205–212.
34. Olmstead J (2011) Feeding the world? Twelve years later, U.S. grain exports are up, so too is hunger. Minneapolis, MN: Institute for Agriculture and Trade Policy. http://www.iatp. org/documents/feeding-the-world. Accessed 2012 Jul 10.
35. Heichel GH, Barnes DK, Vance CP, Henjum KI (1984) N_2 fixation, and N and dry matter partitioning during a 4-year alfalfa stand. Crop Sci 24: 811–815.
36. Heichel GH, Vance CP, Barnes DK, Henjum KI (1985) Dinitrogen fixation, and N and dry matter distribution during 4-year stands of birdsfoot trefoil and red clover. Crop Sci 25: 101–105.
37. Fox RH, Piekielek WP (1988) Fertilizer N equivalence of alfalfa, birdsfoot trefoil, and red clover for succeeding corn crops. J Prod Agric 1: 313–317.
38. Blackmer AM, Voss RD, Mallarino AP (1997) Nitrogen fertilizer recommendations for corn in Iowa. PM-1714. Ames, IA: Iowa State University Extension and Outreach. http://www.extension.iastate.edu/Publications/PM1714.pdf. Accessed 2012 Jul 10.
39. Liebman M, Gibson LR, Sundberg DN, Heggenstaller AH, Westerman PR, et al. (2008) Agronomic and economic performance characteristics of conventional and low-external-input cropping systems in the central corn belt. Agron. J. 100: 600–610.
40. Berthoud A, Maupu P, Huet C, Poupart A (2011) Assessing freshwater ecotoxicity of agricultural products in life cycle assessment (LCA): a case study of wheat using French agricultural practices databases and USEtox model. Int J Life Cycle Assess 16: 841–847.
41. Hauschild MZ, Huijbregts M, Jolliet O, MacLeod M, Margni M, et al. (2008) Building a model based on scientific consensus for life cycle impact assessment of chemicals: The search for harmony and parsimony. Environ Sci Technol 42: 7032–7037.
42. Rosenbaum RK, Bachmann TM, Gold LS, Huijbregts MAJ, Jolliet O, et al. (2008) USEtox-the UNEP-SETAC toxicity model: recommended characterisation factors for human toxicity and freshwater ecotoxicity in life cycle impact assessment. Int J Life Cycle Assess 13: 532–546.
43. Cornell PMEP (2008) Pesticide management education program. http://pmep.cce. cornell.edu/. Accessed 2012 Jul 10.
44. Crawley MJ (2007) The R Book. West Sussex: John Wiley and Sons. 942 pp.
45. R Development Core Team (2009) R: A language and environment for statistical computing (R Foundation for Statistical Computing, Vienna, Austria).

A Built-In Strategy to Mitigate Transgene Spreading from Genetically Modified Corn

Jing Li, Hui Yu, Fengzhen Zhang, Chaoyang Lin, Jianhua Gao, Jun Fang, Xiahui Ding, Zhicheng Shen, Xiaoli Xu*

State Key Laboratory of Rice Biology, Institute of Insect Sciences, School of Agriculture and Biotechnology, Zhejiang University, Hangzhou, China

Abstract

Transgene spreading is a major concern in cultivating genetically modified (GM) corn. Cross-pollination may cause the spread of transgenes from GM cornfields to conventional fields. Occasionally, seed lot contamination, volunteers, mixing during sowing, harvest, and trade can also lead to transgene escape. Obviously, new biological confinement technologies are highly desired to mitigate transgene spreading in addition to physical separation and isolation methods. In this study, we report the development of a built-in containment method to mitigate transgene spreading in corn. In this method, an RNAi cassette for suppressing the expression of the nicosulfuron detoxifying enzyme *CYP81A9* and an expression cassette for the glyphosate tolerant *5-enolpyruvylshikimate-3-phosphate synthase (EPSPS)* gene *G10* were constructed and transformed into corn via *Agrobacterium*-mediated transformation. The GM corn plants that were generated were found to be sensitive to nicosulfuron but resistant to glyphosate, which is exactly the opposite of conventional corn. Field tests demonstrated that GM corn plants with silenced *CYP81A9* could be killed by applying nicosulfuron at 40 g/ha, which is the recommended dose for weed control in cornfields. This study suggests that this built-in containment method for controlling the spread of corn transgenes is effective and easy to implement.

Editor: M. Lucrecia Alvarez, TGen, United States of America

Funding: This work was supported by the National Natural Science Foundation of China (31000855), and Grants from the Department of Agriculture of China (2011ZX08010-003). The funders had no role in study design, data collection and analysis, decision to publish, or preparation of the manuscript.

Competing Interests: The authors have declared that no competing interests exist.

* E-mail: xuxiaoli@zju.edu.cn

Introduction

Corn (*Zea mays.* L) is one of the three most widely grown crops worldwide along with wheat and rice [1]. Genetically modified (GM) corn has been increasing in planting acreage since it was commercialized in 1996. GM corn was the second largest biotech crop, after GM soybean, and accounted for 35% of the total corn-planting acreage in the world in 2012 [2]. Herbicide tolerance and insect resistance are two primary input traits for GM plants. GM corn has also been used to produce pharmaceutical protein by a technique known as molecular pharming [3].

Despite the great benefit of using transgene technology in corn, there have been considerable concerns over transgene spreading, particularly the potential for transgene contamination, which is caused by pollen-mediated gene flow [4,5]. Because corn is a wind-pollinated crop, gene flow via pollen commonly occurs. In spite of the strict management policy in transgenic cornfield tests and in commercial planting, GM corn contamination and unintended transgene spreading still occasionally occurred [6–8]. The contamination by GM corn in conventional cornfields often occurred due to unintended seed lot contamination, volunteers, mixing at sowing, cross-pollination, harvest, and trade [9]. In some countries, there is a demand for labeling GM corn food and feed if a threshold value is achieved, 0.9% in EU, for example. In some instances, such as seeds mixing during the sowing period and the presence of volunteers, the adventitious rates of GMO in corn could exceed the threshold value [10]. Volunteer corn could be common in temperate areas. During harvesting, some cobs, cob fragments or isolated kernels may remain in the fields. Depending on the climatic conditions and crop management, the seeds on these kernels from the previous year might germinate, and these GM plants might flower together with that season's plants [11]. Usually, transgenic plants are difficult to detect, let alone to selectively eliminate the transgenic plants from the non-transgenic ones in a large area of field crops. Furthermore, the gene flow between the transgenic crops and their wild relatives is also an important concern when GM corn is grown in its center of domestication, such as Mexico. Because insect resistant and herbicide tolerant genes confer some advantages under the pressure of insect pests and herbicides, these transgenes might accumulate in the wild relatives over time. To minimize transgene spreading and contamination, new biological confinement technologies are highly desirable in addition to physical separation and isolation measures.

To date, several biological confinement strategies have been developed for containing transgene spreading, such as plastid transformation, male sterility, and gene use restriction technologies (GURTs). Some reviews have evaluated these strategies in detail [12–15]. Plastid transformation has been successfully developed in many plants, such as tobacco, tomato and oilseed rape [16]. However, cereals are particularly recalcitrant to plastid transformation [17]. In recent years, some progress has been reported in rice [18]; however, plastid transformation in corn remains unfeasible. Cytoplasmic male sterility (CMS) was suggested as a biocontainment strategy of GM corn pollen [19]. A blend of cytoplasmic male-sterile hybrids and unrelated male-fertile hybrids, which were called Plus-Hybrids, could minimize the release of GM pollen and simultaneously increase the yield [20]. Gene use restriction technologies (GURTs) were developed to produce sterilized seeds, which germinate only when they are exposed to a specific activator molecule [21] However, these GURTs have not been used in the field for major grain crops, such as corn and rice [12]. Most of the biological confinement mechanisms that were described above were far from being used for commercial production [22]. In addition, some of these confinement mechanisms (such as plastid transformation or male sterility) could not prevent gene spreading that resulted from seed dispersal during cultivation, harvest or transportation.

In previous reports, we described a built-in strategy for the containment of transgenic rice [23–25]. Such transgenic rice was created to be sensitive to bentazon by suppressing the expression of a bentazon degradation enzyme, which was encoded by a cytochrome P450 gene *CYP81A6* [23]. This decontamination method could be incorporated into the rice weed control process, and thus, this method is a simple, reliable and inexpensive way to selectively kill transgenic plants in non-transgenic fields.

Nicosulfuron, which is a sulfonylurea herbicide that was developed by DuPont, has been successfully used for weed control in corn [26]. Sulfonylurea herbicides control a wide range of annual and perennial grasses and broadleaf weeds. This type of herbicide has low application rates and displays low levels of acute and chronic animal toxicity [27]. Nicosulfuron has the widest corn safety margin and the fewest sensitive varieties [28]. Tolerant corn plants metabolize nicosulfuron by hydroxylation similar to cytochrome P450 [29,30]. Previously, the cytochrome P450 enzyme, which is responsible for the detoxification of nicosulfuron in corn, was cloned as *nsf1* (*CYP81A9*, GI: 195612396) by Dam et al. [31] and by our group [32].

In this study, we report a method to create glyphosate tolerant GM corn plants that are sensitive to nicosulfuron. This method was achieved by constructing a T-DNA, which consisted of two functional cassettes: one cassette contained the glyphosate tolerant *5-enolpyruvylshikimate-3-phosphate synthase* (*EPSPS*) gene *G10*, and the other was an RNA interference (RNAi) cassette, which suppressed the expression of the nicosulfuron detoxifying enzyme gene *CYP81A9*. We demonstrated that such transgenic corn could be selectively eliminated by spraying nicosulfuron. Based on laboratory and field test results, we concluded that any hybrid progenies that carry these transgenes could be selectively eliminated by nicosulfuron during the regular weed control process in cornfields.

Results

Construction of a T-DNA Plasmid for Creating Terminable Transgenic Corn

A binary T-DNA transformation plasmid was built that was based on pCAMBIA1300. This plasmid contained an RNAi cassette, which targeted the corn cytochrome P450 gene *CYP81A9*, and an expression cassette for the glyphosate tolerant *5-enolpyruvylshikimate-3-phosphate synthase* (*EPSPS*) gene *G10* (GI: 8469109) (Fig. 1). The RNAi cassette included the cauliflower mosaic virus 35S promoter and an inverted repeat sequence of 712 bp of the corn cytochrome P450 gene *CYP81A9*. The *EPSPS* gene *G10* was originally cloned from *Deinococcus radiodurans* R1 and was then further codon-optimized and synthesized for corn expression. The *G10* expression cassette contained the following sequences: the *Z. mays* polyubiquitin-1 promoter (pZmUbi-1), a chloroplast transit signal peptide from the corn *acetohydroxyacid synthase* gene (GB: X63553.1), the codon-optimized synthetic *G10* gene, and a 3' end terminator fragment from corn *phosphoenolpyruvate carboxylase*. The chloroplast transit signal peptide was used to direct the G10 protein to chloroplasts. The RNAi cassette was constructed in tandem to the *G10* gene expression cassette inside the same T-DNA (Fig. 1).

Corn Transformation and Identification of Transgenic Plants

The corn line "hybrid Hi-II" was used as the recipient of the transgene, and the transformation was conducted using an *Agrobacterium*-mediated approach. A typical protocol for the high-efficiency transformation of maize (*Zea mays* L.), which was mediated by *Agrobacterium tumefaciens*, was used for transformation with only minor modifications [33]. For the transformation, the embryonic tissue from 8–10 day-developing seeds were cultured with the *Agrobacterium tumefaciens* strain LBA4404. To select transgenic events, 2 mM glyphosate was used. In total, 109 independent transgenic events with 375 plants were obtained. All these transgenic plants survived on rooting media, which contained 0.1 mM glyphosate and were all positive in the PCR detection of the *G10* gene, as expected. All T0 transgenic plants were cultivated in a greenhouse and were crossed with the elite corn line Zheng-58 to generate T0×Zheng-58 plants. The T0×Zheng-58 plants were used for further analysis.

Analysis of the T0×Zheng-58 Transgenic Corn Plants

The hybrid seeds that were harvested from T0 plants were germinated, and seedlings were grown in the greenhouse to test their sensitivity to nicosulfuron and glyphosate. Hybrids between Hi-II and Zheng-58 were used as the control plants. The T0×Zheng-58 plants were divided into two groups for the herbicide spray test. One group of the T0×Zheng-58 plants was sprayed with nicosulfuron at 60 mg/L, and the other group was sprayed with 4 g/L glyphosate at the V4 to V5 stage. Among the 109 transgenic events that were generated, 35 were obviously sensitive to nicosulfuron and exhibited symptoms from varying degrees of injuries, such as stunting and malformation, to death 10 days after nicosulfuron spray. The other events showed no obvious damage that was caused by the application of nicosulfuron.

To search for a transgenic event that was sensitive to nicosulfuron and that only had a single copy of the transgene, 5 of the 35 sensitive events were selected for Southern blot analysis. Among the 5 events that were analyzed, events R450-42, R450-58, and R450-93 had a single copy of T-DNA insertions (Fig. 2).

Due to segregation, approximately half of the T0×Zheng-58 plants of events R450-42, R450-58 and R450-93 were expected to have the transgenes, which confer sensitivity to nicosulfuron but tolerance to glyphosate. Indeed, as expected, spraying nicosulfuron severely damaged or killed approximately half of the T0×Zheng-58 plants of events R450-42, R450-58 and R450-93 in 10 days (Table 1, Fig. 3A, 3B and 3C); however, spraying nicosulfuron did not incur any visible damage to the non-transgenic control plants (Fig. 3D). In contrast, spraying glyphosate killed all non-transgenic control plants (Fig. 3H), as well as approximately half of the T0×Zheng-58 plants of events R450-42, R450-58 and R450-93 (Table 1, Fig. 3E, 3F and 3G).

The plants that were killed by nicosulfuron were expected to contain the transgene, whereas the plants that were killed by glyphosate were segregates without the transgene. To verify this prediction, Western blot analysis was performed on the tested plants of the event R450-58 to detect the G10 protein. The results confirmed that the dying plants, which were affected by nicosulfuron, were the transgenic plants, whereas the survived

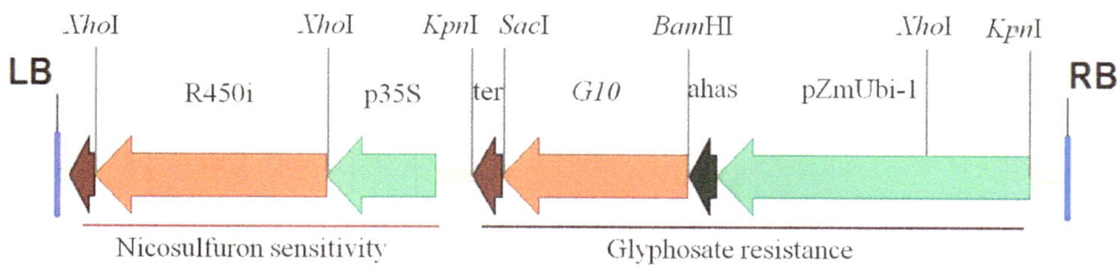

Figure 1. Diagram of T-DNA used for corn transformation. pZmUbi-1, the Z. mays polyubiquitin-1 promoter; ahas, a chloroplast transit signal peptide from a corn acetohydroxyacid synthase gene, which was used to direct the G10 gene to the chloroplast; G10, the corn codon-optimized synthetic glyphosate resistant 5-enolpyruvylshikimate-3-phosphate synthase gene, which was isolated from Deinococcus radiodurans R1;ter, a 3′ end terminator fragment from the corn phosphoenolpyruvate carboxylase (PEPC); p35S, cauliflower mosaic virus 35S promoter; 450i, the inverted repeat sequence of a 712 bp fragment of CYP81A9; RB, right border of T-DNA; LB, left border of T-DNA.

plants were non-transgenic segregates (Fig. 4, upper panel). The Western analysis further indicated that the plants that were killed by glyphosate were non-transgenic segregates, whereas the survived plants were transgenic (Fig. 4, lower panel). These results clearly demonstrated that the transgenic corn plants that were generated in this study could be selectively eliminated by nicosulfuron.

RNAi Suppression of the Nicosulfuron Detoxification Gene *CYP81A9*

To analyze the efficiency of the suppression of *CYP81A9* transcripts in transgenic corn plants, qRT-PCR was performed using the *GAPDH* (*glyceraldehyde-3-phosphate dehydrogenase*) gene as an internal control. qRT-PCR analysis revealed that the transcript levels of *CYP81A9* in the event R450-58, which showed high sensitivity to nicosulfuron, decreased by 97.2% compared with its expression levels in non-transgenic control plants (Fig. 5). This result indicated that the expression of the *CYP81A9* gene was drastically suppressed by transforming this RNAi cassette. Similar results were also observed among other independent events that were sensitive to nicosulfuron (data not shown).

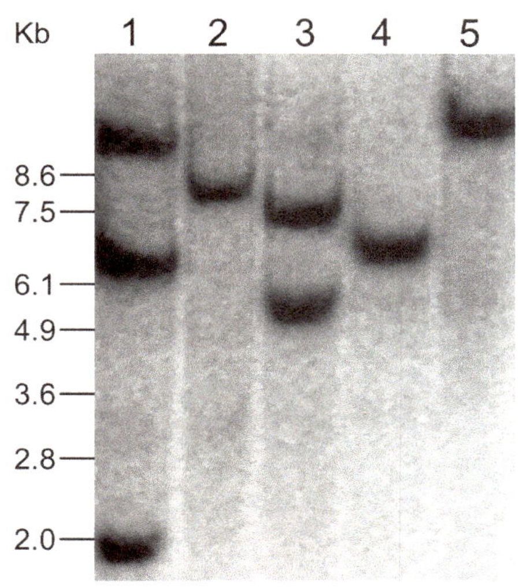

Figure 2. Southern blot analysis of transgenic corn. The genomic DNA of the five independent transgenic events were digested by BamHI and then probed by the synthetic 5-enolpyruvylshikimate-3-phosphate synthase (EPSPS) gene, G10. Lanes 1–5, transgenic events R450-8, R450-42, R450-53, R450-58, and R450-93.

Field Test of the Terminable Transgenic Corn

A field test was performed for the further analysis of the sensitivity of T0×Zheng-58 transgenic plants to nicosulfuron and glyphosate. The plants of the event R450-58 were planted in experimental plots in the summer of 2012 in Hangzhou, Zhejiang Province, China. Non-transgenic corn plants that were produced by crossing Hi-II with Zheng-58 were planted along with the transgenic corn as controls. In total, 98 individual corn seedlings of R450-58 were analyzed by PCR before spraying to identify the non-transgenic segregates, which were removed from the field. The segregation ratio was approximately 1:1 (data not shown), which is consistent with a single T-DNA copy. The planting rows were divided into two parts: one part was sprayed with 40 g/ha nicosulfuron, whereas the other part was sprayed with 4 kg/ha glyphosate at the V5 stage. These application rates were typically used to control weeds under field conditions. Sensitivity was scored 10 days and 20 days after herbicide treatments. All of the transgenic plants were sensitive to nicosulfuron but were resistant to glyphosate. In contrast, the non-transformed control plants were resistant to nicosulfuron but were sensitive to glyphosate (Fig. 6).

Furthermore, there were no significant morphological differences between the transgenic and non-transgenic plants in the field test. The agronomic traits of plant height, ears length, number of kernels per ear and weight per 1000 kernels were measured for the transgenic event R450-58 and the control plants. There were no significant differences ($P>0.05$) between the transgenic and non-transgenic plants in all these parameters, which were analyzed by Student's *t*-test using the DPS statistical software (Table 2). This result suggested that the suppression of *CYP81A9* expression did not have statistically significant side effects on the agronomic performance of the transgenic corn in the field.

Discussion

Previously, we reported a built-in strategy for the containment of transgenes in rice [23]. This method was found to be effective in mitigating the spread of transgenes when this method was used in the development of insect-resistant and herbicide-tolerant GM rice [25].

In the study, we demonstrated that the transgenic corn plants that contained the silenced nicosulfuron-detoxifying gene *CYP81A9* were highly sensitive to nicosulfuron and could be selectively killed by spraying nicosulfuron at a regular application dosage. Because nicosulfuron is widely used for weed control in conventional cornfields, any transgenic corn plants that are developed by this method can be selectively decontaminated without extra effort or cost. Therefore, this transgene spreading

Table 1. Herbicide resistance of segregating T0×Zheng-58 plants.

Events	Nicosulfuron			Glyphosate		
	Total plants	Surviving plants	Survival (%)	Total plants	Surviving plants	Survival (%)
R450-42	19	10	52.6	16	9	56.3
R450-58	20	9	45.0	18	9	50.0
R450-59	17	8	47.1	15	8	53.3
CK	7	7	100.0	8	0	0.0

The T0×Zheng-58 hybrid plants of R450-42, R450-58 and R450-93 along with non-transgenic ones (CK) were cultured in a greenhouse and sprayed with 60 mg/L nicosulfuron or 4 g/L glyphosate. The data were recorded 10 days after spraying.

control method is simple, preventive and could easily be incorporated into the regular weed control process.

In addition to its application in regular transgenic corn, such as transgenic insect-resistant corn, this built-in containment strategy is especially useful for the development of transgenic plants as bioreactors [23,24]. Corn has many advantages for the large-scale production of recombinant pharmaceutical proteins, which makes corn the widest used cereal for molecular pharming [3]. However, the risk of transgene spreading is relatively high because corn is a wind-pollinated plant. Due to the nature of the pharmaceutical or industrial proteins, it is particularly important to develop a reliable method for the containment of such transgenes. The method that is described in this report could be an ideal technology for minimizing the risk of contamination of such transgenic corn into food and feed supplies.

Materials and Methods

Construction of Binary Vector for Corn Transformation

Genomic DNA was extracted from young leaves of corn according to a standard CTAB method [34]. The 712 bp fragment of *CYP81A9* (GI: 195612396) DNA was obtained by PCR using a forward primer 450F (5′-CTCGAGTTCTC-CATGCGCCTGGGGACC-3′, with the *Xho*I recognition site underlined) and a reverse primer 450R1 (5′-AGATCTCAGT-GATCACAGTGTCAGTGTAGAC-3′, with the *Bgl*II recognition site underlined). This fragment represents the 223 to 934 bp of the genomic DNA from the initial codon. The other 923 bp fragment of *CYP81A9* DNA was amplified by PCR using the primer 450F and the reverse primer 450R2 (5′-AGATCTACA-GACTATGTCAACATAAAGCAC-3′, with the *Bgl*II recognition site underlined). This fragment represents the 223 to 1145 bp of the genomic DNA from the initial codon. Both PCR products were separately cloned into the pMD-T vector (Shanghai Sangon,

Figure 3. Sensitivity test of transgenic corn plants to nicosulfuron or glyphosate. The T0×Zheng-58 transgenic corn events (R450-42, R450-58 and R450-93) along with non-transgenic ones (CK) were cultured in a greenhouse and sprayed with 60 mg/L nicosulfuron (A–D) or 4 g/L glyphosate (E–H). The photos were taken 10 days after spraying.

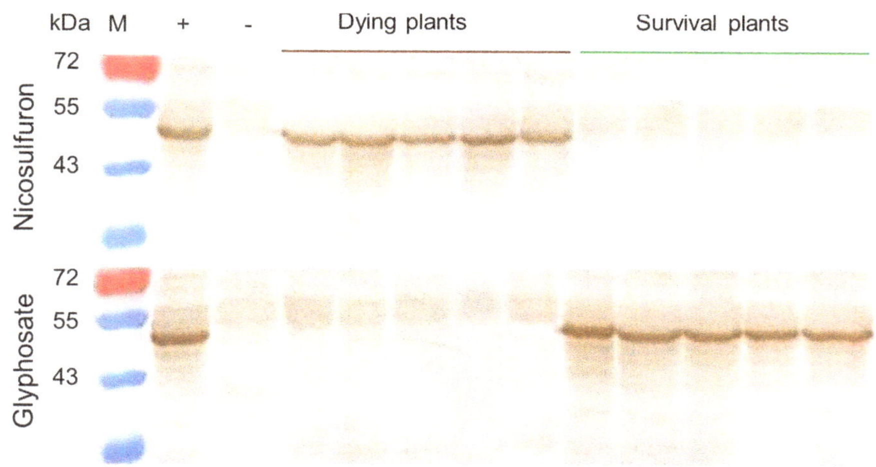

Figure 4. Western blot analysis of G10 (EPSPS) expression in transgenic T0×Zheng-58 plants. The T0×Zheng-58 transgenic corn plants of R450-58 were cultured in a greenhouse and sprayed with 60 mg/L nicosulfuron (upper panel) or 4 g/L glyphosate (lower panel). After 7 days of the treatment, the dying plants and the survival plants were analyzed by Western blotting. −, non-transgenic corn as the negative control; +, G10 protein positive control; M, molecular weight markers.

China) and sequenced. Plasmids were digested with *Xho*I and *Bgl*II to obtain the inserts. A three-way ligation was performed to clone these two fragments into pCAMBIA1300, which was predigested with *Xho*I and dephosphorylated by CIAP. The resulting plasmid, which contained a 712 bp inverted repeat sequence of *CYP81A9* for RNA interference, was designated as p1300-450i. Essentially, the *hpt*II gene (hygromycin resistance) in the pCAMBIA1300 vector was replaced with the 450i gene, which was driven by the CaMV 35S promoter for nicosulfuron sensitivity.

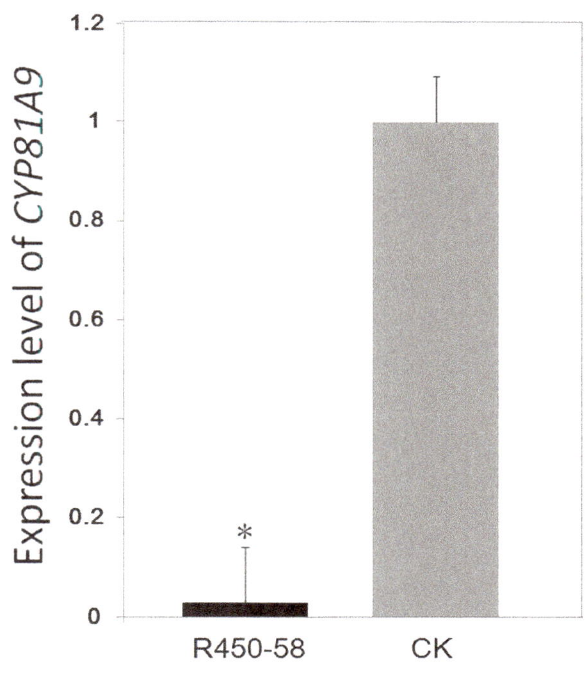

Figure 5. Relative expression of CYP81A9 in transgenic plants (R450-58) and non-transgenic plants (CK). The relative expression levels of CYP81A9 was shown as 2-△△CT, using corn GAPDH as an internal control. The data are expressed as the means of quadruplicates with standard deviations shown by vertical bars. *indicates statistically significant differences between the transgenic corn plants and the non-transgenic control plants under the same conditions (*P<0.05; Student's t-test).

The *Z. mays* polyubiquitin-1 promoter (pZmUbi-1) DNA fragment, including the chloroplast transit signal peptide from the *acetohydroxyacid synthase* of *Z. mays*, was generated by digesting a pMD-T plasmid, which contained the promoter with *Kpn*I and *Bam*HI. The corn codon-optimized synthetic glyphosate resistant *5-enolpyruvylshikimate-3-phosphate synthase* (*EPSPS*) gene *G10* (GI: 8469109), including a 3′ end terminator fragment from the corn *phosphoenolpyruvate carboxylase* (PEPC), was synthesized by Shanghai Sangon Limited, China. To facilitate the cloning, the restriction sites of *Bam*HI and *Kpn*I were introduced into the 5′ and 3′ ends of the *G10* gene, respectively. By a three-way ligation, the restriction fragments of the ZmUbi-1 promoter that was fused with the signal peptide and the *G10* gene were inserted into the plasmid p1300-450i, which was predigested with *Kpn*I and dephosphorylated by CIAP. Finally, the direction of the inserts was identified by *Xho*I. Thus, the RNA interference cassette was constructed in tandem with the glyphosate resistance cassette inside the T-DNA. The final plasmid, p1300-450i-G10, with the two expression cassettes in same orientation, was used for corn transformation.

Agrobacterium-mediated Corn Transformation

The plasmid p1300-450i-G10 was transformed into the *Agrobacterium tumefaciens* strain LBA4404 by electroporation using an Electroporator 2510 (Eppendorf) following the manufacturer's instructions. The hybrid Hi-II corn (*Z. mays* L.) was transformed using a standard *Agrobacterium*-mediated transformation method, which was described previously [33], except that the final concentration of 2 mM glyphosate (Sigma) was used as the selection agent during callus culture and that 0.1 mM glyphosate was used during rooting stage. The independently transformed events were propagated in sterile culture and then planted in soil in the greenhouse for artificial pollinations. T0 plants were crossed with the elite event Zheng-58 to obtain the T0×Zheng-58 seeds.

Greenhouse Test

The resulting T0×Zheng-58 plants from T0 plants, which were crossed with Zheng-58 and seeds were planted in the greenhouse.

Figure 6. Field trial for T0×Zheng-58 population. The T0×Zheng-58 transgenic plants (R450-58), along with the non-transgenic control plants (CK), were sprayed with 40 g/ha of nicosulfuron (A) or 4 kg/ha of glyphosate (B). The photos were taken 10 days after spraying.

Hybrid plants from Hi-II that were crossed with Zheng-58 were used as controls. The amount of the plants that were evaluated for the herbicide resistance was unequal according to the total number of seeds, which were obtained from the events. Briefly, 3–4 plants were planted in one pot (some seeds did not germinate). Plants from an event were planted in 2, 4, 8 or 10 pots for the herbicide spray test. Each of the transgenic events was divided into two groups for the herbicide spray test. One group of the T0×Zheng-58 plants was sprayed with nicosulfuron at 60 mg/L, whereas the other group of the T0×Zheng-58 plants was sprayed with 4 g/L glyphosate at the V4 to V5 leaf stage. A handhold sprayer was used to apply the herbicide formulations. Individual plants were assessed for the nicosulfuron response score 10 days after being sprayed and assigned a visual response score from 1 to 4 (1 = dead plant, 2 = serious damaged plant, 3 = slight damaged plant and 4 = no effect observed). Nicosulfuron (4% suspending solution, Ishihara Sangyo Kalsha Ltd., Japan) and Roundup (41% isopropylamine salt of glyphosate, Monsanto) were used for the herbicide spraying test.

Table 2. Comparison of agronomic traits between the transgenic corn plants (R450-58) and its non-transgenic control plants (CK) in the field trial.

	CK	R450-58	P-value
Plant height (cm)	175.95±1.93	177.90±1.35	0.0707
Ear length (cm)	16.93±0.28	16.36±0.57	0.0521
Kernels per ear	464.33±5.39	454.67±9.48	0.0551
Weight per 1000 kernels (g)	211.62±1.72	213.27±0.95	0.0662

At harvest time, the plant height, ear length and kernels per ear of mature ears were recorded. Mature ears were dried and weighed to determine the weight per 1000 kernels. The values are the means ± SD (n = 6).

P-values indicate statistically insignificant differences between transgenic plants (R450-58) and the non-transgenic control plants (CK) under the same conditions at the 0.05 level, using Student's *t*-test.

Western Blot Analysis

A standard Western blot analysis was performed to detect the expression of G10 in both the survival and the dying transgenic corn plants after being sprayed with 60 mg/L nicosulfuron or 4 g/L glyphosate. The samples from all tested plants were collected 7 days after the treatment. Leaf samples that were collected from transgenic plants as well as from non-transgenic control plants were ground to a powder in liquid nitrogen and then suspended in SDS sample buffer. After lysis and centrifugation, the protein samples were separated by 10% SDS-PAGE and then blotted onto nitrocellulose membranes. The rabbit polyclonal antibodies that were generated against G10 were used as the primary antibodies, and the horseradish peroxidase-conjugated goat anti-rabbit IgG was used as the secondary antibody (Promega) in the analysis.

Southern Blot Analysis

For Southern blot analysis, approximately 100 μg of genomic DNA was digested with the *Bam*HI restriction enzyme. The digested genomic DNA was size-fractionated on a 0.7% (w/v) agarose gel by electrophoresis, transferred onto a positively charged nylon membrane and cross-linked to the membrane at 121°C for 30 min. The hybridization probes, which were specific to the *G10* gene, were prepared as described in the DIG System Manual (Roche). The DNA template that was used for producing the *G10* probe was amplified by PCR using the primers G10-F (5′-CACCTTCGACGTGATCGTGCATCCA-3′) and G10-R (5′-CGAGG TGAGCGAAGAACTGAGGG TAGGA-3′).

Total RNA Isolation and cDNA Synthesis

To minimize dehydration- and wounding-induced gene expression, leaf samples were quickly excised, wrapped in aluminum foil, and immediately frozen in liquid nitrogen. Total RNA was extracted from 100 mg of leaves using the TRIzol reagent (Invitrogen). The resulting RNA was treated with RNase-free DNase I (Promega) to remove all genomic DNA residues. The RNA concentration and purity were determined using a NanoDrop ND-2000 spectrophotometer (Thermo Scientific). The quality of RNA samples were also confirmed by electrophoresis in 1% agarose gels. Only intact RNA was used for the following reactions. Finally, 2.5 μg of the DNase-treated RNA samples were reversing transcribed into cDNA with an oligo (dT) $_{18}$ primer using a Revert Aid First Strand cDNA Synthesis Kit (Fermentas). The synthesized cDNA products were diluted 1:10 in nuclease-free water before being used as templates in the qRT-PCR analysis.

qRT-PCR Analysis

Two-step qRT–PCR was performed using a 2×SYBR Green Master Mix (Applied Biosystems®) on an ABI 7500 Real-time PCR System according to the manufacturer's instructions. The level of *CYP81A9* transcript was normalized using the internal control, the *GAPDH* (*glyceraldehyde-3-phosphate dehydrogenase*) gene. The primers that were used for qRT-PCR were 5′-GCTGGCGACGAGAGCGAAAGTA-3′ and 5′-ATGGCC-CATTCCGTCGTGGT-3′ for *CYP81A9* and were 5′-AG-CAGGTCGAGCATCTTCG-3′ and 5′- CTGTAGCCC-CACTCGTTGTC-3′ for *GAPDH*. The PCR reaction volume was 20 μl, which contained 5.0 μl 10-fold diluted cDNA template, 100 μM of each primer and 10.0 μl of 2×SYBR Mix. The cycling conditions were composed of 10 minutes of polymerase activation at 95°C, followed by 40 cycles of 95°C for 15 seconds, and 60°C for 1 minute. To verify that only a single product was amplified, a dissociation curve analysis was performed using the ABI Prism Dissociation Curve Analysis software. All samples were run on the same plate to avoid between-run variations. For each qRT-PCR experiment, four technical repetitions were performed, and the mean values were calculated. The relative expression levels were calculated using the comparative $2^{-\triangle\triangle CT}$ method [35] with *GAPDH* as the internal control.

Field Trials

The R450-58 plants were planted in the experimental plots in the summer of 2012 in Hangzhou, Zhejiang Province, China. Non-transgenic plants, which were prepared by crossing Hi-II with Zheng-58, were planted along with the transgenic corn as controls. The application rates for nicosulfuron and glyphosate were 40 g/ha and 4 kg/ha, respectively. Testing plots were inspected visually 10 days and 20 days post-herbicide application. The number of dead plants was recorded to determine the segregation ratios. Five major agronomic traits including plant height, ears length, number of kernels per ear and weight per 1000 kernels, were recorded from mature plants. Plant heights were measured from the soil surface to the tip of the tassel of each tested plant. Ears length and kernels number were averaged from 6 randomly selected ears from transgenic and non-transgenic control plants.

Statistical Analysis

All the data were presented as the mean ± SD (standard deviation). Student's *t*-test was performed to compare the differences between the transgenic and the non-transgenic corn plants. $P<0.001$ was considered extremely significant, $P>0.05$ was considered non-significant. All statistical analyses were performed using the DPS statistical software (Refine Information Tech. Inc., Hangzhou, China).

Author Contributions

Conceived and designed the experiments: XX ZS JL. Performed the experiments: JL HY FZ XD. Analyzed the data: JL XX CL JG JF. Wrote the paper: XX JL ZS.

References

1. Shiferaw B, Prasanna B, Hellin J, Bänziger M (2011) Crops that feed the world 6. Past successes and future challenges to the role played by maize in global food security. Food Sec 3: 307–327.
2. James C (2012) Global status of commercialized biotech/GM crops: 2012. ISAAA Brief 44: ISAAA, Ithica, New York.
3. Ramessar K, Sabalza M, Capell T, Christou P (2008) Maize plants: an ideal production platform for effective and safe molecular pharming. Plant Sci 174: 409–419.
4. Stewart CN, Richards HA, Halfhill MD (2000) Transgenic plants and biosafety: science, misconception and public perceptions. Biotechniques 29: 832–836, 838–843.
5. Lu BR (2008) Transgene escape from GM crops and potential biosafety consequences: an environmental perspective. Collection of Biosafety Reviews 4: 66–141.
6. Fox JL (2003) Puzzling industry response to ProdiGene fiasco. Nat Biotechnol 21: 3–4.
7. Dalton R (2009) Mexico's transgenic maize under fire. Nature 462: 404.
8. Mezzalama M, Crouch JH, Ortiz R (2010) Monitoring the threat of unintentional transgene flow into maize gene banks and breeding materials. Electron J Biotechnol 13: 1–6.
9. Bannert Michael(2006). Simulation of transgenic pollen dispersal by use of different grain colour maize. ETH, Institute of Plant Sciences, Group Agronomy and Plant Breeding.
10. Messeguer J, Penas G, Ballester J, Bas M, Serra J, et al. (2006) Pollen-mediated gene flow in maize in real situations of coexistence. Plant Biotechnol J 4: 633–645.
11. Palaudelmas M, Penas G, Mele E, Serra J, Salvia J, et al. (2009) Effect of volunteers on maize gene flow. Transgenic Res 18: 583–594.
12. Breyer D, Goossens M, Herman P, Sneyers M (2009) Biosafety considerations associated with molecular farming in genetically modified plants. J Med Plants Res 3: 825–838.
13. Husken A, Prescher S, Schiemann J (2010) Evaluating biological containment strategies for pollen-mediated gene flow. Environ Biosafety Res 9: 67–73.
14. Lee D, Natesan E (2006) Evaluating genetic containment strategies for transgenic plants. Trends Biotechnol 24: 109–114.
15. Valverde BE, Gressel J (2005) Implications and containment of gene flow from herbicide-resistant rice (*Oryza sativa*). 20th Asian-Pacific Weed Science Society Conference. Ho-Chi Minh City Vietnam 63–84.
16. Maliga P (2012) Plastid transformation in flowering plants.In:Bock R,Knoop V, editors.Genomics of chloroplasts and mitochondria. Dordrecht:Springer. pp, 393–414.
17. Rigano MM, Scotti N, Cardi T (2012) Unsolved problems in plastid transformation. Bioengineered 3: 329–333.
18. Lee SM, Kang K, Chung H, Yoo SH, Xu XM, et al. (2006) Plastid transformation in the monocotyledonous cereal crop, rice (*Oryza sativa*) and transmission of transgenes to their progeny. Mol Cells 21: 401–410.
19. Weider C, Stamp P, Christov N, Husken A, Foueillassar X, et al. (2009) Stability of cytoplasmic male sterility in maize under different environmental conditions. Crop Sci 49: 77–84.
20. Munsch MA, Stamp P, Christov NK, Foueillassar XM, Husken A, et al. (2010) Grain yield increase and pollen containment by Plus-Hybrids could improve acceptance of transgenic maize. Crop Sci 50: 909–919.
21. Hills MJ, Hall L, Arnison PG, Good AG (2007) Genetic use restriction technologies (GURTs): strategies to impede transgene movement. Trends in Plant Sci 12: 177–183.
22. Breyer D, Schrijver AD, Goossens M, Pauwels K, Herman P (2012) Biosafety of molecular farming in genetically modified plants. In: Wang A, Ma S editors. Molecular farming in plants: recent advances and future prospects. Dordrecht: Springer. pp, 259–274.
23. Lin C, Fang J, Xu X, Zhao T, Cheng J, et al. (2008) A built-in strategy for containment of transgenic plants: creation of selectively terminable transgenic rice. PLoS One 3: e1818.
24. Lin C, Nie P, Lu W, Zhang Q, Li J, et al. (2010) A selectively terminable transgenic rice event expressing human lactoferrin. Protein Expr Purif 74: 60–64.
25. Liu C, Li J, Gao J, Shen Z, Lu BR, et al. (2012) A built-in mechanism to mitigate the spread of insect-resistance and herbicide-tolerance transgenes into weedy rice populations. PLoS One 7: e31625.
26. Brown H, Cotterman J (1994) Recent advances in sulfonylurea herbicides.In: Stetter J,editor.Herbicides inhibiting branched-chain amino acid biosynthesis. New York: Springer Berlin Heidelberg. 47–81.
27. Brown H (1990) Mode of action, crop selectivity, and soil relations of the sulfonylurea herbicides. Pesticide Sci 29: 263–281.
28. Green JM, Ulrich JF (1993) Response of corn (*Zea Mays* L.) inbreds and hybrids to sulfonylurea herbicides. Weed Sci 41: 508–516.
29. Kreuz K, Tommasini R, Martinoia E (1996) Old enzymes for a new job-herbicide detoxification in plants. Plant Physiol 111: 349–353.

30. Kang MS (1993) Inheritance of susceptibility to nicosulfuron herbicide in maize. J Hered 84: 216–217.
31. Dam T, Guida AD, Hazel CB, Li B, Williams ME (2010) Polynucleotide encoding a maize herbicide resistance gene and methods for use. US Patent 7,705,200.
32. Shen Z, Lin C, Cheng Y (2006) Herbicide resistant gene and use thereof. CN Patent WO/2012/068966 A1.
33. Ishida Y, Saito H, Ohta S, Hiei Y, Komari T, et al. (1996) High efficiency transformation of maize (*Zea mays* L.) mediated by *Agrobacterium tumefaciens*. Nat Biotechnol 14: 745–750.
34. Dellaporta SL, Wood J, Hicks JB (1983) A plant DNA minipreparation: version II. Plant Mol Biol Reptr 1: 19–21.
35. Livak KJ, Schmittgen TD (2001) Analysis of relative gene expression data using real-time quantitative PCR and the$2^{-\triangle\triangle CT}$ method. Methods 25: 402–408.

16

Annual Glyphosate Treatments Alter Growth of Unaffected Bentgrass (*Agrostis*) Weeds and Plant Community Composition

Collin W. Ahrens, Carol A. Auer*

Agricultural Biotechnology Lab, Department of Plant Science, University of Connecticut, Storrs, Connecticut, United States of America

Abstract

Herbicide resistance is becoming more common in weed ecotypes and crop species including turfgrasses, but current gaps in knowledge limit predictive ecological risk assessments and risk management plans. This project examined the effect of annual glyphosate applications on the vegetative growth and reproductive potential of two weedy bentgrasses, creeping bentgrass (CB) and redtop (RT), where the glyphosate resistance (GR) trait was mimicked by covering the bentgrass plants during glyphosate application. Five field plots were studied in habitats commonly inhabited by weedy bentgrasses including an agricultural hayfield, natural meadow, and wasteland. Results showed that annual glyphosate treatment improved bentgrass survivorship, vegetative growth, and reproductive potential compared with bentgrass in unsprayed subplots. In the second year of growth, RT plants had an 86-fold increase in flower number in glyphosate-treated subplots versus controls, while CB plants had a 20-fold increase. At the end of the three year study, plant community composition had changed in glyphosate-treated subplots in hayfield and meadow plots compared to controls. Soils in subplots receiving glyphosate had higher nitrate concentrations than controls. This is the first study to mimic the GR trait in bentgrass plants with the goal of quantifying bentgrass response to glyphosate selection pressure and understanding the impacts on surrounding plant communities.

Editor: Luis Herrera-Estrella, Centro de Investigación y de Estudios Avanzados del IPN, Mexico

Funding: This project was supported by the Biotechnology Risk Assessment Program Competitive Grant no. 2008-39211-19566 from the USDA National Institute of Food and Agriculture (http://www.csrees.usda.gov/fo/biotechnologyriskassessment.cfm) to CAA. Funds were also provided by the Storrs Agricultural Experiment Station and the University of Connecticut. The funders had no role in study design, data collection and analysis, decision to publish, or preparation of the manuscript.

Competing Interests: The authors have declared that no competing interests exist.

* E-mail: carol.auer@uconn.edu

Introduction

Herbicide resistance has increased in weed species concurrent with the expanding acreage used for herbicide resistant crops in the U.S., but the long-term environmental impact of these traits on plant communities is largely unknown [1]. Crops and weeds have become resistant to the herbicide glyphosate [*N*(phosphono-methyl)glycine] through two mechanisms: 1) genetic engineering with the 5-enolpyruvylshikimate-3-phosphate synthase (EPSPS) transgene or glyphosate oxidoreductase (GOX) transgene, or 2) repeated glyphosate selection pressure coupled with physiological and genetic changes [2,3,4]. Genetic engineering has been used in the U.S. to commercialize glyphosate resistance (GR) in important agronomic crops (e.g. maize, cotton, soybeans, alfalfa, canola, and sugarbeet) and the trait has been found in feral GR crops (e.g. canola, volunteer corn) [5,4]. Genetic engineering has been used to add the GR trait to two perennial turfgrasses, creeping bentgrass (*Agrostis stolonifera*) and Kentucky bluegrass (*Poa pratensis* L.). While GR creeping bentgrass remains under regulatory review in the U.S., the USDA recently declined to regulate GR Kentucky bluegrass because it was generated without using any DNA sequences from an organism deemed a plant pest [6,7].

Unfortunately, the presence of the GR trait in diverse weed species is negatively affecting the utility of GR crops [8,9]. Selection pressure has created GR and glyphosate tolerant ecotypes in diverse weed species such as Canadian horseweed (*Conyza canadensis*), carelessweed (*Amaranthus palmeri*), ryegrass (*Lolium* species), Indian goosegrass (*Eleusine indica*), annual ragweed (*Ambrosia artemisiifolia*), Johnsongrass (*Sorghum halapense*), Mexican fireplant (*Euphorbia heterophylla*) and narrowleaf plantain (*Plantago lanceolata*) [10,1,4]. The mechanism for GR in weeds appears to vary with evidence for EPSPS gene amplification, EPSPS gene mutation, vacuolar sequestration and other mechanisms [11,12,13]. Regardless of the mechanism, glyphosate selection pressure will cause the heritable GR trait to spread in weed populations via pollen-mediated gene flow, seed dispersal, and vegetative propagules. Without selection pressure, the GR trait will be governed by genetic drift and gene flow. However, a study of field mustard (*Brassica rapa*) showed that the GR trait can persist without selection pressure [14].

Plant populations and community assemblages are affected by natural forces such as plant competition, facilitation, and stochastic processes [15]. In addition, many plant communities are also intentionally or unintentionally impacted by management regimes including herbicides and herbicide drift. Glyphosate usage has increased dramatically in the U.S., and recent experiments have shown that glyphosate can be detected in some air and surface water samples [16,17]. Deposition of sub-lethal glyphosate concentrations (drift) can negatively affect plant reproductive fitness [5]. However, there is no clear consensus about the effects of glyphosate on plant community assemblages or biodiversity. While some studies have reported that glyphosate applications had little impact on species richness or diversity in plant communities [18], other studies have reported significant changes [19,20]. A U.K. study by Heard et al. [21] showed that herbicides reduced weed abundance in GR beet (*Beta*) and canola (*Brassica*) fields, but weed species diversity and species richness within the fields was affected.

Risk assessments for the ecological impact of herbicide resistance trait require knowledge about host plant biology, crop-to-wild gene flow, and the ecology of the site for environmental release [22,23,24]. GR creeping bentgrass (*Agrostis stolonifera* L., CB) has generated concern because it is a cold-hardy perennial, a weed in many natural and managed habitats, part of a complex of closely-related species that hybridize through pollen-mediated gene flow, and able to spread vegetatively [25,26,27]. *Agrostis* species (bentgrasses) are common in early to mid-successional grasslands [28], implying that they readily establish in habitats with a variety of plant competion levels [29]. CB is a non-native species in the U.S. and a common weed outside of cultivation [30,31,32,33]. Our previous research showed that that CB is common and abundant as a weed in various managed and disturbed habitats (e.g. roadsides, wastelands) in two ecoregions in the Northeastern U.S. [31]. Furthermore, CB frequently co-occurs with a closely-related, non-native *Agrostis* called redtop (*Agrostis gigantea* Roth., RT). This co-occurrence suggests a potential for pollen-mediated gene flow [31]. In another study, CB, four other bentgrass species, and interspecific CB hybrids were able to establish and compete in residential and roadside habitats [34]. CB has also been shown to persist in sand dune, salt marsh, meadow and polder habitats [35]. In the Western U.S., pollen-mediated transgene flow between CB and RT populations moved the GR trait long distances [36], and removal of the escaped GR CB plants has been problematic along irrigation ditches because glyphosate is the only herbicide approved for these waterways [37]. Thus, the complex of native and non-native *Agrostis* species already present in the U.S. support intraspecific and interspecific gene flow, although the temporal and spatial scales have not been defined [38,37,27,39].

It has been reported that GR CB plants were not significantly different to their non-transformed counterparts with regard to infloresence development, reproductive morphology, pollen longevity, seed-set capacity or growth and establishment [40,41]. However, these studies did not examine CB growth under glyphosate selection pressure which would reduce competition between CB and other plants in natural, cultural or agricultural landscapes. It is important to understand GR CB behavior with glyphosate selection pressure in order to predict its behavior as a weed, create predictive models of gene flow and dispersal, and quantify impacts on plant community assemblages. Therefore, the primary goals of this project were to: 1) examine the impact of annual glyphosate applications on established plant communities in agricultural and natural areas, and 2) determine if one glyphosate application per year could alter the survivorship, growth, and reproductive capacity of introduced CB and RT plants. In this project, the GR trait in the bentgrass plants had to be mimicked due to a variety of factors including the potential for transgene escape during flowering and seed production. Our first hypothesis was that plant community composition would be altered in subplots receiving glyphosate compared to control (unsprayed) subplots. Second, we predicted that CB and RT plants with mimicked GR in glyphosate-treated subplots would have increased growth and reproductive potential compared to control subplots. To our knowledge, this is the first study to examine the effects of glyphosate on weedy bentgrass plants in the context of natural and agricultural plant communities.

Materials and Methods

Field Plots and Glyphosate Treatment

Five field plots (12 m×15 m) were established in Mansfield, Connecticut in the Lower New England Ecoregion, subecoregion

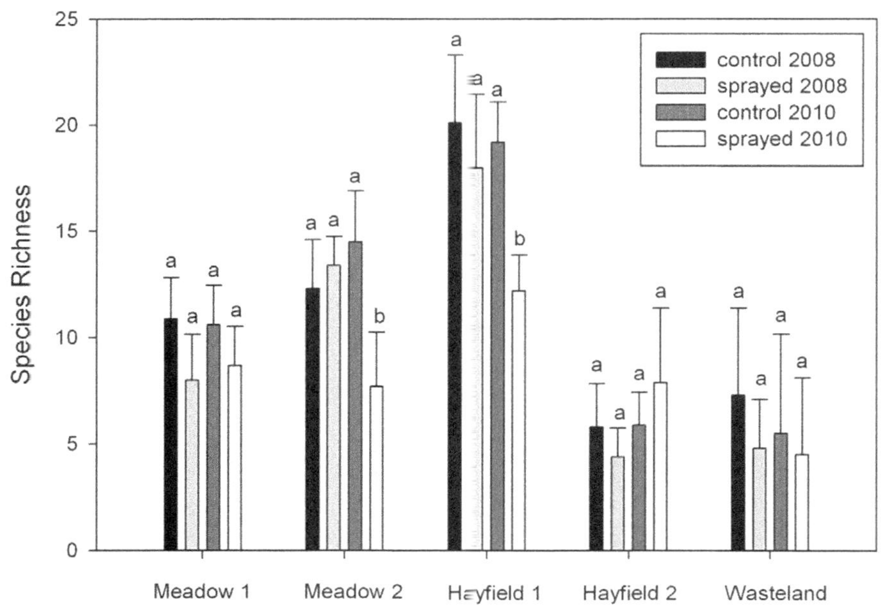

Figure 1. Species richness in subplots before the study (2008) and at the end of the experiment (2010). Error bars indicate SD and letters indicate differences within plot locations according to Fisher's LSD test ($\alpha<0.05$).

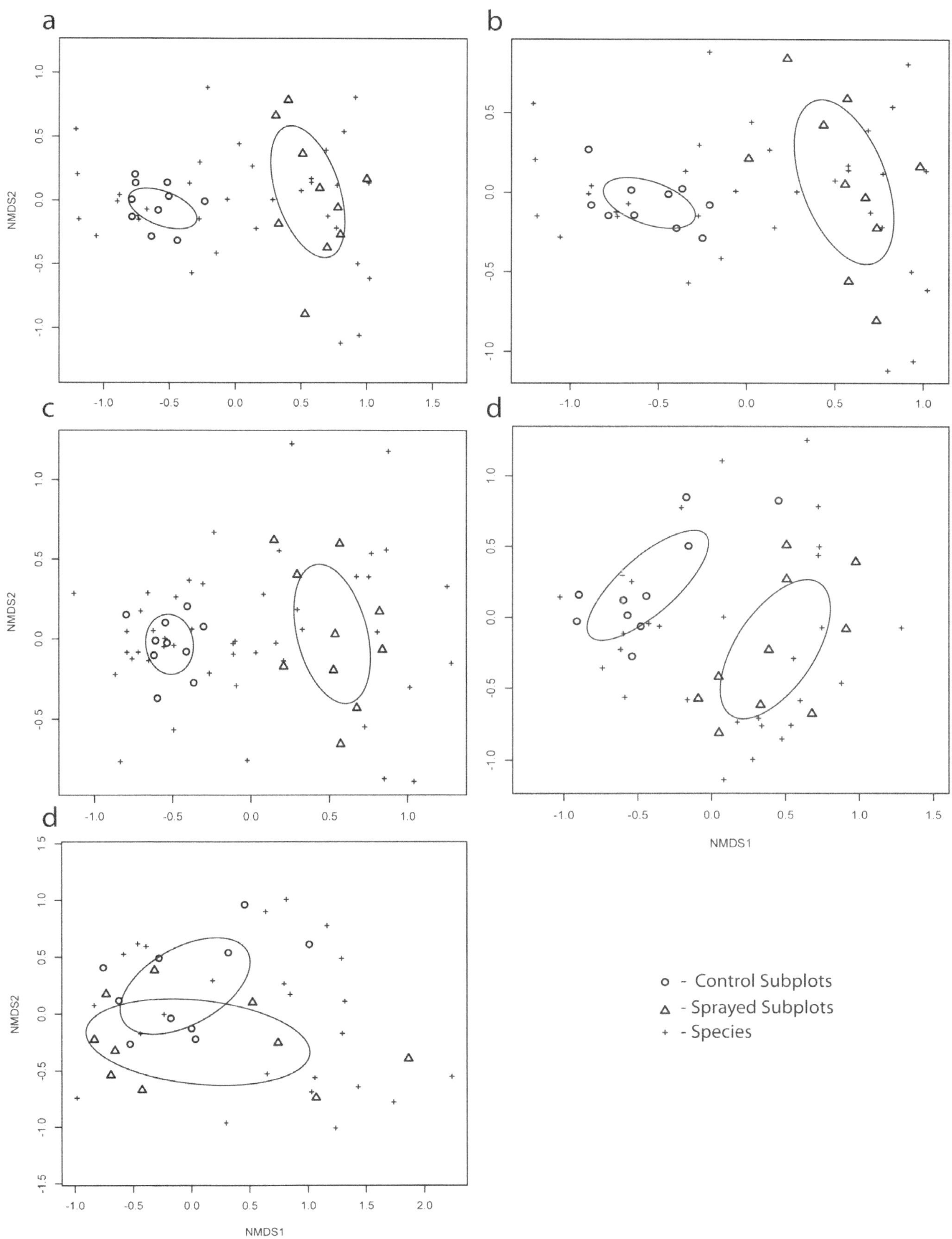
a
b
c
d
d
NMDS2
NMDS1
o - Control Subplots
Δ - Sprayed Subplots
+ - Species

Figure 2. Non-multidimensional scaling of each plot using Bray-Curtis Dissimilarity. a) meadow 1, b) meadow 2, c) hayfield 1, d) hayfield 2, and e) wasteland. The ellipses enclose the 99% confidence intervals.

Southern New England Coastal Hills and Plains. The plot names and locations were: Meadow 1, 41 49′ 30.30″ N, 72 14′ 12.91″ W; Meadow 2, 41 49′ 30.71″ N, 72 14′ 18.43″ W; Hayfield 1, 41 46′ 58.30″ N, 72 13′ 07.59″ W; Hayfield 2, 41 46′ 55.80″ N, 72 13′ 10.56″ W; and Wasteland, 41 49′ 01.25″ N, 72 13′ 10.56″ W. Meadow 1 and 2 were established in a natural meadow along the Fenton River. Hayfield 1 and 2 were placed on the upper and lower edges of an agricultural hayfield. The Wasteland plot was established in an area bordered by a road, cow pasture and wetlands, and dominated by the invasive grass canary reedgrass (*Phalaris arundinaceae* L.). Plots were not fertilized, irrigated or mowed during the 3 years of the study, and there was no known history of herbicide use at these sites. The overall intent was to mimic the localized application of glyphosate to control invasive plants or serious weeds in conservation areas, hayfields, or roadside wastelands. No permits were required for the project. Glyphosate was applied according to label specifications using appropriate personal protective equipment. All field plots were located on property owned by the University of Connecticut and did not contain any species of special concern.

A 2×2 factorial experiment was set up in each plot using a randomized complete block design with two management regimes (annual glyphosate spray or non-sprayed control) and two weedy, non-native bentgrass species (CB or RT). Each treatment group in the factorial design was replicated five times creating 20 randomized subplots with a size of 3 m^2. Soil samples (combination of two soil cores 2.2 cm×10.2 cm depth) were taken in each subplot prior to the study and 3 years later at the same time of year (Oct.-Nov.). Samples were analyzed for pH and nutrients using a modified Morgan test and a soil test for nitrates (University of Connecticut Soils Testing Lab, Storrs, CT). The soil texture for all plots was either sandy loam or loamy sand. Weather data were obtained from the National Oceanic and Atmospheric Administration (NOAA, (www.noaa.gov)). Yearly rainfall totals from November 1 to October 31 were: 2008, 375 cm; 2009, 310 cm; 2010, 304 cm. Mid-summer rainfall was: June 2008, 10.8 cm; July, 2008, 11.5 cm; June, 2009, 14.5 cm; July 2009, 19.5 cm; June, 2010, 11.1 cm; July, 2010, 8.1 cm. Mid-summer mean maximum temperatures (C°) were: June, 2008, 24.2; July, 2008, 26.4; June, 2009, 21.1; July, 2009, 24.1; June, 2010, 24.8; July, 2010, 27.8.

In 2007, conventional (non-genetically engineered) CB 'Penn A-4' and RT 'unspecified' (Des Moines Forage and Turf Seed Corporation, Ankeny, IA) plants were grown at the university research farm with biweekly mowing (1.27 cm height) and two fertilization treatments at a rate of 0.45 kg per 0.0093 ha for N, P and K (19-19-19, Lesco fertilizer, Troy, MI). Genetically engineered GR CB were not used in the study because: 1) we did not have permission to use commercially-developed, patented, transgenic GR CB seeds, 2) the U.S. government would be unlikely to provide permits for experimental field trials since the bentgrass plants needed to flower, release pollen, and produce seed in order to measure reproductive potential, and 3) to our knowledge, RT has not been engineered to contain the GR trait.

Table 1. Soil nutrients and pH in five field plots prior to the experiment.

Plot Name	Ca	Mg	P	K	NO_3	pH
	kg/ha				mg/kg	
Meadow 1	3226^{a}	433^{a}	1^{b}	164^{b}	3.49	5.78
Meadow 2	2839^{a}	519^{a}	1^{b}	147^{b}	4.34	5.64
Hayfield 1	1101^{b}	147^{b}	0^{b}	160^{b}	1.26	5.82
Hayfield 2	1572^{b}	283^{a}	2^{b}	247^{a}	4.11	6.07
Wasteland	1466^{b}	322^{a}	44^{a}	276^{a}	3.92	5.76

Letters indicate significant differences within a column according to Fisher's LSD test ($\alpha<0.05$). No differences were observed for nitrate-N or pH.

In May, 2008, small plugs (3 cm×3 cm) of acclimated CB or RT were transplanted into the center of 100 subplots (20 subplots×5 plots). Measurements of bentgrass growth were taken monthly (May-October) for three years including: surface area covered, maximum leaf height, number of tillers (stolons), number of panicles, and number of flowers. All vascular plant species in each subplot were identified and monitored during the three year study. At the end of the third year (November, 2010), bentgrass survival and above-ground plant biomass were recorded. Subplots with CB or RT plants were randomly assigned to two treatment groups: annual glyphosate treatment or control (no herbicide). Subplots receiving glyphosate (1.69 kg per hectare, Roundup Pro ®, Monsanto Company, St. Louis, MO) were sprayed once per year on June 24–26th using a hand-held wand with a single boom Teejet XR8004VS nozzle (Wheaton, IL) and a backpack sprayer (SP system backpack, Santa Monica, CA). *Agrostis* plants within the sprayed plots were temporarily covered with white plastic bags (0.5 mm thickness) for 10–15 minutes to mimic the herbicide resistance trait. Control subplots received no weed management during the experiment.

At the end of the experiment, dependent variables were examined using analyses of variance (SAS ver. 9). Soil nutrient levels were analyzed as possible covariates. Data from all subplots were combined to analyze treatment, year (treated as a random variable) and bentgrass species interaction effects. Survivorship was analyzed using a probit model to determine the significance of glyphosate application. Probit analysis was chosen because this type of regression analysis was developed for biological systems involving quantal responses (e.g. death) to differing dosages of a toxin or a stress (e.g. glyphosate application) [42,43].

To assess changes in plant community composition, subplots were graphed on a two-axis graph using non-multidimensional scaling. The Bray-Curtis Dissimilarity between each subplot was calculated and the layout was based on the lowest amount of stress between actual dissimilarity and graphed dissimilarity. Confidence intervals (99%) were calculated and drawn for both control and sprayed subplot groups within each plot. If the confidence intervals did not overlap, then the communities in the subplot treatments (control vs. sprayed) were considered to be different [44].

Results

Effects of Glyphosate on Plant Community Composition and Soil Nitrogen

The five plot sites were typical examples of a natural meadow, hayfield or wasteland plant community that frequently contain non-native bentgrasses in our ecoregion. The high plant community diversity in Hayfield 1 could be attributed to cultivation of agronomic species such as smooth brome (*Bromus inermis* Leyss.)

Table 2. Mean bentgrass biomass (g dry weight) and survivorship (percent) for introduced creeping bentgrass (CB) and redtop (RT) plants at the end of the field experiment (2010).

Species	Treatment	Biomass	Meadow 1	Meadow 2	Hayfield 1	Hayfield 2	Wasteland	Total
CB	sprayed	5.85[a] (7.85)	100	60	80	40	0	56
	control	0.0[b] (0)	0	0	0	0	0	0
RT	sprayed	25.30[a] (22.13)	100	20	60	60	40	56
	control	0.1[b] (0)	0	0	20	0	0	4

Standard deviation is given in parentheses for plant biomass and letters indicate differences within bentgrass species according to Fisher's LSD test ($\alpha<0.05$).

and timothy (*Phleum pratense* L.), plus the naturalization of native and non-native species such as RT, CB, velvet bentgrass (*A. canina* L.), colonial bentgrass (*A. capillaris* L.), rough bentgrass (*A. scabra* Willd.), rough goldenrod (*Solidago rugosa* Mill.), white panical aster (*Symphyotrichum lanceolatum* (Willd.) G.L. Nesom), poverty oatgrass (*Danthonia spicata* (L.) P. Beauv. Ex Roem. & Schult.), Kentucky bluegrass, fowl bluegrass (*Poa palustris* L.) and annual bluegrass (*Poa annua* L.). The Hayfield 2 plot had a lower species richness value because it contained mostly cultivated species including smooth brome and oat (*Avena sativa* L.). Meadow 1 was dominated by coastal plain joe pye weed (*Eupatoriadelphus dubius* (Wild. Ex Poir.) King & H. Rob.), rough goldenrod, giant goldenrod (*Solidago gigantea* Aiton), flat top goldenrod (*Euthamia graminifolia* (L.) Nutt.) and arrowleaf tearthumb (*Polygonum sagittatum* L.). Meadow 2 contained plants such as black raspberry (*Rubus occidentalis* L.), American red raspberry (*Rubus idaeus* L.), flat top goldenrod and rough goldenrod. The low species richness in the Wasteland plot was due to the predominance of reed canarygrass, although chicory (*Cichorium intybus*), Canada thistle (*Cirsium arvense*), bull thistle (*Cirsium vulgare*), Kentucky bluegrass and prickly lettuce (*Lactuca serriola*) were also identified.

After three years in the study, the glyphosate-treated subplots in Meadow 2 and Hayfield 1 had lower plant species richness (14.5 to 7.7 species per subplot, and 19.2 to 12.2 species, respectively) and altered plant community assemblages compared to control subplots (Fig. 1). These two plots also had the highest number of species prior to the study ($P<0.0001$). This result contrasts with the meta-analysis performed by Sullivan and Sullivan [18] which found that glyphosate did not change species richness. However, while glyphosate treatment in Meadow 1, Hayfield 2, and Wasteland subplots did not alter the metric of plant species richness, changes in plant community assemblages were observed. Plant community composition shifted in four plot sites due to glyphosate treatment (Fig. 2). This change could largely be attributed to the increased occurrence of early and late annuals such as fall panicum (*Panicum dichotomiflorum* Michx.), witchgrass (*Panicum capillare* L.), large crabgrass (*Digitaria sanguinalis* (L.) Scop.), Virginia pepperweed (*Lepidium virginicum* L.) and yellow rocket (*Barbarea vulgaris* W.T. Aiton) in sprayed subplots. The Wasteland plot was an exception to this trend (Fig. 2) due to the localized dominance of the invasive reed canarygrass, which quickly re-established itself after glyphosate treatment.

Soil samples taken before the experiment showed differences in soil fertility between plots (Table 1). The high concentration of P and K in the Wasteland plot was most likely due to

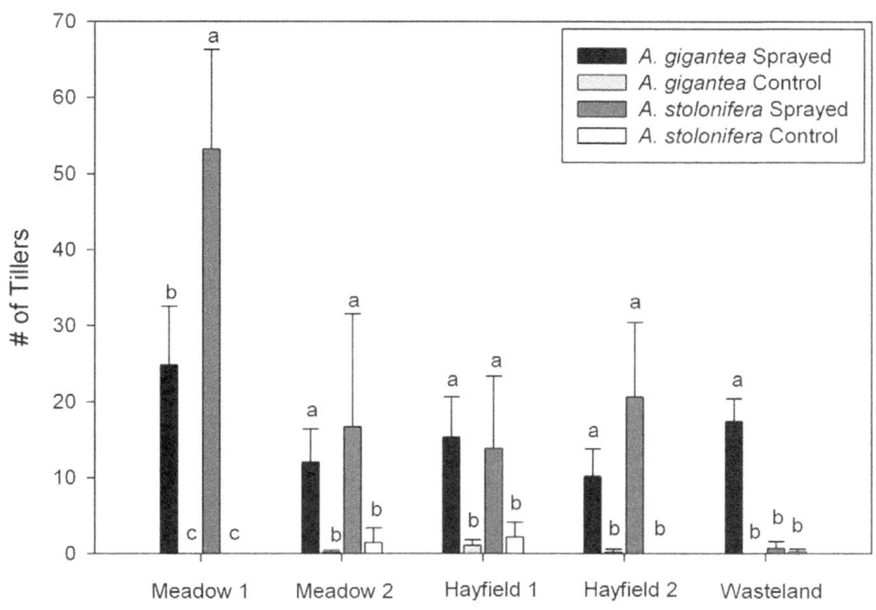

Figure 3. Number of tillers on creeping bentgrass (CB) and redtop (RT) plants in the last year of the field study (May, 2010). Error bars indicate SD and letters indicate differences within plot location at according to Fisher's LSD test ($\alpha<0.05$).

Table 3. Mean number of flowers produced by *A. gigantea* (RT) and *A. stolonifera* (CB) plants in glyphosate-treated subplots or unsprayed controls.

Species	Treatment	2008	2009	2010
RT	Control	0	213^b (92)	0^b (0)
	Glyphosate	0	$18,377^a$ (16,0[illegible]2)	7959^a (8551)
CB	Control	0	104^b (46)	0^b (0)
	Glyphosate	0	$2,042^a$ (1310)	1346^a (789)

Standard deviation is given in parentheses and letters indicate differences within bentgrass species and year according to Fisher's LSD test ($\alpha<0.05$).

previous land use for crop production, manure storage and/or manure application (personnal communication, Tom Morris). Meadow plots 1 and 2 had higher calcium concentrations than other plots, and Hayfield 1 had a lower Mg level (Table 1). However, analysis of covariance indicated that the differences in soil Mg, Ca, P, and K did not affect bentgrass growth or survivorship. At the end of the third year, soils in subplots receiving glyphosate had higher nitrate concentrations as follows (Nitrate-N shown as mg/Kg in sprayed/control subplots): Meadow 1, 11.8/3.5 ($P<0.0001$); Meadow 2, 9.2/3.9 ($P=0.0024$); Hayfield 1, 6.1/1.3 ($P=0.0021$); Hayfield 2, 6.8/4.2 ($P=0.167$); Wasteland, 13.6/3.8 ($P<0.0001$). The nitrate levels in sprayed and control subplots in Hayfield 2 did not change. Other soil nutrients (P, K, Ca, Mg) were not affected by glyphosate application.

Effect of Glyphosate on Bentgrass Survivorship, Vegetative Growth and Reproductive Potential

Introduced CB and RT plants had higher survivorship in subplots receiving glyphosate (56%) than control subplots (4%) (Table 2). This was confirmed by probit model analysis ($P<0.001$). RT and CB plants from sprayed subplots had higher final biomass than bentgrass in control subplots, and plot location did not affect final biomass (Table 2). The number of CB or RT tillers increased in sprayed subplots compared to controls (Fig. 3). The only exception was CB plants in the Wasteland plot where competition with invasive reed canarygrass negatively affected growth in all subplots (Fig. 3). CB plants had significantly more tillers than RT plants in sprayed subplots in Meadow 1; a similar but non-significant trend was observed in Meadow 2 and Hayfield 2.

Reproductive potential of CB and RT plants was measured in each year of the study. There was no significant plot effect ($P=0.524$), so data were combined for analysis. As expected, CB and RT plants did not produce panicles or flowers in the first year (2008). In the second year (2009), RT plants in glyphosate subplots had an 86-fold increase in flower number compared to those in control subplots (Table 3). Thirteen RT plants produced a total of 238,896 flowers compared with 640 flowers on 3 RT control plants. Results in the third year (2010) were similar, although total flower numbers were probably reduced due to dry weather; 2009 had cooler temperatures and more rain during May and June which favored the cool-season grasses. RT seed collected in 2010 had 80% germination in the laboratory. Throughout the study, CB plants had a somewhat lower reproductive potential (flower number) than RT plants, but CB seeds had a relatively high germination rate of 74%.

Discussion and Conclusion

Effects of Glyphosate on Plant Community Composition and Soil Nitrate-N

Annual glyphosate application over a three-year period altered plant community composition in agricultural hayfields and natural meadows supporting our original hypothesis. This result is relevant to long-term risk assessment for HR crops that can escape cultivation because changes in plant community assemblages can produce negative impacts on other trophic levels [45]. Our results contrast with some previous reports that herbicides have little effect on plant species richness [20,46,47,48,18]. This discrepancy may be due, at least in part, to plant community types, differences in glyphosate application, and the metric (species richness) used to characterize changes in plant community composition over time. In our study, species richness declined in glyphosate-treated subplots in two plot locations (Meadow 2 and Hayfield 1). However, species richness data alone failed to capture the glyphosate-induced plant community changes in two other plots (Meadow 1 and Hayfield 2). This was because the number of new plant species appearing in the glyphosate-treated subplots was approximately equal to the number lost due to herbicide treatment. For example, the loss of some perennial plant species was often concurrent with an increase in weeds such as witchgrass (an annual warm-season C4 grass) or yellow rocket (a cool-season biennial), which were not present in subplots without glyphosate treatment. Furthermore, native perennial plants that declined in sprayed subplots included goldenrods such as tall goldenrod (*Solidago altissima* L.), rough goldenrod, giant goldenrod, and narrowleaf goldenrod (*Euthamia graminifolia* (L.) Nutt.), a trend that has been observed by other researchers [20]. Thus, measuring community composition over time is important for describing the response of complex agricultural or natural plant communities to annual glyphosate applications.

The only plot in our study that showed no effect from glyphosate on community composition was the Wasteland plot (Fig. 2). This plot was different because it had high soil P and K from past agricultural use and was dominated by one invasive plant species (reed canarygrass). This invasive regrew quickly after glyphosate application and competed with the introduced bentgrass plants, leading to very low survival rates. Nearly all of the introduced bentgrass plants died except for two RT plants in glyphosate-treated subplots. Based on our observations, it would be difficult for weedy bentgrass species to compete in this particular type of plant community regardless of the GR trait or glyphosate treatment.

The observation of increased nitrate-N in subplots sprayed with glyphosate was unexpected and difficult to explain, in part because the study did not include studies on microbial activity, soil nitrogen fluxes, and other minerals. A study in the Hubbard Brook ecosystem observed an increase in nitrate-N in a watershed where all vegetation was cut and suppressed for two years by herbicide application [49]. A study in young loblolly pine forests showed increased soil nitrogen in response to glyphosate, and the authors postulated the involvement of microbial mineralization [50,51]. Increased soil nitrate-N was observed when turfgrass was removed with glyphosate, but nitrate-N levels returned to normal six weeks after reseeding [52]. Previous research has also suggested that the removal of living plants coupled with the deposition of dead plant material in the sprayed subplots could affect the nitrogen competition dynamics between microbes and plants [53]. In our plots, increased microbial mineralization of dead plant tissues in sprayed subplots may have lead to the increased nitrate-N observed in October-November. Alternatively, nitrate levels in our plots may have been affected by changes in the C:N balance because of shifts in the plant community and the amount or type of plant detritus in the sprayed subplots. In this study, high nitrate-N levels were dependant on glyphosate application and may have contributed to the growth of the introduced bentgrass plants because nitrogen is often the limiting element for plant growth [8]. However, the effect of nitrate-N on *Agrostis* growth and survivorship is difficult to separate from other glyphosate effects such as the removal of plant competitors. Further studies are needed to understand glyphosate effects on GR plant growth, competition and soil nutrient dynamics.

Effect of Glyphosate on Bentgrass Growth and Reproductive Capacity

Our results supported the hypothesis that CB and RT plants with mimicked GR trait would have higher levels of survivorship, vegetative growth and reproductive potential with selection pressure. Transgenic GR bentgrasses could not used in this study for a variety of reasons including: 1) it would be necessary to obtain a Memorandum of Understanding (MOU) and approval from the company holding the intellectual property rights and patents for the plant product, 2) the experimental design required quantifying flower and seed number and this would have allowed transgene escape because bentgrasses were already present in the study sites, and 3) the GR trait has not been inserted into RT. Unintended transgene flow has been well documented in bentgrasses in Oregon [36]. Given our special attention to mimicking the GR trait, we feel confident that the results are applicable to transgenic bentgrasses in the habitats and plant communities studied.

CB and RT plants had more tillers, above-ground biomass, and flowers in sprayed subplots. The most likely explanation for this is that they experienced less competition from neighboring vegetation for resources such as light, soil nutrients, or soil moisture. This contrasts with the inference from some previous bentgrass studies [40,41], but agrees with a study on 'mock feral' GR canola that showed increased crop fitness when exposed to glyphosate drift selection pressure [5] and a study that found glyphosate resistant *Brassica* species persisted as the dominant species after two annual applications of glyphosate drift [54].

CB and RT produced a high number of flowers and viable seeds in the second and third seasons of growth. RT plants in glyphosate-sprayed subplots had an 86-fold higher number of flowers compared to control subplots in 2009 (Table 3). CB plants in glyphosate-treated subplots also produced more flowers than controls. In addition, the germination rates for seed collected from CB and RT plants were above 70%. If the number of viable seeds is proportional to the number of flowers observed in 2009 and 2010, then seeds represent a significant pathway for dispersal of weedy bentgrasses carrying the GR trait. In addition, the large number of flowers suggests a strong potential for pollen-mediated transgene flow through either intraspecific or interspecific hybridization. Forced crosses and landscape-scale experiments have already demonstrated pollen-mediated gene flow between bentgrass species in the U.S. [38,36,39,55]. In our study area, seven bentgrass species can be found in natural and cultural landscapes, and all of these species are able to hybridize [30].

The removal of vegetation by glyphosate is a disturbance that creates gaps in plant communities. Research has shown that, without gaps, seed recruitment in herbaceous plant communities (e.g. tallgrass prairie and pastures) is rare and episodic [56], but disturbed sites have a higher rate of seedling establishment [57]. A positive correlation between gap diameter, grass population growth rate, and panicle production has also been demonstrated [58]. These observations are consistent with our study where subplots with glyphosate disturbance had higher CB and RT survival and growth (Table 2). In addition, the gaps created by glyphosate were postively correlated with increased bentgrass panicle and tiller production. Thus, preventing or reducing gaps in plant communities could be one strategy for preventing the establishment of GR bentgrasses in natural and agricultural habitats in the Northeastern U.S.

Acknowledgments

We thank Thomas Morris (Univ. of Connecticut) for help in analyzing soil fertility data, and Karl Guillard (Univ. of Connecticut) for helpful comments on the manuscript.

Author Contributions

Conceived and designed the experiments: CWA CAA. Performed the experiments: CWA Analyzed the data: CWA. Contributed reagents/materials/analysis tools: CWA CAA. Wrote the paper: CWA CAA.

References

1. Heap I (2012) The International Survey of Herbicide Resistant Weeds. www.weedscience.com. Accessed: May 02, 2012.
2. Cerdeira AL, Duke SO (2006) The current status and environmental impacts of glyphosate-resistant crops: a review. J Environ Qual 35: 1633–1658.
3. Duke SO, Powles SB (2008) Glyphosate: a once-in-a-century herbicide. Pest Manag Sci 64: 319–325.
4. Powles S (2008) Evolved glyphosate-resistant weeds around the world: lessons to be learnt. Pest Manag Sci 64: 360–365.
5. Londo JP, Bautista NS, Sagers CL, Lee EH, Watrud LS (2010) Glyphosate drift promotes changes in fitness and transgene gene flow in canola (*Brassica napus*) and hybrids. Ann Bot-London 106: 957–965.
6. APHIS (2011) Animal and Plant Health Inspection Service. http://www.aphis.usda.gov/brs/aphisdocs/scotts_kbg_resp.pdf Accessed: May 2, 2012.
7. Ledford H (2011) Transgenic grass skirts regulators: Technological advances remove basis for government oversight of genetically modified crops. Nature 475: 274–275.
8. Service RF (2007) Agbiotech: A growing threat down on the farm. Science 316: 1114–1117.
9. Shaner DL, Lindenmeyer RB, Ostlie MH (2011) What have the mechanisms of resistance to glyphosate taught us? Pest Manag Sci. 68: 3–9.
10. Baucom RS, Mauricio R (2008) Constraints on the evolution of tolerance to herbicide in the common morning glory: resistance and tolerance are mutually exclusive. Evolution 62: 2842–2854.
11. Baerson SR, Rodriguez DJ, Tran M, Feng YM, Biest NA, et al. (2002) Glyphosate-Resistant Goosegrass. Identification of a mutation in the target enzyme 5-Enolpyruvylshikimate-3-Phosphate Synthase. Plant Physiol 129: 1265–1275.
12. Gaines TA, Shaner DL, Ward SM, Leach JE, Preston C, et al. (2011) Mechanism of resistance of evolved glyphosate-resistant palmer amaranth (*Amaranthus palmeri*). J Agr Food Chem 59: 5886–5889.
13. Ge X, d'Avignon DA, Ackerman JJH, Sammons RD (2010) Rapid vacuolar sequestration: the horseweed glyphosate resistance mechanism. Pest Manag Sci 66: 345–348.
14. Warwick SI, Légère A, Simard M-J, James T (2008) Do escaped transgenes persist in nature? The case of an herbicide resistance transgene in a weedy *Brassica rapa* population. Mol Ecol 17: 1387–1395.
15. Lortie CJ, Brooker RW, Choler P, Kikvidze Z, Michalet R, et al. (2004) Rethinking plant community theory. Oikos 107: 433–438.
16. Chang F, Simcik MF, Capel P (2011) Occurrence and fate of the herbicide glyphosate and its degradate aminomethylphosphonic acid in the atmosphere. Environ Toxicol Chem 30: 548–555.
17. Coupe RH, Kalkhoff SJ, Capel PD, Gregoire C (2011) Fate and transport of glyphosate and aminomethylphosphonic acid in surface waters of agricultural basins. Pest Manag Sci 68: 16–30.
18. Sullivan TP, Sullivan DS (2003) Vegetation management and ecosystem disturbance: impact of glyphosate herbicide on plant and animal diversity in terrestrial systems. Environ Rev 11: 37–59.
19. Boyd RS, Freeman JD, Miller JH, Edwards MB (1995) Forest herbicide influences on floristic diversity seven years after broadcast pine release treatments in central Georgia, USA. New Forest 10: 17–37.

20. Freedman B, Morash R, MacKinnon D (1993) Short-term changes in vegetation after the silvicultural spraying of glyphosate herbicide onto regenerating clearcuts in Nova Scotia, Canada. Can J Forest Res 23: 2300–2311.
21. Heard MS, Hawes C, Champion GT, Clark SJ, Firbank LG, et al. (2003) Weeds in fields with contrasting conventional and genetically modified herbicide-tolerant crops. I. Effects on abundance and diversity. Philos T R Soc Biol 358: 1819–1832.
22. Marvier M, Van Acker R (2005) Can crop transgenes be kept on a leash? Front Ecol Environ 3: 99–106.
23. NASS (2011) National Agricultural Statistics Service. http://www.ers.usda.gov/data/biotechcrops/Accessed August 2011.
24. Warwick S, Beckie HJ, Hall LM (2009) Gene flow, invasiveness, and ecological impact of genetically modified crops. Ann NY Acad Sci 168: 72–99.
25. Behrendt S, HanfM (1979) *Grass Weeds in World Agriculture*. BASF Aktiengesellschaft, Ludwigshafen am Rhein, Germany, 73–80.
26. IPA (2011) Invasive Plant Atlas. http://www.invasiveplantatlas.org/grass.html Accessed: August 2011.
27. MacBryde B (2006) White Paper: Perspectives on creeping bentgrass, *Agrostis stolonifera* L. USDA/APHIS/BRS, Riverdale, Maryland, USA. http://www.aphis.usda.gov/peer_review/downloads/perspectiveCBG-wp.pdf. Accessed: November 3, 2009.
28. Wedin DA, Tilman D (1990) Species effects on nitrogen cycling: a test with perennial grasses. Oecologia 84: 433–441.
29. Gurevitch J, Scheiner SM, Fox GA (2006) The Ecology of Plants. 2nd ed. Sunderland, MA: Sinauer Associates, Inc.
30. Ahrens C, Chung J, Meyer T, Auer C (2011) Bentgrass distribution surveys and habitat suitability maps support ecological risk assessment in cultural landscapes. Weed Sci 59: 145–154.
31. Ahrens C, Ecker G, Auer C (2011) The intersection of ecological risk assessment and plant communities: an analysis of *Agrostis* and *Panicum* species in northeastern U.S. Plant Ecol 212: 1629–1642.
32. Beard JB (2002) Turf management for golf courses. 2nd ed. Chelsea, MI, USA: Ann Arbor Press,. 137–138, 716–726.
33. Magee DW, Ahles HE (2007) Flora of the northeast: a manual of the vascular flora of New England and adjacent New York. Amherst, MA: University of MA press.
34. Hart SE, Belanger FC, McCullough PE, Rotter D (2009) Competitiveness of *Agrostis* interspecific hybrids in turfgrass swards. Crop Sci 49: 2275–2284.
35. Kik C, Van Andel J, W. Joenjes W (1990) Life-history variation in ecologically contrasting populations of *Agrostis stolonifera*. Ecology 78: 962–973.
36. Watrud LS, Lee EH, Fairbrother A, Burdick C, Reichman JR, et al. (2004) Evidence for landscape-level, pollen-mediated gene flow from genetically modified creeping bentgrass with CP4 EPSPS as a marker. P Natl Acad Sci USA 101: 14533–14538.
37. Charles D (2011) Scientist in the middle of the GM-Organic Wars. Science 332: 168.
38. Belanger FC, Meagher TR, Day PR, Plumley K, Meyer WA (2003) Interspecific hybridization between *Agrostis stolonifera* and related *Agrostis* species under field conditions. Crop Sci 43: 240–246.
39. Wipff JK, Fricker CR (2000) Determining gene flow of transgenic creeping bentgrass and gene transfer to other bentgrass species. Diversity 16: 36–39.
40. Fei S, Nelson E (2004) Greenhouse evaluation of fitness-related reproductive traits in roundup®-tolerant transgenic creeping bentgrass (*Agrostis Stolonifera* L.). In Vitro Cell Dev B 40: 266–273.
41. Gardner DS, Nelson EK, Waldecker MA, Tarter WR (2006) Establishment and lateral growth of glyphosate-resistant creeping bentgrass in bare soil. HortTechnology 16: 590–594.
42. Bliss CL (1934) Methods of probits. Science 79: 38–39.
43. Kuykendall KL, Mahoney CJ, Conroy GC (1992) Probit and survival analysis of tooth emergence ages in mixed-longitudinal sample of chimpanzees (*Pan troglodytes*). Am J Phys Anthropol 89: 379–399.
44. Oksanen J (2009) Ordination and analysis of dissimilarities: tutorial with R and Vegan. www.bio.utk.edu/fesin/MSA2009/R/veganbird.pdf Accessed: April 2012.
45. Flanders AA, Kuvlesky WP, Ruthven DC, Zaiglin RE, Bingham RL, et al. (2006) Effects of invasive exotic grasses on south Texas rangeland breeding birds. Auk 123: 171–182.
46. Krueger-Mangold J, Sheley R, Roos B (2002) Maintaining plant community diversity in a waterfowl production area by controlling canada thistle (*Cirsium arvense*) using glyphosate. Weed Technol 16: 4 57–463.
47. Puricelli E, Tuesca D (2005) Weed density and diversity under glyphosate-resistant crop sequences. Crop Prot 24: 533–542.
48. Rice PM, Toney JC, Bedunah DJ, Carlson CE (1997) Plant community diversity and growth form responses to herbicide applications for control of *Centaurea maculosa*. J Appl Ecol 34: 1397–1412.
49. Likens GE, Bormann FH, Johnson NM, Fisher DW, Pierce RS (1970) Effects of forest cutting and herbicide treatment on nutrient budgets in the Hubbard Brook Watershed-Ecosystem. Ecol Monogr 40: 23–47.
50. Vitousek PM, Matson PA (1985) Disturbance, nitrogen availability, and nitrogen losses in an intensively managed loblolly pine plantation. Ecology 66: 1360–1376.
51. Vitousek PM, Andarlese SW, Matson PA, Morris L, Sanford RL (1992) Effects of harvest intensity, site preperation, and herbicide use on soil nitrogen transformations in a young loblolly pine plantation. Forest Ecol Manag 49: 277–292.
52. Bushoven JT, Jiang Z, Ford HJ, Sawyer CD, Hull RJ, et al. (2000) Stabilization of soil nitrate by reseeding with perenial ryegrass following sudden turf death. J Environ Qual 29: 1657–1661.
53. Schimel JP, Bennet J (2004) Nitrogen mineralization: challenges of a changing paradigm. Ecology 85: 591–602.
54. Watrud LS, King GA, Londo JP, Colasanti R, Smith BM, et al. (2011) Changes in constructed *Brassica* communities treated with glyphosate drift. Ecol Appl 21: 525–538.
55. Zapiola ML, Campbell CK, Butler MD, Mallory-Smith CA (2008) Escape and establishment of transgenic glyphosate-resistant creeping bentgrass *Agrostis stolonifera* in Oregon, U.S.A: a 4-year study. J Appl Ecol 45: 486–494.
56. Benson EJ, Hartnett DC (2006) The role of seed and vegetative reproduction in plant recuritment and demography in tallgrass prairie. Plant Ecol 187: 163–178.
57. Reader RJ, Buck J (1991) Control of seedling density on disturbed ground: role of seedling establishment for some midsuccessional, old-field species. Can J Bot 69: 773–777.
58. Tozer KN, Chapman DF, Quigley PE, Dowling PM, Cousens RD, et al. 2008. Controlling invasive annual grasses in grazed pastures: population dynamics and critical gap sizes. J Appl Ecol 45: 1152–1159.

17

Pharmacological Inhibition of CXCR2 Chemokine Receptors Modulates Paraquat-Induced Intoxication in Rats

Kesiane M. Costa[1], Izaque S. Maciel[1], Luiza W. Kist[1,2], Maria M. Campos[1,3,4], Maurício R. Bogo[1,2,5]*

1 Postgraduate Program in Medicine and Health Sciences, Pontifical Catholic University of Rio Grande do Sul, Porto Alegre/RS, Brazil, **2** Laboratory of Genomics and Molecular Biology, Faculty of Biosciences, Pontifical Catholic University of Rio Grande do Sul, Porto Alegre/RS, Brazil, **3** Institute of Toxicology and Pharmacology, Pontifical Catholic University of Rio Grande do Sul, Porto Alegre/RS, Brazil, **4** Faculty of Dentistry, Pontifical Catholic University of Rio Grande do Sul, Porto Alegre/RS, Brazil, **5** Postgraduate Program in Cellular and Molecular Biology, Pontifical Catholic University of Rio Grande do Sul, Porto Alegre/RS, Brazil

Abstract

Paraquat (PQ) is an agrochemical agent commonly used worldwide, which is a lied to potential risks of intoxication. This herbicide induces the formation of reactive oxygen species (ROS) that ends up compromising various organs, particularly the lungs and the brain. This study evaluated the deleterious effects of paraquat on the central nervous system (CNS) and peripherally, with special attempts to assess the putative protective effects of the selective CXCR2 receptor antagonist SB225002 on these parameters. PQ-toxicity was induced in male Wistar rats, in a total dose of 50 mg/kg, and control animals received saline solution at the same schedule of administration. Separate groups of animals were treated with the selective CXCR2 antagonist SB225002 (1 or 3 mg/kg), administered 30 min before each paraquat injection. The major changes found in paraquat-treated animals were: decreased body weight and hypothermia, nociception behavior, impairment of locomotor and gait capabilities, enhanced *TNF-α* and *IL-1β* expression in the striatum, and cell migration to the lungs and blood. Some of these parameters were reversed when the antagonist SB225002 was administered, including recovery of physiological parameters, decreased nociception, improvement of gait abnormalities, modulation of striatal *TNF-α* and *IL-1β* expression, and decrease of neutrophil migration to the lungs and blood. Taken together, our results demonstrate that damage to the central and peripheral systems elicited by paraquat can be prevented by the pharmacological inhibition of CXCR2 chemokine receptors. The experimental evidence presented herein extends the comprehension on the toxicodynamic aspects of paraquat, and opens new avenues to treat intoxication induced by this herbicide.

Editor: Pedro Fernandez-Funez, University of Florida, United States of America

Funding: This work was supported by grants from the Conselho Nacional de Desenvolvimento Científico e Tecnológico (CNPq), DECIT/SCTIEMS through CNPq and Fundação de Amparo à Pesquisa do Estado do Rio Grande do Sul (FAPERGS) (Proc. FAPERGS 10/0036-5 – PRONEX), and FINEP research grant "Implantação, Modernização e Qualificação de Estrutura de Pesquisa da PUCRS" (PUCRSINFRA) # 01.11.0014-00. The funders had no role in study design, data collection and analysis, decision to publish, or preparation of the manuscript.

Competing Interests: The authors have declared that no competing interests exist.

* Email: mbogo@pucrs.br

Introduction

Paraquat (PQ) dichloride is a fast-acting, non-selective bipyridylium herbicide, widely used in many countries worldwide due to the low costs and effectiveness against a range of weeds. The accidental or intentional poisoning with this agent has been associated with fatal cases since there is a lack of effective treatments to intoxication (for review see: [1]). In fact, the mechanisms implicated in the toxicity caused by paraquat exposure are not yet completely understood [2]. Regardless, it has been demonstrated that paraquat effects are primarily allied with oxidative stress through the production of reactive oxygen species (ROS) [3,4]. After exposure, the herbicide can be detected in several organs including heart, kidneys, liver, thymus, and lungs [5]. The concentration in lungs is commonly highly elevated, as a consequence of its affinity for type 2 pneumocytes. Of note, acute lung injury with fibrosis and massive cell infiltration in the alveolar spaces can be observed after intoxication with a single dose of paraquat [6].

Chemokines are a subset of cytokines that selectively regulate the recruitment of leukocytes to inflammatory sites, mainly through the activation of G-protein-coupled receptors [7,8]. These receptors are classified into different families according to the type of chemokines they bind (CC, CXC, C, or CX_3C) (for review see: [9]). CXC chemokines are especially involved in migration of polymorphonuclear leukocytes (PMNs), and the prototype of this group is interleukin-8 (IL-8/CXCL8), which binds to the CXCR1/2 receptors [10]. Of note, it has been demonstrated that paraquat is able to activate a set of genes related to immune response (IL-10, CXL10, and CXL11), according to evaluation of the human keratinocyte cell line HaCaT [11].

In 1998, White et al. [12] described the first non-peptide, selective, and competitive CXCR2 receptor antagonist, named SB225002, which was found effective in preventing a series of IL-8-mediated inflammatory responses, both *in vitro* and *in vivo*. By using an *in vivo* model of mouse colitis induced by 2,4,6-trinitrobenzene sulfonic acid (TNBS), Bento et al. [13] demonstrated that intraperitoneal (i.p.) administration of SB225002, in

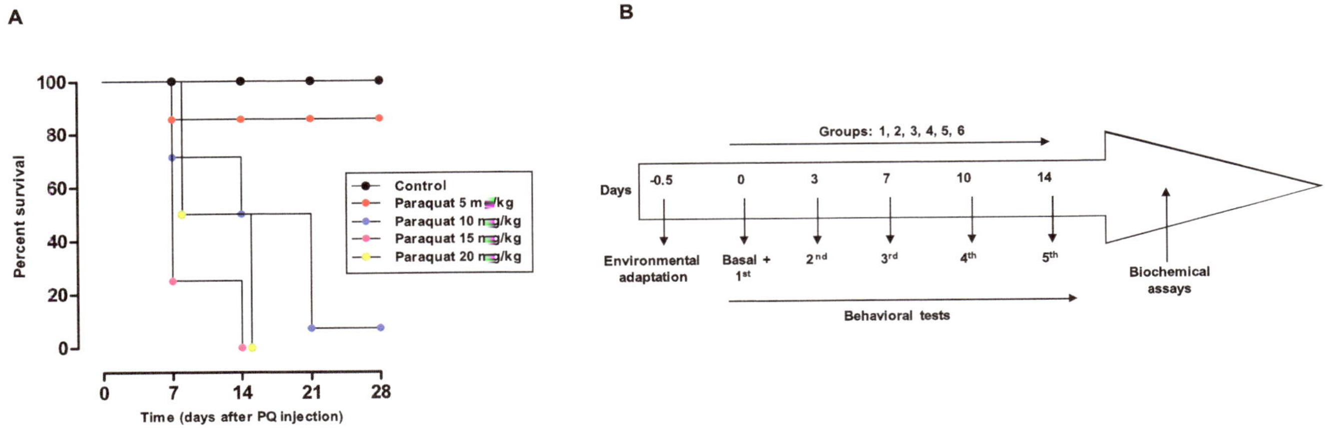

Figure 1. Paraquat survival curve and general schedule of treatment used in the study. (A) Survival curve for cumulative paraquat administration through 28 days, given at different doses [5 (n = 4), 10 (n = 14), 15 (n = 4), and 20 (n = 6) mg/kg], twice a week. (B) The animals received five i.p. applications (at 0, 3, 7, 10, and 14 days) of vehicle/saline (10 ml/kg; group 1), SB225002 (1 and 3 mg/kg; groups 2 and 3)/saline, vehicle/paraquat (10 mg/kg; group 4), or SB225002 (1 and 3 mg/kg)/paraquat (10 mg/kg); groups 5 and 6, respectively. Behavioral tests were performed at days 0, 7, and 14, which corresponded to the basal timepoint, and the 3rd or the 5th application of drugs. In some cases, the behavioral measurements were also carried out under basal conditions (prior to any treatment).

doses as low as 1 mg/kg, was able to prevent several inflammatory parameters, such as neutrophil migration to the colon tissue. Furthermore, i.p. treatment with SB225002 resulted in a marked and long-lasting inhibition of acute or chronic nociception in mice [14]. More recently, SB225002 was demonstrated to be effective in preventing the functional, inflammatory, and nociceptive alterations related to cyclophosphamide-induced cystitis in rats when dosed i.p., alone or in combination with the transient receptor potential vanilloid 1 (TRPV1) antagonist SB266791 [15]. Another recent study demonstrated that spinal injection of SB225002 (5 and 20 µg/site) significantly inhibited the mechanical allodynia caused by CXL1/KC in mice [8].

The present study examined, for the first time, the possible beneficial effects of SB225002 in a rat model of paraquat intoxication, induced by repeated exposure to low doses of this herbicide. Interestingly, our results revealed that pharmacological inhibition of CXCR2 receptors was allied to improvement of both central and peripheral complications induced by paraquat, thus contributing to clarify its mechanisms of toxicity.

Materials and Methods

Ethics Statement

Experiments were conducted in accordance with current guidelines for the care of laboratory animals and ethical guidelines for the investigation of experimental pain in conscious animals [16]. All the experimental procedures were approved by the Animal Ethics Committee of Pontificia Universidade Católica do Rio Grande do Sul (RS) (Protocol Number: CEUA 12/00295).

Animals

Male Wistar rats (weighing 180 g at the beginning of experiments) used in this study were obtained from the Central Animal House of Universidade Federal de Pelotas (UFPEL, Brazil). The animals were housed in groups of four and maintained in controlled temperature (22±1°C) and humidity (60–70%), under a 12 h light-dark cycle (lights on 07:00 AM). Food and water were available *ad libitum*. Animals were acclimatized to the laboratory for at least 1 h before testing and all the tests were performed between 9:00 AM and 5:00 PM. The number of animals and intensity of noxious stimuli used were the minimum necessary to provide consistent effects of the treatments. Analgesics were not administered to avoid any possible interference with the obtained results. In all experimental protocols, the animals were monitored twice daily for signals of severe distress. The criteria adopted to decide on euthanasia included the occurrence of tremors, convulsions, and/or chromodacryorrhea. The euthanasia was performed by deep anesthesia with isoflurane.

General protocols of treatment

Initially, a total of 28 rats received i.p. applications of paraquat at different doses [5 (n = 4), 10 (n = 14), 15 (n = 4), and 20 (n = 6) mg/kg] [17], twice a week for 28 days, in order to determine the appropriate dose for the next experiments. A separate group of animals received saline solution (n = 8) as control (Figure 1A). The number of animals used to determine the experimental design was 36 rats. Next, on the basis of the previous set of experiments and based on the literature [18], animals were provided with paraquat (10 mg/kg), twice a week for 14 days, at a total dose of 50 mg/kg. The animals used in the behavioral analysis were further used in molecular and biochemical analysis. Briefly, rats were randomly divided into six groups comprising: (A) vehicle/saline (1% Tween-80 in 0.9% NaCl solution; 10 ml/kg), with a total number of 15 animals; (B) SB225002 (1 mg/kg)/saline, with a total number of six animals; (C) SB225002 (3 mg/kg)/saline, with a total number of four animals; (D) vehicle/paraquat, with a total number of 20 animals; (E) SB225002 (1 mg/kg)/paraquat, with a total number of nine animals and (F) SB225002 (3 mg/kg)/paraquat, with a total number of 13 animals. The number of animals used in the behavioral, molecular, and biochemical analysis was 67 rats. Therefore, the total number of animals used in this study was 103 rats. The doses of SB225002 were selected on the basis of previous publications [14]. The antagonist (or the vehicle) was administered 30 min before each paraquat or saline application. The protocol of the treatments is outlined in Figure 1B.

Body weight measurement

Rat body weight was recorded (in g) using a digital balance. All the animals were weighed once a day at the basal timepoint, and before the 2nd, 3rd, 4th, and 5th applications. The values are

Table 1. Primer sequences for RT-qPCR experiments.

Gene	Forward primer	Reverse primer
COX-2[a]	5′-CCCCAAGGCACAAATATGATG-3′	5′-CCTCGCTTCTGATCTGTCTTGA-3′
IL-1β[b]	5′-GGACAGAACATAAGCCAACAA-3′	5′-CTTTCATCACACAGGACAGGT-3′
NF-κB1[c]	5′-CAGCTCTTCTCAAAGCAGCA-3′	5′-TCCAGGTCATAGAGAGGCTCA-3′
Rpl13α[d]	5′-ACAAGAAAAAGCGGATGGTG-3′	5′-TTCCGGTAATGGATCTTTGC-3′
TNFα[b]	5′-AGCAGATGGGCTGTACCTTAT-3′	5′-GCTGACTTTCTCCTGGTATGA -3′

According to [a]Hanafy et al., 2012; [b]Singh et al., 2012; [c]Kireev et al., 2012; [d]Bonefeld et al., 2008.

expressed as the difference between the initial weights minus the weight values on the day of measurement.

Body temperature assessment

The rectal temperature was recorded (in °C) using a commercially available thermometer. After recording the initial rectal temperature (basal measurement; in °C), the body temperature was evaluated on days corresponding to the 2nd, 3rd, 4th, and 5th applications at two different timepoints: immediately before, and 1 h after the treatments. The values are expressed as the difference between the temperatures before and after application.

Nociceptive responses

The method adopted in the present study was similar to that described previously [19,20,21,22], with minor modifications. The animals were analyzed in their cages in order to minimize the potential variations in the behavioral responses. After the treatments, the animals were assessed in two different approaches: (i) the spontaneous behavior was measured 25 min after treatment for 5 min; (ii) the spontaneous behavior was measured after 55 min for 5 min, over a total period of a 1 h. The behavioral alterations were scored according to the following scale: 0 = absence of any signal, with normal eyes; 1 – piloerection; 2 – stretching; 1 – hump-backed position; 1 – half-closed eyes; 2 – completely closed eyes. All the experiments were performed in a blinded manner, each experimenter scoring four rats in parallel. The scores were registered over each 5-min period, and summed for an overall pain score (maximum score = 6).

Open-field test

To analyze the locomotor activity, the animals were evaluated in the open-field test, according to the methodology described before [23], with minor modifications. The experiments were conducted in a sound-attenuated room, under low-intensity light. Rats were individually placed in the center of a plywood box (34.5 cm height×249 cm circumference×77.5 cm radius), with the floor divided into 17 squares. Duration of grooming (s) was cumulatively registered over 5 min. The number of rearing behaviors and fecal boluses was also recorded. Additionally, the number of squares crossed with the four paws was registered. All the parameters were measured at the basal timepoint and after the 3rd and 5th applications.

Beam-walking test

To assess the locomotor skills, the animals were evaluated in the beam-walking test. The methodology used was the same described previously [24]. Each rat was positioned in the center of a wooden narrow bridge (3.5 cm wide and 85 cm long), suspended between two platforms at an angle of 8° [25]. The time taken to cross the beam (in seconds, s), and the number of paw slips and falls was registered during 3 min. Each animal was acclimatized on the platform on three consecutive days before the experiments. The measurements were made at the basal timepoint and after the 3rd and the 5th applications.

Footprint test

To determine the gait abnormalities induced by paraquat, the animals were evaluated in the footprint test according to the methodology described by Richter et al. [26] with minor modifications, at the baseline and after the 3rd and the 5th applications of the herbicide. The rats were allowed to walk through an 85-cm-long, 5-cm-wide runway corridor, with a darkened cage at the end of the corridor. The floor of this corridor was covered with a new white absorbing paper for each run. The animals were first trained to pass straight forward through the corridor on three consecutive days before the experiments. After the training, the paws were colored with different colors (red for the forepaws and black for the hindpaws), and the rats were placed into the corridor. For footprint analysis, both the initial and the final 10 cm of the stamped paper was excluded to avoid false-positive results [27]. Two distinct parameters were considered for analysis: (i) the stride length was measured as the average distance between the extremity of third toes of the forepaws and hindpaws, for both left and right paws; (ii) the step frequency was evaluated according to the total number of footprints. The averages obtained by two blinded researchers were considered as a duplicate.

Gene expression analysis by quantitative real-time RT-PCR (RT-qPCR)

Following the behavioral tests assessments, the animals were euthanized by deep inhalation of isoflurane and all the right and left striatum was collected. Total RNA was isolated, quantified by spectrophotometry (A260/280 nm) and after treated with deoxyribonuclease I to eliminate genomic DNA contamination. The cDNA was synthesized from 1 μg total RNA. Quantitative PCR was performed using SYBR Green to detect double-stranded cDNA synthesis Reactions were done in a volume of 25 μL using 12.5 μL of diluted cDNA, containing a final concentration of 0.2x SYBR Green, 100 μM dNTP, 1x PCR Buffer, 3 mM $MgCl_2$, 0.25 U Taq DNA Polymerase, 0.5 M betaine, and 200 nM of *COX-2* [28], *IL-1β*, *TNF-α* [29], *NF-κB1* [30], or *Rpl13α* [31] primers (Table 1). The PCR cycling conditions were: an initial polymerase activation step for 5 min at 95°C, 40 cycles of 15 s at 95°C for denaturation, 35 s at 60°C for annealing, and 15 s at 72°C for elongation. At the end of the cycling protocol, a melting-curve analysis was included and fluorescence measured from 60 to

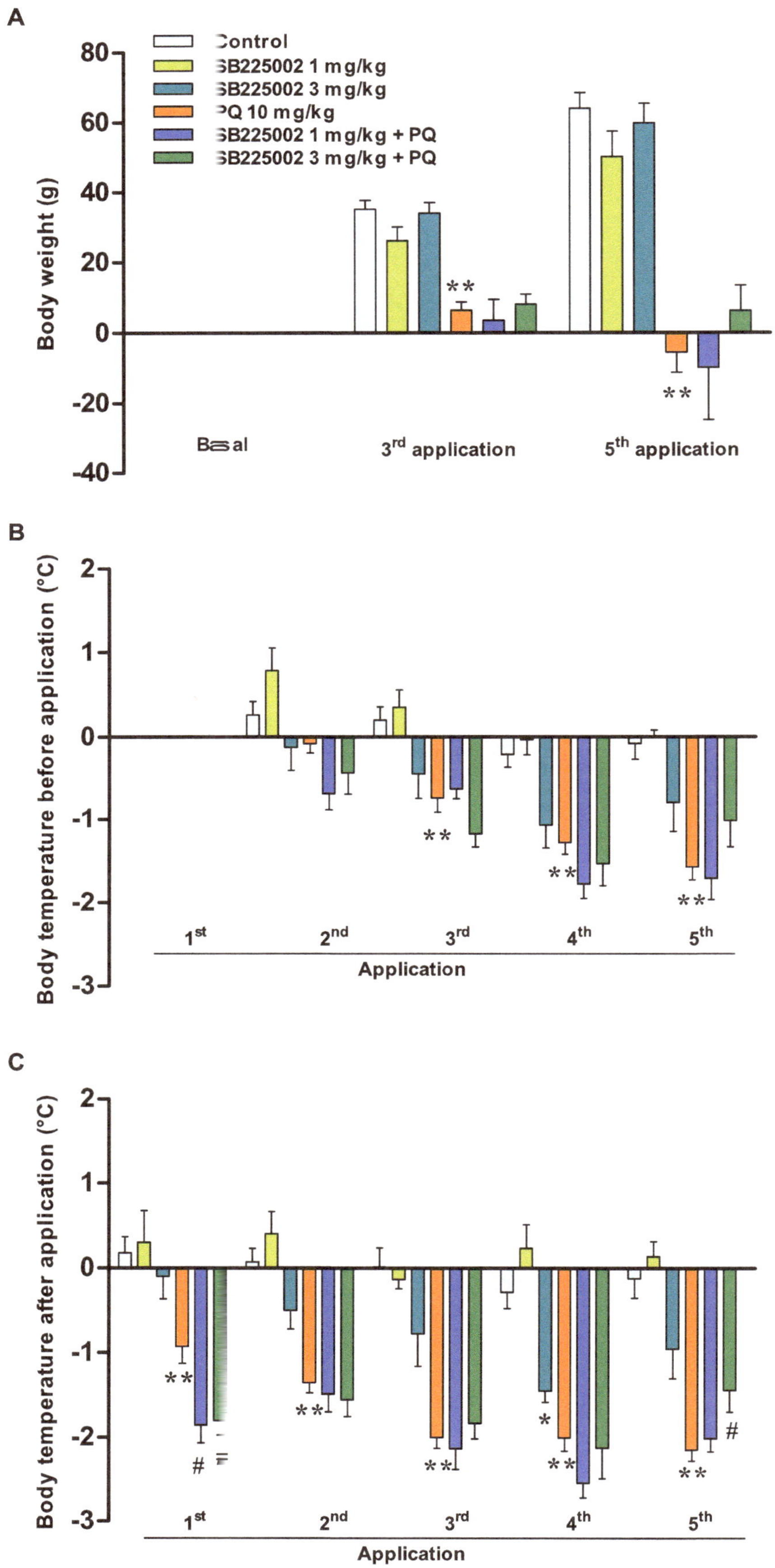

A
Control
SB225002 1 mg/kg
SB225002 3 mg/kg
PQ 10 mg/kg
SB225002 1 mg/kg + PQ
SB225002 3 mg/kg + PQ
Body weight (g)
80
60
40
20
0
-20
-40
Basal
3rd application
5th application
B
Body temperature before application (°C)
2
1
0
-1
-2
-3
1st
2nd
3rd
4th
5th
Application
C
Body temperature after application (°C)
2
1
0
-1
-2
-3
1st
2nd
3rd
4th
5th
Application

Figure 2. Physiological changes elicited by paraquat: effects of SB225002. (A) Changes in body weight (Δ grams), before the 3rd and 5th application; (B) variation of body temperature (Δ °C), measured immediately before treatments; (C) variation of body temperature measured 1 h after each application. Each column represents the mean ± SEM of 4–18 animals per group. *$p<0.05$ and **$p<0.01$ significantly different from control (vehicle/saline) group; #$p<0.05$ significantly different from vehicle/paraquat group (PQ) (ANOVA followed by Bonferroni's post-hoc test).

99°C and showed in all cases one single peak. *Rpl13α* was used as an endogenous control. Relative expression levels were determined with 7500 and 7500 Fast Real-Time PCR Systems Software v.2.0.6 (Applied Biosystems). The efficiency per sample was calculated using LinRegPCR 2012.3 Software (http://LinRegPCR.nl). Relative mRNA expression levels were determined using the $2^{-\Delta\Delta CT}$ method [32].

Hematologic parameters

After the behavioral assessments, the animals were euthanized by deep inhalation of isoflurane and the blood was collected. Immediately after, a small drop of blood was taken for smear evaluation using May-Grunewald-Giemsa staining [33]. Differential counts (neutrophils, eosinophils, basophils, lymphocytes, monocytes, and immature cells) were estimated under a ×40 objective, by counting 100 cells. Representative pictures were captured.

Myeloperoxidase (MPO) activity

Neutrophil recruitment to the rat lung was measured by means of tissue MPO activity, according to the method described by Souza et al. [34], with minor modifications. After euthanasia by deep inhalation of isoflurane, the lungs of animals were removed and immediately stored at −80°C. The tissues were homogenized in 5% (w/v) EDTA/NaCl buffer (pH 4.7) and centrifuged at 4000 rpm for 25 min, at 4°C. The pellet was resuspended in 0.5% hexadecyltrimethyl ammonium bromide buffer (pH 5.4), and the samples were re-centrifuged (4000 rpm, 25 min, 4°C). Twenty-five microliters of the supernatant were used for the MPO assay. The enzymatic reaction was assessed with 1.6 mM tetramethylbenzidine, 80 mM $NaPO_4$, and 0.3 mM hydrogen peroxide. The absorbance was measured at 595 nm, and the results are expressed in optical density (OD) per milligram of tissue.

Statistical analysis

Results are presented as mean ± standard error mean (SEM). The statistical comparison of the data was performed by one-way analysis of variance (ANOVA), followed by Bonferroni's post-hoc test. Comparison of the survival curves was performed using the Log-rank test. *P* values less than 0.05 ($p<0.05$) were considered as indicative of significance.

Drugs and Reagents

The following drugs and reagents were used: paraquat was purchased from Syngenta Limited. N-(2-hydroxy-4-nitrophenyl)-N9-(2-bromophenyl) Urea (SB225002) was synthesized as described before [12] and provided by Universidade Federal de Santa Catarina (UFSC). SB225002 was identified by comparing the 1H NMR data with those published beforehand (Yield 70%, m.p. = 193–194 _C. 1H NMR (400 MHz) (ppm) (DMSO-d6) 11.05 (br. s, 1H), 9.48 (s, 1H), 9.12 (s, 1H), 8.36 (d, 1H), 7.93 (d, 1H), 7.75 (dd, 1H), 7.69 (s, 1H), 7.63 (d, 1H), 7.36 (t, 1H), 7.03 (t, 1H); IR (cm_1) 3364, 3212, 1692, 1588, 1536, 1507, 1430, 1306, 1267, 1212, 1086, 746, 649. LC–MS-ESI (m/z) 350 (M-H)-. SB225002 was dissolved in 1% Tween-80 in 0.9% NaCl solution.

Results

Dose-related toxicity of paraquat and experimental design

This experimental set was designed to characterize the behavioral changes and mortality after treatment with different doses of paraquat in order to define the ideal scheme of treatment

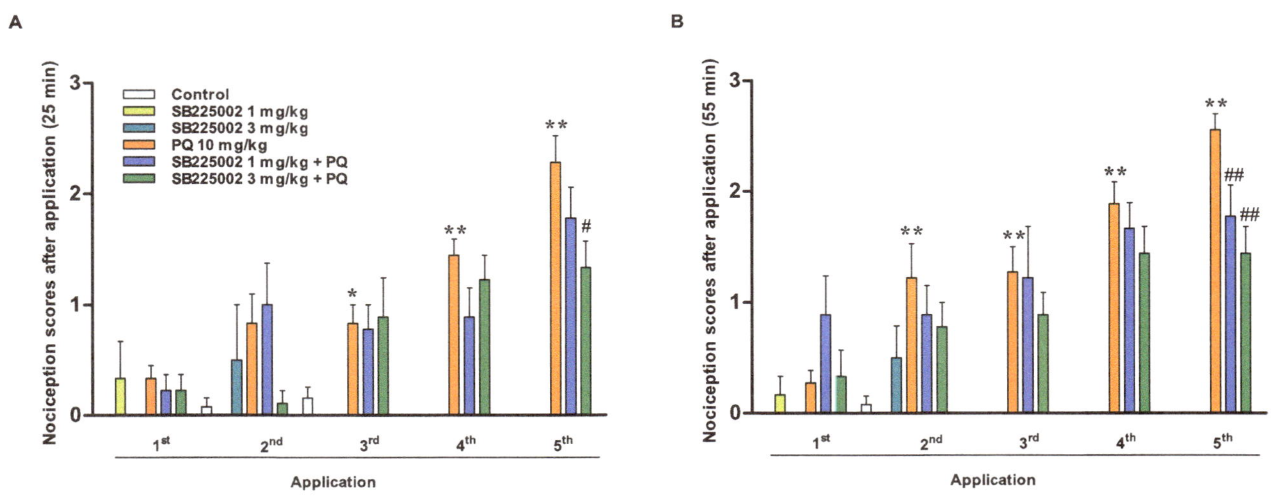

Figure 3. Behavioral scores of nociception. Assessment of behavioral scores at 25 min (A) or 55 min (B) after each treatment. Data is plotted as the cumulative scores of nociception over 5 min. Each column represents the mean ± SEM of 4–18 animals per group. *$p<0.05$ and **$p<0.01$ significantly different from control (vehicle/saline) group; #$p<0.05$ and ##$p<0.01$ significantly different from vehicle/paraquat group (PQ) (ANOVA followed by Bonferroni's post-hoc test).

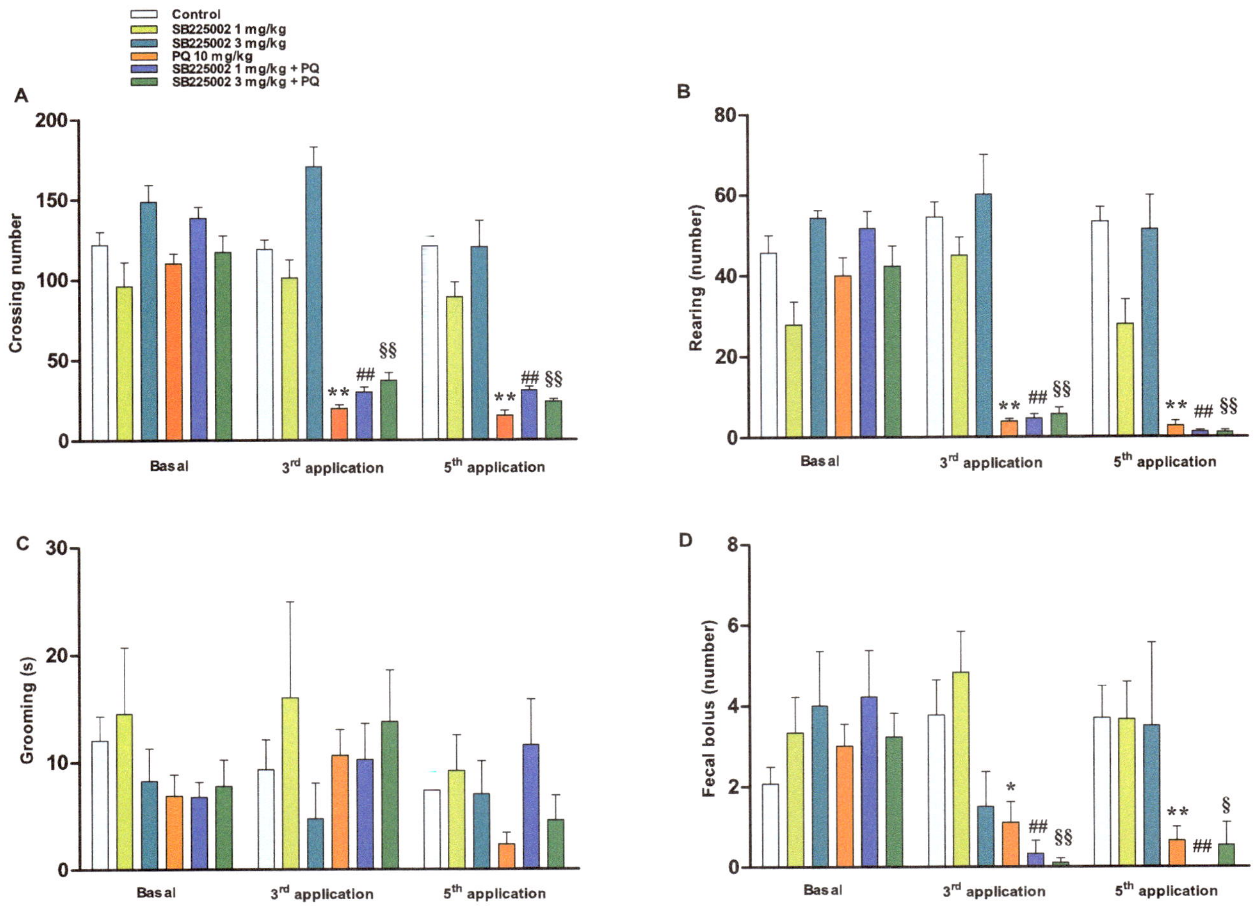

Figure 4. Evaluation of paraquat effects in the open-field test. (A) Number of crossed squares; (B) number of rearings; (C) facial grooming in s; (D) number of fecal boluses. Each column represents the mean ± SEM of 4–18 animals per group. $^{*,\S}p<0.05$; $^{**,\#\#,\S\S}p<0.01$ compared with the basal measurements of the same experimental group (ANOVA followed by Bonferroni's post-hoc test).

for the next experiments. Thus, different doses of paraquat [5 (n = 4), 10 (n = 14), 15 (n = 4), and 20 (n = 6) mg/kg] were administered to rats, twice a week over 28 days and the percentage of survival (Figure 1A) was registered.

The animals treated with 5 mg/kg per dose of paraquat did not show significant changes in behavioral tests, displaying low toxicity, as was observed in the negative control group (treated with saline). In the 10 mg/kg paraquat group significant time-related changes were observed in the behavioral tests, and 7/14 of animals died within the first 14 days. With paraquat doses of 15 mg/kg and 20 mg/kg, all the animals died within the first 14 days of treatment. Therefore, further experiments were performed using the dose of 10 mg/kg (sub-lethal dose), twice a week over 14 days, with a total cumulative dose of 50 mg/kg (Figure 1B).

Physiological changes evoked by paraquat: effects of SB225002

The treatment with 10 mg/kg per dose of paraquat, with a total dose of 50 mg/kg within 14 days, resulted in a significant and time-related reduction of body weight ($p<0.01$) in comparison to the control vehicle/saline group. This effect was partially, although not significantly, prevented by pre-treating animals with the selective CXCR2 antagonist SB225002 (3 mg/kg), as observed at the 5th application (Figure 2A).

The rectal temperature was checked at two distinct timepoints, immediately before and 1 h after treatments. The administration of paraquat resulted in a time-dependent and significant reduction of body temperature in both cases ($p<0.01$). This effect was significant from the 3rd application, when the temperature was checked before each administration of paraquat (Figure 2B), and at all the experimental times following paraquat dosage (Figure 2C). Of note, the treatment with SB225002 (3 mg/kg) was able to significantly reverse the body temperature decrease, according to assessment after the last application of paraquat ($p<0.05$; Figure 2C). SB225002 (either 1 or 3 mg/kg) did not elicit any significant effect (Figures 2A and 2B), except by a significant decrease of rectal temperature after 1 h of antagonist application (4th application, 3 mg/kg SB225002), an action that was reversed in the subsequent measurement (5th application, 3 mg/kg SB225002) (Figure 2C).

Effects of treatment with SB225002 on paraquat-elicited nociception

We also investigated whether nociceptive changes induced by paraquat might be prevented by SB225002. Intoxication with paraquat was associated with time-dependent nociceptive behavior, which was significantly different from vehicle/saline control animals, as recorded at 25 (Figure 3A) or 55 min (Figure 3B)

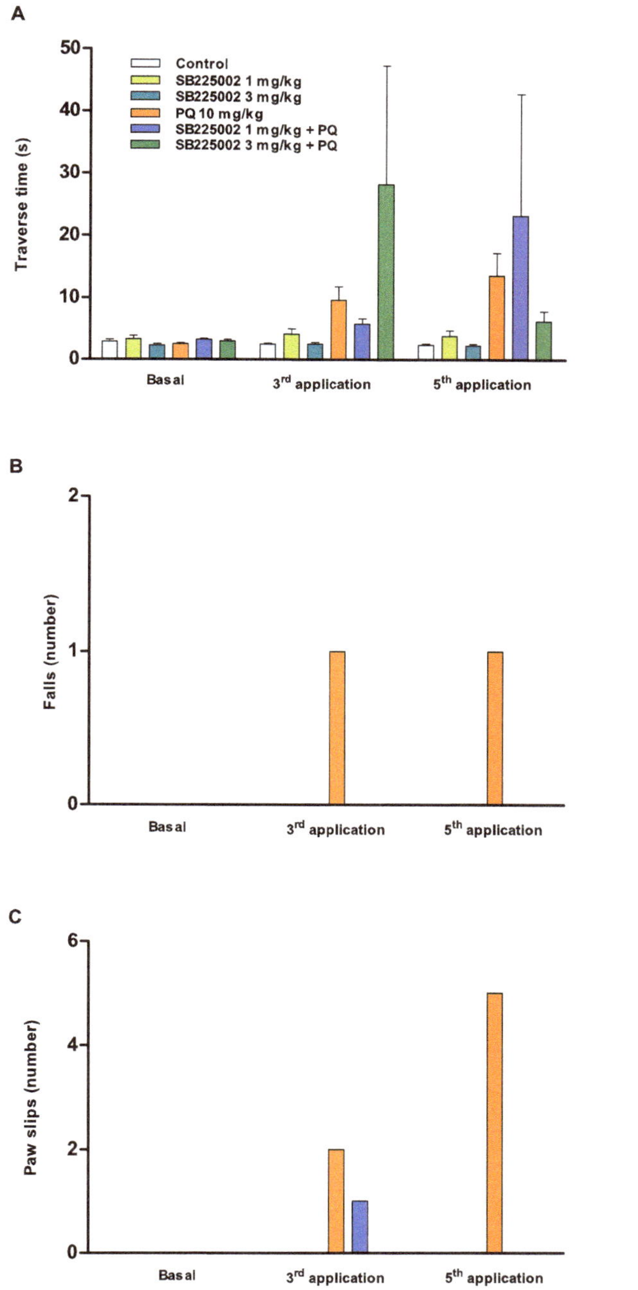

Figure 5. Locomotor skills in the beam-walking test. (A) Time taken to cross the beam in s; (B) number of falls; (C) number of paw slips. Each column represents the mean ± SEM of 4–18 animals per group (ANOVA followed by Bonferroni's post-hoc test was used for analysis of data in Figure 5A).

following the herbicide application. In either protocol, the maximal effects of paraquat was observed at the 5th application ($p<0.01$). Of note, the administration of SB225002 (3 mg/kg) was able to significantly reduce paraquat-evoked nociception, at either 25 ($p<0.05$) or 55 min ($p<0.01$) timepoints (Figure 3A and 3B). However, 1 mg/kg SB225002 displayed a significant effect on paraquat-induced nociception only at 55 min of evaluation ($p<0.01$).

Effects of paraquat administration in the open-field paradigm

In these experiments, we analyzed to what extent the treatment with paraquat might affect locomotor and exploration tasks in the open-field apparatus. Paraquat-intoxicated animals displayed significant reductions of crossing (Figure 4A) and rearing counts (Figure 4B) at both the 3rd and 5th applications ($p<0.01$). Although grooming behavior was not significantly affected (Figure 4C; $p>0.05$), paraquat administration caused a significant decrease in the number of fecal boluses (Figure 4D; $p<0.01$). SB225002 (1 or 3 mg/kg) did not display any unspecific effect, and this antagonist failed to significantly modify the effects of paraquat in the open-field model (Figure 4A–4D; $p>0.05$).

Assessment of motor coordination

The motor coordination was firstly evaluated in the beam-walking test. The traverse time was not significantly affected in either experimental group (Figure 5A). During the analysis of falls, one animal (1/18) fell after the 3rd application and another one (1/18) fell after the 5th application (Figure 5B), both belonging to the vehicle/paraquat group. Regarding paw slips, two animals (2/18) from the vehicle/paraquat group presented paw slips after the 3rd application, and five animals (5/18) of this group presented paw slips after the 5th application. One animal from the 1 mg/kg SB225002 plus PQ group (1/9) presented a paw fault at the 3rd application (Figure 5C).

Next, the animals were evaluated in the footprint model by assessing the stride length mean and the average step frequency. Representative images of footprints after the 5th application are depicted in Figure 6A–D. The administration of paraquat caused a significant reduction of mean stride length (Figure 6E), associated with an increase of step frequency (Figure 6F), after both the 3rd and the 5th applications. Although not significant, effects were observed in the groups pretreated with SB225002 in which the effects of paraquat in motor coordination were virtually prevented by this antagonist, especially at the dose of 3 mg/kg.

The exposure of paraquat increases inflammatory marker expression in the striatum

Figure 7 shows the expression profile of some inflammatory markers in the striatum of rats. The relative gene expression of *TNF-α* was significantly increased in the vehicle/paraquat group ($p<0.05$) in comparison to the control (vehicle/saline) group (Figure 7B). The relative expression of *IL-1β* was increased in the vehicle/SB225002 (3 mg/kg) and vehicle/paraquat ($p<0.05$) groups when compared to the control (vehicle/saline) group (Figure 7C). Of note, the induction of *TNF-α* and *IL-1β* expression in the striatum was abolished by treatment with SB225002 (Figure 7B and C). *NF-κB1* and *COX-2* mRNA levels did not show any changes in the expression profile amongst different experimental groups (Figure 7A and D, respectively).

Hematologic changes and neutrophil migration

At the end of the behavioral experiments, we decided to assess the possible changes of leukocyte populations throughout the different groups. Representative images of different experimental groups are depicted in Figure 8A–F. Total blood cell counts (Figure 8G) revealed a predominance of lymphocytes in the vehicle/saline group and this feature was not significantly modified by SB225002 (1 and 3 mg/kg). Vehicle/paraquat-treated animals showed a predominance of neutrophils, accompanied by a slight increase in the number of immature cells and a reduction of lymphocyte numbers ($p<0.01$). Pretreatment with SB225002 (1

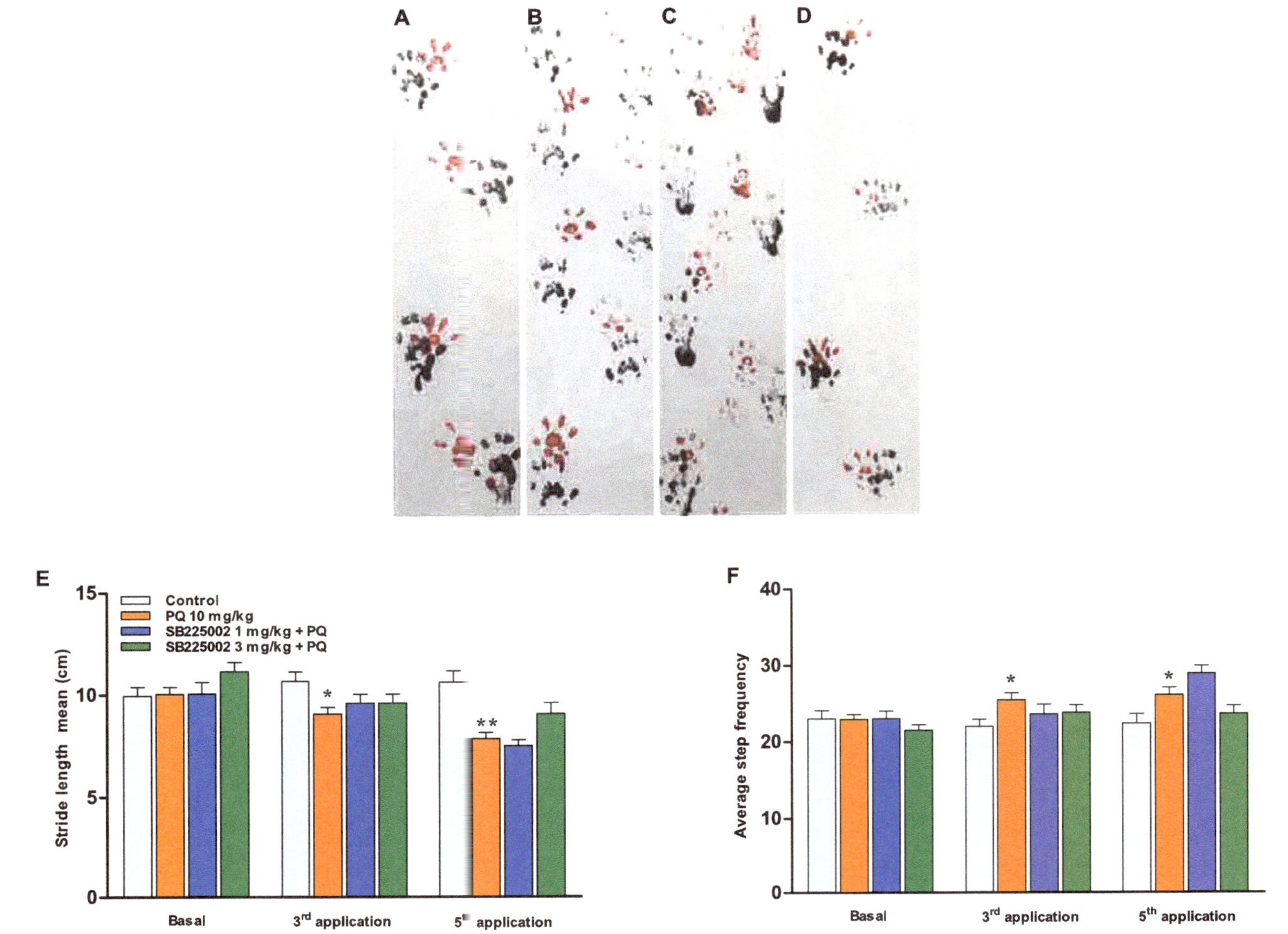

Figure 6. Symptoms of motor dysfunction. Representative walking footprint patterns after the 5th treatment in (A) control (vehicle/saline) group; (B) vehicle/paraquat (PQ)- treated animals; (C) SB225002 (1 mg/kg)/plus PQ group; (D) SB225002 (3 mg/kg)/plus PQ group. The bar graphs depict quantitative analysis of the footprint test regarding the stride length mean (E) and the average step frequency (F). Each column represents mean ± SEM of 9–19 animals per group. *$p<0.05$ and **$p<0.01$ compared with control (vehicle/saline) group (ANOVA followed by Bonferroni's post-hoc test).

and 3 mg/kg) was able to significantly modify the effects of paraquat ($p<0.05$ and $p<0.01$, respectively), as these experimental groups presented the same percentages of lymphocytes and neutrophils.

Neutrophil migration to the rat lungs was assessed by means of MPO activity. The vehicle/paraquat group presented a significant increase of MPO levels ($p<0.05$) in comparison to the control group. This parameter was reversed by the co-administration of SB225002 (3 mg/kg) ($p<0.05$), returning to the control levels (Figure 8H).

Discussion and Conclusions

The herbicide paraquat is widely employed worldwide, although its use has been prohibited in some countries due to the potential toxic effects [35]. Nevertheless, traces of paraquat can be detected in fruits and vegetables, and even in processed products, which might be related to intoxication events [36]. This agent promotes the formation of ROS and leads to toxicity of the central nervous system (CNS) linked to the loss of dopaminergic neurons, inducing Parkinson-like motor alterations in animal models [17,37,38]. The lungs are the major targets of paraquat intoxication, as the herbicide is transported by the polyamine uptake system, accumulating within the alveolar type II epithelial cells, which results in pulmonary fibrosis [39]. In the present study, we evaluated whether the administration of the selective chemokine CXCR2 receptor antagonist SB225002 might prevent either peripheral or central alterations related to paraquat intoxication in rats. We believe that our study contributes to the further understanding of the mechanisms involved in paraquat-induced toxicity, and might open new opportunities to develop potential therapeutic options to treat intoxications caused by this agent.

Initially, we performed a dose-response curve in order to select a sub-lethal scheme of treatment with paraquat using a series of behavioral evaluations. Accordingly, Czerniczyniec et al. [18] demonstrated that administration of paraquat (10 mg/kg, i.p.) in rats, dosed weekly for a month (40 mg/kg overall), resulted in behavioral changes associated with brain mitochondrial dysfunction, which corroborates our choice of protocol.

The schedule of treatment with paraquat adopted by us led to a time-related reduction of body weight and rectal temperature in

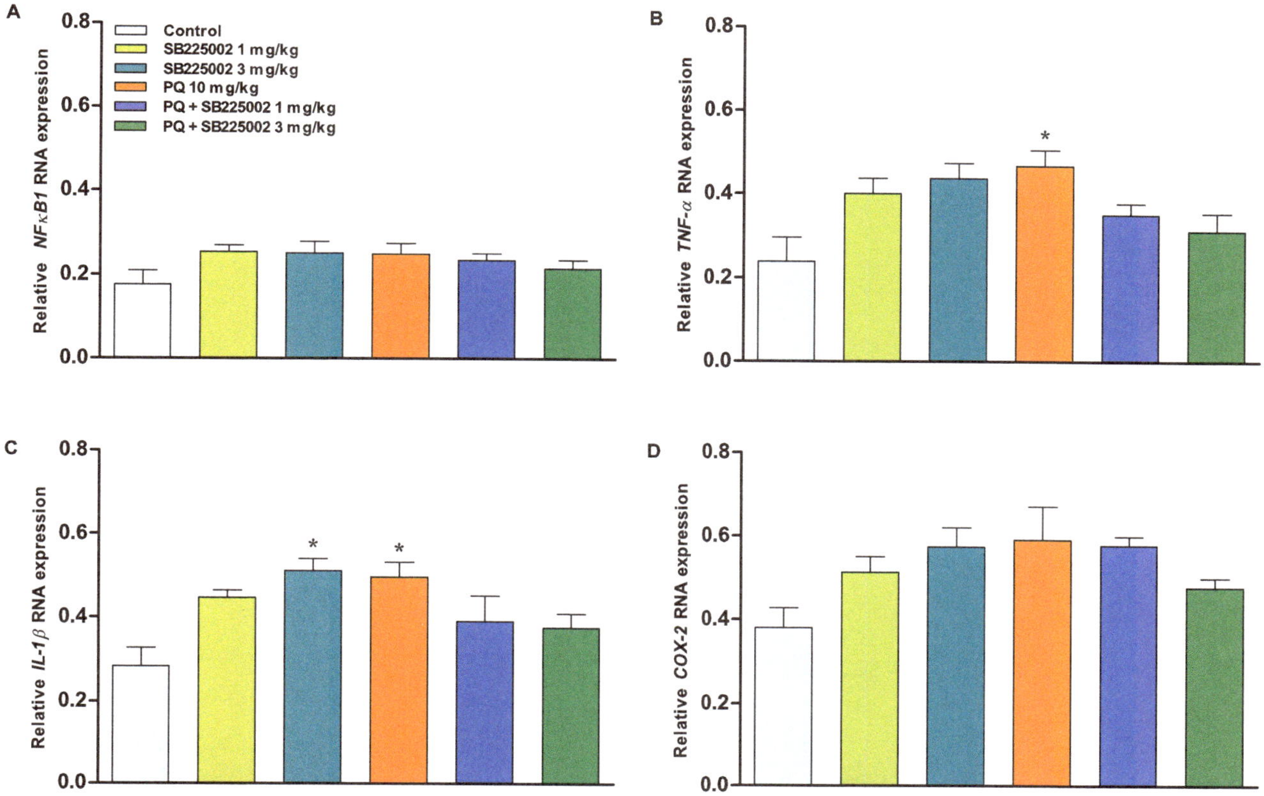

Figure 7. RT-qPCR analysis of inflammatory markers. Relative expression profiles of (A) *NF-κB1*, (B) *TNF-α*, (C) *IL-1β*, and (D) *COX-2*, assessed in the striatum after the 5th paraquat treatment. Each column represents the mean ± SEM, 3–4 samples of striatum. *$p<0.05$ compared with control (vehicle/saline) group (ANOVA followed by Bonferroni's post-hoc test).

rats, which is in accordance with previous literature [40,41,42,43,44]. Although the treatment with SB225002 (3 mg/kg, i.p.) was able to produce only a partial effect on body weight loss, the same dose of this antagonist largely prevented the hypothermia induced by paraquat. This series of results indicates that pharmacological inhibition of CXCR2 receptors by SB225002 can provide protective effects against some complications related to paraquat poisoning. Of high interest, the study conducted by Bento et al. [13] demonstrated that repeated administration of SB225002, at doses as low as 0.3 mg/kg, was effective in markedly preventing the body weight loss in the mouse model of TNBS-induced colitis. Nonetheless, to our knowledge, there is no previous report showing the ability of SB225002 to modulate hypothermia, thus we reveal a novel effect for this antagonist. In spite of that, chemokines and their receptors have been detected in most cell types in CNS [45,46,47] (for review see: [48]), which might help to explain our data.

Among the symptoms of paraquat poisoning, it is relevant to cite the occurrence of intense abdominal pain in clinics [49,50]. The administration of paraquat resulted in a marked and time-related increase of nociception scores, and pretreatment with SB225002 significantly reduced paraquat-elicited nociception at all evaluated timepoints, displaying an apparent dose-related profile. It is well known that paraquat toxicity is likely related to oxidative stress [51], and generation of ROS by this herbicide is able to modulate neutrophil function [52]. Remarkably, prior experimental evidence showed that visceral pain elicited by colorectal distention relies on oxidative stress [21]. Thus, we hypothesize that inhibition of neutrophil migration by SB225002 might prevent the complications originating from paraquat-induced ROS generation, including the abdominal discomfort. Corroborating our data, it was previously demonstrated that treatment with SB225002, in doses ranging from 0.3 to 3 mg/kg, significantly inhibited the visceral nociception induced by acetic acid in mice [14] or by cyclophosphamide in rats [15].

We also aimed to investigate whether the treatment with SB225002 would prevent motor deficits induced by paraquat intoxication in rats. In fact, it has been largely demonstrated that paraquat exposure is associated with serious motor impairments that originate from oxidative stress and neurotransmission deficits in several brain regions (for review see: [37,53]). In this study, the exposure to paraquat led to decreased locomotor and exploratory activities (open-field test), which is in agreement with previous literature [38,44,54,55]. Furthermore, a reduction of fecal bolus number was also observed in paraquat-treated rats, as demonstrated before [35]. Furthermore, the administration of paraquat resulted in worsened performance in beam-walking and footprint paradigms. A series of alterations related to locomotor dysfunction were observed in previous publications dealing with the consequences of paraquat poisoning, mainly via mechanisms involving dopaminergic neurotoxicity [17,35,44,49,51,56]. Concerning the treatment with SB225002, this antagonist presented only partial effects on paraquat-induced motor deficits, although a general improvement of coordination skills was achieved, according to assessment in the beam-walking test or the footprint model. As mentioned beforehand, chemokine receptors are expressed throughout different brain regions [48], which might support the beneficial effects of SB225002 in our experimental paradigm,

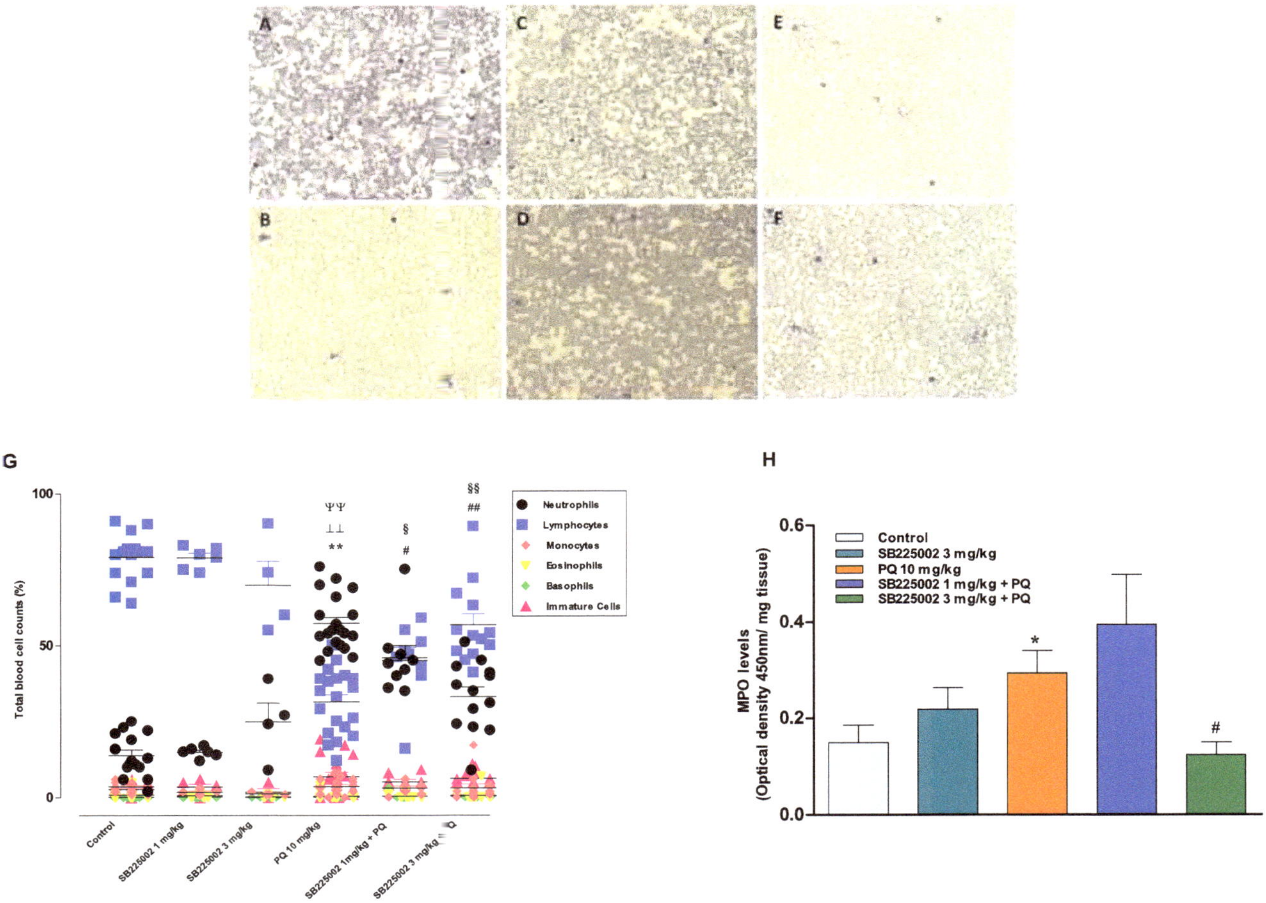

Figure 8. Hematological analysis and lung neutrophil migration. Representative pictures showing the blood smear slides in (A) control (vehicle/saline) groups; (B) vehicle/paraquat (PQ)-treated animals; (C) SB225002 (1 mg/kg)/saline group; (D) SB225002 (3 mg/kg)/saline group; (E) SB225002 (1 mg/kg)/plus PQ group; (F) SB225002 (3 mg/kg)/plus PQ group. Effects of paraquat and SB225002 on the total blood cell counts (G) and neutrophil migration as assessed by MPO activity (H). Each column represents the mean ± SEM of 4–20 animals per group in blood cell counts and 3–12 animals per group in MPO activity. $^{**,\perp\perp,\Psi\Psi}p<0.01$ indicate significant differences of neutrophils, lymphocytes, and immature cells in relation to control (vehicle/saline) group, respectively. $^{\#,\S}p<0.05$; $^{\#\#,\S\S}p<0.01$ indicate significant differences of neutrophils and lymphocytes in relation to vehicle/plus PQ group, respectively. $^{*}p<0.05$ indicates significant difference of MPO activity in relation to control (vehicle/saline) group; $^{\#}p<0.05$ indicates significant difference to vehicle/plus PQ group (ANOVA followed by Bonferroni's post-hoc test).

although the ability of this antagonist to cross the blood-brain barrier remains to be further investigated. Furthermore, it is possible to surmise that prevention of peripheral inflammation by SB225002 might avoid paraquat-induced oxidative stress within the CNS, contributing to the positive effects of this antagonist on motor performance.

It has been previously shown that the striatum is considered the most important site of dopamine action and motor control [57]. For this reason, the striatum was used in the transcriptional analysis of known molecular markers of inflammation. The expression profile of nuclear factor of kappa light polypeptide gene enhancer in B-cells 1 (*NF-κB1*), tumor necrosis factor-α (*TNF-α*), interleukin-1β (*IL-1β*), and cyclooxygenase-2 (*COX-2*) was determined. *TNF-α* and *IL-1β* expression was induced in the striatum after paraquat administration, which, in turn, was abolished by treatment with SB225002. It is important to highlight that *TNF-α* expression is persistent in inflammatory response [29] and *IL-1β* is responsible for recruitment and activation of other immune cells at the inflammatory site (for review see: [58]).

Considering the evidence discussed above, we decided to examine to what extent the administration of SB225002 might modulate peripheral inflammation induced by paraquat exposure. Strikingly, the repeated application of paraquat, resulted in a marked reduction of blood lymphocytes in comparison to saline-treated control animals, accompanied by an increase of neutrophil counts, and the scarce presence of immature cells. More interestingly, SB225002-treated groups (with either 1 or 3 mg/kg) presented the same percentages of lymphocytes and neutrophils, indicating the ability of the tested antagonist to reverse the peripheral inflammation elicited by paraquat. Similar to our data, Aires et al. [59] recently demonstrated that an acute single administration of paraquat (20 mg/kg, i.p.) resulted in marked neutrophilia, an effect that was prevented by the TNF-α inhibitor etanercept. Aligned to these experimental pieces of evidence, an attractive clinical study recently published by Fareed et al. [60] highlighted the gravity of hematological alterations observed in agricultural workers exposed to some pesticides. Furthermore, a recent publication by Kang et al. [61] demonstrated a correlation between the absolute lymphocyte counts and the 30-day mortality rates after paraquat poisoning in humans, suggesting an immunomodulatory action for this herbicide. Whether or not the therapy with anti-inflammatory agents (such as SB225002) might prevent the clinical complications related to paraquat intoxication needs additional investigation.

Lungs are well recognized as the main targets of paraquat intoxication [62]. Paraquat-induced lung toxicity is likely dependent on development of pulmonary edema, inflammatory cell infiltration, and damage of alveolar cells, resulting in fibrosis [63]. A study conducted by Venkatesan [64] demonstrated that single administration of paraquat at a high dose (50 mg/kg, i.p.) led to a significant increase of lung neutrophil migration in rats, as indirectly measured by means of MPO activity. Furthermore, the acute exposure to paraquat (20 mg/kg, i.p.) produced a marked increase of neutrophil influx to the broncoalveolar lavage of mice, and this parameter was reduced by TNF-α inhibition [59]. In the present study, the repeated i.p. administration of paraquat caused a significant increase of lung MPO activity, which was prevented by pretreating animals with SB225002 at 3 mg/kg.

In summary, on the basis of results present herein, it is possible to suggest that pharmacological inhibition of CXCR2 receptors by the antagonist SB225002 is able to interfere, in a dose-dependent manner, with paraquat-related systemic toxicity. Further studies are necessary to reinforce our findings.

Acknowledgments

The authors thank Rodrigo B. M. da Silva and Priscilla B. Pail for their technical support. We also thank Dr. Paulo C. Leal for the kind donation of antagonist SB225002.

Author Contributions

Conceived and designed the experiments: KMC ISM MMC MRB. Performed the experiments: KMC ISM LWK. Analyzed the data: KMC ISM LWK MMC MRB. Contributed reagents/materials/analysis tools: MMC MRB. Wrote the paper: KMC LWK MMC MRB.

References

1. Dinis-Oliveira RJ, Duarte JA, Sanchez-Navarro A, Remiao F, Bastos ML, et al. (2008) Paraquat poisonings: mechanisms of lung toxicity, clinical features, and treatment. Crit Rev Toxicol 38: 13–71.
2. Kornbrust DJ, Mavis RD (1980) The effect of paraquat on microsomal lipid peroxidation in vitro and in vivo. Toxicol Appl Pharmacol 53: 323–332.
3. Castello PR, Drechsel DA, Patel M (2007) Mitochondria are a major source of paraquat-induced reactive oxygen species production in the brain. J Biol Chem 282: 14186–14193.
4. Day BJ, Patel M, Calavetta L, Chang LY, Stamler JS (1999) A mechanism of paraquat toxicity involving nitric oxide synthase. Proc Natl Acad Sci USA 96: 12760–12765.
5. Dinis-Oliveira RJ, de Pinho PG, Santos L, Teixeira H, Magalhaes T, et al. (2009) Postmortem analyses unveil the poor efficacy of decontamination, anti-inflammatory and immunosuppressive therapies in paraquat human intoxications. Plos One 4: e7149.
6. Smith P, Heath D (1975) The pathology of the lung in paraquat poisoning. J Clin Pathol Suppl (R Coll Pathol) 9: 81–93.
7. Holmes WE, Lee J, Kuang WJ, Rice GC, Wood WI (1991) Structure and functional expression of a human interleukin-8 receptor. Science 253: 1278–1280.
8. Zhang ZJ, Cao DL, Zhang X, Ji RR, Gao YJ (2013) Chemokine contribution to neuropathic pain: Respective induction of CXCL1 and CXCR2 in spinal cord astrocytes and neurons. Pain 154: 2185–2197.
9. Murphy PM, Baggiolini M, Charo IF, Hebert CA, Horuk R, et al. (2000) International union of pharmacology. XXII. Nomenclature for chemokine receptors. Pharmacol Rev 52: 145–176.
10. Nasser MW, Raghuwanshi SK, Malloy KM, Gangavarapu P, Shim JY, et al. (2007) CXCR1 and CXCR2 activation and regulation. Role of aspartate 199 of the second extracellular loop of CXCR2 in CXCL8-mediated rapid receptor internalization. J Biol Chem 282: 6906–6915.
11. Paolillo N, Piccirilli S, Giardina E, Rispoli V, Colica C, et al. (2011) Effects of paraquat and capsaicin on the expression of genes related to inflammatory, immune responses and cell death in immortalized human HaCat keratinocytes. Int J Immunopathol Pharmacol 24: 861–868.
12. White JR, Lee JM, Young PR, Hertzberg RP, Jurewicz AJ, et al. (1998) Identification of a potent, selective non-peptide CXCR2 antagonist that inhibits interleukin-8-induced neutrophil migration. J Biol Chem 273: 10095–10098.
13. Bento AF, Leite DF, Claudino RF, Hara DB, Leal PC, et al. (2008) The selective nonpeptide CXCR2 antagonist SB225002 ameliorates acute experimental colitis in mice. J Leukoc Biol 84: 1213–1221.
14. Manjavachi MN, Quintao NL, Campos MM, Deschamps IK, Yunes RA, et al. (2010) The effects of the selective and non-peptide CXCR2 receptor antagonist SB225002 on acute and long-lasting models of nociception in mice. Eur J Pain 14: 23–31.
15. Dornelles FN, Andrade EL, Campos MM, Calixto JB (2013) Role of CXCR2 and TRPV1 in functional, inflammatory and behavioral changes in the rat model of cyclophosphamide-induced hemorrhagic cystitis. Br J Pharmacol. doi: 10.1111/bph.12467.
16. Zimmermann M (1983) Ethical guidelines for investigations of experimental pain in conscious animals. Pain 16: 109–110.
17. Mitra S, Chakrabarti N, Bhattacharyya A (2011) Differential regional expression patterns of alpha-synuclein, TNF-alpha, and IL-1 beta; and variable status of dopaminergic neurotoxicity in mouse brain after Paraquat treatment. J Neuroinflammation 8:163.
18. Czerniczyniec A, Karadayian AG, Bustamante J, Cutrera RA, Lores-Arnaiz S (2011) Paraquat induces behavioral changes and cortical and striatal mitochondrial dysfunction. Free Radic Biol Med 51: 1428–1436.
19. Martins JP, Silva RB, Coutinho-Silva R, Takiya CM, Battastini AM, et al. (2012) The role of P2X7 purinergic receptors in inflammatory and nociceptive changes accompanying cyclophosphamide-induced haemorrhagic cystitis in mice. Br J Pharmacol 165: 183–196.
20. Santos AA, Jr., Leal PC, Edelweiss MI, Lopes TG, Calixto JB, et al. (2010) Effects of the compounds MV8608 and MV8612 obtained from Mandevilla velutina in the model of hemorrhagic cystitis induced by cyclophosphamide in rats. Naunyn Schmiedebergs Arch Pharmacol 382: 399–407.
21. Vaculin S, Franek M, Vejrazka M (2010) Role of oxidative stress in animal model of visceral pain. Neurosci Lett 477: 82–85.
22. Wantuch C, Piesla M, Leventhal L (2007) Pharmacological validation of a model of cystitis pain in the mouse. Neurosci Lett 421: 250–252.
23. Holland HC, Weldon E (1968) A note on a new technique of recording ambulation in the open field test and its validation. Acta Psychol (Amst) 28: 293–300.
24. Wallace JE, Krauter EE, Campbell BA (1980) Motor and reflexive behavior in the aging rat. J Gerontol 35: 364–370.
25. Brooks SP, Dunnett SB (2009) Tests to assess motor phenotype in mice: a user's guide. Nature Reviews Neuroscience 10: 519–529.
26. Richter F, Hamann M, Richter A (2008) Moderate degeneration of nigral neurons after repeated but not after single intrastriatal injections of low doses of 6-hydroxydopamine in mice. Brain Res 1188: 148–156.
27. Dellon ES, Dellon AL (1991) Functional assessment of neurologic impairment: track analysis in diabetic and compression neuropathies. Plast Reconstr Surg 88: 686–694.
28. Hanafy S, El-Kadi AO, Jamali F (2012) Effect of inflammation on molecular targets and drug transporters. J Pharm Pharm Sci 15: 361–375.
29. Singh LP, Devi TS, Nantwi KD (2012) Theophylline regulates inflammatory and neurotrophic factor signals in functional recovery after C2-hemisection in adult rats. Exp Neurol 238: 79–88.
30. Kireev RA, Cuesta S, Ibarrola C, Bela T, Moreno Gonzalez E, et al. (2012) Age-related differences in hepatic ischemia/reperfusion: gene activation, liver injury, and protective effect of melatonin. J Surg Res 178: 922–934.
31. Bonefeld BE, Elfving B, Wegener G (2008) Reference genes for normalization: a study of rat brain tissue. Synapse 62: 302–309.
32. Pereira VM, Bortolotto JW, Kist LW, Azevedo MB, Fritsch RS, et al. (2012) Endosulfan exposure inhibits brain AChE activity and impairs swimming performance in adult zebrafish (Danio rerio). Neurotoxicology 33: 469–475.
33. Pilny AA (2008) Clinical hematology of rodent species. Vet Clin North Am Exot Anim Pract 11: 523–533, vi–vii.
34. Souza DG, Coutinho SF, Silveira MR, Cara DC, Teixeira MM (2000) Effects of a BLT receptor antagonist on local and remote reperfusion injuries after transient ischemia of the superior mesenteric artery in rats. Eur J Pharmacology 403: 121–128.
35. Cicchetti F, Lapointe N, Roberge-Tremblay A, Saint-Pierre M, Jimenez L, et al. (2005) Systemic exposure to paraquat and maneb models early Parkinson's disease in young adult rats. Neurobiol Dis 20: 360–371.
36. Prasad K, Tarasewicz E, Mathew J, Strickland PA, Buckley B, et al. (2009) Toxicokinetics and toxicodynamics of paraquat accumulation in mouse brain. Exp Neurol 215: 358–367.
37. Moretto A, Colosio C (2013) The role of pesticide exposure in the genesis of Parkinson's disease: epidemiological studies and experimental data. Toxicology 307: 24–34.

38. Litteljohn D, Mangano EN, Hayley S (2008) Cyclooxygenase-2 deficiency modifies the neurochemical effects, motor impairment and co-morbid anxiety provoked by paraquat administration in mice. Eur J Neurosci 28: 707–716.
39. Mainwaring G, Lim FL, Antrobus K, Swain C, Clapp M, et al. (2006) Identification of early molecular pathways affected by paraquat in rat lung. Toxicology 225: 157–172.
40. Satomi Y, Tsuchiya W, Mihara K, Ota M, Kasahara Y, et al. (2004) Gene expression analysis of the lung following paraquat administration in rats using DNA microarray. J Toxicol Sci 29: 91–100.
41. Cho JH, Yang DK, Kim L, Ryu JS, Lee HL, et al. (2005) Inhaled nitric oxide improves the survival of the paraquat-injured rats. Vascul Pharmacol 42: 171–178.
42. Hayashi A, Matsunaga N, Okazaki H, Kakimoto K, Kimura Y, et al. (2013) A disruption mechanism of the molecular clock in a MPTP mouse model of Parkinson's disease. Neuromolecular Med 15: 238–251.
43. Dinis-Oliveira RJ, Duarte JA, Remiao F, Sanchez-Navarro A, Bastos ML, et al. (2006) Single high dose dexamethasone treatment decreases the pathological score and increases the survival rate of paraquat-intoxicated rats. Toxicology 227: 73–85.
44. Chanyachukul T, Yoovathaworn K, Thongsaard W, Chongthammakun S, Navasumrit P, et al. (2004) Attenuation of paraquat-induced motor behavior and neurochemical disturbances by L-valine in vivo. Toxicol Lett 150: 259–269.
45. Nguyen D, Stangel M (2001) Expression of the chemokine receptors CXCR1 and CXCR2 in rat oligodendroglial cells. Brain Res Dev Brain Res 128: 77–81.
46. Popivanova BK, Koike K, Tonchev AB, Ishida Y, Kondo T, et al. (2003) Accumulation of microglial cells expressing ELR motif-positive CXC chemokines and their receptor CXCR2 in monkey hippocampus after ischemia-reperfusion. Brain Res 970: 195–204.
47. Valles A, Grijpink-Ongering L, de Bree FM, Tuinstra T, Ronken E (2006) Differential regulation of the CXCR2 chemokine network in rat brain trauma: implications for neuroimmune interactions and neuronal survival. Neurobiol Dis 22: 312–322.
48. Bacon KB, Harrison JK (2000) Chemokines and their receptors in neurobiology: perspectives in physiology and homeostasis. J Neuroimmunol 104: 92–97.
49. Thrash B, Uthayathas S, Karuppagounder SS, Suppiramaniam V, Dhanasekaran M (2007) Paraquat and maneb induced neurotoxicity. Proc West Pharmacol Soc 50: 31–42.
50. Hsieh YW, Lin JL, Lee SY, Weng CH, Yang HY, et al. (2013) Paraquat poisoning in pediatric patients. Pediatr Emerg Care 29: 487–491.
51. Kang MJ, Gil SJ, Lee JE, Koh HC (2010) Selective vulnerability of the striatal subregions of C57BL/6 mice to paraquat. Toxicol Lett 195: 127–134.
52. Mitra S, Abraham E (2006) Participation of superoxide in neutrophil activation and cytokine production. Biochim Biophys Acta 1762: 732–741.
53. Dinis-Oliveira RJ, Remiao F, Carmo H, Duarte JA, Navarro AS, et al. (2006) Paraquat exposure as an etiological factor of Parkinson's disease. Neurotoxicology 27: 1110–1122.
54. Carvalho MM, Campos FL, Coimbra B, Pego JM, Rodrigues C, et al. (2013) Behavioral characterization of the 6-hydroxidopamine model of Parkinson's disease and pharmacological rescuing of non-motor deficits. Mol Neurodegener 8: 14.
55. Choi WS, Abel G, Klintworth H, Flavell RA, Xia Z (2010) JNK3 mediates paraquat- and rotenone-induced dopaminergic neuron death. J Neuropathol Exp Neurol 69: 511–520.
56. Corasaniti MT, Strongoli MC, Rotiroti D, Bagetta G, Nistico G (1998) Paraquat: a useful tool for the in vivo study of mechanisms of neuronal cell death. Pharmacol Toxicol 83: 1–7.
57. Gupta SP, Patel S, Yadav S, Singh AK, Singh S, et al. (2010) Involvement of nitric oxide in maneb- and paraquat-induced Parkinson's disease phenotype in mouse: is there any link with lipid peroxidation? Neurochem Res 35: 1206–1213.
58. Latz E, Xiao TS, Stutz A (2013) Activation and regulation of the inflammasomes. Nat Rev Immunol 13: 397–411.
59. Aires RD, Capettini LS, Silva JF, Rodrigues-Machado Mda G, Pinho V, et al. (2013) Paraquat Poisoning Induces TNF-alpha-Dependent iNOS/NO Mediated Hyporesponsiveness of the Aorta to Vasoconstrictors in Rats. Plos One 8: e73562.
60. Fareed M, Pathak MK, Bihari V, Kamal R, Srivastava AK, et al. (2013) Adverse respiratory health and hematological alterations among agricultural workers occupationally exposed to organophosphate pesticides: a cross-sectional study in North India. Plos One 8: e69755.
61. Kang C, Kim SC, Lee SH, Jeong JH, Kim DS, et al. (2013) Absolute lymphocyte count as a predictor of mortality in emergency department patients with paraquat poisoning. Plos One 8: e78160.
62. Rose MS, Lock EA, Smith LL, Wyatt I (1976) Paraquat accumulation: tissue and species specificity. Biochem Pharmacol 25: 419–423.
63. Tomita M, Okuyama T, Katsuyama H, Miura Y, Nishimura Y, et al. (2007) Mouse model of paraquat-poisoned lungs and its gene expression profile. Toxicology 231: 200–209.
64. Venkatesan N (2000) Pulmonary protective effects of curcumin against paraquat toxicity. Life Sci 66: PL21–28.

18

A Meta-Analysis of the Impacts of Genetically Modified Crops

Wilhelm Klümper, Matin Qaim*

Department of Agricultural Economics and Rural Development, Georg-August-University of Goettingen, Goettingen, Germany

Abstract

Background: Despite the rapid adoption of genetically modified (GM) crops by farmers in many countries, controversies about this technology continue. Uncertainty about GM crop impacts is one reason for widespread public suspicion.

Objective: We carry out a meta-analysis of the agronomic and economic impacts of GM crops to consolidate the evidence.

Data Sources: Original studies for inclusion were identified through keyword searches in ISI Web of Knowledge, Google Scholar, EconLit, and AgEcon Search.

Study Eligibility Criteria: Studies were included when they build on primary data from farm surveys or field trials anywhere in the world, and when they report impacts of GM soybean, maize, or cotton on crop yields, pesticide use, and/or farmer profits. In total, 147 original studies were included.

Synthesis Methods: Analysis of mean impacts and meta-regressions to examine factors that influence outcomes.

Results: On average, GM technology adoption has reduced chemical pesticide use by 37%, increased crop yields by 22%, and increased farmer profits by 68%. Yield gains and pesticide reductions are larger for insect-resistant crops than for herbicide-tolerant crops. Yield and profit gains are higher in developing countries than in developed countries.

Limitations: Several of the original studies did not report sample sizes and measures of variance.

Conclusion: The meta-analysis reveals robust evidence of GM crop benefits for farmers in developed and developing countries. Such evidence may help to gradually increase public trust in this technology.

Editor: emidio albertini, University of Perugia, Italy

Funding: This research was financially supported by the German Federal Ministry of Economic Cooperation and Development (BMZ) and the European Union's Seventh Framework Programme (FP7/2007-2011) under Grant Agreement 290693 FOODSECURE. The funders had no role in study design, data collection and analysis, decision to publish, or preparation of the manuscript. Neither BMZ nor FOODSECURE and any of its partner organizations, any organization of the European Union or the European Commission are accountable for the content of this article.

Competing Interests: The authors have declared that no competing interests exist.

* Email: mqaim@uni-goettingen.de

Introduction

Despite the rapid adoption of genetically modified (GM) crops by farmers in many countries, public controversies about the risks and benefits continue [1–4]. Numerous independent science academies and regulatory bodies have reviewed the evidence about risks, concluding that commercialized GM crops are safe for human consumption and the environment [5–7]. There are also plenty of studies showing that GM crops cause benefits in terms of higher yields and cost savings in agricultural production [8–12], and welfare gains among adopting farm households [13–15]. However, some argue that the evidence about impacts is mixed and that studies showing large benefits may have problems with the data and methods used [16–18]. Uncertainty about GM crop impacts is one reason for the widespread public suspicion towards this technology. We have carried out a meta-analysis that may help to consolidate the evidence.

While earlier reviews of GM crop impacts exist [19–22], our approach adds to the knowledge in two important ways. First, we include more recent studies into the meta-analysis. In the emerging literature on GM crop impacts, new studies are published continuously, broadening the geographical area covered, the methods used, and the type of outcome variables considered. For instance, in addition to other impacts we analyze effects of GM crop adoption on pesticide quantity, which previous meta-analyses could not because of the limited number of observations for this particular outcome variable. Second, we go beyond average impacts and use meta-regressions to explain impact heterogeneity and test for possible biases.

Our meta-analysis concentrates on the most important GM crops, including herbicide-tolerant (HT) soybean, maize, and cotton, as well as insect-resistant (IR) maize and cotton. For these crops, a sufficiently large number of original impact studies have been published to estimate meaningful average effect sizes. We estimate mean impacts of GM crop adoption on crop yield, pesticide quantity, pesticide cost, total production cost, and farmer profit. Furthermore, we analyze several factors that may influence outcomes, such as geographic location, modified crop trait, and type of data and methods used in the original studies.

Materials and Methods

Literature search

Original studies for inclusion in this meta-analysis were identified through keyword searches in relevant literature databanks. Studies were searched in the ISI Web of Knowledge, Google Scholar, EconLit, and AgEcon Search. We searched for studies in the English language that were published after 1995. We did not extend the review to earlier years, because the commercial adoption of GM crops started only in the mid-1990s [23]. The search was performed for combinations of keywords related to GM technology and related to the outcome of interest. Concrete keywords used related to GM technology were (an asterisk is a replacement for any ending of the respective term; quotation marks indicate that the term was used as a whole, not each word alone): GM*, "genetically engineered", "genetically modified", transgenic, "agricultural biotechnology", HT, "herbicide tolerant", Roundup, Bt, "insect resistant". Concrete keywords used related to outcome variables were: impact*, effect*, benefit*, yield*, economic*, income*, cost*, soci*, pesticide*, herbicide*, insecticide*, productivity*, margin*, profit*. The search was completed in March 2014.

Most of the publications in the ISI Web of Knowledge are articles in academic journals, while Google Scholar, EconLit, and AgEcon Search also comprise book chapters and grey literature such as conference papers, working papers, and reports in institutional series. Articles published in academic journals have usually passed a rigorous peer-review process. Most papers presented at academic conferences have also passed a peer-review process, which is often less strict than that of good journals though. Some of the other publications are peer reviewed, while many are not. Some of the working papers and reports are published by research institutes or government organizations, while others are NGO publications. Unlike previous reviews of GM crop impacts, we did not limit the sample to peer-reviewed studies but included all publications for two reasons. First, a clear-cut distinction between studies with and without peer review is not always possible, especially when dealing with papers that were not published in a journal or presented at an academic conference [24]. Second, studies without peer review also influence the public and policy debate on GM crops; ignoring them completely would be short-sighted.

Of the studies identified through the keyword searches, not all reported original impact results. We classified studies by screening titles, abstracts, and full texts. Studies had to fulfill the following criteria to be included:

- The study is an empirical investigation of the agronomic and/or economic impacts of GM soybean, GM maize, or GM cotton using micro-level data from individual plots and/or farms. Other GM crops such as GM rapeseed, GM sugarbeet, and GM papaya were commercialized in selected countries [23], but the number of impact studies available for these other crops is very small.
- The study reports GM crop impacts in terms of one or more of the following outcome variables: yield, pesticide quantity (especially insecticides and herbicides), pesticide costs, total variable costs, gross margins, farmer profits. If only the number of pesticide sprays was reported, this was used as a proxy for pesticide quantity.
- The study analyzes the performance of GM crops by either reporting mean outcomes for GM and non-GM, absolute or percentage differences, or estimated coefficients of regression models that can be used to calculate percentage differences between GM and non-GM crops.
- The study contains original results and is not only a review of previous studies.

In some cases, the same results were reported in different publications; in these cases, only one of the publications was included to avoid double counting. On the other hand, several publications involve more than one impact observation, even for a single outcome variable, for instance when reporting results for different geographical regions or derived with different methods (e.g., comparison of mean outcomes of GM and non-GM crops plus regression model estimates). In those cases, all observations were included. Moreover, the same primary dataset was sometimes used for different publications without reporting identical results (e.g., analysis of different outcome variables, different waves of panel data, use of different methods). Hence, the number of impact observations in our sample is larger than the number of publications and primary datasets (Data S1). The number of studies selected at various stages is shown in the flow diagram in Figure 1. The number of publications finally included in the meta-analysis is 147 (Table S1).

Effect sizes and influencing factors

Effect sizes are measures of outcome variables. We chose the percentage difference between GM and non-GM crops for five different outcome variables, namely yield, pesticide quantity, pesticide cost, total production cost, and farmer profits per unit area. Most studies that analyze production costs focus on variable costs, which are the costs primarily affected through GM technology adoption. Accordingly, profits are calculated as revenues minus variable production costs (profits calculated in this way are also referred to as gross margins). These production costs also take into account the higher prices charged by private companies for GM seeds. Hence, the percentage differences in profits considered here are net economic benefits for farmers using GM technology. Percentage differences, when not reported in the original studies, were calculated from mean value comparisons between GM and non-GM or from estimated regression coefficients.

Since we look at different types of GM technologies (different modified traits) that are used in different countries and regions, we do not expect that effect sizes are homogenous across studies. Hence, our approach of combining effect sizes corresponds to a random-effects model in meta-analysis [25]. To explain impact heterogeneity and test for possible biases, we also compiled data on a number of study descriptors that may influence the reported effect sizes. These influencing factors include information on the type of GM technology (modified trait), the region studied, the type of data and method used, the source of funding, and the type of publication. All influencing factors are defined as dummy variables. The exact definition of these dummy variables is given in Table 1. Variable distributions of the study descriptors are shown in Table S2.

Statistical analysis

In a first step, we estimate average effect sizes for each outcome variable. To test whether these mean impacts are significantly

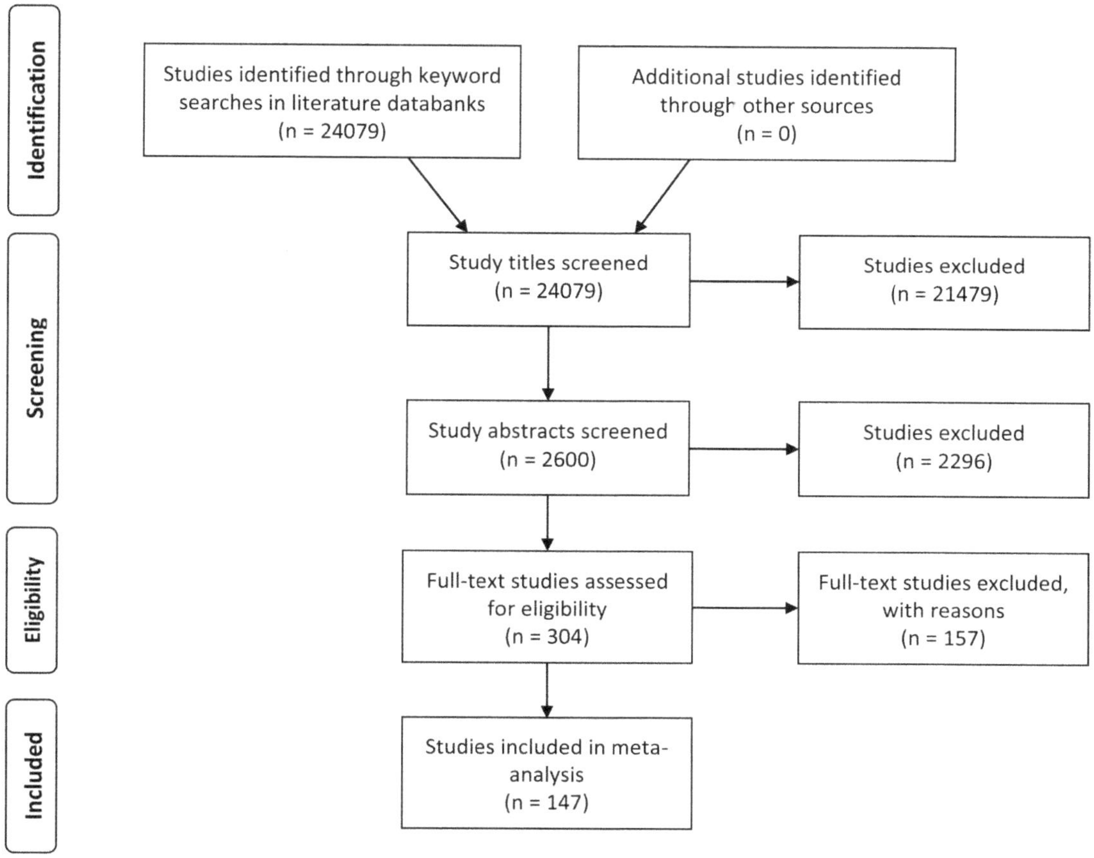

Figure 1. Selection of studies for inclusion in the meta-analysis.

different from zero, we regress each outcome variable on a constant with cluster correction of standard errors by primary dataset. Thus, the test for significance is valid also when observations from the same dataset are correlated. We estimate average effect sizes for all GM crops combined. However, we expect that the results may differ by modified trait, so that we also analyze mean effects for HT crops and IR crops separately.

Meta-analyses often weight impact estimates by their variances; estimates with low variance are considered more reliable and receive a higher weight [26]. In our case, several of the original studies do not report measures of variance, so that weighting by variance is not possible. Alternatively, weighting by sample size is common, but sample sizes are also not reported in all studies considered, especially not in some of the grey literature publications. To test the robustness of the results, we employ a

Table 1. Variables used to analyze influencing factors of GM crop impacts.

Variable name	Variable definition
Insect resistance (IR)	Dummy that takes a value of one for all observations referring to insect-resistant GM crops with genes from *Bacillus thuringiensis* (Bt), and zero for all herbicide-tolerant (HT) GM crops.
Developing country	Dummy that takes a value of one for all GM crop applications in a developing country according to the World Bank classification of countries, and zero for all applications in a developed country.
Field-trial data	Dummy that takes a value of one for all observations building on field-trial data (on-station and on-farm experiments), and zero for all observations building on farm survey data.
Industry-funded study	Dummy that takes a value of one for all studies that mention industry (private sector companies) as source of funding, and zero otherwise.
Regression model result	Dummy that takes a value of one for all impact observations that are derived from regression model estimates, and zero for observations derived from mean value comparisons between GM and non-GM.
Journal publication	Dummy that takes a value of one for all studies published in a peer-reviewed journal, and zero otherwise.
Journal/academic conference	Dummy that takes a value of one for all studies published in a peer-reviewed journal or presented at an academic conference, and zero otherwise.

different weighting procedure, using the inverse of the number of impact observations per dataset as weights. This procedure avoids that individual datasets that were used in several publications dominate the calculation of average effect sizes.

In a second step, we use meta-regressions to explain impact heterogeneity and test for possible biases. Linear regression models are estimated separately for all of the five outcome variables:

$$\%\Delta Y_{hij} = \alpha_h + \mathbf{X}_{hij}\boldsymbol{\beta}_h + \varepsilon_{hij}$$

$\%\Delta Y_{hij}$ is the effect size (percentage difference between GM and non-GM) of each outcome variable h for observation i in publication j, and $\mathbf{X}_{hij}$ is a vector of influencing factors. α_h is a coefficient and $\boldsymbol{\beta}_h$ a vector of coefficients to be estimated; ε_{hij} is a random error term. Influencing factors used in the regressions are defined in Table 1.

Results and Discussion

Average effect sizes

Distributions of all five outcome variables are shown in Figure S1. Table 2 presents unweighted mean impacts. As a robustness check, we weighted by the inverse of the number of impact observations per dataset. Comparing unweighted results (Table 2) with weighted results (Table S3) we find only very small differences. This comparison suggests that the unweighted results are robust.

On average, GM technology has increased crop yields by 21% (Figure 2). These yield increases are not due to higher genetic yield potential, but to more effective pest control and thus lower crop damage [27]. At the same time, GM crops have reduced pesticide quantity by 37% and pesticide cost by 39%. The effect on the cost of production is not significant. GM seeds are more expensive than non-GM seeds, but the additional seed costs are compensated through savings in chemical and mechanical pest control. Average profit gains for GM-adopting farmers are 69%.

Results of Cochran's test [25], which are reported in Figure S1, confirm that there is significant heterogeneity across study observations for all five outcome variables. Hence it is useful to further disaggregate the results. Table 2 shows a breakdown by modified crop trait. While significant reductions in pesticide costs are observed for both HT and IR crops, only IR crops cause a consistent reduction in pesticide quantity. Such disparities are expected, because the two technologies are quite different. IR crops protect themselves against certain insect pests, so that spraying can be reduced. HT crops, on the other hand, are not protected against pests but against a broad-spectrum chemical herbicide (mostly glyphosate), use of which facilitates weed control. While HT crops have reduced herbicide quantity in some situations, they have contributed to increases in the use of broad-spectrum herbicides elsewhere [2,11,19]. The savings in pesticide costs for HT crops in spite of higher quantities can be explained by the fact that broad-spectrum herbicides are often much cheaper than the selective herbicides that were used before. The average farmer profit effect for HT crops is large and positive, but not statistically significant because of considerable variation and a relatively small number of observations for this outcome variable.

Impact heterogeneity and possible biases

Table 3 shows the estimation results from the meta-regressions that explain how different factors influence impact heterogeneity. Controlling for other factors, yield gains of IR crops are almost 7 percentage points higher than those of HT crops (column 1). Furthermore, yield gains of GM crops are 14 percentage points higher in developing countries than in developed countries. Especially smallholder farmers in the tropics and subtropics suffer from considerable pest damage that can be reduced through GM crop adoption [27].

Most original studies in this meta-analysis build on farm surveys, although some are based on field-trial data. Field-trial results are often criticized to overestimate impacts, because farmers may not be able to replicate experimental conditions. However, results in Table 3 (column 1) show that field-trial data do not overestimate the yield effects of GM crops. Reported yield gains from field trials are even lower than those from farm surveys. This is plausible, because pest damage in non-GM crops is often more severe in farmers' fields than on well-managed experimental plots.

Table 2. Impacts of GM crop adoption by modified trait.

Outcome variable	All GM crops	Insect resistance	Herbicide tolerance
Yield	21.57*** (15.65; 27.48)	24.85*** (18.49; 31.22)	9.29** (1.78; 16.80)
n/m	451/100	353/83	94/25
Pesticide quantity	−36.93*** (−48.01; −25.86)	−41.67*** (−51.99; −31.36)	2.43 (−20.26; 25.12)
n/m	121/37	108/31	13/7
Pesticide cost	−39.15*** (−46.96; −31.33)	−43.43*** (−51.64; −35.22)	−25.29*** (−33.84; −16.74)
n/m	193/57	145/45	48/15
Total production cost	3.25 (−1.76; 8.25)	5.24** (0.25; 10.73)	−6.83 (−16.43; 2.77)
n/m	115/46	96/38	19/10
Farmer profit	68.21*** (46.31; 90.12)	68.78*** (46.45; 91.11)	64.29 (−24.73; 153.31)
n/m	136/42	119/36	17/9

Average percentage differences between GM and non-GM crops are shown with 95% confidence intervals in parentheses. *, **, *** indicate statistical significance at the 10%, 5%, and 1% level, respectively. *n* is the number of observations, *m* the number of different primary datasets from which these observations are derived.

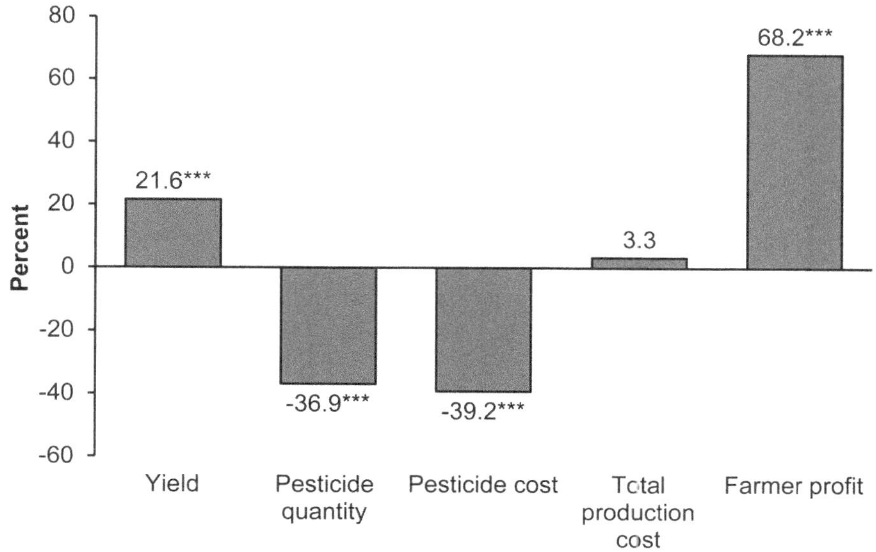

Figure 2. Impacts of GM crop adoption. Average percentage differences between GM and non-GM crops are shown. Results refer to all GM crops, including herbicide-tolerant and insect-resistant traits. The number of observations varies by outcome variable; yield: 451; pesticide quantity: 121; pesticide cost: 193; total production cost: 115; farmer profit: 136. *** indicates statistical significance at the 1% level.

Another concern often voiced in the public debate is that studies funded by industry money might report inflated benefits. Our results show that the source of funding does not significantly influence the impact estimates. We also analyzed whether the statistical method plays a role. Many of the earlier studies just compared yields of GM and non-GM crops without considering possible differences in other inputs and conditions that may also affect the outcome. Net impacts of GM technology can be estimated with regression-based production function models that control for other factors. Interestingly, results derived from regression analysis report higher average yield effects.

Finally, we examined whether the type of publication matters. Controlling for other factors, the regression coefficient for journal publications in column (1) of Table 3 implies that studies published in peer-reviewed journals show 12 percentage points higher yield gains than studies published elsewhere. Indeed, when only including observations from studies that were published in journals, the mean effect size is larger than if all observations are included (Figure S2). On first sight, one might suspect publication bias, meaning that only studies that report substantial effects are accepted for publication in a journal. A common way to assess possible publication bias in meta-analysis is through funnel plots [25], which we show in Figure S3. However, in our case these funnel plots should not be over-interpreted. First, only studies that report variance measures can be included in the funnel plots, which holds true only for a subset of the original studies used here. Second, even if there were publication bias, our mean results would be estimated correctly, because we do include studies that were not published in peer-reviewed journals.

Further analysis suggests that the journal review process does not systematically filter out studies with small effect sizes. The journal articles in the sample report a wide range of yield effects, even including negative estimates in some cases. Moreover, when combining journal articles with papers presented at academic conferences, average yield gains are even higher (Table 3, column 2). Studies that were neither published in a journal nor presented at an academic conference encompass a diverse set of papers, including reports by NGOs and outspoken biotechnology critics. These reports show lower GM yield effects on average, but not all meet common scientific standards. Hence, rather than indicating publication bias, the positive and significant journal coefficient may be the result of a negative NGO bias in some of the grey literature.

Concerning other outcome variables, IR crops have much stronger reducing effects on pesticide quantity than HT crops (Table 3, column 3), as already discussed above. In terms of pesticide costs, the difference between IR and HT is less pronounced and not statistically significant (column 4). The profit gains of GM crops are 60 percentage points higher in developing countries than in developed countries (column 6). This large difference is due to higher GM yield gains and stronger pesticide cost savings in developing countries. Moreover, most GM crops are not patented in developing countries, so that GM seed prices are lower [19]. Like for yields, studies published in peer-reviewed journals report higher profit gains than studies published elsewhere, but again we do not find evidence of publication bias (column 7).

Conclusion

This meta-analysis confirms that – in spite of impact heterogeneity – the average agronomic and economic benefits of GM crops are large and significant. Impacts vary especially by modified crop trait and geographic region. Yield gains and pesticide reductions are larger for IR crops than for HT crops. Yield and farmer profit gains are higher in developing countries than in developed countries. Recent impact studies used better data and methods than earlier studies, but these improvements in study design did not reduce the estimates of GM crop advantages. Rather, NGO reports and other publications without scientific peer review seem to bias the impact estimates downward. But even with such biased estimates included, mean effects remain sizeable.

One limitation is that not all of the original studies included in this meta-analysis reported sample sizes and measures of variance. This is not untypical for analyses in the social sciences, especially when studies from the grey literature are also included. Future

Table 3. Factors influencing results on GM crop impacts (%).

Variables	(1) Yield	(2) Yield	(3) Pesticide quantity	(4) Pesticide cost	(5) Total cost	(6) Farmer profit	(7) Farmer profit
Insect resistance (IR)	6.58** (2.85)	5.25* (2.82)	−37.38*** (11.81)	−7.28 (5.44)	5.63 (5.60)	−22.33 (21.62)	−33.41 (21.94)
Developing country	14.17*** (2.72)	13.32*** (2.65)	−10.23 (8.99)	−19.16*** (5.35)	3.43 (4.78)	59.52*** (18.02)	60.58*** (17.67)
Field-trial data	−7.14** (3.19)	−7.81** (3.08)	–#	−17.56 (11.45)	−10.69* (5.79)	–#	–#
Industry-funded study	1.68 (5.30)	1.05 (5.21)	37.04 (23.08)	−7.77 (10.22)	–#	–#	–#
Regression model result	7.38* (3.90)	7.29* (3.83)	9.67 (10.40)	–#	–#	−11.44 (24.33)	−9.85 (24.03)
Journal publication	12.00*** (2.52)	–	9.95 (6.79)	−3.71 (4.09)	−3.08 (3.30)	48.27*** (15.48)	–
Journal/academic conference	–	16.48*** (2.64)	–	–	–	–	65.29*** (17.75)
Constant	−0.22 (2.84)	−2.64 (2.86)	−4.44 (10.33)	−16.13 (4.88)	−1.02 (4.86)	8.57 (24.33)	−1.19 (24.53)
Observations	451	451	121	193	115	136	136
R^2	0.23	0.25	0.20	0.14	0.12	0.12	0.14

Coefficient estimates from linear regression models are shown with standard errors in parentheses. Dependent variables are GM crop impacts measured as percentage differences between GM and non-GM. All explanatory variables are 0/1 dummies (for variable definitions see Table 1). The yield models in columns (1) and (2) and the farmer profit models in columns (6) and (7) have the same dependent variables, but they differ in terms of the explanatory variables, as shown. *, **, *** indicate statistical significance at the 10%, 5%, and 1% level, respectively. # indicates that the variable was dropped because the number of observations with a value of one was smaller than 5.

impact studies with primary data should follow more standardized reporting procedures. Nevertheless, our findings reveal that there is robust evidence of GM crop benefits. Such evidence may help to gradually increase public trust in this promising technology.

Supporting Information

Figure S1 Histograms of effect sizes for the five outcome variables.

Figure S2 Impacts of GM crop adoption including only studies published in journals.

Figure S3 Funnel plots for the five outcome variables.

Table S1 List of publications included in the meta-analysis.

Table S2 Distribution of study descriptor dummy variables for different outcomes.

Table S3 Weighted mean impacts of GM crop adoption.

Data S1 Data used for the meta-analysis.

Acknowledgments

We thank Sinja Buri and Tingting Xu for assistance in compiling the dataset. We also thank Joachim von Braun and three reviewers of this journal for useful comments.

Author Contributions

Conceived and designed the research: WK MQ. Analyzed the data: WK MQ. Contributed to the writing of the manuscript: WK MQ. Compiled the data: WK.

References

1. Gilbert N (2013) A hard look at GM crops. Nature 497: 24–26.
2. Fernandez-Cornejo J, Wechsler JJ, Livingston M, Mitchell L (2014) Genetically Engineered Crops in the United States. Economic Research Report ERR-162 (United Sates Department of Agriculture, Washington, DC).
3. Anonymous (2013) Contrary to popular belief. Nature Biotechnology 31: 767.
4. Andreasen M (2014) GM food in the public mind–facts are not what they used to be. Nature Biotechnology 32: 25.
5. DeFrancesco L (2013) How safe does transgenic food need to be? Nature Biotechnology 31: 794–802.
6. European Academies Science Advisory Council (2013) Planting the Future: Opportunities and Challenges for Using Crop Genetic Improvement Technologies for Sustainable Agriculture (EASAC, Halle, Germany).
7. European Commission (2010) A Decade of EU-Funded GMO Research 2001–2010 (European Commission, Brussels).
8. Pray CE, Huang J, Hu R, Rozelle S (2002) Five years of Bt cotton in China - the benefits continue. The Plant Journal 31: 423–430.
9. Huang J, Hu R, Rozelle S, Pray C (2008) Genetically modified rice, yields and pesticides: assessing farm-level productivity effects in China. Economic Development and Cultural Change 56: 241–263.
10. Morse S, Bennett R, Ismael Y (2004) Why Bt cotton pays for small-scale producers in South Africa. Nature Biotechnology 22: 379–380.
11. Qaim M, Traxler G (2005) Roundup Ready soybeans in Argentina: farm level and aggregate welfare effects. Agricultural Economics 32: 73–86.
12. Sexton S, Zilberman D (2012) Land for food and fuel production: the role of agricultural biotechnology. In: The Intended and Unintended Effects of US Agricultural and Biotechnology Policies (eds. Zivin, G. & Perloff, J.M.), 269–288 (University of Chicago Press, Chicago).
13. Ali A, Abdulai A (2010) The adoption of genetically modified cotton and poverty reduction in Pakistan. Journal of Agricultural Economics 61, 175–192.
14. Kathage J, Qaim M (2012) Economic impacts and impact dynamics of Bt (*Bacillus thuringiensis*) cotton in India. Proceedings of the National Academy of Sciences USA 109: 11652–11656.
15. Qaim M, Kouser S (2013) Genetically modified crops and food security. PLOS ONE 8: e64879.
16. Stone GD (2012) Constructing facts: Bt cotton narratives in India. Economic & Political Weekly 47(38): 62–70.
17. Smale M, Zambrano P, Gruere G, Falck-Zepeda J, Matuschke I, et al. (2009) Measuring the Economic Impacts of Transgenic Crops in Developing Agriculture During the First Decade: Approaches, Findings, and Future Directions (International Food Policy Research Institute, Washington, DC).
18. Glover D (2010) Is Bt cotton a pro-poor technology? A review and critique of the empirical record. Journal of Agrarian Change 10: 482–509.
19. Qaim M (2009) The economics of genetically modified crops. Annual Review of Resource Economics 1: 665–693.
20. Carpenter JE (2010) Peer-reviewed surveys indicate positive impact of commercialized GM crops. Nature Biotechnology 28: 319–321.
21. Finger R, El Benni N, Kaphengst T, Evans C, Herbert S, et al. (2011) A meta analysis on farm-level costs and benefits of GM crops. Sustainability 3: 743–762.
22. Areal FJ, Riesgo L, Rodríguez-Cerezo E (2013) Economic and agronomic impact of commercialized GM crops: a meta-analysis. Journal of Agricultural Science 151: 7–33.
23. James C (2013) Global Status of Commercialized Biotech/GM Crops: 2013. ISAAA Briefs No.46 (International Service for the Acquisition of Agri-biotech Applications, Ithaca, NY).
24. Rothstein HR, Hopewell S (2009) Grey literature. In: Handbook of Research Synthesis and Meta-Analysis, Second Edition (eds. Cooper, H., Hegdes, L.V. & Valentine, J.C.), 103–125 (Russell Sage Foundation, New York).
25. Borenstein M, Hedges LV, Higgins JPT, Rothstein HR (2009) Introduction to Meta-Analysis (John Wiley and Sons, Chichester, UK).
26. Shadish WR, Haddock CK (2009) Combining estimates of effect size. In: Handbook of Research Synthesis and Meta-Analysis, Second Edition (eds. Cooper, H., Hegdes, L.V. & Valentine, J.C.), 257–277 (Russell Sage Foundation, New York).
27. Qaim M, Zilberman D (2003) Yield effects of genetically modified crops in developing countries. Science 299: 900–902.

19

Evaluation of the Agronomic Performance of Atrazine-Tolerant Transgenic *japonica* Rice Parental Lines for Utilization in Hybrid Seed Production

Luhua Zhang[1☯], Haiwei Chen[1☯¤], Yanan Li[1], Yanan Li[1], Shengjun Wang[2], Jinping Su[2], Xuejun Liu[2], Defu Chen[1*], Xiwen Chen[1*]

1 Laboratory of Molecular Genetics, College of Life Sciences, Nankai University, Tianjin, China, **2** Tianjin Crop Research Institute, Tianjin, China

Abstract

Currently, the purity of hybrid seed is a crucial limiting factor when developing hybrid *japonica* rice (*Oryza sativa* L.). To chemically control hybrid seed purity, we transferred an improved atrazine chlorohydrolase gene (*atzA*) from *Pseudomonas* ADP into hybrid *japonica* parental lines (two maintainers, one restorer), and Nipponbare, by using *Agrobacterium*-mediated transformation. We subsequently selected several transgenic lines from each genotype by using PCR, RT-PCR, and germination analysis. In the presence of the investigated atrazine concentrations, particularly 150 μM atrazine, almost all of the transgenic lines produced significantly larger seedlings, with similar or higher germination percentages, than did the respective controls. Although the seedlings of transgenic lines were taller and gained more root biomass compared to the respective control plants, their growth was nevertheless inhibited by atrazine treatment compared to that without treatment. When grown in soil containing 2 mg/kg or 5 mg/kg atrazine, the transgenic lines were taller, and had higher total chlorophyll contents than did the respective controls; moreover, three of the strongest transgenic lines completely recovered after 45 days of growth. After treatment with 2 mg/kg or 5 mg/kg of atrazine, the atrazine residue remaining in the soil was 2.9–7.0% or 0.8–8.7% respectively, for transgenic lines, and 44.0–59.2% or 28.1–30.8%, respectively, for control plants. Spraying plants at the vegetative growth stage with 0.15% atrazine effectively killed control plants, but not transgenic lines. Our results indicate that transgenic *atzA* rice plants show tolerance to atrazine, and may be used as parental lines in future hybrid seed production.

Editor: Jin-Song Zhang, Institute of Genetics and Developmental Biology, Chinese Academy of Sciences, China

Funding: This work was supported by the Key Project of Tianjin Science and Technology Support Program (11ZCGYNC01000), the Key Program of the Natural Science Foundation of Tianjin (12YFJZJC01700, 14JCZDJC34100), the grant of Natural Science Foundation of China (No. 31070273, 31070717) and the 111 Project (No. B08011). The funders had no role in study design, data collection and analysis, decision to publish, or preparation of the manuscript.

Competing Interests: The authors have declared that no competing interests exist.

* Email: chendefu@nankai.edu.cn (DC); xiwenchen@nankai.edu.cn (XC)

☯ These authors contributed equally to this work.

¤ Current address: College of Life Sciences, Chifeng College, Chifeng, China

Introduction

Rice (*Oryza sativa*) is one of the most important staple food crops globally. According to the National Grain and Oil Information Center, the area of China planted with rice in 2012 was 3.0×10^7 hm^2, including 9.0×10^6 hm^2 of *japonica* rice [1]. *Japonica* rice is mainly planted in the northern region of the Qinling Mountains–Huai River, and its planted area has increased in recent years because of its high quality and good taste. *Japonica* rice production is currently dominated by conventional varieties, with hybrid rice accounting for only 3% of the cultivated area. On the other hand, *indica* hybrid rice represents 70–80% of the total planted area of *indica* rice [2]. Therefore, there is considerable potential for the development of *japonica* hybrid rice. An increase in the annual planted area of *japonica* hybrid rice from 3% to 50%, i.e., to reach 4.0×10^6 hm^2, is estimated to lead to the production of 3.5×10^9 kg of high-quality grain (www.cngrain.com/Publish/qita/200503/207290), thereby contributing considerably to meet consumer's demand for high-quality food both in China and globally.

The three-line system is a traditional and effective production method for hybrid *japonica* rice seed [3]. The most widely used male sterile line in the system is BT-type cytoplasmic male sterile (CMS). However, the panicle of this line is loosely enclosed when heading, and this appearance closely resembles that of the maintainer. This makes it difficult for farmers to distinguish the BT-CMS line when eliminating off-type plants [2]. Furthermore, the BT CMS line has good restorability, and may therefore be easily pollinated with exotic pollens that contaminated during mechanical harvesting and storage of seeds, and also with exotic pollens from other plants [4]. The use of contaminated CMS lines in seed production results in decreased hybrid seed purity. Therefore, off-type contamination must be eliminated as early as possible. This is largely a manual process and requires considerable labor input, particularly in Asia. On the one hand, the need for increased labor will increase the price of hybrid seeds, while on the other hand, the increase in manual procedures may lead to the production of false hybrids. Furthermore, as the Chinese economy develops, increasing numbers of young men are leaving their home towns to seek work in the cities, leaving the elderly and women to work on the farms. The transformation of heavy and complex farming to light and simple farming is therefore becoming increasingly important. Thus, ensuring hybrid seed purity and reducing labor costs are two key issues in hybrid *japonica* rice seed production.

Genetic engineering, especially herbicide resistance engineering, provides an efficient means of controlling purity in hybrid seed production. Yan first proposed a strategy of utilizing herbicide resistance genes to chemically control purity in hybrid seed production [5]. Since then, two-line hybrid rice production has been extensively investigated [6–12] and progress has recently been reviewed [13]. Additionally, some transgenic hybrid rice combinations have been used in field trials [7,8,10,12]. However, the research has mainly focused on the *bar* gene isolated from *Streptomyces hygroscopicus* [6,8,10,12]; other genes, such as the EPSPS (5-enolpyruvylshikimate-3-phosphate synthase) gene from *Agrobacterium* strain CP4, and the protoporphyrinogen oxidase gene from *Bacillus subtilis*, have rarely been investigated [11]. If the strategy is proven to be effective, chemical control of hybrid seed purity will be mainly dependent on an herbicide with a single mode of action, and this will hinder sustainable weed management.

Atrazine (6-chloro-N^2-ethyl-N^4-isopropyl-1,3,5-triazine-2,4-diamine) is a triazine herbicide, and is commonly used in maize, sorghum, and sugarcane fields [14]. By inhibiting electron transport to plastoquinone in the photosystem PSII, atrazine terminates photosynthesis and kills weeds [15]. Atrazine was once the most widely used herbicide worldwide, because of its low cost and high effectiveness [14]. We previously isolated an atrazine chlorohydrolase gene (*atzA*) from a soil bacterium *Pseudomonas* ADP [16], and modified this gene by using directed evolution, to improve the enzymatic activity [17]. In the present study, we transferred the improved *atzA* gene into breeding hybrid *japonica* parental lines. Our results indicate that the transgenic *atzA* rice lines show tolerance to atrazine, and may be used as parental lines to chemically improve seed purity in hybrid seed production.

Materials and Methods

Construction of plant expression vector

Ubiquitin promoter and an improved atrazine chlorohydrolase gene *atzA*-22-4 [17] were respectively amplified from pSTARLING, an RNAi intermediate vector for monocots (a kind gift from the Commonwealth Scientific and Industrial Research Organization, Australia), and *AtzA*-22-4 using primers SL-Ubi-F/SL-Ubi-R and atzA-TJ-F/atzA-TJ-R (Table 1). After respectively inserted into TaKaRa pMD19T-simple vector and confirmed by sequencing, the pAtzA-22-4-19T-simple was restricted with *Eco*RV to collect the 1.4 kb fragment, and then lignated with pUbi-19T-simple. The resulting plasmid was restricted with *Sac*I/*Spe*I to collect the 3.4 kb fragment, and then ligated with pCAMBIA1301. The recombinant plasmid p1301-ubi-22-14 was then introduced into *Agrobacterium tumefaciens* EHA105 by the freeze-thaw method [18].

Genetic transformation and plant regeneration

Oryza sativa L. Nipponbare and *japonica* hybrid rice parental lines in the three-line system, Jindao7 (maintainer), Jindao8 (maintainer) and Jinhui3 (restorer) were used for transformation. Mature seeds were dehulled, surface-sterilized and placed on NB medium (N6 macro elements, B5 micro elements and vitamins) supplemented with 2 g/L proline, 3 mg/L 2, 4-D and 300 mg/L casein hydrolysate in dark at 28°C. After 2–3 weeks, the scutellum-derived calli were excised and subcultured every four weeks on the same medium but with 0.5 g/L proline, 2 mg/L 2, 4-D in dark at 28°C. The highly embryogenic compact calli (3–5 mm in diameter) that subcultured for less than five generations, were selected and co-cultivated with *A. tumefaciens* EHA105 harboring p1301C-ubi-22-14 on the co-cultivation medium (subculture medium but with 100 μM acetosyringone) for 3 days in dark at 28°C. Following that, the explants were then transferred into selection medium (subculture medium but with 50 mg/L hygromycin and 500 mg/L cefotaxime) in dark at 28°C for selection. After two cycles of selection, hygromycin-resistant calli were transferred onto pre-regeneration medium (NB medium with 0.5 g/L proline, 2 mg/L 6-BA, 1 mg/L NAA, 5 mg/L ABA, 300 mg/L casein hydrolysate and 50 mg/L hygromycin) for 14 to 21 days in dark at 28°C, then to regeneration medium (pre-regeneration medium but without 5 mg/L ABA) for 30 days under 54 μmol/m^2/s light at 28°C and finally to the rooting medium (MS medium with 1 mg/L IBA) for 15 days under light at 28°C.

Molecular analysis of transgenic lines

Genomic DNA was isolated from young leaves using a modified CTAB method [19]. PCR was performed to preliminarily select the transformed plants using primers atzA7/atzA8 [20] and SL-Ubi-C-F/SL-Ubi-C-R (Table 1). RT-PCR was performed to further confirm the expression of *atzA* in the PCR-positive transformed plants. Total RNA was extracted from young leaves using an RNAultra Extraction Kit (Qiagen). cDNA was synthesized using atzA8 and oligo(dT) as primers and SuperScript™ II RNase H− (Invitrogen) as reverse transcriptase. RT-PCR was amplified using the cDNA as template and atzA7/atzA2 [20] as primers. *β*-actin (AB047313) was also amplified as the internal control using actin-R/actin-F as primers [21].

Germination and seedling growth in the presence of atrazine

Fifty seeds from each transgenic line and the respective control plant were directly sown on the surface of filter paper in plates containing 0, 75 or 150 μM atrazine. Seeds were placed in a growth chamber at 28°C with 16 h of 54 μmol/m^2/s light per day. Germination (based on radicles >2 mm) was recorded daily and the cumulative values at day 3 and day 7 were calculated to represent as germination potential and germination percentage. The shoot length, root length and their biomass were measured after 7 days. For seedling growth test, seven-day old seedlings germinated in absence of atrazine were placed on filter paper in pots, and incubated in Kimura B nutrition solution [22] containing 0, 75 or 150 μM atrazine in growth chamber described above. During the period, the nutrition solution containing the respective concentration of atrazine was added to keep the filter paper wet. The growth parameters as described above were measured after 10 days.

Soil-grown transgenic T_2 lines in the presence of atrazine

Two-week old seedlings of similar size that germinated in absence of atrazine were transplanted into pots containing 1.12 kg of soil with 0, 2, or 5 mg/kg of atrazine. Nine plants were planted in each pot, and irrigated with the same amount of water every day and with Kimura B nutrition solution [22] twice a week. Plants were incubated in the greenhouse in a 14-h light/10-h dark cycle (28/25°C) at 300 μmol $m^{-2}s^{-1}$ light and 75% relative humidity. Plant growth and chlorophyll content were measured at

Table 1. Primers used in this study.

Primer	Sequence (5′→3′)
SL-Ubi-F	tgagctcctgcagtgcagcgtgacccggtcgt, with *SacI* site
SL-Ubi-R	tgatatcctgcagaagtaacaccaaacaacag, with *EcoRV* site
atzA-TJ-F	tgatatcatgcaaacgctcagcatccg, with *EcoRV* site
atzA-TJ-R	tgatatcactagtctagaggctgcgcaagctg, with *EcoRV* and *SpeI* site
SL-Ubi-C-F	cccgccgtaataaatagacac
SL-Ubi-C-R	accacaccacatcatcacaac

15 -day intervals. For chlorophyll content analysis, approximately 10 mg of leaves was extracted with 5 ml acetone and then quantified the absorbance at 663.6 nm and 646.6 nm, as described by Porra et al. [23]. Chlorophyll a = $12.25 \times A_{663.6} - 2.55 \times A_{646.6}$, Chlorophyll b = $20.31 \times A_{646.6} - 4.91 \times A_{663.6}$.

HPLC analysis of atrazine and its metabolite hydroxyatrazine in the plant and soil

Atrazine and hydroxyatrazine in the rice leaves and soil were also determined in the soil-grown experiment. One gram of leaves, or five grams of soil samples that collected from the soil layer mixture as thick as possible, were extracted with 5 mL dichloromethane. After 30 min of incubation at room temperature, the mixture was centrifuged at 15,294 *g* for 10 min. The supernatant was transferred to a new tube and filtered through a 0.2-µm filter and then subjected to CoM 6000 HPLC system analysis on an analytical C18 column (5 µm, 250 mm×4.6 mm) at 30°C with linear gradients phase as follows: 0 to 6 min, 10% to 25% acetonitrile; 6 min to 21 min, 25% to 65% acetonitrile; 21 min to 23 min, 65% to 100% acetonitrile; and 23 min to 25 min, 100% acetonitrile [24] at the wavelength of 228 nm.

Spraying atrazine to transgenic T2 lines

Two-week old seedlings from the germination experiment were also transplanted into pots with soil. Each pot contained 16 plants. After 30 days' growth, the last second leaves were taken and cut into 2–3 cm section, soaked in 0, 75 and 150 µM atrazine, and incubated at 25°C under light for 2 days. The color of the leaves was observed every day. After 40 days' growth, each pot of plants was sprayed with 20 ml of 0.15% atrazine solution.

Statistical analysis

All the experiments were performed for three times. Significant difference between the specific transgenic line and the respective control was performed using independent-samples *t*-test at 95% or 99% confidence with IBM SPSS Statistics 11.0. Values indicated by * or ** represented significantly difference at $P<0.05$ or $P<0.01$.

Results

Selection of transgenic *atzA* plants

PCR analysis revealed that 18 (100%) Nipponbare transformants, 15 out of 18 (83%) Jindao7 transformants, 14 out of 16 (88%) Jinhui1 transformants, and 17 (100%) Jindao8 transformants were positive. RT-PCR analysis further confirmed the expression of *atzA* in the PCR-positive lines (data not shown). Subsequently, the self-pollinated seeds (T_1 progenies) from each plant were germinated in the presence of hygromycin. Transgenic lines that showed a segregation pattern of 3:1 resistant/sensitive in the germination test were selected to produce the T_2 generations. Individual plants, whose herbicide tolerance did not segregate in the germination test, were considered as homogenous lines and selected. For simplicity, in further research, we used only two or three independent transgenic T_2 lines for each genotype.

Germination of transgenic *atzA* lines in the presence of atrazine

To investigate the atrazine tolerance of transgenic lines during germination, we germinated seeds of three lines of each genotype and the respective controls in the presence of atrazine (Fig. 1A, Table 2). The germination potential of the wild types decreased as the atrazine concentration increased (58.3–98%, 22.4–64.2%, and 0%, respectively, in 0 µM, 75 µM, and 150 µM atrazine); on the other hand, the germination percentage significantly decreased only in presence of 150 µM atrazine (95.2–100%, 72.1–96.7%, and 0–42.2%, respectively, in 0 µM, 75 µM, and 150 µM atrazine). In the presence of the investigated atrazine concentrations, the highest germination potential was determined for Nipponbare, followed by Jinhui1, Jindao8, and Jindao7. In the presence of 0 µM or 75 µM atrazine, the germination potentials and percentages of almost all of the transgenic lines did not differ significantly from those of the respective controls. The exceptions were the germination potential for Jindao8 in the presence of 75 µM atrazine and the germination percentage for Jinhui1 in the presence of 75 µM atrazine. On the other hand, in the presence of 150 µM atrazine, all of the transgenic lines showed significantly higher germination potentials and germination percentages than did the respective controls. Moreover, two of the wild types (Nipponbare and Jinhui1) failed to germinate or only rarely germinated at day 3 or day 7. Interestingly, in the presence of 150 µM atrazine, the highest germination percentage was determined for Jindao7, followed by Jindao8, Jinhui1, and Nipponbare; on the other hand, Jindao7 showed the lowest germination potential. When germinated in the presence of atrazine, all of the transgenic lines produced larger seedlings (with taller shoots and longer roots) than did the respective control plants (Fig. 1A). However, the presence of atrazine significantly inhibited the growth of transgenic lines and wild types (Fig. 1A).

Seedling growth of transgenic *atzA* lines in the presence of atrazine

To investigate the tolerance of transgenic lines to atrazine, we transplanted 7-day old seedlings germinated in the absence of atrazine, to pots of soil containing different concentration of atrazine (Fig. 1B and Table 3). In the absence of atrazine, we determined no differences in shoot or root growth between any of the transgenic lines and the respective controls. However, in the

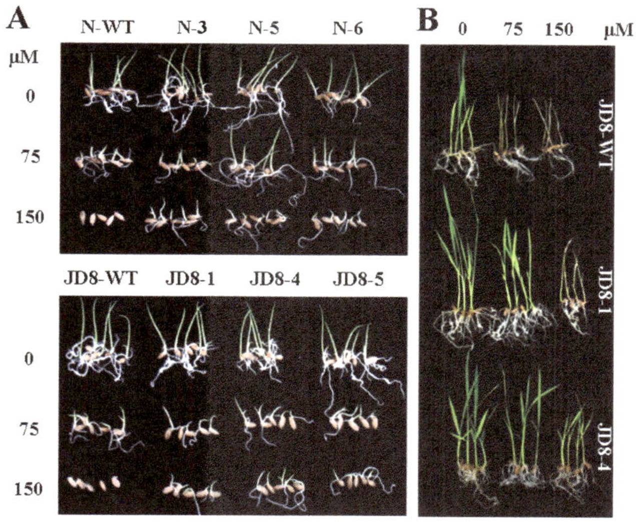

Figure 1. Germination and seedling growth of transgenic *atzA* rice lines in the presence of atrazine. (**A**) Representative images of seeds from the transgenic lines and the respective controls sown on plates containing 0, 75 or 150 μM atrazine for 7 days. (**B**) Representative images from seven-day old seedlings germinated in absence of atrazine were transplanted into pots containing 0, 75 or 150 μM atrazine for 10 days. For simplicity, only images of transgenic Jindao8 (JD8) and/or Nipponbare (N) were shown here.

presence of 75 μM atrazine, all of the transgenic lines (except for Jindao8) were significantly taller than were the respective controls; further, all of the transgenic lines (except for Jinhui1) produced significantly more root biomass than did the respective controls. On the other hand, in the presence of 150 μM atrazine, shoot and root growth of almost all of the transgenic lines and the respective controls were significantly inhibited. The exceptions were the shoot growth of Nipponbare and the root growth of Jinhui1-5. Plants became yellowish, and subsequently rotted and died after 10 days (Fig. 1B).

Tolerance of soil-grown transgenic *atzA* lines to atrazine

To further investigate the tolerance of transgenic lines to atrazine, we transplanted 15-day-old seedlings germinated in the absence of atrazine, to pots of soil containing 0 mg/kg, 2 mg/kg, or 5 mg/kg of atrazine. We measured the plant height (as a non-destructive assay of plant growth) and chlorophyll content at 15-day intervals.

In the absence of atrazine, we determined no significant differences in plant height between the transgenic lines and the respective controls (Fig. 2A). In soil containing 2 mg/kg or 5 mg/kg of atrazine, all the transgenic lines were significantly taller than were the respective control plants. However, the presence of atrazine significantly inhibited the growth of transgenic lines and control plants (by 68.1–111.9% and 46.4–68.2%, respectively). In contrast to the control plants, all of the transgenic lines (except for the Jindao8) were slightly taller in soil containing 2 mg/kg of atrazine than in soil containing 5 mg/kg of atrazine. The growth of transgenic lines gradually recovered with an increase in the growth time. Further, three of the strongest transgenic lines (Jindao7-4, Jinhui1-5, and Jindao8-4) completely recovered after 45 days of growth (data not shown).

Atrazine may disrupt the electron flow of photosystem II and destroy photosynthetic pigments in plant leaves [15]. In the absence of atrazine, we observed no difference in chlorophyll content between transgenic lines and control plants (Fig. 2B). However, in soil containing 2 mg/kg or 5 mg/kg of atrazine, the transgenic lines retained more chlorophyll than did the respective control plants. With a few exceptions, the chlorophyll content of the transgenic lines decreased significantly in soil containing 5 mg/kg of atrazine, but not in soil containing 2 mg/kg of atrazine; on the other hand, the chlorophyll content of the control plants decreased significantly under both conditions. For each genotype grown in soil containing the same atrazine concentration, we observed no difference in chlorophyll content at different growth stages. After 45 days, the chlorophyll contents of the three

Table 2. Germination of transgenic *atzA* rice lines in the presence of atrazine.

Line	0 μM		75 μM		150 μM	
	3 d	7 d	3 d	7 d	3 d	7 d
N[a]-WT	98.0±2.0	100±0.0	54.0±1.6	94.5±1.4	0.0±0.0	0.0±0.0
N-3	98.0±1.6	94.2±0.0	60.0±2.2	98.2±1.1	64.3±1.6**	96.2±1.8**
N-5	96.0±2.0	100±0.0	64.0±1.3	92.1±1.5	62.3±2.6**	92.6±1.5**
JD7-WT	58.3±1.6	100±0.0	24.3±2.1	90.2±2.3	0.0±0.0	42.2±1.1
JD7-2	56.0±1.1	98.1±1.4	22.3±1.7	88.2±2.1	2.3±0.1**	88.5±1.6**
JD7-4	62.3±2.2	98.6±1.6	26.3±1.3	90.2±0.8	2.4±0.2**	86.5±1.3**
JH3-WT	86.3±2.0	95.2±2.8	64.2±2.2	72.1±1.1	0.0±0.0	4.5±0.0
JH-3	82.3±1.2	94.2±1.8	58.7±1.6	90.2±1.2*	32.3±0.9**	78.6±1.2**
JH3-5	88.3±3.2	92.2±1.3	66.3±2.7	94.1±1.0*	38.7±1.1**	80.5±1.2**
JD8-WT	82.0±4.2	98.7±0.8	22.4±1.6	96.7±1.0	0.0±0.0	26.5±1.3
JD8-1	78.3±2.5	99.2±0.8	48.3±1.1**	97.2±0.2	34.3±0.7**	94.5±2.8**
JD8-4	78.0±2.6	100±0.0	54.0±2.6**	98.3±1.0	35.2±1.6**	96.8±1.3**

Note: In each independent treatment 50 seeds were used for the experiment. Values are shown as mean ±SEM ($n = 3$). * or ** indicates significant difference ($P<0.05$) or highly significant difference ($P<0.01$) between the treatment and the control.

[a]N, JD7, JH3 and JD8 were the abbreviations of Nipponbare, Jindao7, Jindao8 and Jinhui3, respectively. For simplicity, only two transgenic lines for each genotype were shown here.

Table 3. Seedling growth performance of transgenic *atzA* rice lines in the presence of atrazine.

Line	0 μM		75 μM		150 μM	
	Shoot length (cm)	Root biomass (mg)	Shoot length (cm)	Root biomass (mg)	Shoot length (cm)	Root biomass (mg)
N [a]-WT	9.90±1.37	32.00±3.16	6.54±0.86 (66.0)[a]	21.00±2.55 (65.6)	5.07±1.17 (51.2)	21.75±3.77 (68.0)
N-3	10.55±1.49	30.25±3.50	9.56±2.08* (90.7)	29.50±3.14** (97.5)	7.55±1.20* (71.6)	21.50±6.40 (71.1)
N-5	8.78±2.42	32.75±4.50	8.94±1.33* (101.8)	29.85±1.83** (91.2)	9.39±0.51** (106.9)	21.00±2.71 (64.1)
JD7-WT	8.47±0.37	28.50±2.56	5.28±0.57 (62.4)	20.00±3.10 (70.2)	6.54±0.61 (77.2)	17.75±5.19 (62.3)
JD7-2	9.40±2.61	31.50±3.00	7.08±0.78* (75.3)	32.50±2.65** (103.2)	5.81±0.56 (61.8)	17.00±3.46 (54.0)
JD7-4	7.99±1.65	27.50±3.35	8.16±1.27* (102.2)	27.50±2.65** (100.0)	6.78±0.53 (84.8)	17.75±4.19 (64.5)
JH3-WT	9.83±1.25	24.00±3.07	5.80±0.43 (59.0)	21.75±3.36 (90.6)	4.62±0.42 (47.0)	13.25±6.24 (55.2)
JH-3	10.44±1.47	26.75±2.75	8.27±0.59* (79.2)	24.25±2.22* (90.7)	5.12±0.61 (49.0)	15.25±2.63 (57.0)
JH3-5	8.89±1.58	25.50±3.57	8.25±0.78* (92.8)	23.25±2.87* (83.8)	6.04±0.19* (67.9)	22.00±3.37** (86.3)
JD8-WT	7.88±0.93	26.75±3.55	7.20±1.85 (91.4)	21.75±2.87 (81.3)	6.43±0.35 (81.5)	12.25±3.46 (45.8)
JD8-1	8.11±1.51	27.50±2.15	7.18±0.40 (88.6)	27.50±1.52** (100.0)	5.65±1.12 (69.6)	11.50±1.91 (41.8)
JD8-4	7.90±0.72	27.75±2.53	7.64±0.69 (96.7)	27.75±1.79** (100.0)	5.64±0.78 (71.3)	11.75±2.99 (42.3)

Note: The parameters were recorded after 10 days. Values are shown as mean ± SEM ($n=3$). * or ** indicates significant difference ($P<0.05$) or highly significant difference ($P<0.01$) between the treatment and the control.
[a]N, JD7, JH3 and JD8 were the abbreviations of Nipponbare, Jindao7, Jindao8 and Jinhui3, respectively.
[b]Data in the bracket are the percentage of the treated values to the mock values.

strongest lines (Jindao7–4, Jinhui1–5, and Jindao8–4) were almost identical when grown in soil containing 5 mg/kg of atrazine and when grown in the absence of atrazine (data not shown).

Ability of soil-grown T_2 transgenic lines to degrade atrazine

To investigate the ability of the soil-grown transgenic plants to degrade atrazine, we determined the atrazine residue in the soil at 15-day intervals, after growth of seedlings in pots of soil containing 0 mg/kg, 2 mg/kg, or 5 mg/kg of atrazine. We observed that the atrazine residue in the soil decreased with an increase in growth time, for transgenic and control plants (Fig. 2C). However, the decrease was greater for transgenic lines than for control plants. After treatment with 2 mg/kg of atrazine, the atrazine residue remaining in the soil was 2.9–7.0% for transgenic lines, and 44.0–59.2% for control plants; after treatment with 5 mg/kg of atrazine, the atrazine residue remaining in the soil was 0.8–8.7% for transgenic lines and 28.1–30.8% for control plant. For each genotype at the same growth stage, the atrazine residue in the soil was higher after growth in 2 mg/kg atrazine than after growth in 5 mg/kg atrazine. We also observed that hydroxyatrazine, not atrazine (only trace) appeared in the leaves of the transgenic plants (Fig. S1), indicating that atrazine was metabolized rather than accumulated in the transgenic plants.

Utilization of transgenic plants in hybrid seed production

To investigate the tolerance of transgenic plants to the atrazine concentration used in weed control, we first examined the tolerance of leaf sections to atrazine solution. In the presence of 75 μM atrazine, the leaf sections of transgenic lines remained green, whereas those of control plants became bleached (Fig. 3A). In the presence of 150 μM atrazine, the leaf sections of transgenic lines became slightly bleached, but remained greener than those of control plants. We subsequently sprayed plants with 0.15% atrazine, and observed that wild-type plants became curled and withered after 6 days, whereas transgenic lines continued to grow well (Fig. 3B). Our results indicate that the transgenic rice plants showed tolerance to the atrazine concentration used in weed control.

Discussion

Current research on the utilization of herbicide resistance genes in hybrid rice is focused on the two-line hybrid production system. Sterile lines are not stable in the two-line system, and therefore most *japonica* hybrid combinations are still produced by using three-line system. In the present study, we have developed atrazine-tolerant transgenic *japonica* rice parental lines with the potential to be used in future hybrid seed production.

Herbicide tolerance during germination is very important when utilizing transgenic *atzA* lines in hybrid seed production, because off-type plants should be eliminated as early as possible. Atrazine is a slightly water-soluble herbicide, with a saturated concentration of 153 μM. Kawahigashi [25] previously showed that an atrazine concentration of 100 μM did not affect the germination of rice plants. Therefore, we used two atrazine concentrations (75 μM and 150 μM) in the germination test, to evaluate the tolerance of transgenic rice lines. In contrast to Kawahigashi [25], we observed that the transgenic lines germinated well in the presence of atrazine whereas the respective controls did not. However, as the seedling growth of transgenic lines were also inhibited by atrazine, chemical control of contamination during germination may not be appropriate.

To evaluate the tolerance of seedlings to atrazine, we used 7-day-old seedlings germinated in the absence of atrazine as a starting material, to avoid cumulative effects derived from germination. We observed that almost all of the transgenic lines grew well in the presence of 75 μM atrazine with no difference from that in absence of atrazine. Therefore, 75 μM atrazine may be used for effective control of off-type plants during the early seedling growth stage. Our data also suggested spraying plants with 0.15% atrazine (equivalent to the standard dosage used in the field for weed control, i.e., 200–250 g 40% suspension concentrate in 30~50 kg of water) at the subsequently vegetative growth stage could serve as an alternative means of chemically controlling

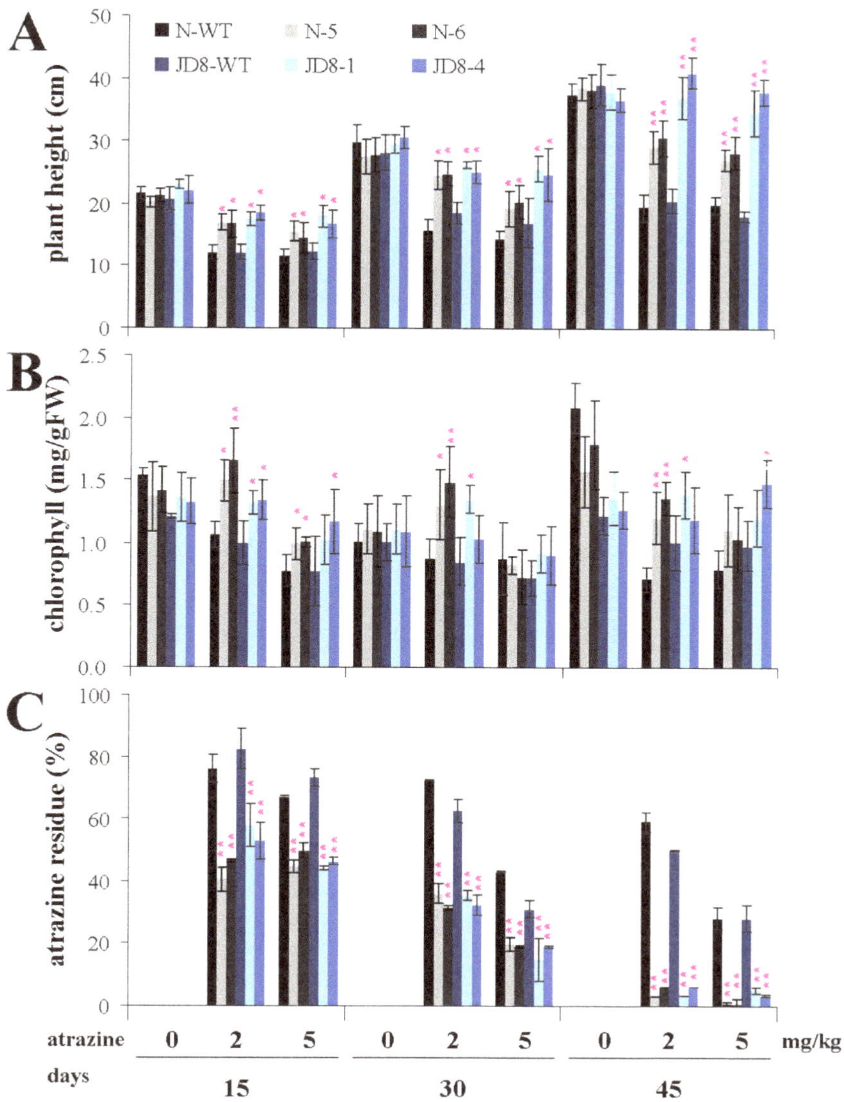

Figure 2. Growth and physiology of transgenic *atzA* rice lines in the presence of atrazine. Fifteen-day old seedlings germinated in absence of atrazine were transplanted into pots with 1.12 kg soil containing 0, 2 or 5 mg/kg of atrazine. (**A**) plant height; (**B**) chlorophyll content; and (**C**) atrazine residue were measured at 15-day intervals. Values are shown as mean ±SEM ($n = 3$). * or ** over the bar indicates significant difference ($P < 0.05$) or highly significant difference ($P < 0.01$) between the treatment and the control. For simplicity, only data of transgenic Nipponbare (N) and Jindao8 (JD8) were shown here.

off-type plants. However, we did not test the tolerance of transgenic *atzA* lines at the reproductive growth stage, which is important for the utilization in mechanical harvesting of hybrid seed production. Further studies are required to investigate whether the lethal dosage for untransformed parental lines is tolerated at the reproductive stage.

Atrazine degrades relatively slowly, with an average half-life of 4–57 weeks [26]. Therefore, application of this herbicide will inevitably lead to accumulation in the field, which may affect current crop growth or damage sensitive succession crops [27]. The transgenic rice lines were taller, and had higher chlorophyll contents than did the respective control plants when growing in the soil of 2 mg/kg or 5 mg/kg atrazine. Three of the transgenic lines (Jindao7–4, Jinhui1–5, and Jindao8–4) completely recovered after 45 days of growth in the presence of atrazine, suggesting that an atrazine concentration up to 5 mg/kg does not affect sustained growth of transgenic lines. Our data also suggested that almost all of the atrazine was degraded after growth of the transgenic rice plants for 45 days. This was also verified by the appearance of hydroxyatrazine in the leaves of transgenic plants. Besides, atrazine metabolite hydroxyatrazine could be directly determined in transgenic plants. However, as the leave, stem and root of the plants all possibly metabolize atrazine [28], it would be more complicated if we want to quantitatively evaluate the atrazine metabolized by plants. That is why we chose to determine the atrazine residue in the soil after growth of the transgenic plants in this study.

Utilization of herbicide resistance genes in hybrid seed production may be achieved by introducing the genes into a maintainer or restorer line in the three-line system. When an herbicide-resistant maintainer is obtained, its corresponding CMS line, which possesses herbicide resistance, may be created by using

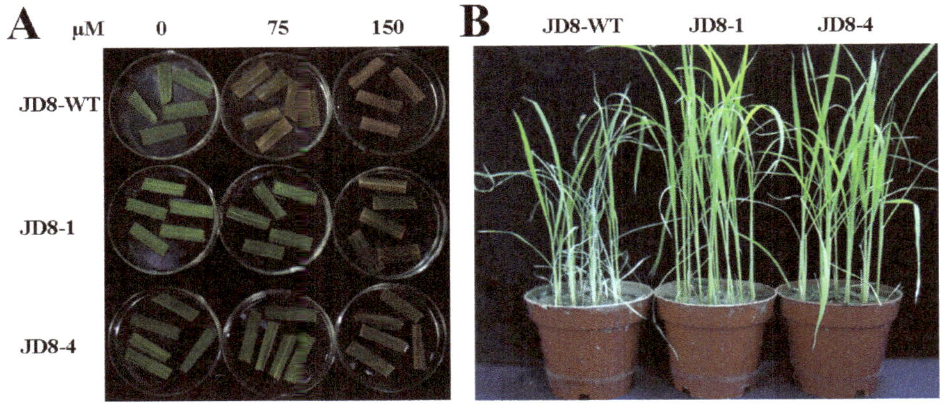

Figure 3. Tolerance of mature transgenic *atzA* plants to atrazine. (**A**) Representative images of leave sections in the presence of atrazine. Fifteen-day old seedlings were transplanted into pots containing soil. After 30 days' growth, the last second leaves were cut into 2–3 cm section, and soaked in 0, 75 and 150 μM atrazine, and incubated at 25°C under light for 2 days. (**B**) Representative images of mature plants after sprayed with atrazine. Fifteen-day old seedlings were transplanted into pots containing soil. After 40 days' growth, each pot of transgenic lines (9 plants) was sprayed with 20 mL of 0.15% atrazine solution. For simplicity, only data of transgenic Jindao8 (JD8) was shown here.

a backcross. The resulting herbicide-resistant CMS breeder seeds may be used to cross with the original maintainer (herbicide sensitive), to multiply the herbicide-resistant foundation seeds. During this process, the selfed maintainer and off-type plants may easily be eliminated by herbicide spraying, thus ensuring the purity of CMS seeds. The herbicide-resistant CMS line may also be used in mechanical harvesting, by spraying herbicide to kill the pollen plants after pollination, and harvesting seeds from the surviving CMS plants. A similar result may be achieved when crossing the herbicide-resistant CMS line with the herbicide-sensitive restorer line. Alternatively, the herbicide-resistant restorer line may be crossed with the herbicide-sensitive sterile line during the hybrid seed production process, by spraying herbicide to kill the off-type plants. Further, the herbicide-resistant maintainer and restorer lines may be useful for eliminating off-type plants, and for weed management when growing hybrid rice.

Field trials with genetically modified rice are strictly controlled in China, and therefore we were unable to evaluate our herbicide-resistant hybrid rice combinations on a field scale. Further studies are required to verify the validity of utilizing atrazine-tolerant transgenic lines in the field, and to determine whether herbicide-resistant hybrid rice combinations retain their original superiority in terms of yield. Nevertheless, the development of atrazine-tolerant transgenic *japonica* rice parental lines represents a valuable tool for future application in hybrid seed production.

Supporting Information

Figure S1 HPLC analysis for determination of atrazine and its metabolite hydroxyatrazine in leaves of transgenic and WT rice plants after grown in 5 mg/kg atrazine soil for 45 days. (**A**) Standard of hydroxyatrazine (I) and atrazine (II). (**B**) Wild type (WT). (**C–E**) Transgenic Nipponbare (N-6), Jindao7 (JD7-4) or Jindao8 (JD8-4), respectively.

Author Contributions

Conceived and designed the experiments: DFC XWC. Performed the experiments: LHZ HWC YLL YNL SJW JPS. Analyzed the data: LHZ DFC XWC. Contributed reagents/materials/analysis tools: SJW XJL DFC XWC. Wrote the paper: DFC XWC.

References

1. Peng C, Zhang H (2013) Analysis of paddy and rice market at home and abroad in the first quarter of 2013 and its prospect. Agric Outlook (4): 4–9 (in Chinese, with English abstract).
2. Shi KB, Deng HS (2004) Chinese *japonica* hybrid rice technology innovation seminar held in Sanya. Hybrid Rice 19: 76 (in Chinese).
3. Li S, Yang D, Zhu Y (2007) Characterization and use of male sterility in hybrid rice breeding. J Integ Plant Biol 49: 791–804.
4. Tang SZ, Zhang HG, Zhu ZB, Liu C, Li P, et al. (2010) Application of HL type male sterile cytoplasm in *japonica* hybrid rice breeding. Chin J Rice Sci 24: 116–124 (in Chinese, with English abstract).
5. Yan W (2000) Crop heterosis and herbicide. US Patent 6066779.
6. Hu G, Xiao H, Yu Y, Zhu Z, Si H, et al. (2000) *Agrobacterium*-mediated transformation of the restorer lines of two-line hybrid rice with *bar* gene. Chin J Appl Environ Biol 6: 511–515 (in Chinese, with English abstract).
7. Li Y, Xu Q, Duan F, Liu G, Yan W (2000) Breeding of herbicide-resistant hybrid rice combinations. Hybrid Rice 15(6): 9–11.
8. Xue S, Zhang W, Gong X, Shen L, Huang D, et al. (2008) Breeding of isotype restorer line with *bar* gene from Miyang 46 and its combinations. J Zhejiang Agric Sci (4): 181–183 (in Chinese).
9. Ramesh S, Nagadhara D, Pasalu IC, Kumari AP, Sarma NP, et al. (2004) Development of stem borer resistant transgenic parental lines involved in the production of hybrid rice. J Biotechnol 111: 131–141.
10. Xiong X, Tang L, Deng X, Xiao G (2004) A preliminary report on the experiments of herbicide-resistant two-line hybrid rice Xiang 125S/Bar 68-1. Hybrid Rice 19(5): 41–43.
11. Wu F, Wang S, Li S, Zhang K, Li P (2006) Research progress on herbicide resistant transgenic rice and its safety issues. Mol Plant Breed 4: 846–852.
12. Xiao G, Yuan L, Sun S (2007) Strategy and utilization of a herbicide resistance gene in two-line hybrid rice. Mol Breed 20: 287–292.
13. Xiao G (2009) Recent advances in development of herbicide resistance transgenic hybrid rice in China. Rice Sci 16: 235–239.
14. Udiković-Kolić N, Scott C, Martin-Laurent F (2012) Evolution of atrazine-degrading capabilities in the environment. Appl Microbiol Biotechnol 96: 1175–1189.
15. Rutherford AW, Krieger-Liszkay A (2001) Herbicide-induced oxidative stress in photosystem II. Trends Biochem Sci 26: 648–653.
16. Cai B, Han Y, Liu B, Ren Y, Jiang S (2003) Isolation and characterization of an atrazine-degrading bacterium from industrial wastewater in China. Lett Appl Microbiol 36: 272–276.
17. Wang Y, Li X, Chen X, Chen D (2013) Directed evolution and characterization of atrazine chlorohydrolase variants with enhanced activity. Biochemistry (Moscow) 78: 1104–1111.
18. Höfgen R, Willmitzer L (1988) Storage of competent cells for *Agrobacterium* transformation. Nucleic Acids Res 16: 9877.

19. Loh JP, Kiew R, Kee A, Gan LH, Gan YY (1999) Amplified fragment length polymorphism (AFLP) provides molecular markers for the identification of *Caladium bicolor* cultivars. Ann Bot 84: 155–161.
20. Wang H, Chen X. Xing X, Hao X, Chen D (2010) Transgenic tobacco plants expressing *atzA* exhibit resistance and strong ability to degrade atrazine. Plant Cell Rep 29: 1391–1399.
21. Chen D, Chen H, Zhang L, Shi X, Chen X (2014) Tocopherol-deficient rice plants display increased sensitivity to photooxidative stress. Planta 239: 1351–1362.
22. Ma J, Takahashi E (1990) Effect of silicon on the growth and phosphorus uptake of rice. Plant Soil 126: 115–119.
23. Porra RJ, Thompson WA, Kriedemann PE (1989) Determination of accurate extinction coefficients and simultaneous equations for assaying chlorophylls *a* and *b* extracted with four different solvents: verification of the concentration of chlorophyll standards by atomic absorption spectroscopy. Biochim Biophys Acta - Bioenergetics 975: 384–394.
24. de Souza ML, Sadowsky MJ, Wackett LP (1996). Atrazine chlorohydrolase from *Pseudomonas* sp strain ADP: gene sequence, enzyme purification, and protein characterization. J Bacteriol 178: 4894–4900.
25. Kawahigashi H, Hirose S, Ohkawa H, Ohkawa Y (2007) Herbicide resistance of transgenic rice plants expressing human CYP1A1. Biotechnol Adv 25: 75–84.
26. Erickson LE, Lee KH, Sumner DD (1989) Degradation of atrazine and related *s*-triazines. Crit Rev Environ Control 19: 1–14.
27. Rhine ED, Fuhrmann JJ, Radosevich M (2003) Microbial community responses to atrazine exposure and nutrient availability: linking degradation capacity to community structure. Microbiol Ecol 46: 145–160.
28. Wang L, Samac DA, Shapir N, Wackett LP, Vance CP, et al. (2005) Biodegradation of atrazine in transgenic plants expressing a modified bacterial atrazine chlorohydrolase (*atzA*) gene. Plant Biotech J 3: 475–486.

Melatonin Reverses Fas, E2F-1 and Endoplasmic Reticulum Stress Mediated Apoptosis and Dysregulation of Autophagy Induced by the Herbicide Atrazine in Murine Splenocytes

Shweta Sharma[1]*, Jayanta Sarkar[1,3], Chandana Haldar[2], Sudhir Sinha[1,3]

1 Biochemistry Division, CSIR-Central Drug Research Institute, Lucknow, Uttar Pradesh, India, 2 Department of Zoology, Banaras Hindu University, Varanasi, Uttar Pradesh, India, 3 Academy of Scientific and Innovative Research, New Delhi, India

Abstract

Exposure to the herbicide Atrazine (ATR) can cause immunotoxicity, apart from other adverse consequences for animal and human health. We aimed at elucidating the apoptotic mechanisms involved in immunotoxicity of ATR and their attenuation by Melatonin (MEL). Young Swiss mice were divided into control, ATR and MEL+ATR groups based on daily (x14) intraperitoneal administration of the vehicle (normal saline), ATR (100 mg/kg body weight) and MEL (20 mg/kg body weight) with ATR. Isolated splenocytes were processed for detection of apoptosis by Annexin V-FITC and TUNEL assays, and endoplasmic reticulum (ER) stress by immunostaining. Key proteins involved in apoptosis, ER stress and autophagy were quantified by immunoblotting. ATR treatment resulted in Fas-mediated activation of caspases 8 and 3 and inactivation of PARP1 which were inhibited significantly by co-treatment with MEL. MEL also attenuated the ATR-induced, p53 independent mitochondrial apoptosis through upregulation of E2F-1 and PUMA and suppression of their downstream target Bax. An excessive ER stress triggered by ATR through overexpression of ATF-6α, spliced XBP-1, CREB-2 and GADD153 signals was reversed by MEL. MEL also reversed the ATR-induced impairment of autophagy which was indicated by a decline in BECN-1, along with significant enhancement in LC3B-II and p62 expressions. Induction of mitochondrial apoptosis, ER stress and autophagy dysregulation provide a new insight into the mechanism of ATR immunotoxicity. The cytoprotective role of MEL, on the other hand, was defined by attenuation of ER stress, Fas-mediated and p53 independent mitochondria-mediated apoptosis as well as autophagy signals.

Editor: Ying-Jan Wang, National Cheng Kung University, Taiwan

Funding: The work was supported by the grant funded by Science and Engineering Research Board, Department of Science and Technology, Government of India under Young Scientist Scheme (Grant no: SR/FT/LS-115/2009). The funders had no role in study design, data collection and analysis, decision to publish, or preparation of the manuscript.

Competing Interests: The authors have declared that no competing interests exist.

* Email: shwetadr8@gmail.com

Introduction

Environmental or occupational exposure to the herbicide Atrazine (ATR, 6-chloro-N^2-ethyl-N^4-isopropyl-1,3,5-triazine-2,4-diamine), which is one of the most commonly used agricultural products worldwide, can produce toxic consequences in animals and humans [1–3]. As a potent endocrine disruptor, ATR targets endocrine, reproductive and nervous systems [4–6]. In addition, it also modulates immunity [7] due to the interaction between endocrine and immune systems [8]. ATR-induced suppression of immunity foreshadows the risk for contracting diseases. Indeed, some ATR applicators have been found to develop chronic bronchitis accompanied by a modulation of their humoral and cellular immune responses [9]. In mice, ATR induced immunotoxicity is manifested by a decline in the number of splenic dendritic cells, selected subpopulations of splenic lymphocytes, thymic T cell populations and Natural Killer cell activity along with an increase in early apoptosis and reduction in weights of spleen and thymus [2,10–12]. To date, Fas-mediated apoptosis has been identified as a mechanism for ATR immunotoxicity in splenocytes of mice [13] though additional mechanisms for immunotoxicity may also exist.

Endoplasmic reticulum (ER) plays a key role in synthesis and maturation of proteins, biosynthesis of lipids, regulation of calcium (Ca^{2+}) and maintenance of cell homeostasis. Disturbance in any of these functions activates the 'ER stress pathway' comprising three different branches: activating transcription factor 6 (ATF6), inositol-requiring enzyme 1 (IRE1) and protein kinase RNA-like ER kinase (PERK); which involve activation of transcription factors ATF-6α, X-box binding protein-1 spliced form (XBP-1s), cAMP-response element binding protein-2 (CREB-2) and growth arrest and DNA damage inducible 153 (GADD153). Activation of ER stress pathway is an adaptive response to restore ER homeostasis. However, a prolonged ER stress triggers apoptosis [14] which can contribute to pathophysiological conditions such as neurodegenerative diseases and diabetes [15,16]. Autophagy, a

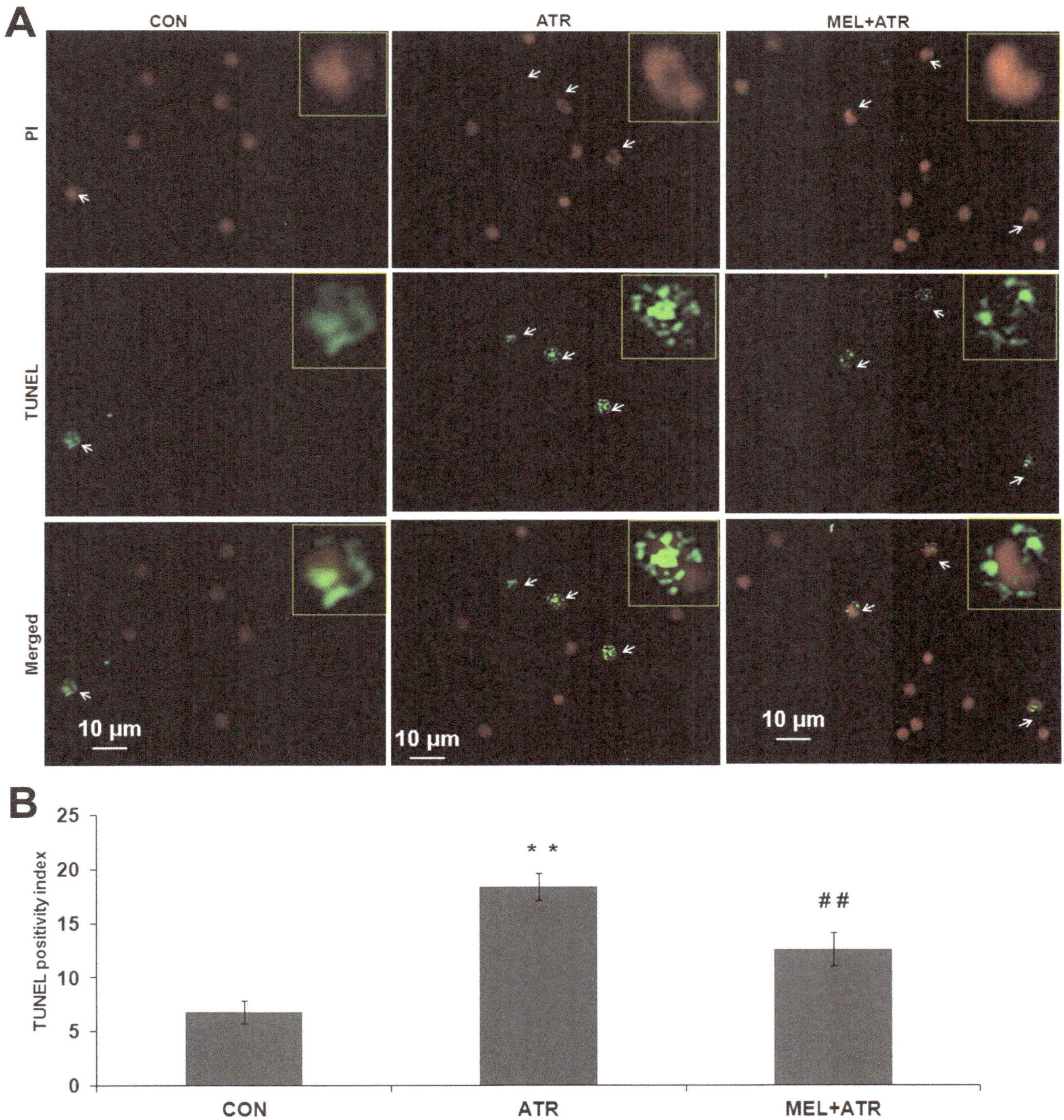

Figure 1. ATR-induced apoptosis in murine splenocytes was ameliorated by MEL. (A) Representative images of TUNEL staining in splenocytes of control (CON), atrazine (ATR) and melatonin with atrazine co-treated (MEL+ATR) mice. All cell nuclei were stained with propidium iodide (red fluorescence). TUNEL (fluorescein-12-dUTP labeled fragmented DNA) positive nuclei, displaying green fluorescence, are arrow-marked. Insets show individual and merged fluorescence of single cells at a higher magnification. (B) Histogram showing TUNEL positivity index. Data are expressed as mean ± SEM of 3 experiments (**$P<0.01$ versus CON; ##$P<0.01$ versus ATR).

dynamic catabolic process that degrades misfolded proteins, organelles and other cytoplasmic contents, serves as a survival mechanism against ER stress [17]. ER stress and autophagy dysregulation have been linked with toxicity induced by certain pesticides and herbicides [18]. However, it is not known whether ATR can also modulate the ER stress and autophagy pathways.

Melatonin (MEL, *N*-acetyl-5-methoxytryptamine) is a ubiquitous indoleamine hormone produced by the pineal gland, retina, gastrointestinal tract, Harderian gland, bone marrow, platelets, lymphocytes, skin, thymus and spleen [19]. Its amphiphilic nature determines its bioavailability as well as action on cells and subcellular organelles [20]. The multifunctional versatility of MEL as a circadian rhythm regulator, antioxidant, free radical scavenger as well as an antiapoptotic (for normal cells), oncostatic and immunomodulatory agent has led to it being considered as a molecule with immense therapeutic potential [21,22]. MEL reduces ER stress and apoptosis caused by the herbicide arsenite by repressing the activation of XBP-1 (IRE-1 branch) and caspase-3 [23]. It is not however clear whether it can regulate all three branches of the ER stress pathway in response to the action of herbicides. Role of MEL in protecting normal cells from apoptosis through modulation of Fas-mediated pathway also remains unresolved [24]. It also remains debatable whether MEL is an inhibitor or stimulator of autophagy [25,26].

This study was aimed at elucidating the mechanism of ATR immunotoxicity in mice and its amelioration by MEL. Since apoptosis, ER stress and autophagy are interrelated, modulation of any of these processes could contribute to immunotoxicity. We therefore monitored changes in expression levels of the key apoptotic, ER stress and autophagic signals in splenocytes in response to administration of ATR, with or without MEL. We found that MEL ameliorated ATR-induced cell death in mouse splenocytes through suppression of Fas, E2F-1 and ER stress mediated apoptosis and also through restoration of impaired autophagy. These findings extend our knowledge of health risks involved in exposure to herbicides such as ATR as well as the therapeutic potential of MEL, particularly against immunotoxicity.

Materials and Methods

Reagents

MEL, ATR, RPMI-1640, fetal bovine serum (FBS), dimethylsulfoxide (DMSO), ammonium chloride (NH_4Cl), bovine serum albumin (BSA), poly-L-lysine, Tween-20, Tris buffer, 30% acrylamide/bis-acrylamide solution, Annexin V-FITC kit, β-actin and HRP-conjugated secondary antibodies were obtained from Sigma-Aldrich (St. Louis, MO, USA). Alexa Fluor 594 antibody and Prolong Gold antifade reagent with DAPI were procured from Invitrogen (Carlsbad, CA, USA). Polyvinylidene fluoride (PVDF) membrane and Immobilon Western chemiluminiscent substrate were purchased from Millipore (Bedford, MA, USA). Mammalian protein extraction reagent (M-PER) and Bichinchonoic acid (BCA) protein assay kit were procured from Pierce Biotechnology (Rockford, IL, USA). DeadEnd Fluorometric TUNEL kit was purchased from Promega (Madison, WI, USA).

Ethics statement

All experimental procedures on mice were approved by the Institutional Animal Ethics Committee (IAEC) of CSIR-Central Drug Research Institute (Permit No. 34/1999/CPCSEA). All ethical guidelines and animal welfare regulations set out by IAEC and Committee for the Purpose of Control and Supervision of Experiments on Animals (CPCSEA), Government of India were strictly followed. Mice were sacrificed by cervical dislocation.

Animals

Swiss mouse is widely used as a model for toxicological or pharmacological evaluation of drugs. *In vivo* protective effects of MEL have also been demonstrated in these mice [27,28]. Young male Swiss albino mice (aged 4 weeks, weighing 20±2 g) used in this study were bred and reared at National Laboratory Animal Centre, CSIR-CDRI. They were housed in a controlled environment (22–26°C; 50–60% humidity; 12 h light/dark cycle) and fed standard rodent food pellet and water *ad libitum*. After 10 days of acclimatization, animals were subjected to the planned treatments and monitored daily for behavioural change, food and water intake, body weight and survival.

Treatments

Taking a cue from earlier reports [10,13,29], the experimental animals were treated intraperitoneally (i.p.) with ATR (100 mg/kg body weight/day), with or without MEL (20 mg/kg b.wt./day) for 14 days. Comparable ATR treatment schedules have been reported to show a significant modulation of immunological parameters, including apoptosis, in splenocytes of young mice [10,13]. Similarly, co-treatment with an identical dose of MEL has been shown to produce significant reduction in the oxidative damage induced by a diquat herbicide in murine splenocytes [29]. ATR stock was made in DMSO and diluted with 0.9% saline to obtain the required dosage whereas MEL was dissolved in minimal amount of ethanol before diluting with saline. Stock solutions were stored at 4°C for one week and working dilutions were prepared fresh daily before use. Animals were randomly divided into 3 groups of 6 mice each. Control (CON) group received the vehicle (0.9% saline, i.p.), ATR group received ATR alone and MEL+ATR group was administered MEL 30 min before ATR. The levels of endogenous MEL and total antioxidants are lowest during the daytime [19] favouring enhanced oxidative damage by toxins [30]. Therefore, treatments were given during daytime, at around 11:00 am. All end points were studied at 24 h after the last treatment. After recording body weights, animals were sacrificed by cervical dislocation and organs were excised aseptically. Weights of spleen and thymus were noted immediately.

Relative organ weight

Relative organ weight was determined as the ratio of weight of the organ (mg) to body weight (g) of a mouse.

Preparation of splenocytes

Spleen was placed on a cell strainer (70 μm pore size) in a petridish containing RPMI-1640 medium (complete with 10% FBS) and pressed gently with the plunger of glass syringe to disperse cells into the medium. Erythrocytes were lysed by hypotonic shock with equal volume of cold 0.84% Tris-NH_4Cl (pH 7.2). The splenocytes were washed, resuspended in complete RPMI medium and counted in a Neubauer's counting chamber.

Annexin V- fluorescein isothiocyanate (FITC)/propidium iodide (PI) double staining assay

Apoptotic cells were detected using Annexin V-FITC kit. Splenocytes (10^6 cells/mL) were resuspended in binding buffer and incubated in dark for 10 min at room temperature (RT) with Annexin V-FITC conjugate and PI. Population of unstained splenocytes were gated to locate quadrant using FACS Calibur flow cytometer (Becton-Dickinson, CA, USA). Later, fluorochrome labelled cells were analysed for FITC+/PI- apoptotic events. The percentage of apoptotic cells (of the total counted cells) was expressed as apoptotic index.

Terminal deoxynucleotidyl transferase mediated dUTP nick end labeling (TUNEL) assay

DNA fragmentation in apoptotic cells was detected using fluorometric TUNEL kit. Briefly, cells were fixed with 4% paraformaldehyde, washed, spotted onto poly-L-lysine coated

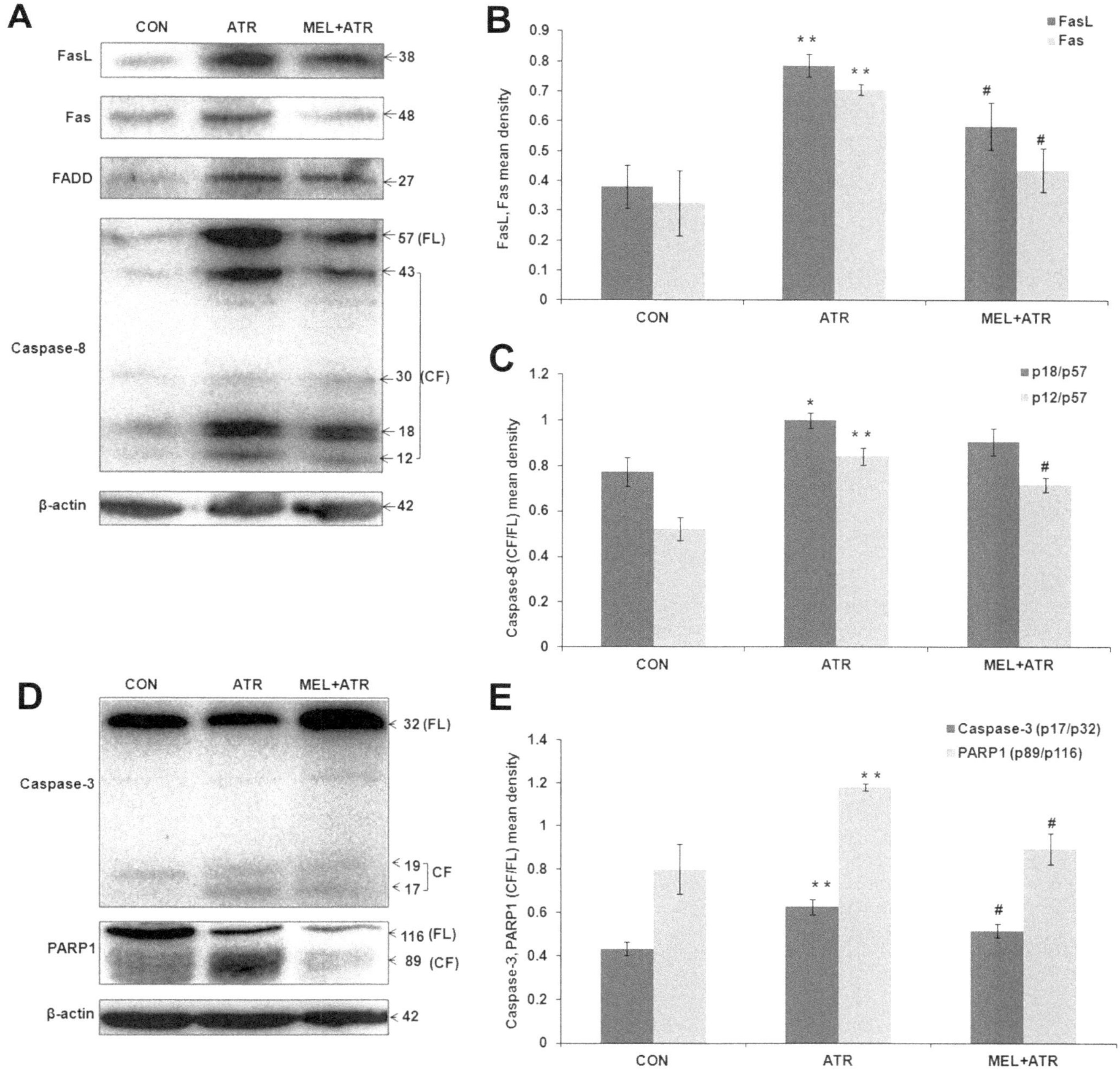

Figure 2. ATR-induced Fas mediated apoptosis was inhibited by MEL. (A) Representative immunoblots of splenocyte lysates from CON, ATR and MEL+ATR mice. FasL, Fas, FADD, Caspase-8 and β-actin proteins were visualized by chemiluminiscence. Corresponding histograms show (B) FasL, Fas and (C) Caspase-8 (CF/FL; p18/p57 and p12/p57) expressions as mean densities normalized by β-actin density. (D) Representative immunoblots of Caspase-3, PARP1 and β-actin. Corresponding histogram (E) shows caspase-3 (CF/FL, p17/p32) and PARP1 (CF/FL, p89/p116) mean densities normalized by β-actin. Data are presented as mean ± SEM of 3 independent experiments (*$P<0.05$, **$P<0.01$ versus CON; $^{\#}P<0.05$ versus ATR). FL, full length; CF, cleaved fragments.

slides and air-dried. Cells were permeabilized with 0.2% Triton X-100, washed and covered with equilibration buffer. Later, rTdT incubation buffer (nucleotide mix and rTdT enzyme in equilibration buffer) was added on to cells and slides were kept in a humidified chamber in dark (37°C, 1 h). Reaction was stopped by adding 2x SSC reagent. After washing, cells were counterstained with freshly diluted PI (1 μg/mL), which stains apoptotic as well as non-apoptotic cell nuclei. Apoptotic cells displaying green fluorescence (due to catalytic incorporation of fluorescein-12-dUTP at 3′OH DNA ends) were determined as TUNEL positive. Over 200 nuclei were counted in a randomly selected field, using Nikon Eclipse Ti inverted microscope (Nikon Instruments, NY, USA) at 200× magnification. The percentage of TUNEL positive nuclei (of the total number of nuclei) was expressed as TUNEL positivity index.

Immunofluorescence

Cells were resuspended in PBS and fixed with equal volume of 4% paraformaldehyde (RT, 15 min). The washed cells were spotted onto poly-L-lysine coated cover glass and air-dried. Cells were permeabilized (4°C, 15 min) with ice-cold 0.3% Triton X-100 in PBS, washed with PBS containing 0.1% Tween-20 (PBST) and incubated (RT, 1 h) in blocking solution (2% BSA in PBST). Subsequently, cells were incubated (37°C, 1 h, in humidified atmosphere) with primary antibodies against ATF-6α, XBP-1 and GADD153 proteins (Table S1). Cells were later washed with PBST, incubated (37°C, 1 h) with Alexa-Fluor 594 secondary antibody (1:300) and washed again with PBST. Cover slips were mounted on antifade reagent (~10 µL). After overnight incubation in dark, cells were observed for expression of signals under Nikon Eclipse Ti microscope (200× magnification).

Preparation of cell lysates

Cell suspension was washed, resuspended in PBS and centrifuged (300 $g\times10$ min, 4°C). Cell sediments were lysed (30 min, on ice) with MPER reagent (containing protease inhibitor cocktail) to extract total protein. Lysates were centrifuged (14000 $g\times10$ min, 4°C) and supernatants collected. After estimating protein by BCA kit, samples were stored in aliquots at −80°C.

Immunoblotting

Samples (20 µg) were subjected to SDS-PAGE on 8% and 12% resolving gels and proteins were transferred electrophoretically onto PVDF membrane. The membrane was blocked with 5% non-fat dry milk in TBST {Tris-buffered saline (10 mM Tris, 100 mM NaCl), containing 0.1% Tween-20} and incubated overnight (4°C) with primary antibodies (FasL, Fas, caspases 8, 3, PARP1, p53, E2F1, PUMA, ATF6, XBP1, CREB2, GADD153, BECN1, LC3B, p62) using indicated dilutions (Table S1). After washing with TBST three times for 5 min each, the membranes were incubated with HRP-conjugated secondary antibody (1:8000, 1 h, RT). After another wash cycle, the membranes were developed using ECL reagent and visualized in ChemiDoc (Bio-Rad Laboratories, USA). Densities of specific bands were quantified using Image J analysis software (version 1.46, NIH, USA). Density of each band was normalized by that of the loading control (β-actin).

Statistical analysis

All data are expressed as mean ± standard error of mean (SEM) of at least 3 independent experiments. Statistical significance was determined using one way ANOVA followed by Student's Newman-Keuls *post hoc* test. $P<0.05$ was considered as statistically significant.

Results

Relative weight loss of lymphoid organs due to ATR treatment was reversed by MEL

We initially measured relative weights of spleen and thymus as a gross sign of immunotoxicity in response to ATR treatment. The 14-day treatment schedule resulted in a significant reduction of relative weights of both organs, with spleen showing greater susceptibility to treatment ($P<0.01$ versus CON; Figure S1). This observation is corroborated by a previous report [10]. Co-treatment with MEL resulted in significant recovery of relative weights of spleen and thymus ($P<0.05$ versus ATR; Figure S1), indicating ameliorating effect of MEL on ATR immunotoxicity.

Apoptosis in splenocytes induced by ATR was inhibited by MEL

The binding of Annexin V-FITC to phosphatidylserine on outer leaflet of cell membrane, which is an early morphological event of apoptotic cells, was detected by flow cytometry. FITC+/PI- single green fluorescence was located in lower right quadrants (Figure S2, a). ATR treatment significantly enhanced ($P<0.01$ versus CON) the apoptotic index, whereas MEL co-treatment led to its significant reduction ($P<0.01$ versus ATR; Figure S2, b). DNA fragmentation was confirmed by TUNEL assay wherein apoptotic cells display green nuclei (marked with arrows; Fig. 1A). ATR treatment produced a significant increase ($P<0.01$ versus CON) in TUNEL positivity which was reduced significantly by MEL ($P<0.01$ versus ATR; Fig. 1B). Altogether these results show that ATR induced apoptosis in splenocytes, which was reversed by MEL.

Fas mediated apoptosis in splenocytes induced by ATR was inhibited by MEL

Expressions of regulatory proteins of the death receptor family were quantified by immunoblot assay (Fig. 2A). Significant enhancement in expressions of Fas ligand (FasL) and Fas ($P<0.01$ versus CON; Fig. 2B) were accompanied by an increase in FADD (Fas-associated protein with death domain) (Fig. 2A). This led to autocatalytic cleavage of inactive full length caspase-8 (57 kDa), a downstream target of Fas pathway, into its 18 and 12 kDa active fragments (p18/p57; $P<0.05$ and p12/p57; $P<0.01$, respectively, versus CON; Fig. 2C). MEL co-treatment significantly inhibited the FasL and Fas expressions ($P<0.05$ versus ATR; Fig. 2B), accompanied by suppression of caspase-8 activation, as demonstrated by a decline in the ratio of cleaved to full length protein (p12/p57; $P<0.05$ versus ATR; Fig. 2C).

The Fas-mediated, Caspase-8 dependent pathway led to activation of effector caspase-3 followed by cleavage of its downstream target PARP1 (Fig. 2D) which execute DNA fragmentation. ATR treatment resulted in activation of caspase-3, releasing its 17 kDa active fragment (p17/p32; $P<0.01$ versus CON; Fig. 2E) which further cleaved full length PARP1 (116 kDa) to its inactive 89 kDa fragment (p89/p116; $P<0.01$ versus CON; Fig. 2E), thereby attenuating the DNA repair function of PARP1 [31]. MEL co-treatment resulted in inhibition of ATR-induced DNA fragmentation, as evident from suppression of the cleavage of caspase-3 (p17/p32; $P<0.05$ versus ATR) and PARP1 (p89/p116; $P<0.05$ versus ATR; Fig. 2E).

Mitochondrial apoptosis induced by ATR was inhibited by MEL

To learn whether ATR treatment also induces mitochondrial apoptosis in splenocytes, we investigated the expressions of p53, E2F-1, PUMA, Bax and Bcl-2 proteins which regulate mitochondria-mediated apoptosis [32,33]. ATR treatment significantly enhanced the expressions of E2F-1 ($P<0.05$ versus CON) and PUMA ($P<0.01$ versus CON), though p53 level remained unaltered when compared with control (Fig. 3A and 3B). Further, E2F-1 and PUMA initiate mitochondrial apoptosis by regulating their downstream targets Bax and Bcl-2. We observed a significant increase in Bax/Bcl-2 ratio ($P<0.05$ versus CON; Fig. 3C) upon ATR treatment. Co-treatment with MEL did also not alter the p53 level (Fig. 3B). However, it suppressed E2F-1 and PUMA expressions ($P<0.01$ versus ATR; Fig. 3B) as well as Bax/Bcl-2 ratio ($P<0.05$ versus ATR; Fig. 3C) thereby inhibiting the mitochondria-mediated apoptosis induced by ATR.

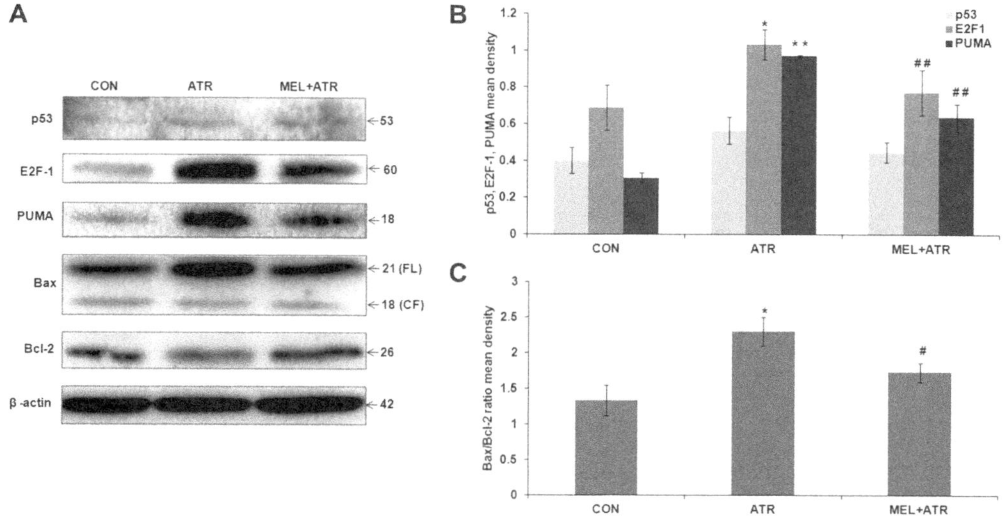

Figure 3. ATR-induced mitochondria mediated apoptosis was inhibited by MEL. (A) Representative immunoblots of p53, E2F-1, PUMA, Bax, Bcl-2 and β-actin in CON, ATR and MEL+ATR groups. Histograms show (B) p53, E2F-1, PUMA and (C) Bax/Bcl-2 ratio mean densities normalized by β-actin density. Data are expressed as mean ± SEM of 3 independent experiments (*$P<0.05$, **$P<0.01$ versus CON; $^{\#}P<0.05$, $^{\#\#}P<0.01$ versus ATR).

ER stress induced by ATR was inhibited by MEL

Along with expression of full length Bax (21 kDa), we also noted the occurrence of its cleaved 18 kDa fragment, even though p18 expression levels remained constant (Fig. 3A). Since p18 is known to be released in response to Calpain1 activation, we went on to check for Calpain1 expression as a marker for ER stress [34]. ATR treatment resulted in cleavage of inactive Calpain1 (80 kDa) to its catalytically active 76 kDa fragment (p76/p80; $P<0.05$ versus CON). This cleavage was significantly inhibited by MEL co-treatment ($P<0.01$ versus ATR; Fig. 4A).

Since Calpain1 activation also suggested the involvement of ER stress in induction of apoptosis, we performed immunostaining of marker proteins for each of the three branches of ER stress pathway: ATF-6α (for ATF6 branch), XBP-1s (for IRE1 branch) and GADD153 (for PERK branch). Signal positive cells were observed in ATR as well as MEL co-treated groups (Figure S3). Further, we quantified the expression levels of these ER stress markers by immunoblot assays (Fig. 4B). ATR treatment caused excessive ER stress, as evident from the activation of ATF-6α ($P<0.01$ versus CON; Fig. 4C) and overexpression of XBP-1s (p56/p32; $P<0.05$ versus CON; Fig. 4D), CREB-2 and GADD153 proteins ($P<0.01$ versus CON; Fig. 4E). MEL co-treatment, on the other hand, inhibited the formation of ATF-6α ($P<0.05$ versus ATR; Fig. 4C) and reduced the XBP-1s/XBP-1u ratio ($P<0.05$ versus ATR; Fig. 4D). It also downregulated CREB-2 expression, followed by inhibition of the expression of its downstream target GADD153 ($P<0.01$ and $P<0.05$, respectively, versus ATR; Fig. 4E). These results reflected that MEL is capable of suppressing all three branches of ER stress response generated by ATR in splenocytes.

ATR-mediated impairment of autophagy was ameliorated by MEL

Autophagy plays an important role in the recovery from ER stress [17] hence we investigated whether ATR and MEL could also modulate the key autophagic signals BECN-1, LC3B-II and p62/SQSTM1. BECN-1 initiates the formation of autophagosomes whereupon LC3B-II gets recruited. LC3B-II, which remains localized to autophagic structures throughout the process of autophagy, binds to p62 laden with the cargo of poly-ubiquitinated proteins and targets the cargo (together with p62) for degradation at the autolysosome. Thus, elevated levels of BECN-1 and LC3B-II and a reduced level of p62 proteins signify an effective autophagy process [35]. However, presently we saw downmodulation of BECN-1 (versus CON, though statistically not significant; Fig. 5A and 5B) and enhancement of LC3B-II ($P<0.01$ versus CON; Fig. 5A and 5C) as well as p62 ($P<0.01$ versus CON; Fig. 5A and 5D) expressions in splenocytes following ATR treatment suggesting an impairment of autophagy. Co-treatment with MEL ameliorated this impairment by way of a significantly elevated expression of BECN-1 ($P<0.05$ versus ATR; Fig. 5B) along with decline in LC3B-II and p62 levels ($P<0.05$ and $P<0.01$, respectively, versus ATR; Fig. 5C and 5D).

Discussion

In this study, we aimed at elucidating the apoptotic mechanisms contributing to immunotoxicity of the herbicide Atrazine and exploring whether and how Melatonin could attenuate them. The short-course treatment of mice with ATR resulted in apoptosis of splenocytes as evident from the results of Annexin V binding and TUNEL assays. An enhanced expression of FasL and Fas in these cells suggested activation of the Fas mediated pathway, as also reported previously [13]. The FasL-Fas interaction further led to DISC formation with recruitment of FADD and procaspase-8, culminating in the release of active fragments of caspase-8 (p43, p30, p18 and p12). Caspase-8 activation at DISC is a complex process occurring via autocatalysis. During the first cleavage, p43 and p30 products are released which further generate their active fragments p12 and p18, respectively [36]. ATR treatment significantly enhanced expression of both the fragments. The active caspase-8 cleaves caspase-3 which executes DNA fragmentation [36]. In addition, the ATR-induced inactivation of PARP1, which plays a role in DNA repair [31], could also have contributed to apoptosis.

Induction of the mitochondria-mediated pathway of apoptosis by ATR was indicated by an increase in the ratio of proapoptotic Bax to antiapoptotic Bcl-2 proteins. Increase in Bax/Bcl-2 ratio

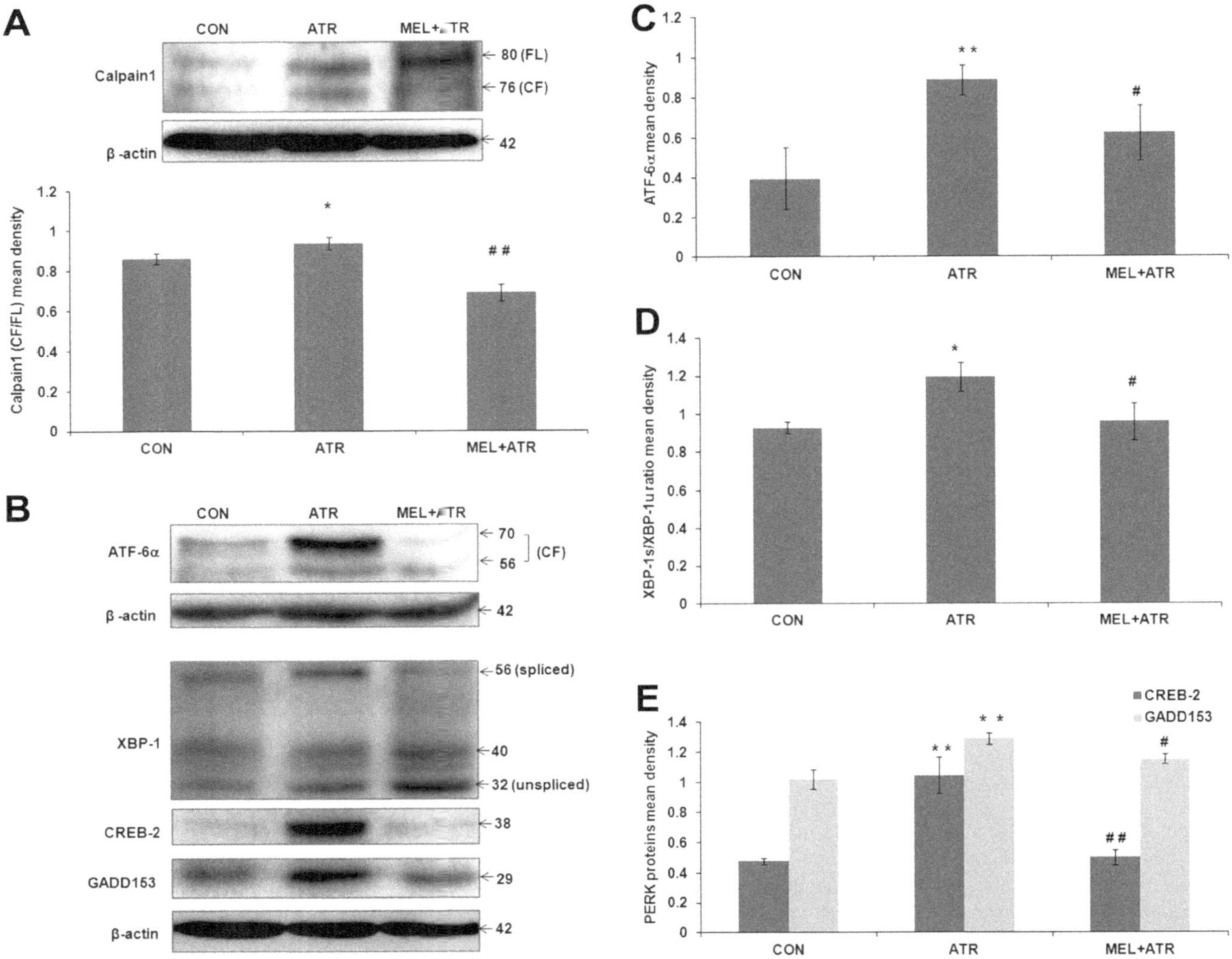

Figure 4. ATR-induced ER stress response in splenocytes was ameliorated by MEL. (A) Representative immunoblot showing Calpain1 cleavage and histogram showing ratio of active to inactive Calpain1 (CF/FL, p76/p80) mean density normalized by β-actin. (B) Representative immunoblots of ATF6α, XBP-1, CREB-2, GADD153 and β-actin. Histograms show mean densities of (C) ATF-6α (CF, 70), (D) XBP-1s/XBP-1u ratio (56/32), and (E) PERK proteins (CREB-2 and GADD153) normalized by β-actin density. Data are expressed as mean ± SEM of 3 independent experiments (*$P<0.05$, **$P<0.01$ versus CON; $^{\#}P<0.05$, $^{\#\#}P<0.01$ versus ATR).

suggests translocation of Bax to mitochondria, permeabilization of mitochondrial membrane and activation of caspases. Mitochondria-mediated apoptosis proceeds primarily through activation of the transcription factor p53 which further activates its downstream targets Bax, PUMA (p53 upregulated modulator of apoptosis) and represses Bcl-2 [33]. The p53 dependent mitochondrial apoptosis has previously been reported in case of herbicides [37]. However, presently a uniformly low expression of p53 in control as well as experimental groups suggested induction of apoptosis through the p53 independent pathway involving E2F-1. E2F-1 is a transcription factor of E2F family having binding sites within the promoters of p53, Puma, Apaf-1, p73 and GRAMD4. Hence it is capable of causing mitochondrial apoptosis through p53 dependent as well as independent mechanisms [32,38]. Both E2F-1 and PUMA proteins were overexpressed after treatment with ATR. This, to our knowledge, is the first documentation of p53 independent mitochondrial apoptosis as a mechanism for toxicity of herbicides.

The 18 kDa cleaved fragment (p18) of Bax was seen invariably in control as well as experimental splenocytes. Though it appeared as a faint band due perhaps to a short half-life (2 h) [39], its occurrence prompted us to look for Calpain1 activity which is known to cleave Bax (21 kDa) into its p18 fragment [34]. Importantly, overexpression of Calpain1 in response to ATR treatment also suggested the involvement of ER stress pathway in induction of apoptosis [37]. ER stores Ca^{2+} and maintains Ca^{2+} homeostasis for optimal protein folding. Under conditions of stress, Ca^{2+} is released from the lumen of ER into cytosol and triggers activation of Calpains [40] leading to activation of one or more of the three branches of ER stress. Activation of ATF6 branch causes production of the cleaved product ATF-6α (56 kDa, 70 kDa) which induces transcription of XBP-1u mRNA. Under the IRE1 branch, XBP-1u mRNA is processed by IRE1 to generate its spliced form XBP-1s (56 kDa) which upregulates ER chaperones and ERAD (endoplasmic reticulum stress associated protein degradation) components [41]. In case of mammals, XBP-1u

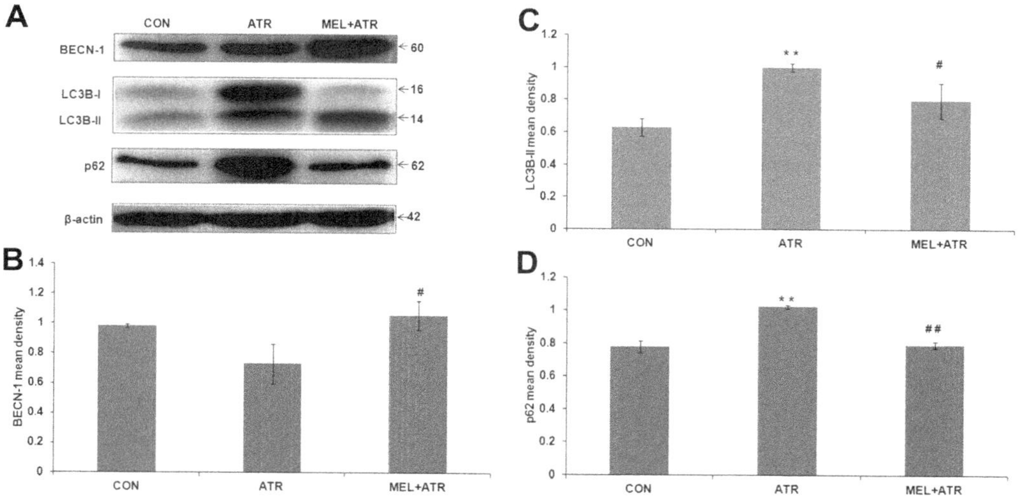

Figure 5. ATR-induced dysregulation of autophagy in splenocytes was ameliorated by MEL. (A) Representative immunoblots of autophagy markers BECN-1, LC3B (I and II), p62 and loading control β-actin in CON, ATR and MEL+ATR groups. Histograms show (B) BECN-1, (C) LC3B-II and (D) p62 mean densities normalized by β-actin. Data are presented as mean ± SEM of 3 experiments (**$P<0.01$ versus CON; #$P<0.05$, ##$P<0.01$ versus ATR).

mRNA also gets translated to the corresponding protein (32 kDa) in cytosol [42]. The PERK branch of ER stress involves activations of CREB-2 followed by GADD153 which attenuate protein translation [43]. As evident from the expression profile of target proteins (cleaved ATF-6α, XBP-1s, CREB-2 and GADD153), ATR treatment activated all three branches of ER stress. Though ER stress normally helps in cell survival by removing misfolded proteins, an elevated and prolonged ER stress level can cause apoptosis [14]. Presently, ER stress-mediated apoptosis induced by ATR was evident from overexpression of the proapoptotic signals GADD153 and PUMA. GADD153 is known to bind to Puma promoter and transactivate PUMA expression [44]. In a related study, the herbicide arsenite has also been shown to induce ER stress and apoptosis [23].

Autophagy serves to ameliorate the ER stress in normal cells [17]. Onset of autophagy is signalled by BECN-1, which forms a 'core complex' (BECN1-Vps34-Vps15) that gets localized on the 'pre-autophagosomal' structures [45]. LC3B-II gets integrated into autophagosomes and selects the cargo of p62-associated protein aggregates for degradation by autolysosomes. Thus enhanced LC3B-II, together with reduced p62 levels are hallmarks of efficient autophagy [35]. However, expression levels of both markers were enhanced in splenocytes after ATR treatment, suggesting an impairment of autophagy. A possible reason for this impairment could be the low expression level of BECN-1which resulted in a low turnover of autophagosomes leading to accumulation of LC3B-II and p62. In addition, ATR-induced caspases 3 and 8 (discussed above) could also have degraded BECN-1 and caused its paucity in splenocytes [46]. Alternatively, a 'block' in autophagic flux at the level of autolysosome generation or activation could also cause accumulation of LC3B-II and p62 [47,48]. Though such a possibility can be explored with the help of specific inhibitors [35], there are certain caveats to this approach, particularly when applied to *in vivo* studies. As the rate of basal autophagic flux for most tissues is unknown, short treatments with the inhibitors may not be effective and long treatments could produce toxicity [49,50]. Moreover, protracted treatments with inhibitors can also lead to 'off-target' effects [35]. For example, 3-methyladenine (3-MA) can, in long term experiments, promote autophagy as well as reduce cell survival.

Melatonin is known to regulate oxidative stress, apoptosis and mitochondrial homeostasis through its free radical scavenging action and interaction with receptors and intracellular targets involved in signal transduction [51]. The protective action of MEL against the death receptor as well as mitochondria-mediated apoptosis induced by ATR in splenocytes was evident from inhibition of FasL, Fas, FADD, caspase-8 and suppression of Bax/Bcl-2 ratio. Suppression of caspase-8 activity through Fas pathway provides a new insight into the cytoprotective action of MEL. In an earlier study, MEL was found to abrogate caspase-8 activity in rabbit liver by modulating the TNF-mediated (rather than Fas-mediated) pathway [24]. MEL is also known to manifest its antiapoptotic effect through suppression of p53 dependent mitochondrial apoptosis [52]. However, the p53 independent (E2F-1 and PUMA dependent) pathway [32,53] was apparently involved in the present case. In addition, cytoprotective action of MEL also involved suppression of caspase-3 cleavage and activation of PARP1 as reported previously [24].

MEL appeared to protect against ER stress through inhibition of Calpain1 activation. Samantaray *et al.* [54] have also demonstrated that MEL attenuates activation of both Calpain and caspase-3 in spinal cord injury of rats. In addition, MEL co-treatment attenuated the activation of ER stress markers ATF-6α, XBP-1s, CREB-2 and GADD153 suggesting suppression of all three branches of ER stress which were activated by ATR. In a relevant report, Zhao *et al.* [55] have also demonstrated down-modulation of the three branches of ER stress by MEL in the mouse model of bleomycin-induced pulmonary fibrosis. During recovery from ER stress, the expression of XBP-1u protein is enhanced which acts as a feedback inhibitor for XBP-1s protein [56]. Thus, the increased expression of XBP-1u in response to MEL co-treatment suggested ER stress recovery in splenocytes.

Enhancement of autophagy signaling in splenocytes after MEL co-treatment was evident from a significant increase in BECN-1 and decrease in LC3B-II and p62 expression levels (compared with levels after ATR treatment). In particular, the decrease in p62 expression, which is a key indicator of normal autophagy process

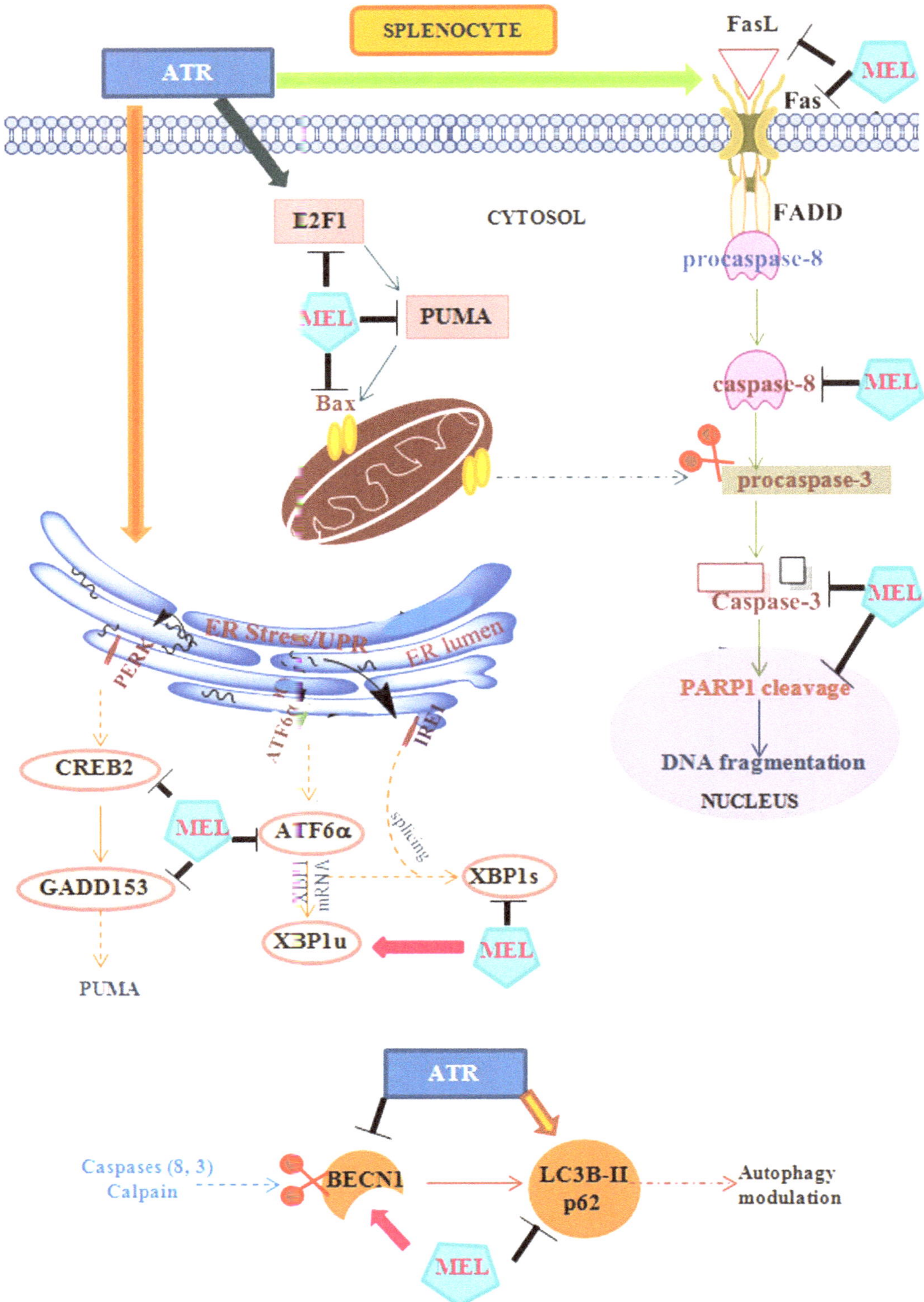

Figure 6. Schematic diagram showing protective action of MEL against ATR immunotoxicity. ATR treatment activates death receptor (FasL, Fas, FADD, Caspase-8) and mitochondrial (E2F-1, PUMA, Bax) apoptosis (Caspase-3 and cleaved PARP1) signals. In addition, ATR induces ER stress (ATF-6α, XBP-1s, CREB-2, GADD153) signals. MEL inhibits the Fas and mitochondrial apoptosis as well as ER stress. ATR treatment also impairs autophagy by suppressing BECN-1 and upregulating LC3B-II and p62 proteins; whereas MEL restores autophagy by reversing this dysregulation. Dotted line arrows indicate known connecting pathways that were not a part of the present study. Line arrows indicate stimulatory effect and sign T indicates inhibitory effect on the expression of corresponding proteins. Scissor symbol indicates the cleavage of target proteins.

[26], signalled recovery from the ATR-induced dysregulation of autophagy. The downmodulation of caspases by MEL could also have contributed to this recovery by preventing the caspase-mediated degradation of BECN-1 [47]. Though we have not looked for it, MEL could possibly have favourably modulated certain other components of the autophagy flux. For instance, in case of 'cytoprotective autophagy' the cleaved 43 kDa fragment of caspase-8 co-localizes with LC3B-II and gets degraded by autolysosomes [57]. This could have been the reason behind the observed low expression of p12 (a product of p43) in the MEL co-treated group.

We have summarized our results in the form of a diagrammatic pathway (Fig. 6). ATR triggers the death receptor (Fas) mediated as well as mitochondrial (E2F-1, PUMA, Bax) apoptosis in mouse splenocytes. It also induces an excessive ER stress response, as evident from activation of ATF-6α (ATF6 branch), XBP-1s (IRE1 branch), CREB-2 and GADD153 (PERK branch). Further, it dysregulates the expression of key autophagy signals BECN-1, LC3B-II and p62. Exogenous MEL, on the other hand, attenuates both Fas-mediated and E2F-1 dependent mitochondria- mediated apoptosis, suppress all three branches of ER stress and enhances autophagy to protect the splenocytes.

In conclusion, our study highlights multiple targets of cell death which could play a role in immunotoxicity of ATR. MEL imparted protection against this apoptosis through, hitherto unreported, 'Fas-mediated' and 'p53 independent E2F-1 dependent' pathways. The latter finding opens a possibility that some other p53 independent pathways (such as p73 and Noxa) could also be involved in anti- apoptotic activity of MEL. MEL also protected splenocytes from ATR-induced ER stress and promoted autophagy. A concerted regulation of ER stress, apoptosis and autophagy targets by exogenous MEL in this manner adds a new dimension to its protective role against immunotoxicity.

Supporting Information

Figure S1 ATR-induced toxicity in lymphoid organs (spleen and thymus) of mice and its reversal by MEL. Histogram shows effect of ATR and MEL treatments on relative spleen and thymus weights (CON, control; ATR, atrazine; MEL+ATR, melatonin and atrazine co-treated group). Data are expressed as mean ± SEM (n = 6) ($*P<0.05$, $**P<0.01$ versus CON; $^{\#}P<0.05$ versus ATR).

Figure S2 MEL inhibited ATR induced early apoptosis. (a) Representative dot plots of Annexin V-FITC and PI stained apoptotic cells were analyzed by flow cytometry in CON, ATR and MEL+ATR groups. Lower right quadrant displays apoptotic cells with FITC+/PI- stains. (b) Histogram shows apoptotic index. Data are expressed as mean ± SEM of 3 experiments ($**P<0.01$ versus CON; $^{\#\#}P<0.01$ versus ATR).

Figure S3 Effect of ATR and MEL on ER stress response in splenocytes. Representative photomicrographs show immunoreactivity of (a) ATF-6α of ATF6 branch, (b) XBP-1 of IRE1 branch and (c) GADD153 of PERK branch. Nuclei stained with DAPI fluoresced blue (Lane 1), cells expressing protein signals fluoresced red (white arrow, Lane 2). Lane 3 shows merged photomicrographs (Scale bar = 10 μM).

Table S1 Sources and dilutions of antibodies used for immunoblot (IB) and immunofluorescence (IF) assays.

Acknowledgments

We are grateful to the Director, CSIR-CDRI for providing laboratory and animal facilities. We are thankful to Neetu Singh for initial assistance in immunoblotting, and to A. L. Vishwakarma for assistance in interpretation of the flow cytometry data. This paper has CSIR-CDRI communication number 8788.

Author Contributions

Conceived and designed the experiments: S. Sharma CH. Performed the experiments: S. Sharma. Analyzed the data: S. Sharma. Contributed reagents/materials/analysis tools: S. Sharma JS S. Sinha. Contributed to the writing of the manuscript: S. Sharma. Reviewed and edited the manuscript: S. Sinha JS.

References

1. Hayes TB, Anderson LL, Beasley VR, de Solla SR, Iguchi T, et al. (2011) Demasculinization and feminization of male gonads by atrazine: consistent effects across vertebrate classes. J Steroid Biochem Mol Biol 127: 64–73.
2. Rowe AM, Brundage KM, Barnett JB (2008) Developmental immunotoxicity of atrazine in rodents. Basic Clin Pharmacol Toxicol 102: 139–145.
3. Rinsky JL, Hopenhayn C, Golla V, Browning S, Bush HM (2012) Atrazine exposure in public drinking water and preterm birth. Public Health Rep 127: 72–80.
4. Kucka M, Pogrmic-Majkic K, Fa S, Stojilkovic SS, Kovacevic R (2012) Atrazine acts as an endocrine disrupter by inhibiting cAMP-specific phosphodiesterase-4. Toxicol Appl Pharmacol 265: 19–26.
5. Cragin LA, Kesner JS, Bachand AM, Barr DB, Meadows JW, et al. (2011) Menstrual cycle characteristics and reproductive hormone levels in women exposed to atrazine in drinking water. Environ Res 111: 1293–1301.
6. Coban A, Filipov NM (2007) Dopaminergic toxicity associated with oral exposure to the herbicide atrazine in juvenile male C57BL/6 mice. J Neurochem 100: 1177–1187.
7. Rajkovic V, Matavulj M, Johansson O (2010) Combined exposure of peripubertal male rats to the endocrine-disrupting compound atrazine and power-frequency electromagnetic fields causes degranulation of cutaneous mast cells: a new toxic environmental hazard? Arch Environ Contam Toxicol 59: 334–341.
8. Clayton EM, Todd M, Dowd JB, Aiello AE (2011) The impact of bisphenol A and triclosan on immune parameters in the U.S. population, NHANES 2003–2006. Environ Health Perspect 119: 390–396.
9. Kluciński P, Kossmann S, Tustanowski J, Friedek D, Kamińska-Kołodziej B (2001) Humoral and cellular immunity rates in chemical plant workers producing dust pesticides. Med Sci Monit 7: 1270–1274.
10. Filipov NM, Pinchuk LM, Boyd BL, Crittenden PL (2005) Immunotoxic effects of short-term atrazine exposure in young male C57BL/6 mice. Toxicol Sci 86: 324–332.
11. Karrow NA, McCay JA, Brown RD, Musgrove DL, Guo TL, et al. (2005) Oral exposure to atrazine modulates cell-mediated immune function and decreases host resistance to the B16F10 tumor model in female B6C3F1 mice. Toxicology 209: 15–28.
12. Chen JY, Song Y, Zhang LS (2013) Immunotoxicity of atrazine in Balb/c mice. J Environ Sci Health B 48: 637–645.
13. Zhang X, Wang M, Gao S, Ren R, Zheng J, et al. (2011) Atrazine-induced apoptosis of splenocytes in BALB/C mice. BMC Med 9: 117.
14. Logue SE, Cleary P, Saveljeva S, Samali A (2013) New directions in ER stress-induced cell death. Apoptosis 18: 537–546.
15. Lindholm D, Wootz H, Korhonen L (2006) ER stress and neurodegenerative diseases. Cell Death Differ 13: 385–392.
16. van der Kallen CJ, van Greevenbroek MM, Stehouwer CD, Schalkwijk CG (2009) Endoplasmic reticulum stress-induced apoptosis in the development of diabetes: is there a role for adipose tissue and liver? Apoptosis 14: 1424–1434.
17. Deegan S, Saveljeva S, Gorman AM, Samali A (2013) Stress-induced self-cannibalism: on the regulation of autophagy by endoplasmic reticulum stress. Cell Mol Life Sci 70: 2425–2441.
18. Mostafalou S, Abdollahi M (2013) Pesticides and human chronic diseases: evidences, mechanisms, and perspectives. Toxicol Appl Pharmacol 268: 157–177.
19. Reiter RJ, Tan DX, Rosales-Corral S, Manchester LC (2013) The universal nature, unequal distribution and antioxidant functions of melatonin and its derivatives. Mini Rev Med Chem 13: 373–384.
20. Venegas C, García JA, Escames G, Ortiz F, López A, et al. (2012) Extrapineal melatonin: analysis of its subcellular distribution and daily fluctuations. J Pineal Res 52: 217–227.

21. Pandi-Perumal SR, Srinivasan V, Maestroni GJ, Cardinali DP, Poeggeler B, et al. (2006) Melatonin: Nature's most versatile biological signal? FEBS J 273: 2813–2838.
22. Seko LM, Moroni RM, Leitao VM, Teixeira DM, Nastri CO, et al. (2014) Melatonin supplementation during controlled ovarian stimulation for women undergoing assisted reproductive technology: systematic review and meta-analysis of randomized controlled trials. Fertil Steril 101: 154–161. e154.
23. Lin AM, Fang SF, Chao PL, Yang CH (2007) Melatonin attenuates arsenite-induced apoptosis in rat brain: involvement of mitochondrial and endoplasmic reticulum pathways and aggregation of alpha-synuclein. J Pineal Res 43: 163–171.
24. Tuñón MJ, San Miguel B, Crespo I, Jorquera F, Santamaría E, et al. (2011) Melatonin attenuates apoptotic liver damage in fulminant hepatic failure induced by the rabbit hemorrhagic disease virus. J Pineal Res 50: 38–45.
25. Chang CF, Huang HJ, Lee HC, Hung KC, Wu RT, et al. (2012) Melatonin attenuates kainic acid-induced neurotoxicity in mouse hippocampus via inhibition of autophagy and α-synuclein aggregation. J Pineal Res 52: 312–321.
26. Zaouali MA, Boncompagni E, Reiter RJ, Bejaoui M, Freitas I, et al. (2013) AMPK involvement in endoplasmic reticulum stress and autophagy modulation after fatty liver graft preservation: a role for melatonin and trimetazidine cocktail. J Pineal Res 55: 65–78.
27. Rehman H, Mohan A, Tabassum H, Ahmad F, Rahman S, et al. (2011) Deltamethrin Increases Candida albicans infection susceptibility in mice. Scand J Immunol 73: 459–464.
28. Carrillo-Vico A, Lardone PJ, Naji L, Fernández-Santos JM, Martín-Lacave I, et al. (2005) Beneficial pleiotropic actions of melatonin in an experimental model of septic shock in mice: regulation of pro-/anti-inflammatory cytokine network, protection against oxidative damage and anti-apoptotic effects. J Pineal Res 39: 400–408.
29. Xu J, Sun S, Wei W, Fu J, Qi W, et al. (2007) Melatonin reduces mortality and oxidatively mediated hepatic and renal damage due to diquat treatment. J Pineal Res 42: 166–171.
30. Benot S, Molinero P, Soutto M, Goberna R, Guerrero JM (1998) Circadian variations in the rat serum total antioxidant status: correlation with melatonin levels. J Pineal Res 25: 1–4.
31. Schult C, Dahlhaus M, Ruck S, Sawitzky M, Amoroso F, et al. (2010) The multikinase inhibitor Sorafenib displays significant antiproliferative effects and induces apoptosis via caspase 3, 7 and PARP in B- and T-lymphoblastic cells. BMC Cancer 10: 560.
32. Hao H, Dong Y, Bowling MT, Gomez-Gutierrez JG, Zhou HS, et al. (2007) E2F-1 induces melanoma cell apoptosis via PUMA up-regulation and Bax translocation. BMC Cancer 7: 24.
33. Kuribayashi K, Finnberg N, Jeffers JR, Zambetti GP, El-Deiry WS (2011) The relative contribution of pro-apoptotic p53-target genes in the triggering of apoptosis following DNA damage in vitro and in vivo. Cell Cycle 10: 2380–2389.
34. Toyota H, Yanase N, Yoshimoto T, Moriyama M, Sudo T, et al. (2003) Calpain-induced Bax-cleavage product is a more potent inducer of apoptotic cell death than wild-type Bax. Cancer Lett 189: 221–230.
35. Klionsky DJ, Abdalla FC, Abeliovich H, Abraham RT, Acevedo-Arozena A, et al. (2012) Guidelines for the use and interpretation of assays for monitoring autophagy. Autophagy 8: 445–544.
36. Lavrik IN, Krammer PH (2012) Regulation of CD95/Fas signaling at the DISC. Cell Death Differ 19: 36–41.
37. Chen YW, Yang YT, Hung DZ, Su CC, Chen KL (2012) Paraquat induces lung alveolar epithelial cell apoptosis via Nrf-2-regulated mitochondrial dysfunction and ER stress. Arch Toxicol 86: 1547–1558.
38. John K, Alla V, Meier C, Pützer BM (2011) GRAMD4 mimics p53 and mediates the apoptotic function of p73 at mitochondria. Cell Death Differ 18: 874–886.
39. Cao X, Deng X, May WS (2003) Cleavage of Bax to p18 Bax accelerates stress-induced apoptosis, and a cathepsin-like protease may rapidly degrade p18 Bax. Blood 102: 2605–2614.
40. Matsumura Y, Saeki E, Otsu K, Morita T, Takeda H, et al. (2001) Intracellular calcium level required for calpain activation in a single myocardial cell. J Mol Cell Cardiol 33: 1133–1142.
41. Yoshida H, Matsui T, Yamamoto A, Okada T, Mori K (2001) XBP1 mRNA is induced by ATF6 and spliced by IRE1 in response to ER stress to produce a highly active transcription factor. Cell 107: 881–891.
42. Uemura A, Oku M, Mori K, Yoshida H (2009) Unconventional splicing of XBP1 mRNA occurs in the cytoplasm during the mammalian unfolded protein response. J Cell Sci 122: 2877–2886.
43. de la Cadena SG, Hernández-Fonseca K, Camacho-Arroyo I, Massieu L (2014) Glucose deprivation induces reticulum stress by the PERK pathway and caspase-7- and calpain-mediated caspase-12 activation. Apoptosis 19: 414–427.
44. Galehdar Z, Swan P, Fuerth B, Callaghan SM, Park DS, et al. (2010) Neuronal apoptosis induced by endoplasmic reticulum stress is regulated by ATF4-CHOP-mediated induction of the Bcl-2 homology 3-only member PUMA. J Neurosci 30: 16938–16948.
45. Kang R, Zeh HJ, Lotze MT, Tang D (2011) The Beclin 1 network regulates autophagy and apoptosis. Cell Death Differ 18: 571–580.
46. Djavaheri-Mergny M, Maiuri MC, Kroemer G (2010) Cross talk between apoptosis and autophagy by caspase-mediated cleavage of Beclin 1. Oncogene 29: 1717–1719.
47. Chen D, Fan W, Lu Y, Ding X, Chen S, et al. (2012) A mammalian autophagosome maturation mechanism mediated by TECPR1 and the Atg12-Atg5 conjugate. Mol Cell 45: 629–641.
48. Herzog C, Yang C, Holmes A, Kaushal GP (2012) zVAD-fmk prevents cisplatin-induced cleavage of autophagy proteins but impairs autophagic flux and worsens renal function. Am J Physiol Renal Physiol 303: F1239–1250.
49. Barth S, Glick D, Macleod KF (2010) Autophagy: assays and artifacts. J Pathol 221: 117–124.
50. Ju JS, Varadhachary AS, Miller SE, Weihl CC (2010) Quantitation of "autophagic flux" in mature skeletal muscle. Autophagy 6: 929–935.
51. Hardeland R, Madrid JA, Tan DX, Reiter RJ (2012) Melatonin, the circadian multioscillator system and health: the need for detailed analyses of peripheral melatonin signaling. J Pineal Res 52: 139–166.
52. Jang SS, Kim WD, Park WY (2009) Melatonin exerts differential actions on X-ray radiation-induced apoptosis in normal mice splenocytes and Jurkat leukemia cells. J Pineal Res 47: 147–155.
53. Gurzov EN, Germano CM, Cunha DA, Ortis F, Vanderwinden JM, et al. (2010) p53 up-regulated modulator of apoptosis (PUMA) activation contributes to pancreatic beta-cell apoptosis induced by proinflammatory cytokines and endoplasmic reticulum stress. J Biol Chem 285: 19910–19920.
54. Samantaray S, Sribnick EA, Das A, Knaryan VH, Matzelle DD, et al. (2008) Melatonin attenuates calpain upregulation, axonal damage and neuronal death in spinal cord injury in rats. J Pineal Res 44: 348–357.
55. Zhao H, Wu QQ, Cao LF, Qing HY, Zhang C, et al. (2014) Melatonin Inhibits Endoplasmic Reticulum Stress and Epithelial-Mesenchymal Transition during Bleomycin-Induced Pulmonary Fibrosis in Mice. PLoS One 9: e97266.
56. Yoshida H, Oku M, Suzuki M, Mori K (2006) pXBP1(U) encoded in XBP1 pre-mRNA negatively regulates unfolded protein response activator pXBP1(S) in mammalian ER stress response. J Cell Biol 172: 565–575.
57. Hou W, Han J, Lu C, Goldstein LA, Rabinowich H (2010) Autophagic degradation of active caspase-8: a crosstalk mechanism between autophagy and apoptosis. Autophagy 6: 891–900.

Pesticide Exposure as a Risk Factor for Myelodysplastic Syndromes

Jie Jin[1,2¶], Mengxia Yu[1,2¶], Chao Hu[1,2¶], Li Ye[1,2], Lili Xie[1,2], Jin Jin[1,2], Feifei Chen[1,2], Hongyan Tong[1,2,3*]

1 Department of Hematology, the First Affiliated Hospital of Zhejiang University, Hangzhou, People's Republic of China, **2** Institute of Hematology, Zhejiang University School of Medicine, Hangzhou, People's Republic of China, **3** Myelodysplastic syndromes diagnosis and therapy center, Zhejiang University School of Medicine, Hangzhou, People's Republic of China

Abstract

Objective: Pesticide exposure has been linked to increased risk of cancer at several sites, but its association with risk of myelodysplastic syndromes (MDS) is still unclear. A meta-analysis of studies published through April, 2014 was performed to investigate the association of pesticide exposure with the risk of MDS.

Methods: Studies were identified by searching the Web of Science, Cochrane Library and PubMed databases. Summary odds ratios (ORs) with corresponding 95% confidence intervals (CIs) were calculated using random- or fixed-effect models.

Results: This meta-analysis included 11 case-control studies, all of which demonstrated a correlation between pesticide exposure and a statistically significant increased risk of MDS (OR = 1.95, 95% CI 1.23–3.09). In subgroup analyses, patients with pesticide exposure had increased risk of developing MDS if they were living in the Europe or Asia and had refractory anemia (RA) or RA with ringed sideroblasts (RARS). Moreover, in the analysis by specific pesticides, increased risk was associated with exposure to insecticides (OR = 1.71, 95% CI 1.22–2.40) but not exposure to herbicides or fungicides.

Conclusion: This meta-analysis supports the hypothesis that exposure to pesticides increases the risk of developing MDS. Further prospective cohort studies are warranted to verify the association and guide clinical practice in MDS prevention.

Editor: Shawn Hayley, Carleton University, Canada

Funding: This study was supported by grants from the Key Innovation Team Foundation of Zhejiang Province (2011R50015), major program fund of the Science Technology Department of Zhejiang Province (2013c03043-2), Zhejiang Province Fund for Distinguished Young Scholars (LR12H08001), National Public Health Grand Research Foundation (201202017) and the National Natural Science Foundation of China (No.30870914, No.81270582). The funders had no role in study design, data collection and analysis, decision to publish, or preparation of the manuscript.

Competing Interests: The authors have declared that no competing interests exist.

* Email: zjuhongyantong@163.com

¶ These authors are joint first authors on this work.

Introduction

Myelodysplastic syndromes (MDS) are a heterogeneous group of stem cell malignancies, characterized by ineffective hematopoiesis, and peripheral blood cytopenias. With disease progression, the risk of transformation into acute myeloid leukemia (AML) increased [1,2]. Despite development of new therapeutic methods in recent years, treatment of MDS is still limited and MDS remains incurable except in the case of the younger patients with good performance status, allogeneic stem cell transplantation eligibility, and adequate donor access [3]. As the whole population ages, MDS will become one of the most common myeloid malignancies. The societal impact and burden of the disease, measured in terms of the number of people affected yearly with a new diagnosis or who are living with the disease, is enormous and will continue to increase in the future. Therefore, a better comprehension of the etiology and further investigation of risk factors may significantly improve MDS prevention measures and reduce MDS incidence.

Since 1950, pesticide use has risen over 50% and pesticide toxicity has increased ten-fold [4]. Pesticide exposure is thought to increase cancer risk by promoting oxidative stress, chromosomal aberrations, cell signaling disturbances or gene mutations [5,6,7].

Over the past few decades, some epidemiological studies have analyzed the association between pesticide exposure and risk of MDS, but the findings are controversial. Five studies showed a positive association between incidence of MDS and pesticide exposure [8,9,10,11,12], and six studies illustrated no association [13,14,15,16,17,18]. Hence, the present meta-analysis was undertaken to further examine the potential involvement of pesticide exposure in MDS etiology.

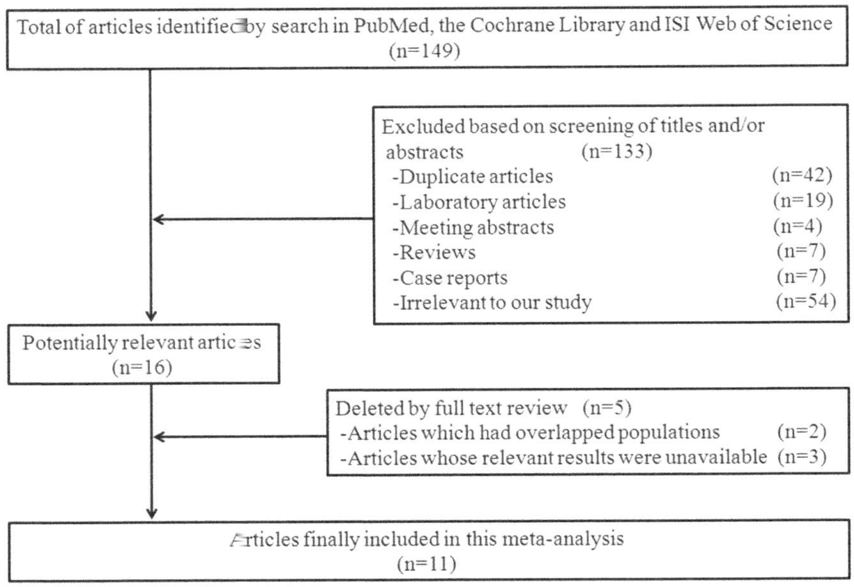

Figure 1. Process of study selection.

Materials and Methods

Literature research

A systematic literature search of the Web of Science, Cochrane Library and PubMed was executed by two independent reviewers (Chao Hu and Mengxia Yu). The following search strategy was used: (myelodysplastic syndrome OR MDS OR myelodysplastic OR myelodysplasia OR preleukemia) AND (pesticides OR herbicides OR fungicides OR insecticides). All relevant titles or abstracts were screened (see Study selection) to determine the suitability of each publication, and full-text articles were retrieved. We also checked the references from retrieved articles for additional studies not identified by database search.

Study selection

Studies included in this meta-analysis had to meet all the following criteria: (a) one of the exposures of interest was pesticide exposure; (b) one of the outcomes of interest was incidence of MDS; (c) a cohort design or case-control design; (d) providing the risk and corresponding 95% confidence intervals (CIs) or data to calculate these; (e) written in the English language. If there were multiple publications from the same study or overlapping study populations, the most recent and detailed study was eligible for inclusion in the meta-analysis.

Data extraction

Data were collected independently by two reviewers using a predefined data collection form. The following data were extracted from each study and included in the final analysis: the study name (together with the first author's name and year of publication), country of origin, gender, age, study design, source of patients, number of cases/controls, risk factor assessment, matching covariates, and adjusted covariates. We contacted the corresponding authors of the primary studies to acquire missing or insufficient data (when necessary), used group consensus and consulted a third reviewer to resolve discrepancies, and assigned scores of <7 and ≥ 7 for low- and high-quality studies, respectively, on the nine-score Newcastle-Ottawa Scale (NOS) [19,20].

Statistical analysis

To determine whether to use the fixed- or random-effects model, we measured statistical heterogeneity [21]. A fixed-effects model was used to calculate a pooled odds ratio (OR) with 95% CI when there was no heterogeneity. Otherwise, we calculated pooled ORs and confidence intervals assuming a random-effects model. The homogeneity of ORs across individual studies was quantified by the Q statistic and the I^2 score. P>0.05 for the Q-test was considered as a lack of heterogeneity among the studies. The I^2 values of 25%, 50%, and 75% represented mild, moderate, and severe heterogeneity, respectively [22]. Potential publication bias was assessed by using Begg's funnel plots (rank correlation method where an asymmetrical plot suggested possible publication bias) [23] and Egger's bias test (linear regression method where P<0.05 indicated the presence of statistically significant publication bias) [24]. Sensitivity analysis was conducted, in which the meta-analysis estimates were calculated by sequential omission of every study in turn, so as to reflect the influence of the data from individual studies on the pooled ORs and evaluate the stability of the results. Cumulative meta-analysis was also conducted by sorting the studies based on publication time. Subset analyses were performed by source of patients, disease subtype, geographic region, study quality, and type of pesticide. All of the statistical analyses were performed with STATA 11.0 (Stata Corporation, College Station, TX) using two-sided P-values, where P<0.05 was considered statistically significant.

Results

Literature search and study characteristics

The results of our literature search strategy and study selection process were detailed in Figure 1. We identified 11 case-control studies on the association of pesticide exposure with risk of MDS published between 1990 and 2011 [8,9,10,11,12,13,14,15,16,17,18]. A total of 1,942 MDS patients and 5,359 controls were included in the present meta-analysis. Among the 5,359 controls, 3853 persons were hospitalized patients without conditions related to hematological diseases and the remaining 1506 were recruited from healthy

Table 1. Main characteristics of studies evaluating the association between pesticides exposure and MDS.

Study	Country	Gender	Age	Study Design	Source of patients	Number of cases	Number of controls	Risk factor Assessment	Study Quality	Matching and Adjustments
Kokouva (2011)[13]	Greece	M/F	27–73	Case-control	Hospital-based	78	455	Questionnaire	5	Gender, age, smoking, family history
Lv (2011)[14]	China	M/F	20–88	Case-control	Hospital-based	403	806	Face-to-face Interview	6	Age, sex, anti-tb drugs, D860, traditional Chinese medicine, alcohol intake, benzene, gasoline, glues, hair dye, education, new building
Pekmezovic (2006)[8]	Serbia Montenegro	M/F	18–85	Case-control	Hospital-based	80	160	Interview	6	Age, sex
Strom (2005)[9]	United States	M/F	24–89	Case-control	Hospital-based	354	452	Mailed questionnaire	7	Age, sex, ethnicity, education, family history of hematopoietic cancer, alcohol intake, benzene, solvent, gasoline
Nisse (2001)[10]	France	M/F	NR	Case-control	Population-based	204	204	Interview	8	Agricultural workers, textile operators, health professionals, living next to an industrial plant, commercial and technical sales representatives, machine operators, oil use, smoking
Rigolin (1998)[11]	Italy	M/F	17–85	Case-control	Hospital-based	178	178	Interview and questionnaire	5	Age, sex
West (1995)[15]	UK	M/F	≥15	Case-control	Hospital-based	400	400	Interview and questionnaire	6	Age, sex, area of residence and hospital, year of diagnosis
Mele (1994)[16]	Italy	M/F	≥15	Case-control	Hospital-based	111	1161	Interview	6	Age, sex, education, and residence outside study town
Ciccone (1993)[12]	Italy	M/F	15–74	Case-control	Hospital-based and population-based	19	246	Interview	5	Sex, area of residence, age
Brown (1990)[17]	United States	M	≥30	Case-control	Population-based	63	1245	Interview	6	Vital status, age, state, tobacco use, family history of lymphopoietic cancer, high-risk occupations and high-risk exposure
Goldberg (1990)[18]	United States	NR	28–88	Case-control	Hospital-based	52	52	Interview	6	Age and sex

M: male; F: female; NR: not reported; tb: tuberculosis.

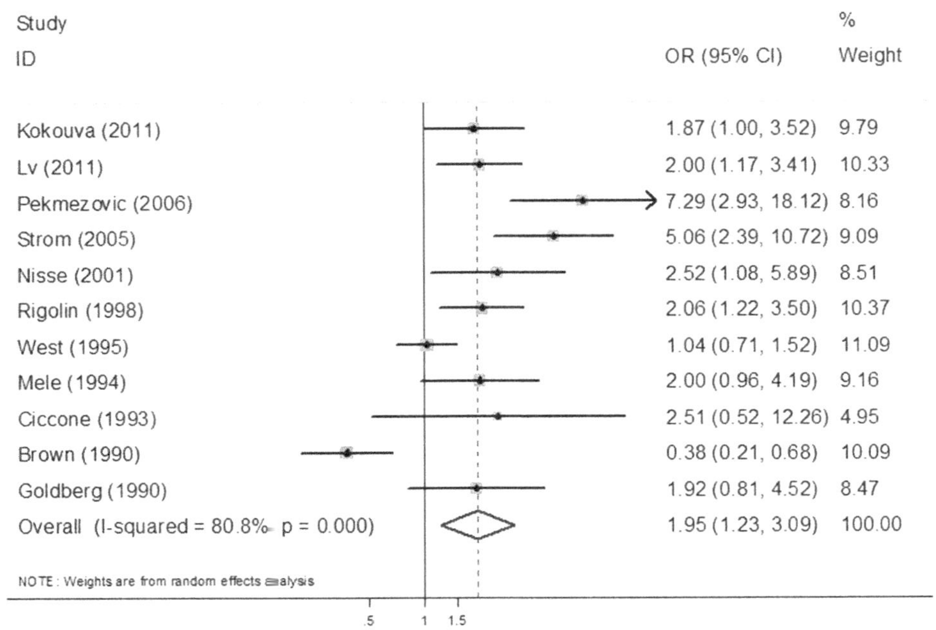

Figure 2. A forest plot illustrating risk estimates from included studies on the relationship between pesticide exposure and MDS risk.

people. A total of 1456 participants have exposed to pesticide, of whom 323 suffered from MDS. Among the chosen studies, seven were conducted in Europe, three in United States, and one in Asia. Pesticide exposure was ascertained by interview or questionnaire or both. The study quality was graded by the Newcastle-Ottawa Quality Assessment Scale, ranged from 5 to 8 (with a mean of 6). The main characteristics of the included articles were listed in Table 1.

Risk estimation

Our analysis demonstrated a significant adverse association between pesticide exposure (exposed vs. non-exposed status) and incidence of MDS (OR = 1.95, 95%CI 1.23–3.09) (Figure 2). Due to a statistically significant heterogeneity across studies ($I^2 = 80.8\%$, $P<0.001$), the summary OR we estimated using the DerSimonian and Laird random effects model [25]. A Galbraith plot identified four studies as major sources of heterogeneity (Figure 3A). After excluding these four studies [8,9,15,17], there was no study heterogeneity existed ($P = 0.999$, $I^2 = 0.0\%$) and the overall association became stronger (OR = 2.04, 95% CI 1.57–2.66).

Stratified analysis

Next, we pooled the OR estimates by patient source (population-based or hospital-based), MDS subtypes refractory anemia (RA) and RA with ringed sideroblasts (RARS) or RA with excess blasts (RAEB) and RAEB in transformation (RAEBt)), geographic region (United States, Europe, or Asia), study quality (low or high), and type of pesticide (insecticide, herbicide or fungicide) (Table 2). When separated by patient source, the ORs (95% CI) were 2.26 (1.49–3.42) for hospital-based studies and 0.95 (0.15–6.06) for population-based studies. When stratified by MDS subtype, the associations were more positive for the RA/RARS sbutype (OR = 1.63, 95%CI 1.06–2.51) than the RAEB/RAEBt subtype (OR = 1.49, 95%CI 0.78–2.84). In the subset analyses stratified by geographic region, a statistically significant adverse effect of pesticide exposure on MDS was observed in Europe (OR = 2.13, 95%CI 1.35–3.36) and Asia (OR = 2.00, 95%CI 1.17–3.41), but not in United States (OR = 1.52, 95%CI 0.30–7.73). Furthermore, when stratified by study quality, the relationship was more significant in high quality studies (OR = 2.19, 95%CI 1.40–3.42) than in low quality studies (OR = 1.90, 95%CI 1.09–3.33). In addition, when analyzed by type of pesticide, the ORs (95% CI) for insecticides, herbicides, and fungicides were 1.71 (1.22–2.40), 1.16 (0.55–2.43) and 0.70 (0.20–3.20), respectively.

Sensitivity analysis

We also carried out sensitivity analysis by sequentially excluding one study at a time to detect the influence of a single study on the overall estimate. The results displayed that no study disproportionately affected the summary risk estimates in this meta-analysis (Figure 3B). The eleven study-specific ORs ranged from a low of 1.72 (95%CI 1.11–2.68) to 2.27 (95%CI 1.58–3.27) via the omission of the study by Pekmezovic et al. [8] and the study by Brown et al. [17], respectively.

Cumulative meta-analysis

Cumulative meta-analysis of the relationship between pesticide exposure and risk of MDS was also implemented by sorting the studies based on publication time. Figure 4 showed the results from the cumulative meta-analysis of this connection in chronologic order. The 95% CIs became increasingly narrower with each addition of more data, suggesting the precision of each estimate was progressively increasing with the addition of more cases.

Publication bias

As reflected by the funnel plot (Begg's test, $P = 0.350$) and the Egger's test ($P = 0.113$), there was no publication bias being discovered. The data witnessed our result was statistically robust.

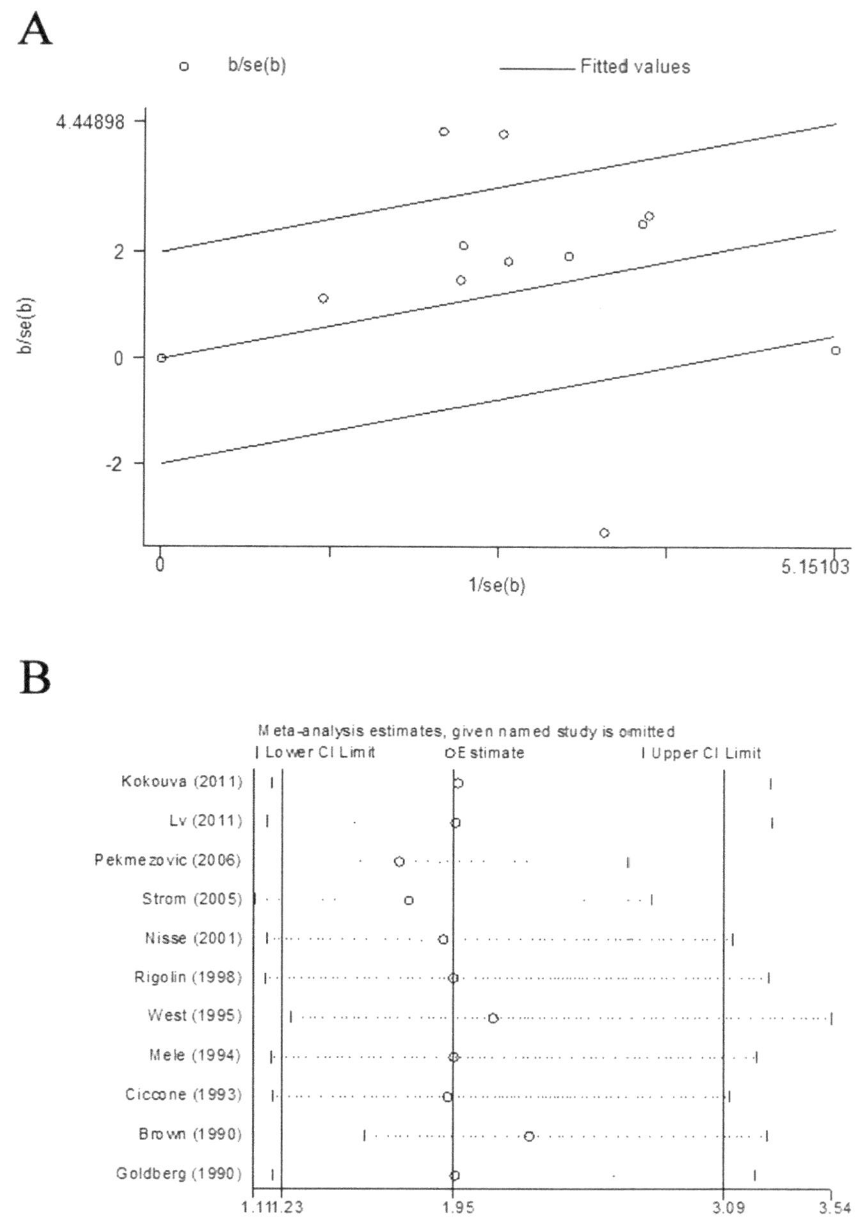

Figure 3. Evaluating the heterogeneity and the stability of the results. (A) Galbraith plot evaluating the heterogeneity; (B) Sensitivity analyses by sequential omission of individual studies in our study.

Discussion

The first myelodysplastic syndromes (MDS) case series was reported about 40 years ago [26]. Thus, the recognition of MDS is approximately 100 years behind the recognition of other hematologic malignancies. Similarly, the level of epidemiologic knowledge of MDS is far below that of other cancers. Therefore, further investigation of risk factors in MDS patients is needed to improve MDS prevention.

Pesticides are widely applied in agriculture all over the world. Three million cases of acute severe pesticide poisoning and over 200,000 deaths are reported annually [4]. Pesticides are thus considered a risk factor for some cancers. Moreover, two previous meta-analyses have been performed in hematological malignancies, which indicated that pesticide exposure could increase risk of non-Hodgkin lymphoma, leukemia and multiple myeloma [27,28]. However, the result about MDS and pesticide exposure was limited. Recent epidemiological studies have examined the potential association between pesticide exposure and the risk of MDS, but none of the results has been conclusive. We attempted to clarify this possible relationship through a meta-analysis of eleven case-control studies.

To the best of our knowledge, this is the first meta-analysis assessing the relationship between pesticide exposure and MDS. Several interesting points raised by our analysis are worth discussing. Firstly, our research demonstrated a significantly positive correlation between pesticide exposure and MDS, which indicated pesticide exposure was associated with a 95% increased

Table 2. Stratified pooled odds ratios of the relationship between pesticide exposure and risk of MDS.

Variables	Number of studies	Pooled OR (95%CI)	Q-test for heterogeneity *P* value (I^2 score)	Egger's test P value	Begg's test P value
Total	11 (8, 9, 10, 11, 12, 13, 14, 15, 16, 17, 18)	1.95 (1.23–3.09)	<0.001 (80.8%)	0.350	0.113
Source of patients					
Population based	2 (10, 17)	0.95 (0.15–6.06)	<0.001 (92.3%)	–	1.000
Hospital based	8 (8, 9, 11, 13, 14, 15, 16, 18)	2.26 (1.49–3.42)	0.001 (71.7%)	0.098	0.266
Disease subtype					
RA/RARS	3 (9, 11, 18)	1.63 (1.06–2.51)	0.258 (25.6%)	0.413	0.734
RAEB/RAEBt	4 (9, 11, 14, 16)	1.49 (0.78–2.84)	0.005 (70.4%)	0.734	0.452
Geographic region					
Europe	7 (8, 10, 11, 12, 13, 15, 16)	2.13 (1.35–3.36)	0.006 (66.8%)	0.133	0.057
Asia	1 (14)	2.00 (1.17–3.41)	–	–	–
United States	3 (9, 17, 18)	1.52 (0.30–7.73)	<0.001 (93.4%)	0.407	1.000
Study quality					
High	2 (10, 11)	2.19 (1.40–3.42)	0.698 (0.0%)	–	1.000
Low	9 (8, 9, 12, 13, 14, 15, 16, 17, 18)	1.90 (1.09–3.33)	<0.001 (83.9%)	0.155	0.348
Type of pesticides					
Insecticides	9 (10, 11, 12, 13, 14, 15, 16, 17, 18)	1.71 (1.22–2.40)	0.009 (60.8%)	0.147	0.348
Herbicides	4 (14, 15, 16, 17)	1.16 (0.55–2.43)	0.056 (60.3%)	0.203	0.089
Fungicides	1 (17)	0.70 (0.20–3.20)	–	–	–

RA: refractory anemia; RARS: RA with ringed sideroblasts; RAEB: RA with excess blasts (RAEB); RAEBt: RAEB in transformation.

risk of MDS. Sensitivity analysis and cumulative analysis confirmed the robustness of our outcomes. In addition, subgroup analyses showed a stronger effect of pesticide exposure on RA/RARS than on RAEB/RAEBt (i.e., exposed MDS patients had 63% increased risk of RA/RARS and 49% increased risk of RAEB/RAEBt, respectively). Our study also illustrated that exposure to insecticides can the increase risk of MDS by 71%, while exposure to herbicides (OR = 1.16, 95%CI 0.55–2.43) and fungicides (OR = 0.70, 95%CI 0.20–3.20), respectively, add no risk. Our subset analysis according to geographical region noted

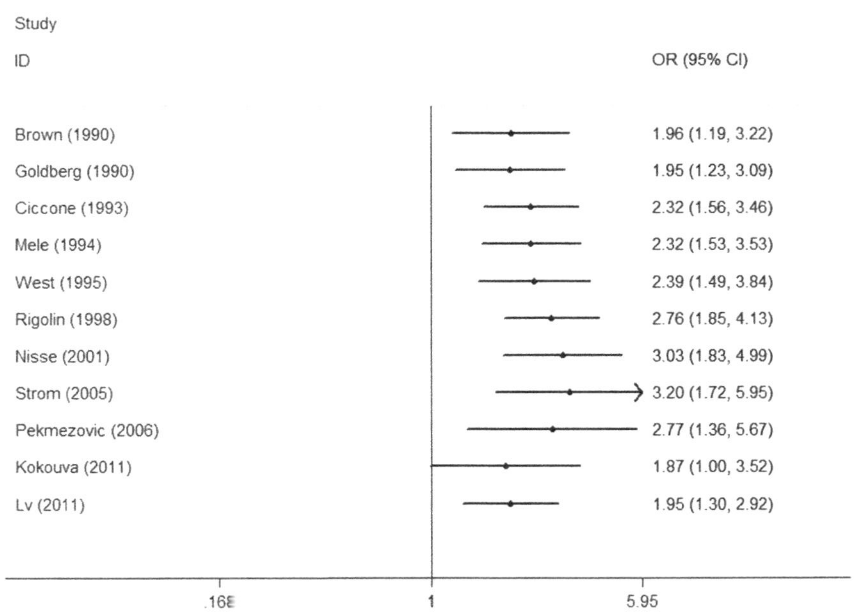

Figure 4. Forest plots showing the result of the cumulative meta-analysis.

higher risk of MDS in Europe (113%) than in Asia (100%) and the United States (52%).

The biological mechanism underlying the linkage of pesticide exposure to the pathogenesis of MDS remains largely unknown. However, several mechanisms are conceivable. Exposure to pesticides might cause overexpression of reactive oxygen species (ROS) sufficient to overwhelm antioxidant defense mechanisms and thereby lead to extensive DNA damage, protein damage, and hematopoietic irregularities [29]. On the other hand, pesticides might bind to and displace endogenous ligands of steroid nuclear receptors, including estrogen and androgen receptors, thus aberrantly activating receptor function and inducing changes in gene expression networks [30]. Recent in vitro mechanistic studies offer novel insight. For example, Boros and Williams reported that exposure of leukemic cell lines (K562) to increasing doses of an organophosphate pesticide (isofenphos) resulted in dose-dependent leukemic cell proliferation [31]. In addition, some previous studies demonstrating that pesticide exposure could induce chromosomal defects [32,33], might also suggest that pesticides could increase the risk of developing MDS. Further research is warranted to elucidate the likely biological mechanisms.

As a meta-analysis of previously published observational studies, our research has some limitations that influence interpretation of the results. First, although the present results seemed to suggest the absence of publication bias, our meta-analysis was vulnerable to publication bias, because only studies published in English were included. Limited resources prevented us from including articles published in other languages and databases. Second, no prospective studies of the association between pesticide exposure and MDS risk were available, and all included studies had a retrospective case-control design. Thus, owing to the limitations of case-control design, the possibility of undetected bias could not be excluded. Third, it is known farmers in China use large amounts of pesticides and this is an ideal population to study their effect on health. However, only one study from China was included in this meta-analysis. Fourth, too few original studies have separated biocides into insecticides, herbicides or fungicides to justify concluding the potential existence of a relationship between exposure to one or several categories of biocides and MDS. Significantly increased risks were observed, with an apparently higher increase when the exposure was to insecticides.

In summary, our findings support that pesticide exposure is associated with the increased risk of MDS, and this association varies widely across disease subtype, geographic region and specific biocide category. Larger and more rigorous analytical studies will be warranted to generate more robust conclusion to guide clinical practice in MDS prevention in the future.

Author Contributions

Conceived and designed the experiments: Jie Jin HT. Performed the experiments: CH MY. Analyzed the data: LY LX Jin Jin FC. Contributed reagents/materials/analysis tools: CH. Contributed to the writing of the manuscript: CH MY.

References

1. Kasner MT, Luger SM (2009) Update on the therapy for myelodysplastic syndrome. Am J Hematol 84: 177–186.
2. Newman K, Maness-Harris L, El-Hemaidi I, Akhtari M (2012) Revisiting use of growth factors in myelodysplastic syndromes. Asian Pac J Cancer Prev 13: 1081–1091.
3. Warlick ED, Smith BD (2007) Myelodysplastic syndromes: review of pathophysiology and current novel treatment approaches. Curr Cancer Drug Targets 7: 541–558.
4. Malek AM, Barchowsky A, Bowser R, Youk A, Talbott EO (2012) Pesticide exposure as a risk factor for amyotrophic lateral sclerosis: a meta-analysis of epidemiological studies: pesticide exposure as a risk factor for ALS. Environ Res 117: 112–119.
5. Infante-Rivard C, Weichenthal S (2007) Pesticides and childhood cancer: an update of Zahm and Ward's 1998 review. J Toxicol Environ Health B Crit Rev 10: 81–99.
6. Lafiura KM, Bielawski DM, Posecion NC, Jr., Ostrea EM, Jr., Matherly LH, et al. (2007) Association between prenatal pesticide exposures and the generation of leukemia-associated T(8;21). Pediatr Blood Cancer 49: 624–628.
7. Agopian J, Navarro JM, Gac AC, Lecluse Y, Briand M, et al. (2009) Agricultural pesticide exposure and the molecular connection to lymphomagenesis. J Exp Med 206: 1473–1483.
8. Pekmezovic T, Suvajdzic Vukovic N, Kisic D, Grgurevic A, Bogdanovic A, et al. (2006) A case-control study of myelodysplastic syndromes in Belgrade (Serbia Montenegro). Ann Hematol 85: 514–519.
9. Strom SS, Gu Y, Gruschkus SK, Pierce SA, Estey EH (2005) Risk factors of myelodysplastic syndromes: a case-control study. Leukemia 19: 1912–1918.
10. Nisse C, Haguenoer JM, Grandbastien B, Preudhomme C, Fontaine B, et al. (2001) Occupational and environmental risk factors of the myelodysplastic syndromes in the North of France. Br J Haematol 112: 927–935.
11. Rigolin GM, Cuneo A, Roberti MG, Bardi A, Bigoni R, et al. (1998) Exposure to myelotoxic agents and myelodysplasia: case-control study and correlation with clinicobiological findings. Br J Haematol 103: 189–197.
12. Ciccone G, Mirabelli D, Levis A, Gavarotti P, Rege-Cambrin G, et al. (1993) Myeloid leukemias and myelodysplastic syndromes: chemical exposure, histologic subtype and cytogenetics in a case-control study. Cancer Genet Cytogenet 68: 135–139.
13. Kokouva M, Bitsolas N, Hadjigeorgiou GM, Rachiotis G, Papadoulis N, et al. (2011) Pesticide exposure and lymphohaematopoietic cancers: a case-control study in an agricultural region (Larissa, Thessaly, Greece). BMC Public Health 11: 5.
14. Lv L, Lin G, Gao X, Wu C, Dai J, et al. (2011) Case-control study of risk factors of myelodysplastic syndromes according to World Health Organization classification in a Chinese population. Am J Hematol 86: 163–169.
15. West RR, Stafford DA, Farrow A, Jacobs A (1995) Occupational and environmental exposures and myelodysplasia: a case-control study. Leuk Res 19: 127–139.
16. Mele A, Szklo M, Visani G, Stazi MA, Castelli G, et al. (1994) Hair dye use and other risk factors for leukemia and pre-leukemia: a case-control study. Italian Leukemia Study Group. Am J Epidemiol 139: 609–619.
17. Brown LM, Blair A, Gibson R, Everett GD, Cantor KP, et al. (1990) Pesticide exposures and other agricultural risk factors for leukemia among men in Iowa and Minnesota. Cancer Res 50: 6585–6591.
18. Goldberg H, Lusk E, Moore J, Nowell PC, Besa EC (1990) Survey of exposure to genotoxic agents in primary myelodysplastic syndrome: correlation with chromosome patterns and data on patients without hematological disease. Cancer Res 50: 6376–6881.
19. Xu X, Cheng Y, Li S, Zhu Y, Zheng X, et al. (2014) Dietary carrot consumption and the risk of prostate cancer. Eur J Nutr.
20. Wells G, Shea B, O'Connell D (2009) Ottawa Hospital Research Institute: The Newcastle-Ottawa Scale (NOS) for assessing the quality of nonrandomized studies in meta-analyses. Available: http://www.ohri.ca/programs/clinical epidemiology/oxford.asp.
21. Hu ZH, Lin YW, Xu X, Chen H, Mao YQ, et al. (2013) Genetic polymorphisms of glutathione S-transferase M1 and prostate cancer risk in Asians: a meta-analysis of 18 studies. Asian Pac J Cancer Prev 14: 393–398.
22. Castillo JJ, Dalia S, Shum H (2011) Meta-analysis of the association between cigarette smoking and incidence of Hodgkin's Lymphoma. J Clin Oncol 29: 3900–3906.
23. Begg CB, Mazumdar M (1994) Operating characteristics of a rank correlation test for publication bias. Biometrics 50: 1088–1101.
24. Egger M, Davey Smith G, Schneider M, Minder C (1997) Bias in meta-analysis detected by a simple, graphical test. BMJ 315: 629–634.
25. DerSimonian R, Laird N (1986) Meta-analysis in clinical trials. Control Clin Trials 7: 177–188.
26. Saarni MI, Linman JW (1973) Preleukemia. The hematologic syndrome preceding acute leukemia. Am J Med 55: 38–48.
27. Van Maele-Fabry G, Duhayon S, Lison D (2007) A systematic review of myeloid leukemias and occupational pesticide exposure. Cancer Causes Control 18: 457–478.
28. Merhi M, Raynal H, Cahuzac E, Vinson F, Cravedi JP, et al. (2007) Occupational exposure to pesticides and risk of hematopoietic cancers: meta-analysis of case-control studies. Cancer Causes Control 18: 1209–1226.

29. Alavanja MC, Ross MK, Bonner MR (2013) Increased cancer burden among pesticide applicators and others due to pesticide exposure. CA Cancer J Clin 63: 120–142.
30. Schug TT, Janesick A, Blumberg B, Heindel JJ (2011) Endocrine disrupting chemicals and disease susceptibility. J Steroid Biochem Mol Biol 127: 204–215.
31. Boros LG, Williams RD (2001) Isofenphos induced metabolic changes in K562 myeloid blast cells. Leuk Res 25: 883–890.
32. Smith MT, McHale CM, Wiemels JL, Zhang L, Wiencke JK, et al. (2005) Molecular biomarkers for the study of childhood leukemia. Toxicol Appl Pharmacol 206: 237–245.
33. Chiu BC, Blair A (2009) Pesticides, chromosomal aberrations, and non-Hodgkin's lymphoma. J Agromedicine 14: 250–255.

22

Invasive Parasites, Habitat Change and Heavy Rainfall Reduce Breeding Success in Darwin's Finches

Arno Cimadom[1], Angel Ulloa[2], Patrick Meidl[3], Markus Zöttl[4], Elisabet Zöttl[1], Birgit Fessl[2], Erwin Nemeth[5], Michael Dvorak[5], Francesca Cunninghame[2], Sabine Tebbich[1]*

1 Department of Behavioural Biology, University of Vienna, Vienna, Austria, **2** Charles Darwin Foundation, Puerto Ayora, Santa Cruz Island, Galápagos, Ecuador, **3** Institute for Science and Technology, Klosterneuburg, Austria, **4** Department of Zoology, University of Cambridge, Cambridge, United Kingdom, **5** BirdLife Austria, Vienna, Austria

Abstract

Invasive alien parasites and pathogens are a growing threat to biodiversity worldwide, which can contribute to the extinction of endemic species. On the Galápagos Islands, the invasive parasitic fly *Philornis downsi* poses a major threat to the endemic avifauna. Here, we investigated the influence of this parasite on the breeding success of two Darwin's finch species, the warbler finch (*Certhidea olivacea*) and the sympatric small tree finch (*Camarhynchus parvulus*), on Santa Cruz Island in 2010 and 2012. While the population of the small tree finch appeared to be stable, the warbler finch has experienced a dramatic decline in population size on Santa Cruz Island since 1997. We aimed to identify whether warbler finches are particularly vulnerable during different stages of the breeding cycle. Contrary to our prediction, breeding success was lower in the small tree finch than in the warbler finch. In both species *P. downsi* had a strong negative impact on breeding success and our data suggest that heavy rain events also lowered the fledging success. On the one hand parents might be less efficient in compensating their chicks' energy loss due to parasitism as they might be less efficient in foraging on days of heavy rain. On the other hand, intense rainfalls might lead to increased humidity and more rapid cooling of the nests. In the case of the warbler finch we found that the control of invasive plant species with herbicides had a significant additive negative impact on the breeding success. It is very likely that the availability of insects (i.e. food abundance)is lower in such controlled areas, as herbicide usage led to the removal of the entire understory. Predation seems to be a minor factor in brood loss.

Editor: Csaba Moskát, Hungarian Academy of Sciences, Hungary

Funding: The study was funded by the University of Vienna (Focus of Excellence grant), the Galápagos Conservation Trust, and the Ethologische Gesellschaft e.V. During the study Sabine Tebbich was funded by the FWF (project number V95-B17) and Arno Cimadom by a scholarship from the University of Vienna. The publication was supported by the publication fund of the University of Vienna. The funders had no role in study design, data collection and analysis, decision to publish, or preparation of the manuscript.

Competing Interests: The authors have declared that no competing interests exist.

* Email: sabine.tebbich@univie.ac.at

Introduction

Invasive alien pathogens and parasites are a major and growing threat to biodiversity worldwide. Small host populations of endemic species are particularly vulnerable because extinction can occur before the hosts have a chance to evolve effective defences [1,2]. This is especially true if the parasite is a generalist because it can switch to another host if it has driven one close to extinction [3]. The adverse effect of introduced parasites has been documented in a range of avian species (reviewed in [2]). A famous example is the extinction of most of the endemic Hawaiian honeycreepers because of the introduction of avian malaria and its vector [4]. The avifauna of almost all islands of the Pacific Ocean has been dramatically altered by introduced species or other human impact [5]. In this respect the Galápagos archipelago is one of the exceptions, as no bird species has become extinct in modern times. This archipelago hosts the endemic Darwin's finches which have provided the inspiration for some of the most important ideas in evolutionary biology. However, the introduction of predators, pathogens, and parasites have led to increasing pressures affecting several Darwin's finch species [6–10].

One of the biggest threats to the Galápagos avifauna is the obligate bird parasite *Philornis downsi* [11], which was first discovered on the archipelago in 1997 [12]. Adult flies lay their eggs in bird nests where the parasitic larvae then hatch and suck blood from the nestlings [13]. Correlative as well as experimental studies have shown that *P. downsi* has a negative impact on nestling growth, haemoglobin levels, and fledgling success [14–18]. The influence of parasitism on breeding success in Darwin's finches is highly variable from year to year, and mortality ranges from 16 to 95% in all finch nests (reviewed in [18,19]). The reasons for this high variation are poorly understood so far but could stem from differences in precipitation between years [20], but also see [21]. The Galápagos archipelago is characterised by a highly seasonal climate with pronounced wet and dry seasons, as well as extraordinary yearly variation in rainfall [22]. Years with intense rainfalls (El Niño years) and severe drought years reoccur at irregular intervals. However, within the wet season there is also variation in precipitation, which has received little attention so far. Rainfall can reach 50 ml and more on some days, whereas on other days no precipitation is measured. An interaction between rainfall and *P. downsi* prevalence and intensity has been found in

Philornis sp. which affect various bird species in Puerto Rico [23] and Argentina [24].

The highest prevalence and intensity (larvae per nest or chick; [25]) of *P. downsi* infestation on the Galápagos Islands was found on Santa Cruz Island [26]. On this island, several insectivorous passerines declined between 1997 and 2010 [27]. The greatest decline was observed in the warbler finch (*Certhidea olivacea*). This insectivorous species is an arboreal finch and is the smallest of the Darwin's finches. It is restricted to the highlands, which were previously covered by the endemic humid *Scalesia* forest. During the last century this forest has to a large extent been transformed into agricultural areas. The warbler finch population has shrunk by 50% in the *Scalesia* forest and by up to 75% in the agricultural areas [27]. The closely related grey warbler finch (*Certhidea fuscua*) has already gone extinct on Floreana Island as there are no records of this species since more than sixty years [28]. Other than parasitism, the loss of primary habitat could be a possible reason for the population decline of the warbler finch. By 2009, the *Scalesia* forest, which holds the highest density of warbler finches [27], was reduced to only 2% of its original area [29]. Additionally, introduced trees and shrubs have invaded these remnant *Scalesia* forest patches [30,31]. The *Scalesia* forests are most affected by the introduced plant species *Rubus niveus*. To control this invasive neophyte, the National Park uses strong herbicides which lead to the temporary removal of the entire understory. We hypothesize that this dramatic habitat change may lead to changes in plant communities and species composition of invertebrates, which could also negatively influence the insectivorous bird species. Finally, introduced predators, especially rats (*Rattus rattus*), may negatively affect bird species [32]. Black rats (*Rattus rattus*) and Norway rats (*Rattus norvegicus*) have been introduced on all inhabited Galápagos Islands [33]. Due to their climbing skills and omnivorous diet, the black rats in particular are considered a threat to the Galápagos avifauna [10,19,34,35]. A study on mangrove finches found that rats can cause significant brood loss [10]. Furthermore, rats may have a stronger influence on the warbler finch than the other finch species, as warbler finch nests are built lower down in the canopy [17].

A population decline may be caused by high adult mortality or low recruitment. Therefore, one possible factor in the decline of the warbler finch on Santa Cruz Island is low breeding success. Thus, the aim of the study was to measure the breeding success of the warbler finch and to identify the reasons for brood loss. We compared the breeding success of the warbler finch, whose populations declined, with the breeding success of the sympatric small tree finch, whose population remained stable during the respective time period [27]. We predicted that warbler finches suffer a higher brood loss due to *P. downsi* than small tree finches because they showed a higher parasite intensity than expected for their body size and nest size in previous studies [17]. Furthermore, we predicted that warbler finches suffer a higher brood loss due to predation (by rats) because their nests are built lower in the vegetation than those of small tree finches (warbler finch: 4.2 m, small tree finch: 6.2 m [17]). In addition, we predicted lower breeding success in areas where herbicides were used to control invasive plant species because both species may be negatively affected by reduced insect availability during chick rearing.

Methods

Study site

The study was conducted during the main breeding season (January to the end of March 2010 and 2012) in the highland area around Los Gemelos (0°37′34″ S, 90°23′10″ W) on Santa Cruz Island, Galápagos. This area is dominated by the endemic tree *Scalesia pedunculata* and is thus called the *Scalesia* zone. On 11 ha of our 30 ha study site, the National Park applied strong herbicides to control the invasion of *R. niveus*, which results in the removal of the whole forest understory. The procedure of *Rubus* control consisted in manually cutting down all introduced and invasive plant species and then regularly treating the area with herbicides to prevent re-growth of neophytes.

Population survey

Our data collection is part of a population survey of the two focal species, warbler finch and small tree finch, that started in the *Scalesia* zone in 1997. We conducted point counts with unlimited distance following Dvorak et al. [27]. Data of the population survey in 2010 are already published (see [27]). In 2012, 26 points were counted once from February 18 to 21 by Michael Dvorak and Birgit Fessl. Each point count lasted 5 minutes and was conducted between 06:30 and 10:30 am, which is when Darwin's finches show the highest singing activity. The 26 points were spaced at least 70 m apart. The number of all singing birds (presumed to be territory holding males) and the distance of each singing individual to the observer were recorded for each point. However, in order to measure the population trends over the whole period from 1997 to 2012, only the number of singing males per point was used.

In 2012, we also recorded the age structure of the small tree finch population in the *Scalesia* forest. The brownish head coloration of young small tree finch males turns black in a specific moulting pattern as the subjects age [36]. Although six distinct moulting patterns were identified, interobserver reliability of plumage classification was low for some categories [36].Thus, each small tree finch male found displaying near its nest (irrespective if it was already paired or not) was assigned to one of two categories following Kleindorfer's [36] coloration categories: (i) young males, whose black plumage is maximally extended from around the beak to the eye region, and (ii) old males, whose black plumage has already extended to the throat and the back part of the head and nape.

Parasite life cycle

While *P. downsi* is an obligate bird parasite in its three larval stages, adult flies feed on organic matter [13,38]. The first instar larvae usually develop within the chicks' nares, although they were also found freely moving in the nest material ([18], Cimadom personal observation). First instar larvae cause beak malformations in the developing chick, which can persist into adulthood [37]. The second and third instar larvae are found in the bottom layer of the nest and feed on nestling blood and tissue during the night [13,38,39]. The larvae pupate at the bottom of the nest and emerge as adult flies approximately 10 to 14 days later ([38], Cimadom personal observation). Bird nests are often subject to multiple infections by *P. downsi* [40].

Nest monitoring and parasite collection

We monitored the nests of both Darwin's finch species – warbler finch and small tree finch – in an area of approximately 30 ha within the *Scalesia* forest. Nest status was checked at different intervals depending on the current breeding status to determine onset of breeding, number of eggs, hatching day, number of nestlings and date of failure or fledging. Intervals of nest observations varied depending on breeding status to minimize disturbance and optimize accuracy of information. During nest building nests were checked at 5 days interval, during incubation at 3 days interval, during feeding at 2 days interval and close to fledging daily. We inspected the inside of the nests with a video camera in 2010 and a small endoscopic video camera (dnt Findoo 3.6) in 2012, which allowed recordings of clutch as well as brood size (only in 2012). After each nest inspection, we waited until the parents resumed incubation or feeding, which occurred in all instances. Therefore we are confident that filming did not lead to abandonment of nests. Successful nests (breeding success) were defined as nests that produced at least one fledgling. Nests that failed were assigned to the following categories: (1) abandoned (pair stopped incubating, no chicks but eggs still present in the nest), (2) dead chicks (all chicks dead but still in the nest), (3) empty nest (previously active nest, chicks ≤6 days), These nests showed no clear signs of predation but predation cannot be excluded. Alternatively dead chicks may have been thrown out by parents [14]. (4) Predated (nest destroyed, with clear signs of predation or intact and empty, previously contained chicks ≥7 days – assumed to be too big to be thrown out by parents), and (5) others (e.g. nesting tree collapsed, nest fallen down due to wind).

Because of the smaller size of the endoscopic camera, the quality of the pictures was much better and allowed reliable age estimates. Therefore, we included only the data set of 2012 in analysis that included chick age, date of starting incubation and hatching date. The chicks' age (in days) in successful and failed nests was calculated by combining information from nest inspections with a series of images of growing nestlings (when inspections coincided with hatching). The hatching date was calculated as the last observation date minus chick age. Where possible, the calculated hatching date was cross-checked with observational data from nest inspections.

In total, 43 nests were found before incubation started and were successfully incubated until the chicks' hatched. The incubation time was calculated for each nest as the time period (in days) from date of first incubating observation to hatching date. The mean incubation time was 13.7±2.0 SD days (n = 27, median 14) for the warbler finch and 13.6±2.0 SD days (n = 16, median 14) for the small tree finch. Thus, for all nests, the date when incubation began was defined as the hatching date minus 14 days. For nests that had been found before incubation started and were abandoned during incubation, the first observation of incubation was used as starting date of incubation.

All monitored nests were collected in separate sealed plastic bags after nesting failure or after the chicks fledged. Eggs found in failed nests were inspected to identify the developmental stage of the embryo (undeveloped: no signs of developed embryo, developed embryo: first tissues of small embryo visible, and developed embryo close to hatching: embryo showing feathers and nest monitoring suggested end of incubation). Nests were later dismantled in the laboratory in order to count *P. downsi* larvae, pupae and empty puparia. *P. downsi* intensity per nest was defined as the total number of *P. downsi* specimens per nest (after [25]). *P. downsi* intensity per chick was calculated by dividing *P. downsi* intensity per nest by the number of chicks of a given nest. However, as brood size was not known for all nests in 2010, *P. downsi* intensity per nest was used for species and year comparisons of *P. downsi* infestation. We are well aware that brood size plays an important role as the detrimental impact of *P. downsi* parasitism increases with decreasing brood size, but (1) we are interested in the general infestation pattern of the two species over the years and (2) there is no difference in clutch sizes between the two species (warbler finch: 2.43±0.5, mean ± SD, n = 56; small tree finch: 2.46±0.5, n = 48; Mann-Whitney test: $Z_{102} = -0.303$, $p = 0.844$) and (3) there is no indication that clutch size differs between the two breeding seasons. As the number of parasites can only be counted by destroying the nest, parasite abundance was assigned to the age of the chicks at the time breeding activity terminated at a given nest. In our data parasite intensities increase with chick age (but see [20]) because of multiple infections of nests with *P. downsi* [40]. This leads to the contradictory finding that successful nests have the highest *P. downsi* intensities. Since we cannot asses *P. downsi* intensity of successful nests prior to fledging and thus cannot compare them to failed nests of the same age, *P. downsi* intensity could not predict breeding success and thus was excluded from further analysis.

Habitat parameters and daily rainfall

In 2012 we estimated nesting height, maximum and average canopy height, as well as vegetation cover of herb-layer, bush-layer (including *R. niveus*), *R. niveus* separately and canopy-layer (vegetation>3 m), of the surrounding area (5 m radius) for each active nest encountered. Height levels were estimated in one-meter intervals and vegetation covers in 10 percent intervals. Furthermore, the nesting location was assigned to (i) *Rubus* controlled sites (area of 5 m radius around the nest,still showed clear signs of herbicide usage) or (ii) not controlled sites (no signs of herbicide usage detected, usually dense understory).

Data of daily rainfall were provided by the Charles Darwin Station from the meteorological station near Santa Rosa (0°39′16,45″ S, 90°24′12,96″ W, elevation 500 m a.s.l., about 3.5 km from our study site) for 2012. Within the relevant study period of 101 days in 2012 (date of starting incubation at the first nest to the date of the last nest inspection), it rained more than 10 mm per day on 26 days. These days were classified as "heavy rain days".

Estimation of rat predation

In March 2012, we conducted an artificial nest experiment to identify potential nest predators and estimate nest predation rates following the method developed by Fessl et al. [10]. The artificial nests were made from coconut fibre. Each nest contained two Plasticine eggs (Acrilex modelling clay) resembling finch eggs dipped in egg albumen, which were tied to the nest. All nests and eggs were handled with surgical gloves to avoid them being tainted with human odour. A total of 30 artificial nests were systematically placed along two transects (15 nests per transect), with a minimum distance of at least 30 m between nests, and about 3 m height, mainly in *Scalesia* trees. To examine the effect of *Rubus* control (understory removal), one transect was located in a controlled area with a 5 m bush cover around the artificial nest of 21%±27 (mean ± SD), whilst the other was in a non-controlled area with a bush cover of 87%±15 (mean ± SD). Nests were revisited after 3, 6 and 9 days before being removed. Eggs with possible signs of predation were removed from nests and replaced with new Plasticine eggs, resulting in a total of 180 possible egg predation events. Traces on artificial eggs were compared to marks left by a rat's dental impression or by a bird's beak and assigned to (1) rats, if clear rat tooth marks were present, (2) birds, if single punctures were visible, or (3) unidentified. We calculated the percentage of predated individual nests over the whole 9-day period for each transect. Furthermore, we measured the number of predated eggs and, of these eggs, calculated the percentage that had been predated by rats.

Analysis

As breeding pairs did re-nest within one breeding season, especially after a breeding failure, it is likely that we monitored different nests from the same breeding pair. Nests found in close proximity of a monitored nest, where a breeding attempt had recently terminated, were considered to belong to the same

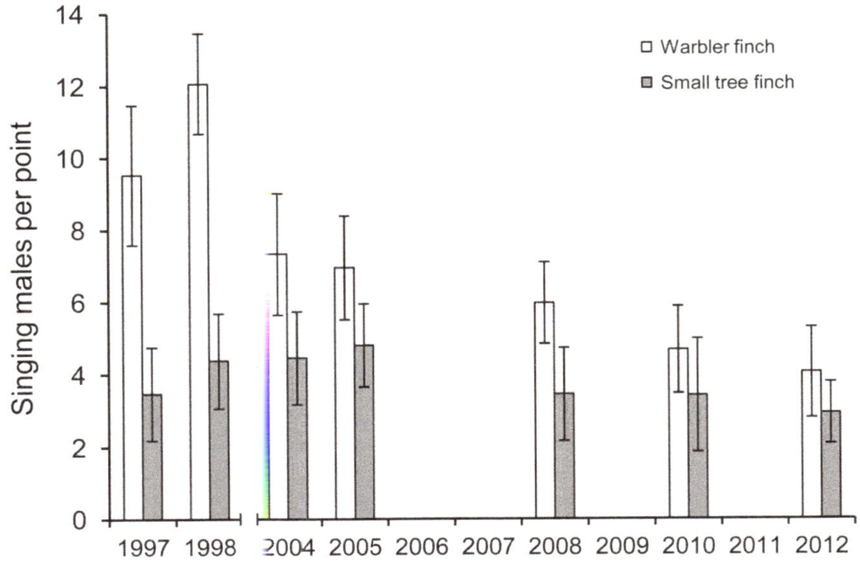

Figure 1. Population trends of warbler finches and small tree finches. Mean (± SD) number of singing warbler finch males and small tree finch males per point count in the *Scalesia* zone on Santa Cruz, Galápagos, for the years 1997, 1998, 2008, 2010 (data from [27]), 2004 and 2005 (Dvorak et al. unpublished data) and 2012.

breeding pair. To reduce such pseudo-replication, only one randomly chosen nest per breeding pair was included in the analysis. We monitored a total of 204 active nests and excluded 13 nests because of possible re-nesting, resulting in a dataset of 191 nests. In 2010 we monitored 30 warbler finch nests and 32 small tree finch nests, and in 2012, 74 warbler finch nests and 55 small tree finch nests were observed. To assess variation between years, we conducted a separate analysis for breeding success and *P. downsi* infestation for 2010 and 2012.

To assess population trends for both species, non-parametric correlations and t-tests were calculated to compare bird numbers over specific years. To compare different distributions (e.g. age structure of the small tree finch population at different time periods, species differences in breeding success, breeding success and predation of artificial nests in *Rubus* controlled compared with uncontrolled sites), we either computed Chi^2-tests or Fisher exact tests (when the sample size was too small for Chi^2-tests).

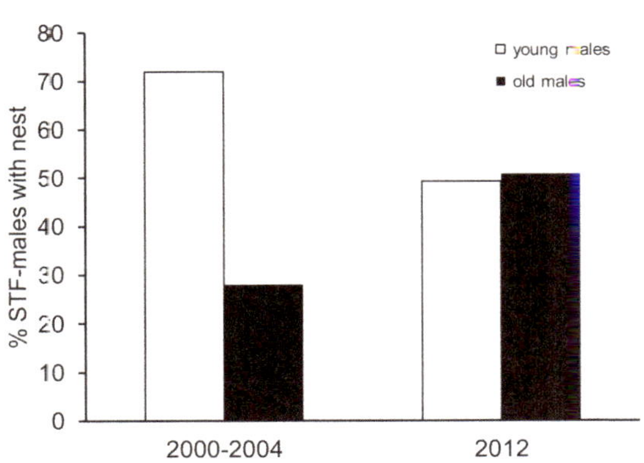

Figure 2. Age structure of the small tree finch population. Percentage of young and old displaying small tree finch (STF) males of the *Scalesia* zone population in 2000–2004 (n = 132, [36]) and 2012 (n = 63) on Santa Cruz, Galápagos.

To compare *P. downsi* abundance between species and years, we calculated a two-way ANOVA. Species differences in *P. downsi* intensity between age classes were tested with two-tailed Student t-tests. To make the data set comparable to previous studies, we distinguished two age classes according to Dudaniec et al. [20]. For the multivariate analyses of the breeding success in 2012, we constructed Generalized Linear Models (GLMs) with binomial error structure and a logit-link function. Successful nests were defined as nests with at least one fledged young, while all other nesting attempts were counted as failures. Since stepwise methods may produce inflated Type I errors, we present full models with all variables entered [41] and reduced models that retained only the significant predictor variables. To avoid multi-colinearity [42], we chose among predictor variables that were highly correlated (Spearman rank correlation $r > 0.4$), the one that seemed to be of higher biological relevance. Predicted variables were breeding success (fledged young or not), predictor variables were height of the nest (m), cover of canopy (%), the day-of-year of the start of incubation (ordinal date), and the percentage of heavy rain days during the incubation and nestling period. We tested the effect of *Rubus* control on the breeding success separately using Chi^2-tests because of inter-correlation with other habitat variables and low sample size. Statistical analyses were done with R 3.0.1 (R Development Core Team 2013).

Ethics statement

The study was conducted in the protected areas of the Galápagos National park. Permission to conduct this study was granted by the Galápagos National Park and the Charles Darwin research station (Project: PC-54-11, Permit Nr. PR.CDS.A-CI.P01.R02). As our study was purely descriptive, strictly non-invasive, and based exclusively on behavioural observations, they are classified as non-animal experiments in accordance with the Austrian Animal Experiments Act (§ 2. Federal Law Gazette No. 501/1989).

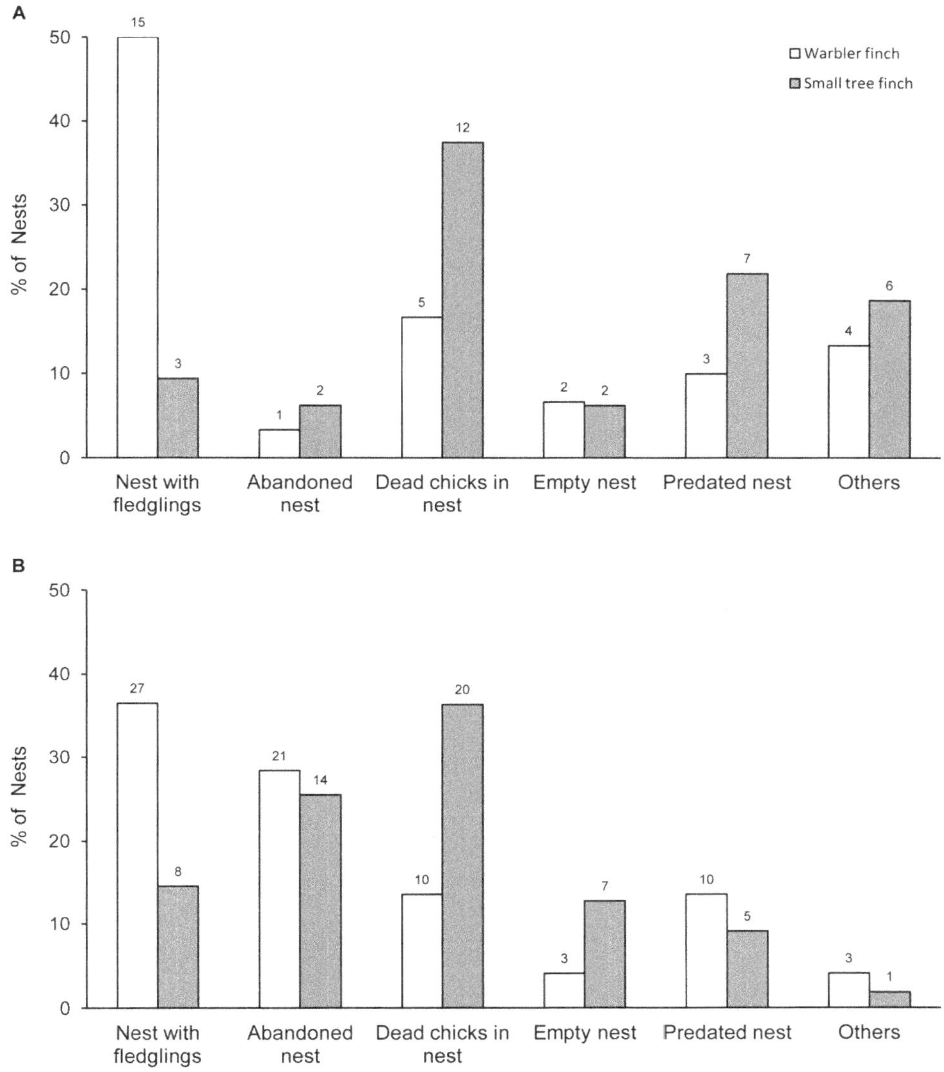

Figure 3. Breeding success and types of nesting failure. Proportional nesting outcome for the breeding season 2010 (A) and 2012 (B) of warbler finch nests and small tree finch nests of the *Scalesia* zone population on Santa Cruz, Galápagos. Numbers above bars indicate total numbers of cases.

Results

Population survey

Point counts showed that the warbler finch population decreased significantly over the last 15 years (Spearman's $Rho_5 = -0.964$, $n = 7$, $p<0.01$, data from [27] and Dvorak et al. unpublished data). Compared to 1997, the number of singing males dropped significantly by 58% (two tailed Student t-test: $n_{1997} = 17$, $n_{2012} = 26$, $t_{41} = 11.318$, $p<0.001$; Figure 1). In contrast to the warbler finch, the small tree finch population did not decrease significantly over the whole period from 1997 to 2012 (Spearman's $Rho_5 = -0.643$, $n = 7$, $p = 0.12$). Instead, there was an initial significant increase from 1997 to 2005 by 38% (two-tailed Student t-test: $n_{1997} = 17$, $n_{2005} = 20$, $t_{35} = 3.324$, $p = 0.002$, data from [27] and Dvorak et al. unpublished data) followed by a significant decrease by 39% until 2012 (two-tailed Student t-test: $n_{2005} = 20$, $n_{2012} = 26$, $t_{44} = 6.380$, $p<0.001$, Figure 1). Additionally, the age structure of the small tree finch population significantly changed between 2000–2004 [36] and 2012 (Chi^2-test: $\chi^2_{1,\ n=195} = 10.396$, $p = 0.001$; Figure 2). The population from 2000–2004 consisted of 72% "young" males and 28% "old" males, whereas in 2012, the proportion of "old" males increased to 51%. However, this comparison is based on pooled data. Unfortunately no data on yearly age structure from 2000–2004 is available and thus, the observed difference might be still within the year to year variation.

Breeding success

In 2010, at least one chick fledged in 9% of the small tree finch nests, compared to 50% of the warbler finch nests (Figure 3a). The breeding success of small tree finches in 2012 was 16% compared to 37% of the warbler finches (Figure 3b). Breeding success was significantly higher for warbler finches than small tree finches in both years (Chi^2-test, 2010: $\chi^2_{5,\ n=62} = 12.403$, $p = 0.001$; 2012: $\chi^2_{5,\ n=129} = 7.682$, $p = 0.009$; corrected α-level after Bonferroni: $\alpha = 0.025$) and did not differ between the two years in both species (Chi^2-test, warbler finch: $\chi^2_{1,\ n=104} = 1.619$, $p = 0.270$; Fisher

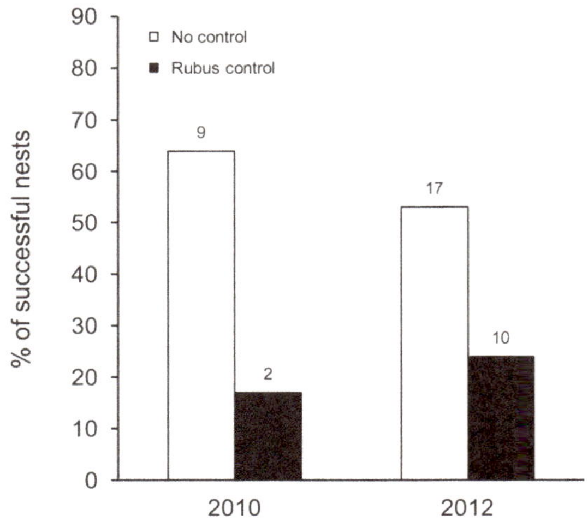

Figure 4. Effect of herbicide use on the breeding success of warbler finches. Percentage of successful warbler finch nests in areas with no control measures by the National Park (2010: n = 14, 2012: n = 32) and in areas where the National Park recently sprayed herbicides to control the invasive *Rubus niveus* (2010: n = 12, 2012: n = 42). Numbers above bars indicate total numbers of cases.

exact test, small tree finch: 3 out of 32 successful nests in 2010 and 8 out of 55 in 2012, p = 0.739).

Breeding success was significantly influenced by *Rubus* control measures. In areas where the National Park had sprayed herbicides, breeding success of warbler finches was significantly lower than in areas that were not managed in both years (Chi^2-test:, 2010: $\chi^2_{1,\ n=26} = 6.003$, p = 0.021; 2012: $\chi^2_{1,\ n=74} = 6.735$, p = 0.014, Figure 4). In small tree finches, 18% (n = 20) of the nests in the not controlled and none of the 15 nests of the controlled area were successful in 2010 and 11% (n = 21) of the nests in the not controlled and 17% (n = 42) of the nests in the controlled area were successful in 2012. Due to the overall low breeding success, a statistical analysis was not run for the small tree finch.

For both species, the majority of failed breeding attempts were assigned to the following three categories: "abandoned nest", "predated nest" and "dead chicks in the nest" (Figure 3). However, reasons for failed breeding attempts differed between years (warbler finch: $\chi^2_{5,\ n=104} = 12.253$, p = 0.029, small tree finch: $\chi^2_{5,\ n=87} = 15.253$, p = 0.007). The post-hoc analysis revealed that this difference was due to a higher proportion of abandoned nests in 2012 (warbler finch: $\chi^2_{1,\ n=104} = 8.188$, p = 0.007, tendency in small tree finch: $\chi^2_{1,\ n=87} = 4.813$, p = 0.042, all other post-hoc comparisons n.s., corrected α-level after Bonferroni: $\alpha = 0.017$).

In 2012, most of the small tree finch nests that lost their total brood did so before the chicks reached 7 days (71%, Figure 5). In the warbler finch, two distinct age classes showed the highest percentage of nests with total brood loss (Figure 5): 38% failed when chicks were 1–3 days old and 33% failed when chicks were 7–9 days old. For both species, the main categories of breeding failure in the age class 1–3 days was "dead chicks in the nest" and "empty nest" (18 nests out of 20).

Abandoned nests

In 2012, 21 warbler finch nests and 13 small tree finch nests were abandoned before the chicks hatched, thus still containing eggs (Figure 3). We found undeveloped eggs in 33% of abandoned warbler finch nests and in 19% of small tree finch nests, and chicks close to hatching, which did not manage to hatch, in 19% of abandoned warbler finch nests and in 15% of small tree finch nests. In both species, abandoned nests experienced a significantly higher percentage of heavy rain days during incubation than those in which chicks hatched (two tailed Student t-test: warbler finch $t_{67} = 2.698$, p = 0.009; small tree finch $t_{51} = 2.129$, p = 0.038).

Predation and artificial nest experiment

Predation as a cause for nesting failure was observed in 22% of small tree finch nests in 2010 and 9% in 2012 as well as 10% of warbler finch nests in 2010 and 14% in 2012 (Figure 3). Potential predators observed in the study area were the short-eared owl (*Asio flammeus galapagoensis*, endemic), the smooth-billed ani

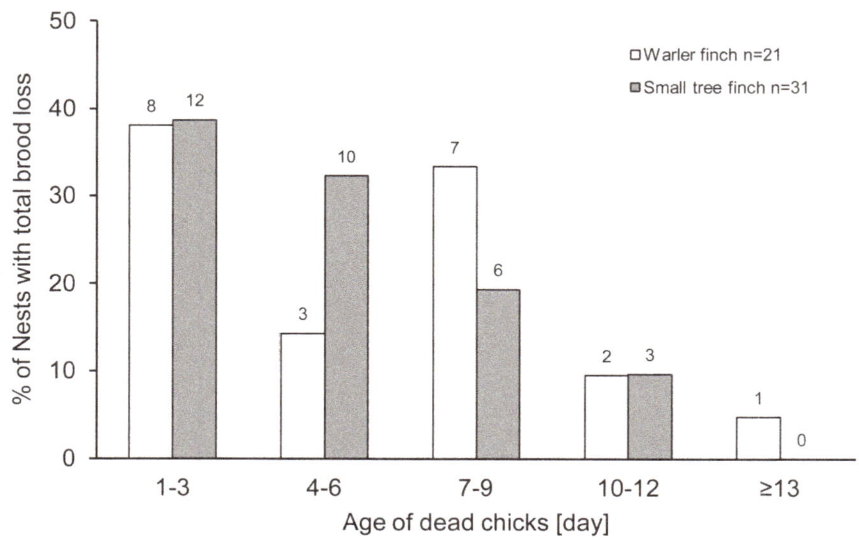

Figure 5. Age of chicks at brood loss. Percentage of nests with total brood loss depending on the chicks' age. Warbler finch (n = 21), small tree finch (n = 31). Total brood loss included the following types of nesting failures: dead chicks in the nest, empty nest and predated nests. Numbers above bars indicate total numbers of cases.

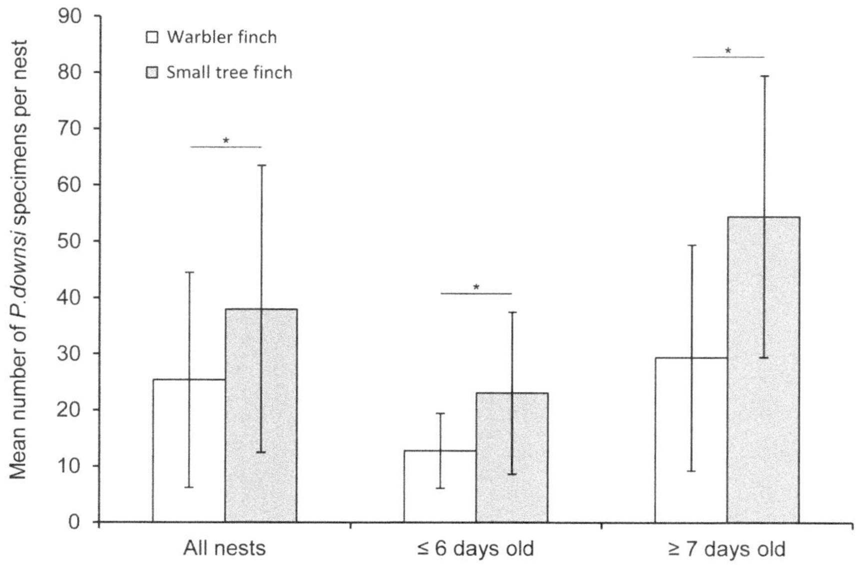

Figure 6. *Philornis downsi* intensity in warbler finch and small tree finch nests. Mean (± SD) number of *P. downsi* specimens (larvae, pupae and puparia) per nest of the breeding season 2012 of warbler finches and small tree finches for all nests with chicks (warbler finch n = 44, small tree finch n = 38), nests with chicks of six days and younger (warbler finch n = 11, small tree finch n = 20) and nests with chicks of seven days and older (warbler finch n = 33, small tree finch n = 18).

(*Crotophaga ani*, introduced), and the black rat (*Rattus rattus*, introduced). The artificial nest experiment revealed similar results. Over the whole experimental period (9 days), eggs from 7 of the 30 artificial nests (23%) showed signs of predation at least once. There was no difference in predation rate between the *Rubus* controlled area (3 of 15 predated artificial nests) and the non-controlled area (4 of 15 predated artificial nests, Fisher exact test: p = 1.000). In total, 20 Plasticine eggs of the possible 180 eggs showed signs of predation. Of these eggs, 18 had clear rat tooth marks.

Nesting failure and *P. downsi* infestation

A high percentage of nests contained dead chicks, though they showed no signs of predation (2010: warbler finch 17%, small tree finch 38%; 2012: warbler finch 14% and small tree finch 30% of the nests; Figure 3). Since all of these nests suffered high parasite loads and no other cause of death was found, we attributed these nesting failures to *P. downsi* parasitism. If we include the fledging nests that also contained dead chicks (partial brood loss), then 56% of small tree finch chicks and 37% of warbler finch chicks most likely died due to *P. downsi* parasitism in 2012 (data for 2010 not available).

We found 100% prevalence of *P. downsi* for nests which contained nestlings. However, the two finch species were affected differently. Small tree finches, the species with significantly lower breeding success, had significantly more *P. downsi* specimens per nest than the warbler finch (two-way ANOVA: species, $F_{1,125} = 7.349$, $p = 0.008$). There was no year effect ($F_{1,125} = 1.912$, $p = 0.169$) nor an interaction of year and species ($F_{1,125} = 0.11$, $p = 0.916$).

Because warbler finches had a higher breeding success than the small tree finch, mean nestling age was higher, too. As *P. downsi* intensity increases with nestling age [40], this could bias the data. Thus, we analysed nests with chicks of 6 days and younger separately to reduce the effect of nestling age. In 2012, significant differences in *P. downsi* intensity were already present in nests with chicks of 6 days and younger (Student t-test: *P. downsi*/nest, $t_{29} = -2.222$, $p = 0.034$; *P. downsi*/chick, $t_{29} = -2.631$, $p = 0.014$, Figure 6). They were also still present in nests with chicks of 7 days and older (two tailed Student t-test: *P. downsi*/nest, $t_{50} = -3.934$, $p<0.001$; *P. downsi*/chick, $t_{50} = -4.295$, $p<0.001$, Figure 6).

Because of the 100% prevalence of *P. downsi* larvae, we are not able to test whether a lower infection rate at early nestling stage results in a higher probability of a successful nest. However, parasitism may not be the only reason for breeding failure, and so we analysed whether other factors may also influence breeding success. In 2012, we found that the only factor to negatively influence breeding success significantly in both species was the percentage of heavy rain days during the nestling period (full and reduced model, Table 1). In warbler finches breeding success also increased later in the season (full and reduced model, Table 1).

Discussion

In summary, breeding success was decreased by *P. downsi* parasitism and heavy rain days in warbler finches and small tree finches while predation had only a minor influence. Additionally, the use of herbicides on the surrounding habitat had a significant negative effect on the reproductive performance of the warbler finch.

Although the direct cause of death could not be identified, it is very likely that parasitism by *P. downsi* played a major role in the death of 56% of small tree finch chicks and 37% of warbler finch chicks. Indirect evidence for the role of *P. downsi* comes from the comparison of breeding success in relation to parasite intensity between the small tree finch and the warbler finch: the nests of the small tree finch, the species which had significantly lower breeding success, had significantly higher *P. downsi* abundance. This difference was already present in the early nestling phase, in which mortality was particularly high. However, only experimentally manipulating parasite intensity can reveal the impact of *P. downsi* on mortality. In an experimental study on Darwin's ground finches where parasites were eliminated in the nests, mortality was reduced from 66% to only 14% [15] (but see [21]). In a similar

Table 1. Variables that explained variation in breeding success of warbler finches and small tree finches in 2012 by GLMs with a binomial error structure.

Dependent variable: Breeding success		Warbler finch (n = 44)				Small tree finch (n = 33)			
	Predictors	Estimate	SE	z	P	Estimate	SE	z	P
Full model	Intercept	−4.79	4.07	−1.18	0.24	−2.94	10.54	−0.28	0.78
	% Heavy rain days in nestling phase	**−0.21**	**0.07**	**−3.15**	**0.002**	**−0.21**	**0.10**	**−2.19**	**0.03**
	% Heavy rain days in incubation phase	6.41	4.18	1.53	0.13	3.32	8.93	0.37	0.71
	Nest height	0.22	0.40	0.54	0.59	−0.12	0.32	−0.37	0.71
	Canopy cover	0.04	0.02	1.64	0.10	0.01	0.03	0.34	0.73
	Date of incubation start	**0.13**	**0.06**	**2.06**	**0.04**	0.14	0.19	0.75	0.46
Reduced model	Intercept	3.05	1.54	1.98	0.05	3.05	1.54	1.98	0.05
	% Heavy rain days in nestling phase	**−13.73**	**4.58**	**−2.99**	**0.003**	**−14.21**	**6.02**	**−2.36**	**0.02**
	Date of incubation start	**0.08**	**0.03**	**2.33**	**0.02**	-	-	-	-

Significant variables are marked in bold.

study by Koop et al. [18], mortality decreased from 96% to 67% as parasite intensity was reduced.

The higher parasite loads and lower breeding success of small tree finch nests compared with warbler finch nests were contradictory to our predictions and to previous findings, which report higher *P. downsi* intensities in the warbler finch than in the small tree finch [17,20]. In the small tree finch, the mean *P. downsi* intensity per chick (26.0±3.1, n = 18) for nests with nestlings of ≥6 days old was also higher than previously reported (mean *P. downsi*/chick: 20.5±2.3, n = 29 [20]). However, these comparisons are based on pooled data from 1998 to 2005 [20]. Since the Galápagos Islands are subject to massive environmental variation between years, and breeding density as well as parasite intensity might be affected by that, the average could smooth out annual variability. The pooled data from Dudaniec et al. [20] included one El Niño year (1998) and five years with low precipitation. Our data were collected in very humid years, which could explain the reversed infestation pattern. Findings on the relationship between parasite intensity, precipitation and host density are mixed (reviewed in [21]) and thus this relationship is not well understood so far.

The low breeding success of small tree finches may additionally be explained by limiting ecological conditions such as availability of suitable prey for chick rearing. In Darwin's ground finches, gonadal activity is triggered by rain [43] as insect abundance starts to increase about 10 days after the first rains on the Galápagos Islands [22]. It seems plausible that warbler finches and small tree finches similarly start to breed when the rain starts, but that they depend on prey [44] with different phenology. Preliminary data on breeding phenology suggests that small tree finches start breeding later and are more synchronic than the opportunistic warbler finches. Future studies focusing on breeding phenology and detailed foraging observations may highlight some key factors, such as special food sources, which have been overlooked so far.

In 2012, a high percentage of nests (small tree finch 25%, warbler finch 28%) were abandoned during incubation (before the chicks hatched). Studies on Darwin's finches on several other islands have previously reported much lower values (e.g. [12]: 1.8–6.7 %; [36]: 7%; [10]: up to 10%). However, O'Connor et al. [45] found 30.4% (data of 2004 and 2006 combined) of small ground finch nests abandoned in the lowlands, and in 2005, 25% abandoned nests in the highlands of Floreana Island. For 2005, it was suggested that extreme drought conditions lead to this high percentage of abandoned nests. Our data suggest that nest abandonment during the incubation stage may also be increased by heavy rain days. Intense rains may negatively affect nest temperature and humidity, thus leading to nest abandonment. This may be linked to higher energy costs for the incubating female, since clutches will cool more rapidly in a wet nest (reviewed in [46]). Thus, in both cases, extreme weather conditions are related to abandonment of the nest in the incubating stage. Heavy rains also affected breeding success of both species during chick rearing. During intense rain, parents may be less efficient in foraging or even stop searching for food as insects are more difficult to find. This temporary lower food abundance probably leads to lower feeding rates by the parents and thus poses stress on chicks, especially those that are very young and require regular feeding. This hypothesis should be investigated in future studies.

Rubus control with herbicides also negatively affected breeding success. Warbler finches had significantly lower breeding success in areas where herbicides had recently been applied compared with uncontrolled areas. The use of strong herbicides by the National Park to control the invasion of *R. niveus* leads to removal of more or less the entire understory (Cimadom personal observation). This may have lead to changes in abundance and composition of invertebrate species, which could negatively affect chick rearing, especially in insectivorous bird species. However, more data on chick food is needed, as it is very likely that more vegetarian bird species also depend on invertebrates for chick rearing. Additionally, the likely decrease in invertebrate abundance triggered by the use of herbicides might also negatively affect adult survival of insectivorous bird species, especially during the dry season, when arthropods are less common in general. The fact that four of the six declining species are insectivorous [27] supports this notion. It has already been shown that the use of herbicides especially glyphosate has a negative effect on abundance of several bird species [47,48] and Betts et al. [49] found that especially leaf-gleaning bird species declined after herbicide treatment. Specifically open-cup nesting species had a significantly lower breeding success in herbicide-treated forest areas than in manually thinned areas [50]. However some management to restore the highly endangered *Scalesia* forest is indispensable: The high cover of the invasive *Rubus niveus* has lead to significantly lower native plant species richness and cover, as well as to changes in the *Scalesia* forest structure [51] which is likely to affect flora and fauna. The restoration of the *Scalesia* forest is an example of the conservation of a highly fragile ecosystem. Restoration measures might be positive for some species but negative for others. Thus, it is important to consider the whole ecosystem when assessing the costs and benefits of different management strategies.

The effect of predation on breeding success was lower than expected and lower than in previous studies from Santa Cruz (pooled data from 1998, 2000, 2001, 2002, 2004 and 2005, 50% of small tree finch and 37.1% of warbler finch nests were predated, [20,36,52]). These findings could either be attributed to differing definitions of depredated nests or to a decline in the main predators (rats and short eared owls). Possibly, these predators are affected directly or indirectly by intense plant control with herbicides. The results of the artificial nest experiment support the notion that predation in the *Scalesia* forest was low.

Conclusions and future plans

We found three factors that negatively influenced the breeding success of both finch species: *P. downsi* parasitism, habitat change due to control of invasive plant species and adverse climatic conditions (days of heavy rain). We hypothesize that the combination and interaction of these factors, rather than one single factor, leads to breeding failures. For instance, the negative role played by *P. downsi* parasitism may be increased by occurrence of intense rainfall (natural stressor) and major habitat changes, leading to low breeding success. While parents might be able to compensate for energy losses due to parasitism under optimal conditions, they might be unable to overcome the negative effect of parasitism when additional stressors (e.g. heavy rain events, *Rubus* control) come into play. This hypothesis should be investigated in future studies on breeding success, for example by selectively manipulating parasite abundance through use of insecticides in areas with invasive plant species control and in uncontrolled areas. These data could then be combined with daily weather conditions.

It is still unclear whether the low breeding success measured in this study can explain the dramatic decline of the warbler finch population in the *Scalesia* forest on Santa Cruz (down 58% compared to 1997), as reliable population models are still missing. To date, there is no data on the age structure, juvenile survival, and dispersal behaviour of the warbler finch. Future studies should aim to collect this relevant information in order to be able to develop meaningful population models, which are helpful tools for evaluating possible conservation measures. In small tree finches, our data on a change in age structure and the extremely low breeding success in 2010 and 2012 go in parallel with a significant population decrease since 2005. Whether this decrease is due to natural fluctuation in population size or is the result of parasitism and habitat change needs to be assessed in future studies. Furthermore, more information regarding breeding ecology and foraging behaviour are needed to further explore the relationship between the three identified factors and their negative influence on breeding success.

Acknowledgments

We thank Sophia Stankewitz for field assistance and Odessa Madakacherry and Gesche Westphal-Fitch for valuable comments on the manuscript. Permission for this study was granted by the Galápagos National Park. We thank the Charles Darwin Station for logistical support.

Author Contributions

Conceived and designed the experiments: AC AU MD BF ST EZ MZ PM. Performed the experiments: AC AU EZ MZ MD BF FC ST PM. Analyzed the data: AC EN ST MD. Contributed reagents/materials/analysis tools: PM FC. Wrote the paper: AC ST EN FC BF.

References

1. Cleaveland S, Hess GR, Dobson AP, Laurenson MK, McCallum HI, et al. (2002) The role of pathogens in biological conservation. In: Hudson PJ, Rizzoli A, Grenfell BT, Heesterbeek H, Dobson AP: The Ecology of Wildlife Diseases. Oxford University Press, Oxford, pp 139–150.
2. Wikelski M, Foufopoulos J, Vargas H, Snell H (2004) Galápagos birds and diseases: Invasive pathogens as threats for island species. Ecol Soc 9: 5 [online: http://www.ecologyandsociety.org/vol9/iss1/art5].
3. De Castro F, Bolker B (2005) Mechanisms of disease-induced extinction. Ecol Lett 8: 117–126.
4. Warner RE (1968) The role of introduced diseases in the extinction of the endemic Hawaiian avifauna. Condor 70: 101–120.
5. Steadman DW (1995) Prehistoric extinctions of Pacific island birds: biodiversity meets zooarchaeology. Science 267: 1123–1131.
6. Thiel T, Whiteman NK, Tirapé A, Baquero MI, Cedeno V, Walsh T, Jiménez-Uzcátegui G, Parker PG (2005) Characterization of canarypox-like viruses infecting endemic birds in the Galápagos Islands. J Wildl Dis 41: 432–453.
7. Kleindorfer S, Dudaniec R (2006) Increasing prevalence of avian poxvirus in Darwin's finches and its effect on male pairing success. J Avian Biol 37: 69–76.
8. Deem SL, Cruz M, Jiménez-Uzcátegui G, Fessl B, Miller RE, Parker PG (2008) Pathogens and parasites: an increasing threat to the conservation of Galápagos avifauna. In: Galápagos Report 2007–2008. Puerto Ayora, Galápagos, Ecuador. pp. 125–130.
9. Wiedenfeld DA, Jiménez-Uzcátegui G (2008) Critical problems for bird conservation in the Galápagos Islands. Cotinga 29: 22–27.
10. Fessl B, Young HG, Young RP, Rodriguez-Matamoros J, Dvorak M, Tebbich S, Fa JE (2010) How to save the rarest Darwin's finch from extinction: the mangrove finch on Isabela Island. Philos Trans R Soc Lond B Biol Sci 365: 1019–1030.
11. Causton CE, Peck SB, Sinclair BJ, Roque-Albedo L, Hodgson CJ, Landry B (2006) Alien insects: threats and implications for the conservation of the Galápagos Islands. Ann Entomol Soc Am 99: 121–143.
12. Fessl B, Tebbich S (2002) *Philornis downsi* – a recently discovered parasite on the Galápagos archipelago – a threat for Darwin's finches? Ibis 144: 445–451.
13. Fessl B, Sinclair BJ, Kleindorfer S (2006) The life cycle of *Philornis downsi* (Diptera: Muscidae) parasitizing Darwin's finches and its impacts on nestling survival. Parasitology 133: 739–747.
14. Dudaniec R, Kleindorfer S, Fessl B (2006) Effects of the introduced ectoparasite *Philornis downsi* on haemoglobin level and nestling survival in Darwin's Small Ground Finch (*Geospiza fuliginosa*). Austral Ecol 31: 88–94.
15. Fessl B, Kleindorfer S, Tebbich S (2006) An experimental study on the effects of an introduced parasite in Darwin's finches. Biological Conservation 127: 55–61.
16. Huber SK (2008) Effects of the introduced parasite *Philornis downsi* on nestling growth and mortality in the medium ground finch (*Geospiza fortis*). Biological Conservation 141: 601–609.

17. Kleindorfer S, Dudaniec R (2009) Love thy neighbour? Social nesting pattern, host mass and nest size affect ectoparasite intensity in Darwin's finches. Behav Ecol Sociobiol 63: 731–739.
18. Koop JAH, Huber SK, Laverty SM, Dale HC (2011) Experimental demonstration of the fitness consequences of an introduced parasite of Darwin's finches. PLoS ONE 6: e19706, doi: 10.1371/journal.pone.0019706.
19. O'Connor JA, Sulloway FJ, Robertson J, Kleindorfer S (2010) *Philornis downsi* parasitism is the primary cause of nestling mortality in the critically endangered Darwin's medium tree finch (*Camarhynchus pauper*). Biodivers Conserv 19: 853–866.
20. Dudaniec R, Fessl B, Kleindorfer S (2007) Interannual and interspecific variation in intensity of the parasitic fly, *Philornis downsi*, in Darwin's finches. Biological Conservation 139: 325–332.
21. Koop JAH, Le Bohec C, Dale HC (2013) Dry year does not reduce invasive parasitic fly prevalence or abundance in Darwin's finch nests. Reports Parasitol 3: 11–17.
22. Grant BR, Grant PR (1989) Evolutionary dynamics of a natural population: the large cactus finch of the Galápagos. Chicago: University of Chicago Press.
23. Arendt WJ (2000) Impact of nest predators, competitors, and ectoparasites on Pearly-eyed thrashers, with comments on the potential implications for Puerto Rican Parrot recovery. Ornitol Neotrop 11: 13–63.
24. Antoniazzi LR, Manzoli DE, Rohrmann D, Saravia MJ, Silvestri L, Beldomenico PM (2011) Climate variability affects the impact of parasitic flies on Argentinean forest birds. J Zool 283: 126–134.
25. Bush AO, Lafferty KD, Lotz JM, Shostak AW (1997) Parasitology meets ecology on its own terms: Margolis, et al. revisited. J Parasitol 83: 575–583.
26. Wiedenfeld DA, Jimenez U, Gustavo A, Fessl B, Kleindorfer S, Valarezo JC (2007) Distribution of the introduced parasitic fly *Philornis downsi* (Diptera, Muscidae) in the Galápagos Islands. Pac Conserv Biol 13: 14–19.
27. Dvorak M, Fessl B, Nemeth E, Kleindorfer S, Tebbich S (2012) Distribution and abundance of Darwin's finches and other land birds on Santa Cruz Island, Galápagos: evidence for declining populations. Oryx 46: 78–86.
28. P.R. Grant PR, Grant BR, Petren K, Keller LF (2005) Extinction behind our backs: the possible fate of one of the Darwin's finch species on Isla Floreana, Galápagos. Biological Conservation 122: 499–503.
29. Mauchamp A, Atkinson R (2010) Pérdida de habitat rábida reciente e irreversible: los bisques de Scalesia en las islas Gálapagos. In: Informe Galápagos 2009–10 Puerto Ayora, Galápagos, Ecuador. pp. 111–114.
30. Rentería JL, Buddenhagen CE (2006) Invasive plants in the *Scalesia pedunculata* forest at Los Gemelos, Santa Cruz, Galápagos. Noticias de Galápagos - Galápagos Research 64: 31–35.
31. Jäger H, Tye A, Kowarik I (2007) Tree invasion in naturally treeless environments: Impacts of quinine (*Cinchona pubescens*) trees on native vegetation in Galapagos. Biological Conservation 140: 297–307.
32. Towns DR, Atkinson IAE, Daugherty CH (2006) Have the harmful effects of introduced rats on islands been exaggerated? Biol Invasions 8: 863–891.
33. Jiménez-Uzcátegui G, Carrión V, Zabala J, Buitrón P, Milstead B (2007) Status of introduced vertebrates in Galapagos. In: Galápagos Report 2006–2007. Puerto Ayora, Galápagos, Ecuador. pp. 136–141.
34. Grant PR, Grant BR (1997) The rarest of Darwin's finches. Conserv Biol 11: 119–126.
35. Dvorak M, Vargas H, Fessl B, Tebbich S (2004) On the verge of extinction: a survey of the mangrove finch *Cactospiza heliobates* and its habitat on the Galápagos islands. Oryx 38: 1–9.
36. Kleindorfer S (2007) Nesting success in Darwin's small tree finch, *Camarhynchus parvulus*: evidence of female preference for older males and more concealed nests. Anim Behav 74: 795–804.
37. Galligan TH, Kleindorfer S (2009) Naris and beak malformation caused by the parasitic fly, *Philornis downsi* (Diptera: Muscidae), in Darwin's small ground finch, *Geospiza fuliginosa* (Passeriformes: Emberizidae). Biol J Linn Soc Lond 98: 577–585.
38. Dudaniec RY, Kleindorfer S (2006) The effects of the parasitic flies of the genus *Philornis* (Diptera: Muscidae) on birds. Emu 10: 13–20.
39. O'Connor JA, Robertson J, Kleindorfer S (2010) Video analysis of host-parasite interactions in nests of Darwin's finches. Oryx 44: 588–594.
40. Dudaniec R, Gardner M, Kleindorfer S (2010) Offspring genetic structure and nest infestation behaviour of an invasive parasitic fly (*Philornis downsi*) of Galápagos birds. Biol Invasions 12: 581–592.
41. Mundry R, Nunn CL (2009) Stepwise model fitting and statistical inference: turning noise into signal pollution. Am Nat 173: 119–123. doi: 10.1086/593303.
42. Tabachnick BG, Fidell LS (2007) Using Multivariate Statistics. 5th edition, Boston: Allyn and Bacon. 980p.
43. Hau M, Wikelski M, Gwinner H, Gwinner E (2004) Timing of reproduction in a Darwin's finch: temporal opportunism under spatial constraints. Oikos 106: 489–500.
44. Tebbich S, Taborsky M, Fessl B, Dvorak M, Winkler H (2004) Feeding behavior of four arboreal Darwin's Finches: adaptations to spatial and seasonal variability. Condor 106: 95–105.
45. O'Connor JA, Dudaniec RY, Kleindorfer S (2010) Parasite infestation and predation in Darwin's small ground finch: contrasting two elevational habitats between islands. J Trop Ecol 26: 285–292.
46. Heenan CB (2013) An overview of the factors influencing the morphology and thermal properties of avian nests. Avian Biol Res 6: 104–118.
47. Santillo DJ, Brown PW, Leslie DM (1989) Response of songbirds to glyphosate-induced habitat changes on clearcuts. J Wildl Manage 53: 64–71.
48. Stolseon SH, Ristau TE, de Calesta DS, Horsley SB (2011) Ten-year response of bird communities to an operational herbicide–shelterwood treatment in a northern hardwood forest. Forest Ecol Manag 262: 1205–1214.
49. Betts MG, Verschuyl J, Giovanini J, Stokely T, Kroll AJ (2013) Initial experimental effects of intensive forest management on avian abundance. Forest Ecol Manag 310: 1036–1044.
50. Easton WE, Martin K (1998) The effect of vegetation management on breeding bird communities in British Columbia. Ecol Appl 8: 1092–1103.
51. Rentería JL, Gardener MR, Panetta FD, Atkinson R, Crawley MJ (2012) Possible impacts of the invasive plant *Rubus niveus* on the native vegetation of the *Scalesia* forest in the Galápagos Islands. PLoS ONE 7: e48106. doi:10.1371/journal.pone.0048106.
52. Kleindorfer S, Sulloway FJ, O'Connor J (2009) Mixed species nesting associations in Darwin's tree finches: nesting pattern predicts predation outcome. Biol J Linn Soc Lond 98: 313–324.

Differential Responses of the Antioxidant System of Ametryn and Clomazone Tolerant Bacteria

Leila Priscila Peters[1], Giselle Carvalho[1], Paula Fabiane Martins[1], Manuella Nóbrega Dourado[1], Milca Bartz Vilhena[1], Marcos Pileggi[2], Ricardo Antunes Azevedo[1]*

1 Departamento de Genética, Escola Superior de Agricultura Luiz de Queiroz, Universidade de São Paulo, Piracicaba, Brazil, **2** Departamento de Biologia Estrutural, Molecular e Genética, Universidade Estadual de Ponta Grossa, Ponta Grossa, Brazil

Abstract

The herbicides ametryn and clomazone are widely used in sugarcane cultivation, and following microbial degradation are considered as soil and water contaminants. The exposure of microorganisms to pesticides can result in oxidative damage due to an increase in the production of reactive oxygen species (ROS). This study investigated the response of the antioxidant systems of two bacterial strains tolerant to the herbicides ametryn and clomazone. Bacteria were isolated from soil with a long history of ametryn and clomazone application. Comparative analyses based on 16S rRNA gene sequences revealed that strain CC07 is phylogenetically related to *Pseudomonas aeruginosa* and strain 4C07 to *P. fulva*. The two bacterial strains were grown for 14 h in the presence of separate and combined herbicides. Lipid peroxidation, reduced glutathione content (GSH) and antioxidant enzymes activities were evaluated. The overall results indicated that strain 4C07 formed an efficient mechanism to maintain the cellular redox balance by producing reactive oxygen species (ROS) and subsequently scavenging ROS in the presence of the herbicides. The growth of bacterium strain 4C07 was inhibited in the presence of clomazone alone, or in combination with ametryn, but increased glutathione reductase (GR) and glutathione S-transferase (GST) activities, and a higher GSH concentration were detected. Meanwhile, reduced superoxide dismutase (SOD), catalase (CAT) and GST activities and a lower concentration of GSH were detected in the bacterium strain CC07, which was able to achieve better growth in the presence of the herbicides. The results suggest that the two bacterial strains tolerate the ametryn and clomazone herbicides with distinctly different responses of the antioxidant systems.

Editor: Guillermo López Lluch, Universidad Pablo de Olavide, Centro Andaluz de Biología del Desarrollo-CSIC, Spain

Funding: This work was funded by grant 09/54676-0 from the Fundação de Amparo à Pesquisa do Estado de São Paulo to RAA. The authors thank Conselho Nacional de Desenvolvimento Científico e Tecnológico (RAA and MBV) and the Fundação de Amparo à Pesquisa do Estado de São Paulo (LPP, PFM, GC and MND) for the fellowship and scholarships granted. The funders had no role in study design, data collection and analysis, decision to publish, or preparation of the manuscript.

Competing Interests: Dr. Paula F. Martins current address is: Souza Cruz S/A, Unidade Cachoeirinha, Avenida Frederico Augusto Ritter, 8000, Industrial, 94930-000 - Cachoeirinha, RS - Brazil. Dr. Paula F Martins was not an employee of the company when the study was being conducted. The Company is not involved in any way with this study.

* Email: raa@usp.br

Introduction

Pesticides are powerful tools in modern agriculture to minimize economic losses caused by weeds, insects and diseases, and to ensure adequate food production [1,2]. It is estimated that nearly 3 billion tons of pesticides are released into the environment each year [3], more than 35% of which are herbicides [4]. The intensive use of these xenobiotics in agroecosystems can result in the contamination of water and soils [5–7].

The herbicides ametryn and clomazone are widely employed in crops such as sugarcane, soybeans, corn, cotton, and are often detected in the environment [8–11]. In plants, ametryn toxicity is related to the blockage of the electron transport chain binding specifically to D1 proteins of photosystem II, thereby preventing photosynthesis [12]. In contrast, the mode of action of clomazone consists of inducing lipid peroxidation in cells by blocking carotenoid synthesis [13,14]. These two herbicides in the soil can affect microbial activity and induce a selection pressure, which in turn allows the identification of tolerant microorganisms [15,16].

Previous studies have shown that many herbicides are redox-cycling agents able to alter the aerobic metabolism of microorganisms culminating in an oxidative stress condition [17–19]. This process induced by herbicides in bacteria results in the increased production and subsequent accumulation of reactive oxygen species (ROS), such as the superoxide radical ($O_2^{\bullet-}$), hydrogen peroxide (H_2O_2) and hydroxyl radical ($OH^{\bullet}$) [17]. These products of aerobic cell metabolism are toxic and may lead to enzyme inactivation, protein denaturation, lipid peroxidation and DNA mutation [20,21]. Therefore, any excess ROS that is produced has to be eliminated if a microbe is to survive [22].

Many bacteria can increase the rate of synthesis and accumulate non-enzymatic antioxidant compounds in response to excessive production of ROS (e.g. reduced glutathione (GSH) and ascorbic acid), as well as increase the activity of antioxidant enzymes [18,23]. The enzymes superoxide dismutase (SOD, EC 1.15.1.1) and catalase (CAT, 1.11.1.6) play crucial roles in the detoxification process of $O_2^{\bullet-}$ and H_2O_2, respectively [24–26]. Furthermore, the enzyme glutathione reductase (GR, EC 1.6.4.2) carries out the reduction of oxidized glutathione (GSSG), which is a fundamental reaction for maintaining the homeostasis between GSH/GSSG levels [27]. GSH is an antioxidant capable of directly neutralizing $OH^{\bullet}$ [28] and thus, is considered a key compound, in the stress tolerance process [29]. Glutathione S-transferase (GST, EC 2.5.1.18) is another important enzyme that is required for the degradation of pollutants, since it is primarily involved in cellular detoxification and redox biochemical mechanisms [30,31].

In this study we have examined the effects of the herbicides ametryn and clomazone on the antioxidant stress responses of two bacteria isolated from agricultural soils, previously treated with herbicides.

Materials and Methods

Ethics statement

The bacteria used in this work were isolated from soil samples collected in Fazenda Areão, Escola Superior de Agricultura Luiz de Queiroz (47°38′00″W; 22°42′30″S), Piracicaba, São Paulo State, Brazil. The location is an experimental area of the University and no specific permissions were required for sampling soils in this location. This field sampling did not involve or cause any harm to endangered or protected species.

Herbicides

Relevant characteristics of the two herbicides, ametryn (2-ethylamino-4-isopropylamino-6-methyl-thio-s-triazine) and clomazone (2-(2-chlorophenyl) methyl-4,4-dimethyl-3-isoxazolidinone), are listed in Table 1. Ametryn, which is a selective herbicide (Gesapax 500, Ciba Geigy) applied at 6 L ha^{-1} (3 kg ha^{-1}) to control narrow leaved weeds and broad leaved weeds, was used at 500 g L^{-1} (active ingredient-ai). Clomazone, which is a selective herbicide (Gamit 360 CS) that can be applied at a recommended dose of 1.8 L ha^{-1} (650 g ai ha^{-1}) in sugarcane and maize, was used at 360 g ai L^{-1}. However, ametryn and clomazone can be applied in combination in sugarcane at a recommended dose of 1 L ha^{-1} (300 g ai ha^{-1}), was used at 300 g ai L^{-1} and 1 L ha^{-1} (200 g ai ha^{-1}), was used at 200 g ai L^{-1}, respectively.

Bacterial strains isolation and growth conditions

The bacteria used in this work were isolated from soil samples collected in Fazenda Areão, Escola Superior de Agricultura Luiz de Queiroz (47°38′00″W; 22°42′30″S), Piracicaba, São Paulo State, Brazil. The soils were classified as Oxisol [32] of medium texture and had a history of ametryn and clomazone applications for five consecutive years.

The initial bacterial isolation was carried out using a plating technique with a serial dilution in 0.85% NaCl at concentrations of 10^{-3} and 10^{-5} inoculated in Minimal Salts medium containing 1.0 g $(NH_4)_2SO_4$, 1.0 g NaCl, 1.5 g K_2HPO_4, 0.5 g KH_2PO_4, and 0.2 g $MgSO_4.7H_2O$, per L of distilled water, at 30°C (pH 7.0) [33] in the absence and presence of the two herbicides. The concentrations of 25 mM ametryn, 9 mM clomazone and 20 mM of each herbicide were used based on the recommendations on the spray tank solution for each herbicide (5 g L^{-1} for ametryn, 1.8 g L^{-1} for clomazone and 5 g L^{-1} each, in combination).

The tolerant bacterial strains, CC07 and 4C07, were selected based on faster growth rates (compared to other bacteria isolates) and halo formation observed around the bacterial colony, indicating possible herbicide degradation, as observed by Nie et al. [33] and Martins et al. [34].

The bacterial strains were grown aerobically in nutrient Agar (Biobrás - Brazil) containing 5 g peptone, 3 g yeast extract and 15 g agar per L of distilled water, at 30°C (pH 7.0) both in the absence and presence of the herbicides. The herbicides concentrations were added as described above.

Bacterial identification

Bacterial DNA was extracted as previously described by Araújo et al. [35] and a partial sequence of the 16S rRNA gene was amplified with primers R1387 [36] and P027F [37]. PCR products were purified and sequenced with primers R1387, 519R and P027F for 16S rRNA (MegaBACE 1000). The sequences of the bacterial strains CC07 and 4C07 were retrieved from databases and used for alignment and phylogeny analyses [38,39] with MEGA 4.0 software package [40] based on the maximum parsimony (MP). The sequences obtained were deposited in GenBank[T] under the accession numbers JX109938 and JX109935 for strains CC07 and 4C07, respectively.

Growth determination

Bacterial growth was monitored by measuring the number of colony-forming units (CFUs) mL^{-1} as described by Sangali and Brandelli [41]. Cultures inoculated with 0.1% of the original (Absorbance = 1.0 at 600 nm) were grown in 250 mL Erlenmeyer flasks containing 50 mL of nutrient medium and incubated in the dark on a rotary shaker (140 rpm) at 30°C for 14 h. The bacterial suspension was diluted to 10^{-6} in a saline solution containing 0.85% NaCl and then homogenized. At 2 h intervals the samples (20 μL) were loaded in triplicate for each treatment onto nutrient agar plates, which were further incubated at 30°C for 24 h. At the end of this period the CFUs were determined.

Table 1. Characteristics of the herbicides ametryn and clomazone.

	Ametryn	Clomazone
Manufacture	Syngenta, BASF, Bayer, Servatis, Sipcam Isagro Brasil	FMC Corporation
Agrochemical formulation	500 g ai L^{-1}, EC	360 g ai L^{-1}, EC
Molecular formula	$C_9H_{17}N_5S$	$C_{12}H_{14}ClNO_2$
Molecular weight	227.3	239.7
Vapor pressure	2.74×10^{-6} mm Hg at 25°C	1.4×10^{-4} mm Hg at 25°C
Water solubility	185 mg L^{-1} at 25°C	1,100 mg L^{-1} at 25°C

Physiological and biochemical measurements

Lipid peroxidation. Lipid peroxidation was determined by estimating the content of thiobarbituric acid reactive substance (TBARS) following the method of Heath and Packer [42]. Malondialdehyde (MDA) was monitored by measuring at 535 and 600 nm in a Perkin Elmer Lambda 40 spectrophotometer, and the concentration was calculated using an extinction coefficient of 155 mM^{-1} cm^{-1}.

Quantification of reduced glutathione (GSH). Bacterial cells (100 mg) were homogenized in 1.5 mL of 5% sulfosalicylic acid in a mortar and pestle at 4°C. The homogenate was centrifuged at 12,000×*g* for 20 min at 4°C. The content of GSH and GSSG was determined as described by Anderson [43] at 25°C in a mixture consisting of 1.75 mL 100 mM potassium phosphate buffer (pH 7.5) containing 0.5 mM ethylenediaminetetraacetic acid (EDTA) and 100 µL of 3 mM 5,5′-dithiobis(2-nitrobenzoic acid) (DTNB). The reaction was started by the addition of 250 µL of the bacterial cell homogenate. After 5 min, the absorbance for the determination of GSH was read at 412 nm using a spectrophotometer. A standard curve was prepared with known concentrations of GSH and the results were expressed in nmol g^{-1} FW.

Enzyme extraction and protein determination. Cultures were centrifuged at 12,000×*g* for 20 min at 4°C and the pellets macerated with liquid nitrogen in a mortar with a pestle. The extracts were homogenized (5:1, buffer volume: fresh weight) in 100 mM potassium phosphate buffer (pH 7.5) containing 1 mM EDTA, 3 mM DL-dithiothreitol (DTT) and 5% (w/w) polyvinyl-polypyrrolidone [44]. The homogenates were centrifuged at 12,000×*g* for 30 min at 4°C and the supernatants were stored in separate aliquots at −80°C prior to enzymatic analysis. The concentration of protein was determined by the method of Bradford [45] using bovine serum albumin as standard.

Polyacrylamide gel electrophoresis (PAGE). Electro-fphoretic analyses of antioxidant enzymes were carried out under non-denaturing condition in 12% polyacrylamide gels as described by Gratão et al. [44]. For denaturing SDS-PAGE, the gels were rinsed in distilled deionized water and incubated overnight in 0.05% Coomassie blue R-250 in a water/methanol/acetic acid 45:45:10 (v/v/v) solution and destained by successive washings in the same water/methanol/acetic acid 45:45:10 (v/v/v) solution.

SOD activity staining. SOD activity staining was carried out as described by Garcia et al. [46]. After non-denaturing PAGE separation, the gel was rinsed in distilled deionized water and incubated in the dark in 50 mM potassium phosphate buffer (pH 7.8) containing 1 mM EDTA, 0.05 mM riboflavin, 0.1 mM nitroblue tetrazolium, and 0.3% N,N,N′,N′-tetramethylethylene-diamine. One unit of bovine liver SOD (Sigma, St. Louis, USA) was used as a positive control of activity. After 30 min, the gels were rinsed with distilled deionized water and then illuminated in water until the development of achromatic bands of SOD activity on a purple-stained gel. SOD isoenzyme characterization was performed as described by Azevedo et al. [47]. Briefly, SOD isoenzymes were distinguished by their sensitivity to inhibition by 2 mM potassium cyanide and 5 mM hydrogen peroxide. The relative intensities of the stained bands were determined by an ImageScanner III (GE Healthcare, Little Chalfont, UK) and the ImageQuant^{TM} TL software (GE Healthcare, Uppsala, Sweden).

CAT activity staining. CAT activity following non-denaturing PAGE was determined as described by Boaretto et al. [48]. Gels were incubated in 0.003% hydrogen peroxide (H_2O_2) for 10 min and subsequently in a 1% (w/v) ferric chloride ($FeCl_3$) and 1% (w/v) potassium hexacyanoferrate III ($K_3Fe(CN_6)$) solution for additional 10 min. One unit of bovine liver CAT (Sigma, St. Louis, USA) was used as a positive control.

GR activity staining. GR activity following non-denaturing PAGE was determined as described by Gomes-Junior et al. [49]. The gels were rinsed in distilled deionized water and incubated in the dark for 30 min at room temperature in the reaction solution contained 250 mM Tris (pH 7.5), 0.5 mM 3-(4,5-dimethyl-2-thiazolyl)-2,5-diphenyl-2H-tetrazolium bromide (MTT), 0.7 mM 2,6-dichloro-*N*-(4-hydroxylphenyl)-1,4-benzoquinoneimine sodium salt (DPIP), 3.4 mM GSSG (oxidized glutathione) and 0.5 mM NADPH. One unit of bovine liver GR (Sigma, USA) was used as a positive control of activity.

GR total activity determination. Total GR activity was assayed as described by Gratão et al. [44] at 30°C in a mixture consisting of 1.7 mL 100 mM potassium phosphate buffer (pH 7.5) containing 1 mM 5,5′-dithiobis(2-nitrobenzoic acid) (DTNB), 1 mM GSSG and 0.1 mM NADPH. The reaction was started by the addition of 50 µL of protein extract. The rate of reduction of oxidized glutathione was followed in a spectrophotometer by monitoring the change in absorbance at 412 nm for 1 min. GR activity was expressed as µmol min^{-1} mg^{-1} protein.

GST total activity determination. GST activity was assayed spectrophotometrically at 30°C in a mixture containing 900 µL 100 mM potassium phosphate buffer (pH 6.5), 25 µL 40 mM 1-chloro-2,4-dinitrobenzene (CDNB), 50 µL 1 mM GSH and 25 µL enzyme extract. The reaction mixture was followed by monitoring the increase absorbance at 340 nm over 5 min [50]. GST activity was expressed as µmol min^{-1} mg^{-1} protein.

Experimental design and statistical analysis. Total protein content and enzyme activity determinations were conducted on three replicates of each treatment, which were performed in a completely randomized design. The significance of the observed differences was verified by using a one-way analysis of variance (ANOVA) followed by the Tukey's test ($p<0.05$). All statistical analyses were carried out by using R software (URL http://www.r-project.org).

Results

Phylogenetic identity and bacterial growth

An almost-complete 16S rRNA gene sequence (1312 nts) was determined for both the CC07 and 4C07 strains and a phylogenetic tree was built up (Fig. 1). *Burkhloderia cariophilli* and *B. plantarii* were used as outgroups. Comparative analyses based on 16S rRNA gene sequences revealed that the CC07 strain is phylogenetically related to *Pseudomonas aeruginosa*, whereas the 4C07 strain exhibited homology with *P. fulva* (Fig. 1).

The growth of the strains in the presence of the herbicides is shown in Fig. 2. The two strains exhibited very distinct growth curves. Growth of both strains was not greatly affected in the presence of the herbicide ametryn, whilst clomazone strongly inhibited the growth of both strains. Strain CC07 exhibited a long (10 h) lag phase as an adaptation period before the exponential growth (12 h) (Fig. 2A), while strain 4C07 grew only for the first six hours. When the herbicides were used in combination, growth of the strain CC07 was only slightly inhibited, mainly during the early period, whereas for strain 4C07 there was considerable inhibition of growth, similar to that shown by clomazone alone.

Lipid peroxidation (MDA)

Lipid peroxidation was determined as the MDA content after 14 h of growth. Although similar trends in MDA content were detected in both strains (Fig. 3), statistically there was only a higher MDA content in strain CC07 in the presence of clomazone

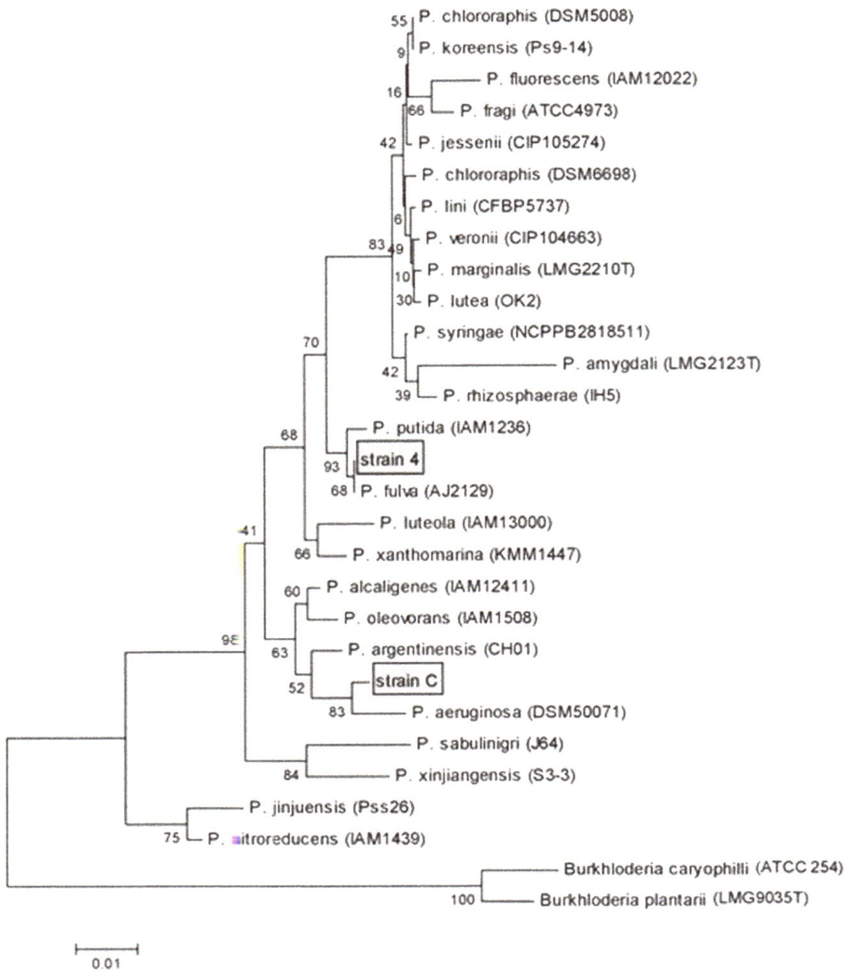

Figure 1. Maximum-parsimony phylogenetic tree constructed from the 16S rRNA gene. A total of 600 bp nucleotide of *Pseudomonas* spp. from RDP database were used. *Burkhloderia cariophilli* and *B. plantarii* served as outgroup. Bootstrap values were 1000 repetitions. Bars indicate the number of evolutionary steps with diverging sequences. Strains CC07 and 4C07 are shown inside the boxes.

(30.3%) and when exposed to the combination of the two herbicides (110.7%) (Fig. 3), when compared to growth in the control herbicide-free medium.

The content of reduced glutathione (GSH)

Different amounts of GSH were detected in the two bacterial strains, following exposure to the herbicides (Fig. 4). There was a significant increase in the GSH content of CC07 (16.5%) in the presence of ametryn, but GSH was reduced below control levels in the presence of clomazone or the herbicide mixture. Meanwhile in strain 4C07, the GSH content increased, following exposure to all the herbicide treatments (ametryn 39.4%, clomazone 60.7% and the mixture 50.5%), when compared to the untreated control (Fig. 4).

SDS-PAGE protein profile

Analysis following SDS-PAGE revealed clearly different protein profiles between the two bacterial strains, both with and without herbicide treatments (Fig. 5). There was a general reduction in intensity of the majority of the protein bands following electrophoresis of extracts of CC07 that had been subjected to herbicide treatment (Fig. 5, lanes 3, 4 and 5). In addition, a number of protein bands varied in intensity or appearance/disappearance, depending on the treatment as indicated by the arrows in Fig. 5. For instance, a 225 kDa protein band (strain CC07) was reduced following treatment with the herbicide clomazone (Fig. 5, lane 4, band I), and a 58 kDa protein band was absent in strain 4C07 exposed to clomazone (Fig. 5, lane 8, band II), and greatly reduced following treatment by both herbicides (Fig.5, lane 9, band II), among other changes (arrows).

Effects of the herbicides on antioxidant enzymes

There were different responses in the antioxidant enzymes (SOD, CAT, GR and GST) isolated from the two bacterial strains, when treated with the herbicides alone or in combination. SOD activity was determined by activity staining following non-denaturing PAGE (Fig. 6). The analysis revealed that there were two distinct isoenzymes present in each bacterium, but that they had different electrophoretic mobilities (Fig. 6A and B). Both of the SOD isoenzymes isolated from strain CC07 were characterized as Mn/SODs (Fig. 6A, SOD I and II, lanes 1–4; and Fig. 6C 1, 2 and 3), whereas the two SOD isoenzymes, isolated from strain 4C07 comprised a Cu-Zn/SOD (I) and a Fe/SOD (II) (Fig. 6B, SOD I and II, lanes 5–8; and Fig. 6C 4, 5 and 6).

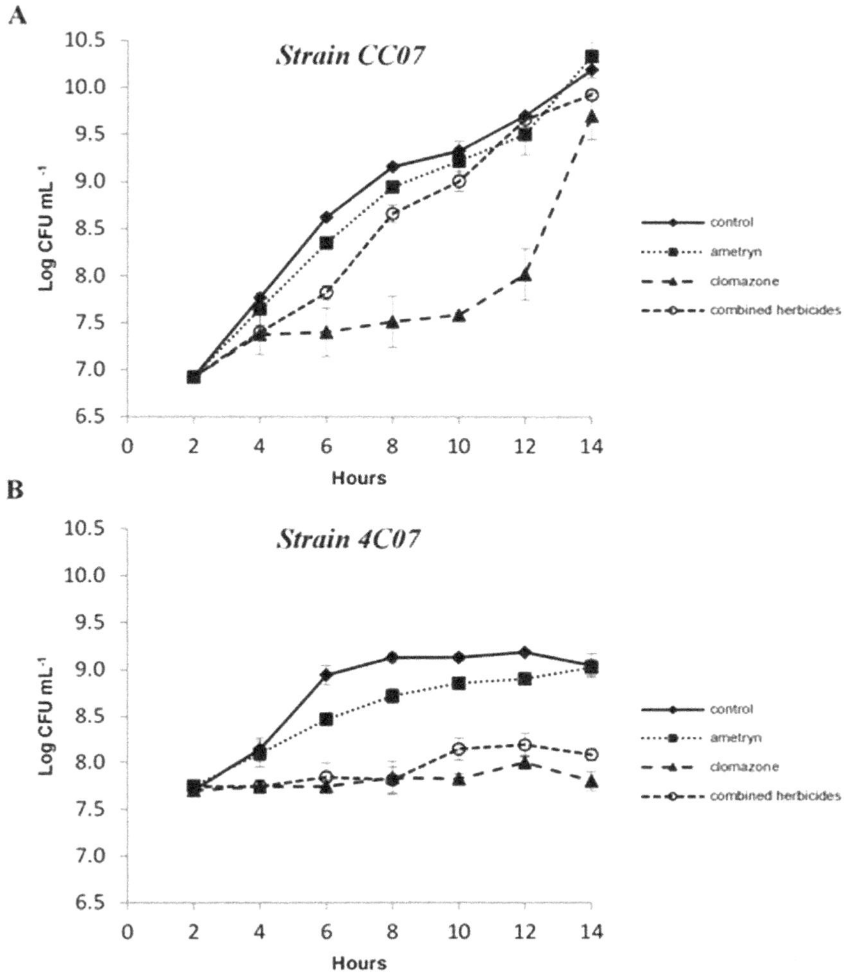

Figure 2. Growth curve of bacterial strains in the presence of 0 mM (control), 25 mM ametryn, 9 mM clomazone and 20 mM of each the herbicide. (A) Strain CC07 and (B) Strain 4C07. Values represent the means from three replicates ±SEM.

The activity of SOD I in strain CC07, was reduced following incubation with ametryn and the two combined herbicides, whilst SOD II was unaffected (Fig. 6A, lanes 2 and 4, respectively). A different SOD isoenzyme pattern was detected following native PAGE of extracts of 4C07, in which SOD II activity was increased following treatment with clomazone and the two combined herbicides (Fig. 6B, lanes 7 and 8, respectively), whereas the activity of SOD I was also stimulated in the presence of clomazone, but drastically inhibited following the combined herbicide treatment (Fig. 6B, lanes 7 and 8, respectively). Ametryn did not produce any major change in SOD I and II activity when compared to the control (Fig. 6B, lanes 5 and 6).

CAT activity staining following non-denaturing PAGE revealed the presence of three isoenzymes for strain CC07 (CAT I, II and III; Fig. 7, lanes 1–4) and four isoenzymes for strain 4C07 (CAT I, II, III and IV; Fig. 7, lanes 5–8), with CAT isoenzymes II (strain CC07), I (strain 4C07) and III (both bacteria) with the same relative mobility, suggesting they may be the same isoenzymes in both bacteria. The activity of CAT isoenzyme II was higher in 4C07 and was further increased when the strain was subjected to the combined herbicide treatment (Fig. 7, lane 8).

Furthermore, the CAT isoenzyme activity profiles were in a way similar to the enzyme pattern also observed for SOD (Fig. 6A and B). For instance, in strain CC07, the activity of CAT I was reduced in the ametryn and combined herbicide treatments (Fig. 7, lanes 2 and 4) as observed for SOD I (Fig. 6, lanes 2 and 4). On the other hand, the activity of the CAT II isoenzyme was slightly reduced in the clomazone and combined herbicide treatments (Fig. 7, lanes 3 and 4), whereas the CAT III isoenzyme was unaltered in all treatments. For strain 4C07, apart from the unique CAT II isoenzyme (Fig. 7, lane 8), the activity of CAT isoenzyme I was clearly increased following treatment with clomazone and the combined herbicides (Fig. 7, lanes 7 and 8).

GR activity was determined as total specific activity (Fig. 8A and B) and by non-denaturing PAGE for isoenzyme identification (Fig. 8C). Total GR activity was not altered in strain CC07 regardless of the treatment (Fig. 8A), but was increased in strain 4C07 when exposed to clomazone (30.8% increase) and the combined herbicides (55.3% increase) (Fig. 8B). GR activity staining revealed the existence of 6 GR isoenzymes for strain CC07 (Fig. 8C, lanes 1–4), all present in all treatments tested, and up to 5 GR isoenzymes in the strain 4C07 depending on the treatment tested (Fig. 8C, lanes 5–8). In general, the GR isoenzyme activities were in accordance with the results obtained for total GR activity (Fig. 8A and B). Slight variations in band intensity were observed among the treatments, but all GR

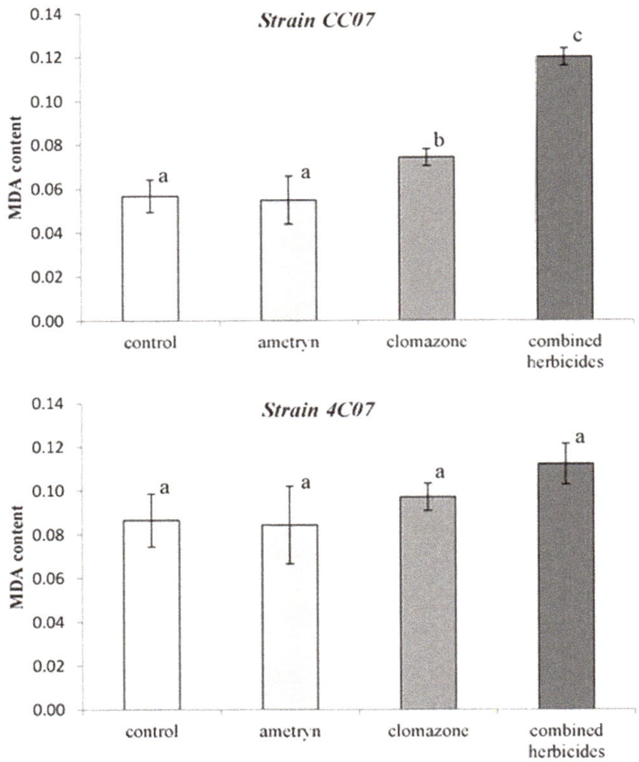

Figure 3. Lipid peroxidation (MDA content) of bacteria exposed to the herbicides. Values of MDA content (nmol g^{-1} fr. wt) represent the means from three replicates ±SEM. Means with different letters are significantly different ($P<0.05$) by one-way analysis of variance (ANOVA) and Tukey's test.

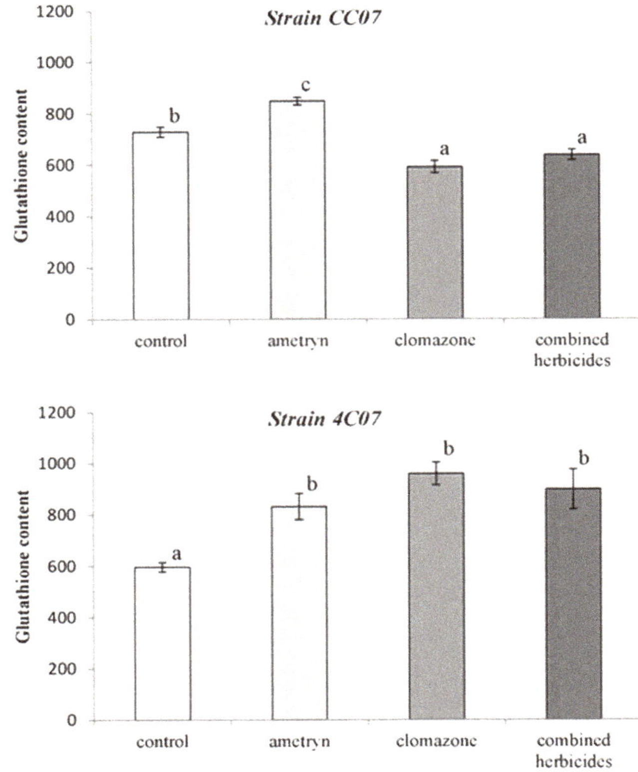

Figure 4. GSH content (nmol g^{-1} fr. wt) of bacteria exposed to the herbicides. Values represent the means from three replicates ±SEM. Means with different letters are significantly different ($P<0.05$) by one-way analysis of variance (ANOVA) and Tukey's test.

isoenzymes were present in the strain CC07 (Fig. 8C, lanes 1–4). On the other hand, there was increased total GR activity in strain 4C07, when subjected to clomazone and combined herbicide treatments, which is most likely due to the increased activity of GR isoenzyme II (Fig. 8C, lanes 7 and 8).

Total specific GST activity was also determined and the results revealed a distinct response for each bacterium when exposed to the herbicides (Fig. 9). Total GST activity increased by 51.5%, 41.5% and 105% when strain 4C07 was grown in the presence of ametryn, clomazone or the combination of herbicides, respectively (Fig. 9). On the other hand, there was a reduction in GST activity when strain CC07 was exposed to the herbicides ametryn or clomazone, but the activity was similar to the control when CC07 was grown in the presence of the combined herbicides (Fig. 9).

Discussion

Microorganisms present in soils must quickly adapt to environmental changes. This biochemical and physiological adaptation process play an important role in microbial survival, especially under stressful conditions [51]. Strains CC07 and 4C07 were selected for this study due to their ability to tolerate the high concentrations of herbicides tested and to the fact that bacteria of the *Pseudomonas* genus play a major role in degradation of xenobiotic compounds [52–55].

Generally, herbicides can alter the growth of degrading or tolerant bacteria [56]. Growth of the strains CC07 and 4C07 exhibited distinct responses when exposed to the herbicides

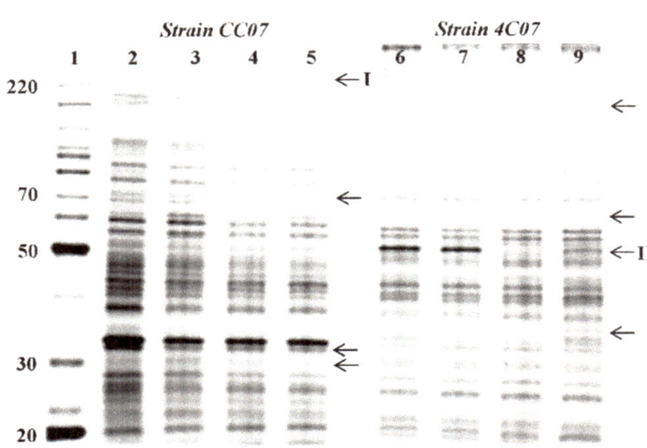

Figure 5. SDS-PAGE protein profiles of bacteria exposed to the herbicides. Lane 1, protein molecular mass markers (220 to 20 kDa). Lanes 2, 3, 4 and 5, strain CC07 grown in the presence of 0 mM (control), 25 mM ametryn, 9 mM clomazone and 20 mM of each the herbicide, respectively. Lanes 6, 7, 8 and 9, strain 4C07 grown in the presence of 0 mM (control), 25 mM ametryn, 9 mM clomazone and 20 mM of each the herbicide, respectively. Arrows indicate selected variations in intensity or appearance/disappearance depending on the treatment tested. I and II indicate protein bands of 225 kDa and 58 KDa that are depleted specifically in the presence of clomazone.

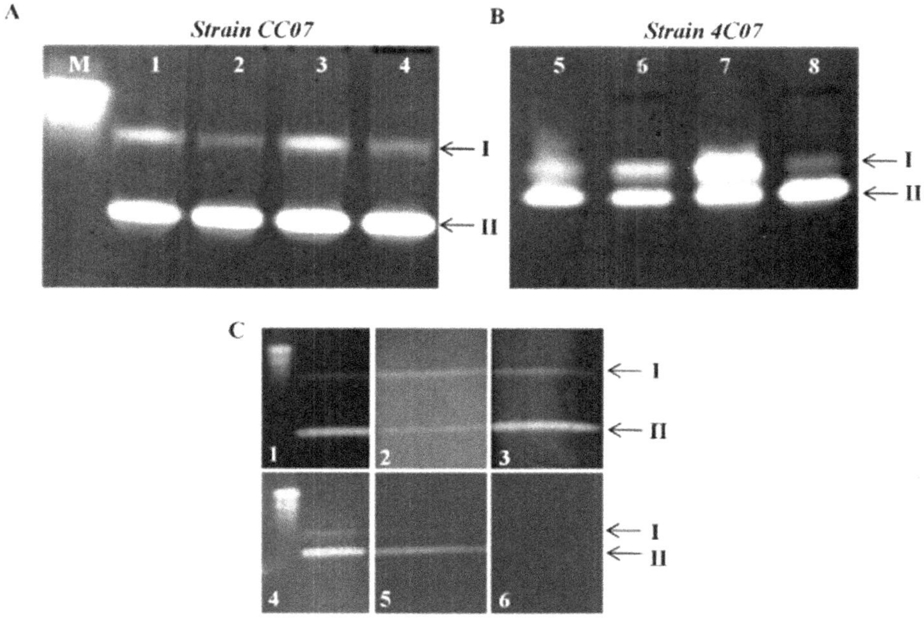

Figure 6. Activity staining for SOD following non-denaturing PAGE of extracts from cultured bacterial cells. (A) First lane is a bovine SOD standard, lanes 1, 2, 3 and 4 are strain CC07 grown in the presence of 0 mM (control), 25 mM ametryn, 9 mM clomazone and 20 mM of each herbicide, respectively. (B) Lanes 5, 6, 7 and 8 are strain 4C07 grown in the presence of 0 mM (control), 25 mM ametryn, 9 mM clomazone and 20 mM of each herbicide, respectively. Arrows indicate sequentially numbered SOD bands (I–II) that are independent of the bacterial strain. (C) Activity staining for SOD of strain CC07 (1, 2 and 3) and strain 4C07 (4, 5 and 6), used for classification of SOD isoenzymes. Lanes 1 and 4, control SOD activity. Lanes 2 and 5, SOD activity with 2 mM potassium cyanide treatment; lanes 3 and 6; 5 mM H_2O_2 treatment. Arrows indicate SOD bands that are sequentially numbered (I–II) according to Fig. 6A and B.

ametryn and clomazone, separately or in combination. The number of bacterial cells for all treatments was lower when compared to the control, indicating that the ROS generated by the herbicides and its metabolites can damage bacteria cells and decrease bacterial growth. However, clomazone was the treatment that markedly affected growth of both strains. These results may be associated with the higher toxicity of the clomazone molecule, due the presence of chloride (Table 1). García-Cruz et al. [57] demonstrated the toxicity of the chlorine atoms present in the herbicide 2,4-dichlorophenoxyacetic acid (2,4-D) and possible intermediates in bacterial biofilms. Another possible cause of the aggressiveness of clomazone towards bacteria cells may be due to its mechanism of action, since it is an inhibitor of the deoxy-D-xylulose 5-phosphate (DXP) synthase enzyme [14,58], which plays a key role in isoprenoid biosynthesis and is also required for thiamine and pyridoxal phosphate production in prokaryotes [59].

Previous studies have shown that when bacteria are exposed to herbicides or their metabolites, a significant increase in ROS generation may occur and, consequently, induce an oxidative stress condition [22,23]. Lipid peroxidation is one of the best predictors of ROS level inducted under stress conditions [60], and can be used as a marker of oxidative stress [61]. An increased content of MDA was detected when strain CC07 was exposed to clomazone alone or in mixture with ametryn (Fig. 3), which

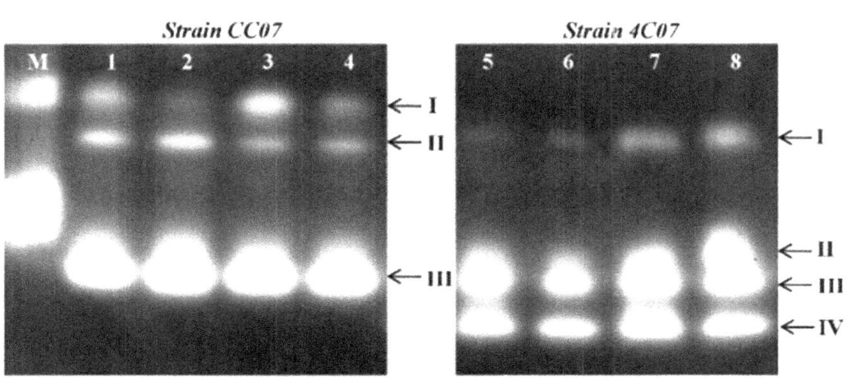

Figure 7. Activity staining for CAT following non-denaturing PAGE of extracts from cultured bacterial cells. Lane M is a bovine CAT standard, lanes 1, 2, 3 and 4, strain CC07 grown in the presence of 0 mM (control), 25 mM ametryn, 9 mM clomazone and 20 mM of each herbicide, respectively. Lanes 5, 6, 7 and 8, strain 4C07 grown in the presence of 0 mM (control), 25 mM ametryn, 9 mM clomazone and 20 mM of each herbicide, respectively. Arrows indicate sequentially numbered CAT bands for strains CC07 (I–III) and 4C07 (I–IV).

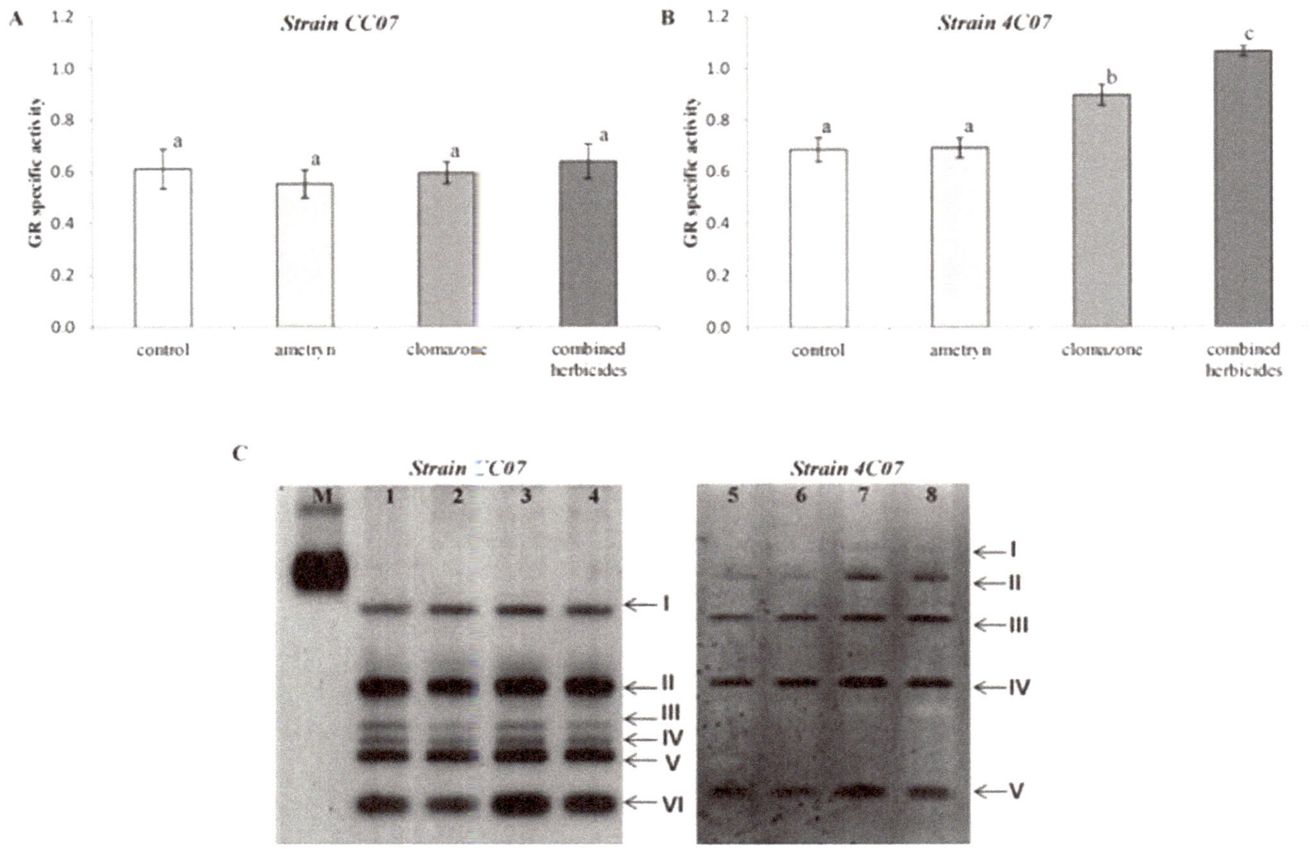

Figure 8. GR specific activity. (A) and (B) Specific activity of GR, expressed as μmol min^{-1} mg^{-1} protein. Values are the means of three replicates ±SEM. Means with different letters are significantly different ($P<0.05$) by one-way analysis of variance (ANOVA) and Tukey's test. (C) Activity staining for GR following non-denaturing PAGE of extracts of cultured bacterial cells. Lane M is a bovine GR standard, lanes 1, 2, 3 and 4, strain CC07 grown in the presence of 0 mM (control), 25 mM ametryn, 9 mM clomazone and 20 mM of each herbicide, respectively. Lanes 5, 6, 7 and 8, strain 4C07 grown in the presence of 0 mM (control), 25 mM ametryn, 9 mM clomazone and 20 mM of each herbicide, respectively. Arrows indicate sequentially numbered GR bands (I–VI) that are independent of the bacterial strain.

suggests that the high concentration used in the combined treatment may have increased the amount of lipid peroxidation. Despite this, the CC07 strain was able to respond and tolerate the peroxidation stress, since following all the herbicide treatments, the growth observed after 14 h was very similar to that of the control (Fig. 2A). High concentrations of MDA have been found in strains of *Enterobacter asburiae* and *E. amnigenus* in the presence of the herbicides metolachlor and acetochlor, respectively [23], which correlated the oxidative stress with the mode of action of the herbicides tested, which involved the inhibition of the elongation of C18 and C16-fatty acids. On the other hand, changes in MDA content were not observed for strain 4C07 in the presence of the herbicides, suggesting that strain 4C07 could possess an effective antioxidant system to avoid the damage caused by ROS. Although no changes in MDA content were observed, the 4C07 strain exhibited limited growth, thus indicating sensitivity to the herbicides dose (s) in the medium.

To maintain ROS under the baseline levels bacteria utilize a complex antioxidant defense system [20]. SOD is able to detoxify $O_2^{-\bullet}$, one of the two substrates of the Haber-Weiss reaction, which generates $OH^{\bullet}$ radicals, and is therefore plays a key role in the central defense mechanism of living organisms [62]. Following analysis of SOD activity by non-denaturing PAGE, two isoenzymes were identified in each strain, which were classified as Mn-SOD (SODs I and II) for strain CC07, and as Cu/Zn-SOD (SOD I) and Fe-SOD (SOD II) for strain 4C07. Fe-SOD and Mn-SOD are present in the bacterial cytoplasm [63], whilst Cu/Zn-SOD isoenzymes are present in the periplasm and are more sensitive to endogenous oxidative stress. However, Cu/Zn-SODs are also responsible for protection against oxidative damage to DNA. Studies with mutants of *Escherichia coli* revealed that Cu/Zn-SOD activity increased bacterial resistance to the herbicide paraquat [64]. Thus, the presence of Cu/Zn-SOD in strain 4C07, in comparison to CC07, may have favored its antioxidant response to the $O_2^{-\bullet}$ increase in the presence of the herbicides. In contrast, following exposure to the herbicide quinclorac, there was an increase in both Fe-SOD and Mn-SOD activity in the bacterium *B. cepacia* WZ [17], which appeared to protect *B. cepacia* from the redox action of quinclorac [17].

Catalases are widely distributed and are considered important components of detoxification routes that prevent the formation of hydroxyl radicals [65]. Three active isoenzymes (CAT I, III and IV) were detected in extracts of the untreated strain 4C07, however in the presence of both herbicides a new isoenzyme (CAT II) was induced. This may have occurred due to bacterial sensitivity to both herbicides, thereby, the expression of a new isoform would contribute to maintain the CAT activity at a standard level.

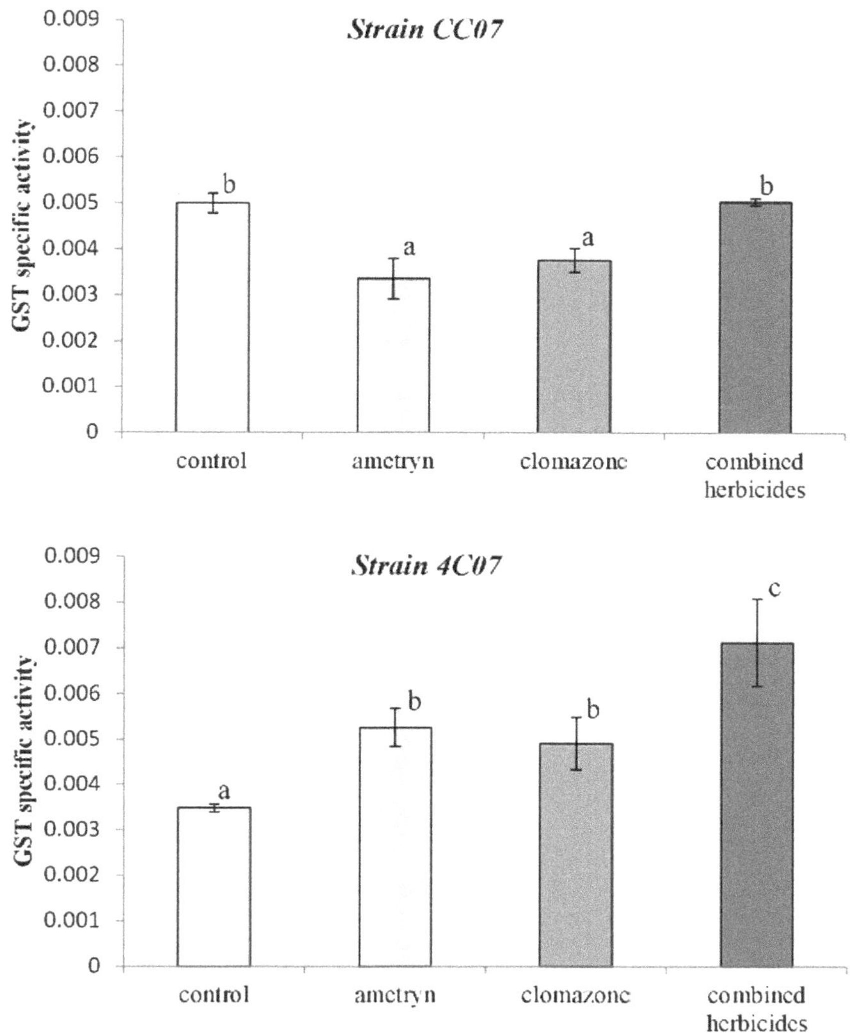

Figure 9. GST specific activity, expressed as units mg^{-1} protein. Values represent the means from three replicates ±SEM. Means with different letters are significantly different ($P<0.05$) by one-way analysis of variance (ANOVA) and Tukey's test.

By comparing the CAT and SOD results for strain CC07, it can be seen that the herbicide ametryn inhibited the activity of both enzymes, (isoenzymes SOD I and CAT I in particular), whilst the herbicide clomazone had a greater effect on the enzymes of strain 4C07. There is no consensus on the effect of herbicides on CAT and SOD; both enzymes are essential in *Bacillus subtilis* B19, *B. megaterium* and *E. coli* K12 for survival in the presence of the herbicide bensulfuron-methyl [19], and both enzymes also played an important role in copper resistance in *Amycolatopsis* spp species [66]. In contrast, Martins et al. [23] reported that SOD activity was not significantly changed when strains of *E. asburiae* and *E. amnigenus* were exposed to the herbicides metolachlor and acetochlor, respectively.

Reduced glutathione is the most abundant thiol in cells and its main function is to maintain the cytoplasm redox state in equilibrium [29] There was a significant increase in the GSH content of strain CC07 in the presence of ametryn, possibly minimizing the occurrence of lipid peroxidation. In the presence of clomazone alone or in mixture with ametryn, however the GSH content was reduced. Hultberg [67] observed decreases in the concentration of GSH when the bacterium *P. fluorescens* was exposed to cadmium and suggested that GSH may be used as a marker for the intensity of environmental stress. On the other hand, the GSH concentrations in strain 4C07 increased significantly in the presence of the herbicides. This result suggests a direct participation of GSH in ROS detoxification, particularly of the $OH^{\bullet}$ radical, consequently preventing lipid peroxidation and thus, contributing to the protection against oxidative stress. Generally in gram-negative bacteria, including *Pseudomonas*, the concentration of GSH is high [68] and it can react directly with a free radical resulting in a thiol radical, $GS^{\bullet}$, which reacts with another $GS^{\bullet}$ radical producing GSSG [69].

Under many intracellular stressful conditions the GSSG concentration increases due to GSH oxidation, however, a high GSH/GSSG rate is necessary to maintain the role of GSH as antioxidant and reducing agent [70], therefore, GR acts as a fundamental link between the two redox metabolites within a cell [29]. In strain CC07, total GR activity remained unchanged in the presence of the herbicides, which was reconfirmed with non-denaturing PAGE analysis. Veremeenko and Maksimova [68] showed that in *P. aurantiaca*, GSH and GR activity increased significantly in the presence of antibiotics. In this study, the increase in total GR activity in strain 4C07 in the presence of

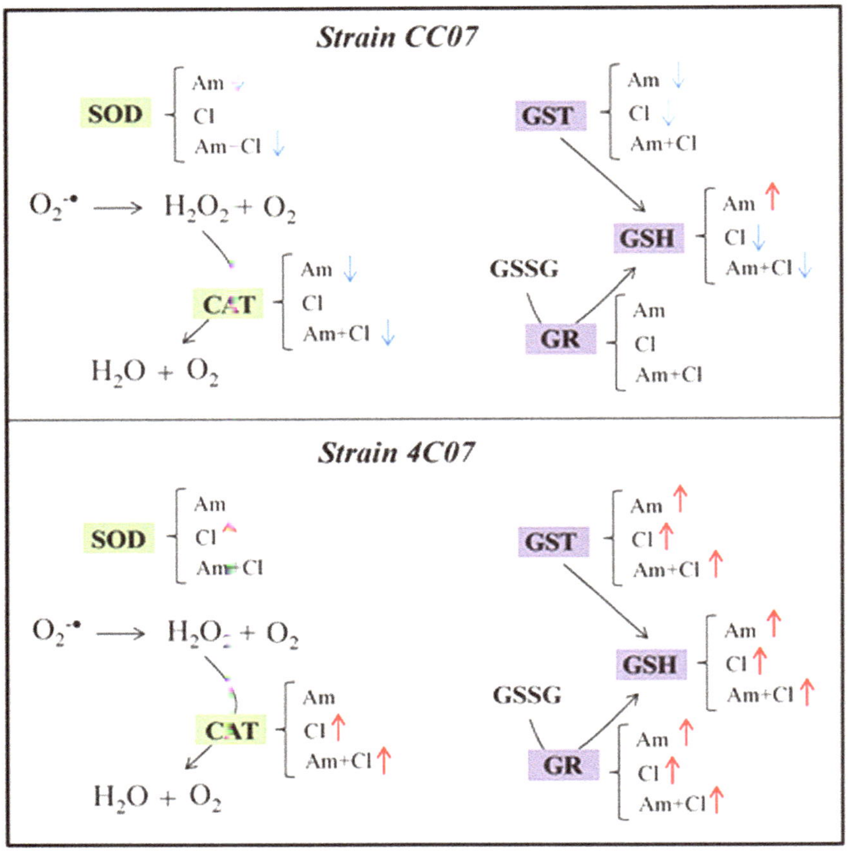

Figure 10. General view of the antioxidant system involving enzymatic and non-enzymatic components for the strains CC07 and 4C07 based on the data obtained in the present work. SOD (superoxide dismutase); CAT (catalase); GST (glutathione S-transferase); GR (glutathione reductase); GSSG (oxidized glutathione); GSH (reduced glutathione); Am (ametryn); Cl (clomazone); Am+Cl (combined herbicides). Blue arrows indicate decreases in enzymatic activity and glutathione content; red arrows indicate increases in enzymatic activity and glutathione content; the absence of symbols indicates no alterations. All changes are relative to the untreated control.

clomazone alone or in mixture with ametryn is associated with the increase in isoenzymes II and to a lesser extent I and V. These data suggest that the different isoenzymes may play a specific role in the antioxidant response of strain 4C07 to oxidative stress induced by the herbicides. Martins et al. [23] reported that when *E. asburiae* was exposed to the herbicide s-metolachlor (34 mM), two new GR isoforms were induced in response to the oxidative stress condition, suggesting a role for GR in herbicide tolerance. Similar responses were also found in plant species. For instance, in coffee cell suspension cultures subjected to the oxidative stress induced by heavy metals (cadmium, nickel and selenium) a new GR isoenzyme was induced during the first hours of stress. The authors also suggested that GR activity could be used as an early stress marker for this plant species [49,71,72].

GST is another enzyme with activity closely related to GSH. Its function is to conjugate GSH to xenobiotics, and therefore has a fundamental central role in detoxification [31]. GSTs are capable of detoxifying numerous classes of pesticides in bacterial cells including s-triazines and are involved in the first stage of the biodegradation of the herbicide atrazine, when the removal of the chlorine atom from the atrazine-GSH conjugate occurs [73]. The increase in GST activity in strain 4C07 may also be associated with the increase in GSH concentration, and could indicate the involvement of GST in the detoxification of the two herbicides.

Our results indicate that both strains are tolerant to the herbicides ametryn and clomazone or to their mixture. However, the herbicides caused an imbalance in the redox potential and metabolism of the bacterial cells. A summary of the key alterations observed are presented in Fig 10. The enzymes GR and GST, together with the antioxidant compound GSH, may play a major role in the tolerance of strain 4C07 to the herbicides ametryn and clomazone supplied individually or combined. In contrast, there was a decrease in the activity of enzymes SOD, CAT, GST and the GSH content in CC07, whilst GR activity remained unchanged. Compensatory mechanisms for the reduction of these enzymes activities and the depletion of GSH content may occur and account for the induction of another defense mechanism, since the strain CC07 managed to adjust its metabolism in response to the stress. These results indicate a different antioxidant response of the two bacterial strains to the herbicides; however, additional studies are required in order to understand the tolerance mechanisms. Nevertheless, strain CC07 grew at a higher rate, indicating that this bacterium was able to adapt better to the stressful environment, which could be useful for bioremediation strategies of environments contaminated with herbicides.

Author Contributions

Conceived and designed the experiments: LPP GC PFM RAA. Performed the experiments: LPP GC PFM MND. Analyzed the data: LPP GC PFM

MND MP RAA. Contributed reagents/materials/analysis tools: LPP GC PFM MND RAA. Wrote the paper: LPP GC PFM MND MBV MP RAA.

References

1. Juraske R, Antón A, Castells F, Huijbregts MAJ (2007) PestScreen: A screening approach for scoring and ranking pesticides by their environmental and toxicological concern. Environ Int 33: 886–893.
2. Tejada M, Gomez I, Del Toro M (2011) Use of organic amendments as a bioremediation strategy to reduce the bioavailability of chlorpyrifos insecticide in soils. Effects on soil biology. Ecotox Environ Safe 74: 2075–2081.
3. Pimentel D, Peshin R, Dhawan AK (2009) Pesticides and pest control integrated pest management: Innovation-development process. Netherlands: Springer. 83–87 p.
4. Vercraene-Eairmal M, Lauga B, Saint Laurent S, Mazzella N, Boutry S, et al. (2010) Diuron biotransformation and its effects on biofilm bacterial community structure. Chemosphere 81: 837–843.
5. Mishra K, Sharma RC, Kumar S (2012) Contamination levels and spatial distribution of organochlorine pesticides in soils from India. Ecotox Environ Safe 76: 215–225.
6. McKnight US, Rasmussen JJ, Kronvang B, Bjerg PL, Binning PJ (2012) Integrated assessment of the impact of chemical stressors on surface water ecosystems. Sci Total Environ 427–428: 319–331.
7. Pileggi M, Pileggi SAV, Olchanheski LR, da Silva PAG, Munoz Gonzalez AM, et al. (2012) Isolation of mesotrione-degrading bacteria from aquatic environments in Brazil. Chemosphere 86: 1127–1132.
8. Andrade SRB, Silva AA, Lima CF, D'Antonino L, Queiroz MELR, et al. (2010) Lixiviação do ametryn em Argissolo Vermelho-Amarelo e Latossolo Vermelho-Amarelo, com diferentes valores de pH. Planta Daninha 28: 655–663.
9. Cumming JP, Doyle RB, Brown PH (2002) Clomazone dissipation in four tasmanian topsoils. Weed Sci 50: 405–409.
10. Monquero PA, Binha DP, Amaral LR, Silva PV, Silva AC, et al. (2008) Lixiviação de clomazone + ametryn, diuron + hexazinone e isoxaflutole em dois tipos de solo. Planta Daninha 26: 685–691.
11. Carlomagno M, Matho C, Cantou G, Sanborn JR, Last JA, et al. (2010) A clomazone immunoassay to study the environmental fate of the herbicide in rice (*Oryza sativa*). J Agr Food Chem 58: 4367–4371.
12. Vieira VC, Alves P, Picchi SC, Lemos MVF, Sena JAD (2010) Molecular characterization of accessions of crabgrass (*Digitaria nuda*) and response to ametryn. Acta Sci Agron 32: 255–261.
13. Yasuor H, TenBrook PL, Tjeerdema RS, Fischer AJ (2008) Responses to clomazone and 5-ketoclomazone by *Echinochloa phyllopogon* resistant to multiple herbicides in Californian rice fields. Pest Manag Sci 64: 1031–1039.
14. Ferhatoglu Y, Barrett M (2006) Studies of clomazone mode of action. Pestic Biochem Physiol 85: 7–14.
15. Navaratna D, Elliman J, Cooper A, Shu L, Baskaran K, et al. (2012) Impact of herbicide ametryn on microbial communities in mixed liquor of a membrane bioreactor (MBR). Bioresource Technol 113: 181–190.
16. Mervosh TL, Sims GK, Stoller EW (1995) Clomazone fate in soil as affected by microbial activity, temperature, and soil-moisture. J Agr Food Chem 43: 537–543.
17. Lü ZM, Min H, Xia Y (2004) The response of *Escherichia coli*, *Bacillus subtilis*, and *Burkholderia cepacia* WZ1 to oxidative stress of exposure to quinclorac. J Environ Sci Health Part B Pestic Food Contam Agric Wastes 39: 431–441.
18. Lü ZM, Sang LY, Li ZM, Min H (2009) Catalase and superoxide dismutase activities in a *Stenotrophomonas maltophilia* WZ2 resistant to herbicide pollution. Ecotox Environ Safe 72: 136–143.
19. Lin X, Xu X, Yang C, Zhao Y, Feng Z, et al. (2009) Activities of antioxidant enzymes in three bacteria exposed to bensulfuron-methyl. Ecotox Environ Safe 72: 1899–1904.
20. Gratão PL, Polle A, Lea PJ, Azevedo RA (2005) Making the life of heavy metal-stressed plants a little easier. Funct Plant Biol 32: 481–494.
21. Monteiro CC, Carvalho RF, Gratão PL, Carvalho G, Tezotto T, et al. (2011) Biochemical responses of the ethylene-insensitive *Never ripe* tomato mutant subjected to cadmium and sodium stresses. Environ Exp Bot 71: 306–320.
22. Zhang Y, Meng DF, Wang ZG, Guo HS, Wang Y, et al. (2012) Oxidative stress response in atrazine-degrading bacteria exposed to atrazine. J Hazard Mater 229: 434–438.
23. Martins PF, Carvalho G, Gratão PL, Dourado MN, Pileggi M, et al. (2011) Effects of the herbicides acetochlor and metolachlor on antioxidant enzymes in soil bacteria. Process Biochem 46: 1186–1195.
24. Gratão PL, Monteiro CC, Carvalho RF, Tezotto T, Piotto FA, et al. (2012) Biochemical dissection of *diageotropica* and *Never ripe* tomato mutants to Cd-stressful conditions. Plant Physiol Biochem 56: 79–96.
25. Monteiro CC, Rolão MB, Franco MR, Peters LP, Cia MC, et al. (2012) Biochemical and histological characterization of tomato mutants. An Acad Bra Cienc 84: 573–585.
26. Cia MC, Guimarães ACR, Medici LO, Chabregas SM, Azevedo RA (2012) Antioxidant responses to water deficit by drought-tolerant and -sensitive sugarcane varieties. Ann Appl Biol 161: 313–324.
27. Carvalho RF, Piotto FA, Schmidt D, Peters LP, Monteiro CC, et al. (2011) Seed priming with hormones does not alleviate induced oxidative stress in maize seedlings subjected to salt stress. Sci Agric 68: 598–602.
28. Lushchak VI (2011) Adaptive response to oxidative stress: Bacteria, fungi, plants and animals. Comp Biochem Physiol Part C Toxicol Pharmacol 153: 175–190.
29. Masip L, Veeravalli K, Georgioui G (2006) The many faces of glutathione in bacteria. Antioxid Redox Signaling 8: 753–762.
30. Ghelfi A, Gaziola SA, Cia MC, Chabregas SM, Falco MC, et al. (2011) Cloning, expression, molecular modelling and docking analysis of glutathione transferase from *Saccharum officinarum*. Ann Appl Biol 159: 267–280.
31. Allocati N, Federici L, Masulli M, Di Ilio C (2009) Glutathione transferases in bacteria. Febs Journal 276: 58–75.
32. Camargo AO, Moniz AC, Jorge JA, Valadares JMA (2009) Métodos de análise química, mineralógica e física de solos do instituto agronômico de Campinas. Instituto Agronômico 106, 77.
33. Nie Z-J, Hang B-J, Cai S, Xie X-T, He J, et al. (2011) Degradation of Cyhalofop-butyl (CyB) by *Pseudomonas azotoformans* strain QDZ-1 and cloning of a novel gene encoding CyB-hydrolyzing esterase. J Agric Food Chem 59: 6040–6046.
34. Martins PF, Martinez CO, Carvalho G, Carneiro PIB, Azevedo RA, et al. (2007) Selection of microorganisms degrading S-metolachlor herbicide. Braz Arch Biol Technol 50: 153–159.
35. Araujo WL, Marcon J, Maccheroni W, Van Elsas JD, Van Vuurde JWL, et al. (2002) Diversity of endophytic bacterial populations and their interaction with *Xylella fastidiosa* in citrus plants. Appl Environ Microbiol 68: 4906–4914.
36. Heuer H, Krsek M, Baker P, Smalla K, Wellington EMH (1997) Analysis of actinomycete communities by specific amplification of genes encoding 16S rRNA and gel-electrophoretic separation in denaturing gradients. Appl Environ Microbiol 63: 3233–3241.
37. Lane DJ, Pace B, Olsen GJ, Stahl DA, Sogin ML, et al. (1985) Rapid-determination of 16S ribosomal RNA-sequences for phylogenetic analyses. Proc Natl Acad Sci USA 82: 6955–6959.
38. Konstantinidis KT, Tiedje JM (2005) Towards a genome-based taxonomy for prokaryotes. J Bacteriol 187: 6258–6264.
39. Saitou N, Nei M (1987) The neighbor-joining method - a new method for reconstructing phylogenetic trees. Mol Biol Evol 4: 406–425.
40. Tamura K, Dudley J, Nei M, Kumar S (2007) MEGA4: Molecular evolutionary genetics analysis (MEGA) software version 4.0. Molecular Mol Biol Evol 24: 1596–1599.
41. Sangali S, Brandelli A (2000) Feather keratin hydrolysis by a *Vibrio* sp strain kr2. J Appl Microbiol 89: 735–743.
42. Heath RL, Packer L (1968) Photoperoxidation in isolated chloroplasts I. Kinetics and stoichiometry of fatty acid peroxidation. Arch Biochem Biophys 125: 189–198.
43. Anderson ME (1985) Dertermination of glutathione and glutathione disulfide in biological samples. Methods Enzymol 113: 548–555.
44. Gratão PL, Monteiro CC, Antunes AM, Peres LEP, Azevedo RA (2008) Acquired tolerance of tomato (*Lycopersicon esculentum* cv. Micro-Tom) plants to cadmium-induced stress. Ann Appl Biol 153: 321–333.
45. Bradford MM (1976) A rapid and sensitive method for the quantitation of microgram quantities of protein utilizing the principle of protein-dye binding. Anal Biochem 72 248–254.
46. Garcia JS, Gratão PL, Azevedo RA, Arruda MAS (2006) Metal contamination effects on sunflower (*Helianthus annuus* L.) growth and protein expression in leaves during development. J Agric Food Chem 54: 8623–8630.
47. Azevedo RA, Alas RM, Smith RJ, Lea PJ (1998) Response of antioxidant enzymes to transfer from elevated carbon dioxide to air and ozone fumigation, in the leaves and roots of wild-type and a catalase-deficient mutant of barley. Physiol Plant 104: 280–292.
48. Boaretto LF, Carvalho G, Borgo L, Creste S, Landell MGA, et al. (2013) Water stress reveals differential antioxidant responses of tolerant and non-tolerant sugarcane genotypes. Plant Physiol Biochem 74: 165–175.
49. Gomes-Junior RA, Gratão PL, Gaziola SA, Mazzafera P, Lea PJ, et al. (2007) Selenium-induced oxidative stress in coffee cell suspension cultures. Funct Plant Biol 34: 449–456.
50. Zablotowicz RM, Hoagland RE, Locke MA, Hickey WJ (1995) Glutathione-S-transferase activity and metabolism of glutathione conjugates by rhizosphere bacteria. Appl Environ Microbiol 61: 1054–1060.
51. Mongkolsuk S, Dubbs J, Vattanaviboon P (2005) Chemical modulation of physiological adaptation and cross-protective responses against oxidative stress in soil bacterium and phytopathogen, *Xanthomonas*. J Ind Microbiol Biotechnol 32: 687–690.
52. Dwivedi S, Singh BR, Al-Khedhairy AA, Musarrat J (2011) Biodegradation of isoproturon using a novel *Pseudomonas aeruginosa* strain JS-11 as a multi-functional bioinoculant of environmental significance. J Hazard Mater 185: 938–944.
53. Viegas CA, Costa C, André S, Viana P, Ribeiro R, et al. (2012) Does S-metolachlor affect the performance of *Pseudomonas* sp. strain ADP as bioaugmentation bacterium for atrazine-contaminated soils? PLoS ONE 7(5): e37140.

54. Ramu S, Seetharaman B (2014) Biodegradation of acephate and methamidophos by a soil bacterium *Pseudomonas aeruginosa* strain Is-6. Environ Sci Heal B 49: 23–34.
55. aMattos MLT, Thomas R (1996) Degradation of the herbicide clomazone by *Pseudomonas fluorescens*. In: W Sand editor.Biodeterioration and Biodegradation.Weinheim: V C H Verlagsgesellschaft. pp.623–630.
56. Tironi SP, Belo AF, Fialho CMT, Galon L, Ferreira EA, et al. (2009) Efeito de herbicidas na atividade microbiana do solo. Planta Daninha 27: 995–1004.
57. García-Cruz U, Celis LB, Poggi P, Meraz M (2010) Inhibitory concentrations of 2,4D and its possible intermediates in sulfate reducing biofilms. J Hazard Mater 179: 591–595.
58. Ferhatoglu Y, Avdiushko S, Barrett M (2005) The basis for the safening of clomazone by phorate insecticide in cotton and inhibitors of cytochrome P450s. Pestic Biochem Physiol 81: 59–70.
59. Matsue Y, Mizuno H, Tomita T, Asami T, Nishiyama M, et al. (2010) The herbicide ketoclomazone inhibits 1-deoxy-D-xylulose 5-phosphate synthase in the 2-C-methyl-D-erythritol 4-phosphate pathway and shows antibacterial activity against *Haemophilus influenzae*. J Antibiot 63: 58[illegible]–588.
60. Imlay JA (2008) Cellular defenses against superoxide and hydrogen peroxide. Annu Rev Biochem 77: 755–776.
61. Shao TJ, Yang GQ, Wang MZ, Lu ZM, Min H, et al. (2010) Reduction of oxidative stress by bioaugmented strain *Pseudomonas* sp HF-1 and selection of potential biomarkers in sequencing batch reactor treating tobacco wastewater. Ecotoxicology 19: 1117–1123.
62. Ryan KC, Johnson OE, Cabelli DE, Brunold TC, Maroney MJ (2010) Nickel superoxide dismutase: structural and functional roles of Cys2 and Cys6. J Biol Inorg Chem 15: 795–807.
63. Lushchak VI (2001) Oxidative stress and mechanisms of protection against it in bacteria. Biochemistry-Moscow 66: 476–489.
64. Goulielmos GN, Arhontaki K, Eliopoulos E, Tserpistali K, Tsakas S, et al. (2003) Drosophila Cu,Zn superoxide dismutase gene confers resistance to paraquat in *Escherichia coli*. Biochem Biophys Res Commun 308: 433–438.
65. Zeng HW, Cai YJ, Liao XR, Qian SL, Zhang F, et al. (2010) Optimization of catalase production and purification and characterization of a novel cold-adapted Cat-2 from mesophilic bacterium *Serratia marcescens* SYBC-01. Ann Microbiol 60: 701–708.
66. Dávila Costa JS, Albarracín VH, Abate CM (2011) Responses of environmental *Amycolatopsis* strains to copper stress. Ecotox Environ Safe 74: 2020–2028.
67. Hultberg M (1998) Rhizobacterial glutathione levels as affected by starvation and cadmium exposure. Curr Microbiol 37: 301–305.
68. Veremeenko EG, Maksimova NP (2010) Activation of the antioxidant complex in *Pseudomonas aurantiaca*-producer of phenazine antibiotics. Microbiology 79: 439–444.
69. Smirnova GV, Oktyabrsky ON (2005) Glutathione in bacteria. Biochemistry-Moscow+ 70: 1199–1211.
70. Foyer CH, López-Delgado H, Dat JF, Scott IM (1997) Hydrogen peroxide and glutathione associated mechanisms of acclamatory stress tolerance and signaling. Physiol Plant 100: 241–254.
71. Gomes-Junior RA, Moldes CA, Delite FS, Gratão PL, Mazzafera P, et al. (2006) Nickel elicits a fast antioxidant response in *Coffea arabica* cells. Plant Physiol Biochem 44: 420–429.
72. Gomes-Junior RA, Moldes CA, Delite FS, Pompeu GB, Gratão PL, et al. (2006) Antioxidant metabolism of coffee cell suspension cultures in response to cadmium. Chemosphere 65: 1330–1337.
73. Labrou NE, Karavangeli M, Tsaftaris A, Clonis YD (2005) Kinetic analysis of maize glutathione S-transferase I catalysing the detoxification from chloroacetanilide herbicides. Planta 222: 91–97.

24

Non-Hodgkin Lymphoma Risk and Insecticide, Fungicide and Fumigant Use in the Agricultural Health Study

Michael C. R. Alavanja[1]*, Jonathan N. Hofmann[1], Charles F. Lynch[2], Cynthia J. Hines[3], Kathryn H. Barry[1], Joseph Barker[4], Dennis W. Buckman[4], Kent Thomas[5], Dale P. Sandler[6], Jane A. Hoppin[6], Stella Koutros[1], Gabriella Andreotti[1], Jay H. Lubin[1], Aaron Blair[1], Laura E. Beane Freeman[1]

1 Division of Cancer Epidemiology and Genetics, National Cancer Institute, Rockville, Maryland, United States of America, **2** College of Public Health, University of Iowa, Iowa City, Iowa, United States of America, **3** National Institute for Occupational Safety and Health, Cincinnati, Ohio, United States of America, **4** IMS, Inc, Calverton, Maryland, United States of America, **5** National Exposure Research Laboratory, U.S. Environmental Protection Agency, Research Triangle Park, North Carolina, United States of America, **6** Epidemiology Branch, National Institute for Environmental Health Sciences, Research Triangle Park, North Carolina, United States of America

Abstract

Farming and pesticide use have previously been linked to non-Hodgkin lymphoma (NHL), chronic lymphocytic leukemia (CLL) and multiple myeloma (MM). We evaluated agricultural use of specific insecticides, fungicides, and fumigants and risk of NHL and NHL-subtypes (including CLL and MM) in a U.S.-based prospective cohort of farmers and commercial pesticide applicators. A total of 523 cases occurred among 54,306 pesticide applicators from enrollment (1993–97) through December 31, 2011 in Iowa, and December 31, 2010 in North Carolina. Information on pesticide use, other agricultural exposures and other factors was obtained from questionnaires at enrollment and at follow-up approximately five years later (1999–2005). Information from questionnaires, monitoring, and the literature were used to create lifetime-days and intensity-weighted lifetime days of pesticide use, taking into account exposure-modifying factors. Poisson and polytomous models were used to calculate relative risks (RR) and 95% confidence intervals (CI) to evaluate associations between 26 pesticides and NHL and five NHL-subtypes, while adjusting for potential confounding factors. For total NHL, statistically significant positive exposure-response trends were seen with lindane and DDT. Terbufos was associated with total NHL in ever/never comparisons only. In subtype analyses, terbufos and DDT were associated with small cell lymphoma/chronic lymphocytic leukemia/marginal cell lymphoma, lindane and diazinon with follicular lymphoma, and permethrin with MM. However, tests of homogeneity did not show significant differences in exposure-response among NHL-subtypes for any pesticide. Because 26 pesticides were evaluated for their association with NHL and its subtypes, some chance finding could have occurred. Our results showed pesticides from different chemical and functional classes were associated with an excess risk of NHL and NHL subtypes, but not all members of any single class of pesticides were associated with an elevated risk of NHL or NHL subtypes. These findings are among the first to suggest links between DDT, lindane, permethrin, diazinon and terbufos with NHL subtypes.

Editor: Suminori Akiba, Kagoshima University Graduate School of Medical and Dental Sciences, Japan

Funding: This work was supported by the Intramural Research Program of the NIH, National Cancer Institute, Division of Cancer Epidemiology and Genetics (Z01CP010119) and The National Institutes of Environmental Health Sciences (Z01ES049030). The funders had no role in study design, data collection and analysis, decision to publish, or preparation of the manuscript. IMS, Inc, provided support in the form of salaries for authors Joseph Barker and Denis W Buckman, but did not have any additional role in the study design, data collection and analysis, decision to publish, or preparation of the manuscript. The specific roles of these authors are articulated in the 'author contributions' section.

Competing Interests: The authors have the following interests. Joseph Barker and Dennis W. Buckman are employed by IMS, Inc. There are no patents, products in development or marketed products to declare.

* Email: alavanjm@mail.nih.gov

Introduction

Since the 1970s, epidemiologic studies of non-Hodgkin lymphoma (NHL) and multiple myeloma (MM) have shown increased risk among farmers and associations with the type of farming practiced [1–6]. While farmers are exposed to many agents that may be carcinogenic [7]; there has been a particular focus on pesticides. Studies from around the world have suggested increased risk of NHL or MM [8,9] and other NHL subtypes [10] in relation to the use of specific pesticides in different functional classes (i.e., insecticides, fungicides, fumigants and herbicides). A meta-analysis of 13 case-control studies published between 1993–2005 observed an overall significant meta-odds ratio (OR) between occupational exposure to pesticides and NHL (OR = 1.35; 95% CI: 1.2–1.5) [11]. This risk was greater among individuals with more than 10 years of exposure (OR = 1.65; 95% CI: 1.08–1.95) [11], but the meta-analysis lacked details about the use of specific pesticides and other risk factors [11]. Although the International Agency for Research on Cancer (IARC) has classified "Occupational exposures in spraying and application of non-arsenical insecticides" as "probably carcinogenic to humans", the human evidence for the 17 individual pesticides evaluated in this monograph was determined to be inadequate for nine and there were no epidemiological studies for eight pesticides [12]. Since then, more studies have focused on cancer risk from specific pesticides, although the information is still relatively limited for many cancer-pesticide combinations [8,9].

To help fill the current information gap we evaluated the relationships between the use of specific insecticides, fungicides and fumigants and NHL in the Agricultural Health Study (AHS), a prospective cohort of licensed private (i.e., mostly farmer) and commercial pesticide applicators. Because the etiology of NHL and its B and T cell subtypes may differ by cell type[13], we also evaluated risk by subtype while controlling for potential confounding factors suggested from the literature [1?], and the AHS data.

Novelty and Impact

These findings on occupationally exposed pesticide applicators with high quality exposure information are among the first to suggest links between DDT, lindane, permethrin, diazinon and terbufos and specific NHL subtypes in a prospective cohort study.

Materials and Methods

Study Population

The AHS is a prospective cohort study of 52,394 licensed private pesticide applicators (mostly farmers) in Iowa and North Carolina and 4,916 licensed commercial applicators in Iowa (individuals paid to apply pesticides to farms, homes, lawns, etc.), and 32,346 spouses of private applicators. Only applicators are included in this analysis. The cohort has been previously described in detail [14,15] and study questionnaires are available on the AHS website (www.aghealth.nih.gov). Briefly, individuals seeking licenses to apply restricted use pesticides were enrolled in the study from December 1993 through December 1997 (82% of the target population enrolled). At enrollment, subjects did not sign a written informed consent form. However, the cover letter of the questionnaire booklet informed subjects of the voluntary nature of participation, the ability to not answer any question, and it provided an assurance of confidentiality (including a Privacy Act Notification statement). The letter also included a written summary of the purpose of research, time involved, benefits of research, and a contact for questions about the research. The cover letter to the take-home questionnaire included all of the above and also informed the participant that they had the right to withdraw at any time. Finally, subjects were specifically informed that their contact information (including Social Security Number) would be used to search health and vital records in the future. The participants provided consent by completing and returning the questionnaire booklet. These documents and procedures were approved in 1993 by all relevant institutional review boards (i.e., National Cancer Institute Special Studies Institutional Review Board, Westat Institutional Review Board, and the University of Iowa Institutional Review Board-01).

Excluded from this analysis were study participants who had a history of any cancer at the time of enrollment (n = 1094), individuals who sought pesticide registration in Iowa or North Carolina but did not live in these states at the time of registration (n = 341) and were thus outside the catchment area of these cancer registries and individuals that were missing information on potential confounders (i.e., race or total herbicides application days [n = 1,569]). This resulted in an analysis sample of 54,306. We obtained cancer incidence information by regular linkage to the population-based cancer registry files in Iowa and North Carolina. In addition, we linked cohort members to state mortality registries of Iowa and North Carolina and the nation-wide National Death Index to determine vital status, and to the nation-wide address records of the Internal Revenue Service, state-wide motor vehicle registration files, and pesticide license registries of state agricultural departments to determine residence in Iowa or North Carolina. The current analysis included all incident primary NHL, as well as CLL and MM (which are now classified as NHL) [13] ($n = 523$) diagnosed from enrollment (1993–1997) through December 31, 2010 in North Carolina and from enrollment (1993–1997) through December 31, 2011 in Iowa, the last date of complete cancer incidence reports in each state. We ended follow-up and person-year accumulation at the date of diagnosis of any cancer, death, movement out of state, or December 31, 2010 in North Carolina and December 31, 2011 in Iowa, whichever was earlier.

Tumor Characteristics

Information on tumor characteristics was obtained from state cancer registries. We followed the definition of NHL and six subtypes of NHL used by the Surveillance Epidemiology and End Results (SEER) coding scheme [16] which was based on the Pathology Working Group of the International Lymphoma Epidemiology Consortium (ICD-O-3 InterLymph modification) classification (Table S1 in File S1, [17], i.e., 1. Small B-cell lymphocytic lymphomas (SLL)/chronic B-cell lymphocytic lymphomas (CLL)/mantle-cell lymphomas (MCL); 2. Diffuse large B-cell lymphomas; 3. Follicular lymphomas; 4. 'Other B-cell lymphomas' consisting of a diverse set of B-cell lymphomas; 5. Multiple myeloma; and 6. T-cell NHL and undefined cell type). There were too few T-cell NHL cases available for analysis [n = 19] so this cell type was not included in the subtype analysis). The ICD-O-3 original definition (used in many earlier studies of pesticides and cancer) of NHL [18] was also evaluated in relation to pesticide exposure to allow a clearer comparison of our results with previous studies.

Exposure Assessment

Initial information on lifetime use of 50 specific pesticides (Table S2 in File S1), including 22 insecticides, 6 fungicides and 4 fumigants was obtained from two self-administered questionnaires [14,15] completed during cohort enrollment (Phase 1). All 57,310 applicators completed the first enrollment questionnaire, which inquired about ever/never use of 50 pesticides, as well as duration (years) and frequency (average days/year) of use for a subset of 22 pesticides including 9 insecticides, 2 fungicides and 1 fumigant. In addition, 25,291 (44%) of the applicators returned the second (take-home) questionnaire, which inquired about duration and frequency of use for the remaining 28 pesticides, including 13 insecticides, 4 fungicides and 3 fumigants.

A follow-up questionnaire, which ascertained pesticide use since enrollment, was administered approximately 5 years after enrollment (1999–2005, Phase 2) and completed by 36,342 (63%) of the original participants. The full text of the questionnaires is available at www.aghealth.nih.gov. For participants who did not complete the Phase 2 questionnaire (20,968 applicators, 37%), a data-driven multiple imputation procedure which used logistic regression and stratified sampling [19] was employed to impute use of specific pesticides in Phase 2. Information on pesticide use from Phase 1, Phase 2 and imputation for Phase 2 was used to construct three cumulative exposure metrics: (i) lifetime days of pesticide use (i.e., the product of years of use of a specific pesticide and the number of days used per year); (ii) intensity-weighted lifetime days of use (i.e., the product of lifetime days of use and a measure of exposure intensity) and (iii) ever/never use data for each pesticide. Intensity was derived from an exposure-algorithm, which was based on exposure measurements from the literature and individual information on pesticide use and practices (e.g., whether or not they mixed pesticides, application method, whether or not they repaired equipment and use of personal protective equipment) obtained from questionnaires completed by study participants [20].

Statistical Analyses

We divided follow-up time into 2-year intervals to accumulate person-time and update time-varying factors, such as attained age and pesticide use. We fit Poisson models to estimate rate ratios (RRs) and 95% confidence intervals (95% CI) to evaluate the effects of pesticide use on rates of overall NHL and the five NHL subtypes.

We evaluated pesticides with 15 or more exposed cases of total NHL, thereby excluding aluminum phosphide, carbon tetrachloride/carbon disulfide, ethylene dibromide, trichlorfon, and ziram leaving 26 insecticides, fungicides and fumigants for analysis (permethrin for animal use and crop use were combined into one category, all insecticides, fungicides and fumigants are listed in Table S2 in File S1). For each pesticide, we evaluated ever vs. never exposure, as well as tertiles of exposure which were created based on the distribution of all NHL exposed cases and compared to those unexposed. In the NHL subtype analysis and in circumstances where multiple pesticides were included in the model we categorized exposure for each pesticide into unexposed (i.e., never users) and two exposed groups (i.e., low and high) separated at the median exposure level. The number of exposed cases included in the ever/never analysis and in the trend analysis can differ because of the lack of information necessary to construct quantitative exposure metrics for some individuals.

Several lifestyle and demographic factors associated with NHL in the AHS cohort or previously suggested as possible confounders in the NHL literature[13] were evaluated as potential confounders in this analysis. These included: age at enrollment, gender, race, state, license type, education, autoimmune diseases, family history of lymphoma in first-degree relatives, body mass index, height, cigarette smoking history, alcohol consumption per week and several occupational exposures[1–13] including number of livestock, cattle, poultry, whether they raised poultry, hogs or sheep, whether they provided veterinary services to their animals, number of acres planted, welding, diesel engine use, number of years lived on the farm, total days of any pesticide use, and total days of herbicide use. However, since most of these variables did not change the risk estimates for specific pesticides, we present results adjusted for age, race, state and total days of herbicide use, which impacted risk estimates by more than 10% for some subtypes. We also performed analyses adjusting for specific insecticides, fungicides and fumigants shown to be associated with NHL or a specific NHL subtype in the current analysis. Tests for trend used the median value of each exposure category. All tests were two-sided and conducted at $\alpha = 0.05$ level. Analysis by NHL subtype was limited to insecticides, fungicides, and fumigants with 6 or more exposed cases.

We also fit polytomous logit models, where the dependent variable was a five-level variable (i.e., five NHL subtypes) and a baseline level (i.e., no NHL) to estimate exposure-response odds ratios (ORs) and 95% confidence intervals (CIs) for each subtypes of NHL. We then used polytomous logit models to estimate exposure-response trend while adjusting for age, state, race and total days of herbicide use, as in the Poisson models, and tested homogeneity among the 5 NHL subtypes.

Poisson models were fit using the GENMOD procedure and polytomous logit models were fit using the LOGISTIC procedure of the SAS 9.2 statistical software package (SAS Institute, Cary, NC). Summary estimates of NHL and NHL subtype risks for both Poisson models and polytomous logit models incorporated imputed data and were calculated along with standard error estimates, confidence intervals, and p-values, using multiple imputation methods implemented in the MIANALYZE procedure of SAS 9.2.

We also evaluated the impact of the additional pesticide exposure information imputed for Phase 2 on risk estimates. We compared risk estimates for those who completed both the phase 1 enrollment and take-home questionnaires and the phase 2 questionnaires (n = 17,545) with risk estimates obtained from the combined completed questionnaire data plus the imputed phase 2 data (n = 54,306). We also explored the effect of lagging exposure data 5 years because recent exposures may not have had time to have an impact on cancer development. For comparison to previous studies, we also assessed the exposure-response association for NHL using the original ICD-O-3 definition of NHL [18] and the new definition [16] in Table S3 in File S1. Unless otherwise specified, reported results show un-lagged exposure information from both Phase 1 and Phase 2 including Phase 2 imputed data for lifetime exposure-days and intensity-weighted lifetime days of use and NHL defined by the InterLymph modification of ICD-O-3 [17]. Data were obtained from AHS data release versions P1REL201005.00 (for Phase 1) and P2REL201007.00 (for Phase 2).

Results

The 54,306 applicators in this analysis contributed 803,140 person-years of follow-up from enrollment through December 31, 2010 in North Carolina and December 31, 2011 in Iowa (Table 1). During this period, there were 523 incident cases of NHL, including 148 SLL/CLL/MCL, 117 diffuse large B-cell lymphomas, 67 follicular lymphomas, 53 'other B-cell lymphomas' (consisting of a diverse set of B-cell lymphomas) and 97 cases of MM. Another 41 cases consisting of T-cell lymphomas (n = 19) and non-Hodgkin lymphoma of unknown lineage (n = 22) were excluded from cell type-specific analyses because of small numbers of cases with identified cell types. Between enrollment and the end of follow-up, 6,195 individuals were diagnosed with an incident cancer other than NHL, 4,619 died without a record of cancer in the registry data, and 1,248 cohort members left the state and could not be followed-up for cancer. Person-years of follow-up accumulated for all of these study participants after enrollment until they were censored for the incident cancer, death or moving out of the state (data not shown). The risk of NHL increased significantly and monotonically with age in the AHS cohort in this analysis (p = 0.001) and age-adjusted risks were significant for state and NHL overall and race for multiple myeloma (data not shown). Total days of herbicide use had a small but significant effect on the risk of some NHL subtypes, but not on NHL overall. No other demographic or occupational factors showed evidence of confounding so they were not included in the final models.

In Table 2 we present ever/never results for 26 insecticides, fungicides and fumigants by total NHL and by NHL subtype adjusted for age, race, state and herbicide use (total life-time days). Terbufos was the only pesticide associated with an increased risk of total NHL in the ever/never use analysis (RR = 1.2 [1.0–1.5]), although the trend for increasing use and risk of total NHL was not significant (p trend = 0.43) (Table 3). In contrast, there were a few chemicals that were not associated with ever/never use, but

Table 1. Baseline characteristics of AHS study participants in the NHL incidence analysis[1,2].

Variables	All NHL cases (%)	Cohort Person-years.
Age at Enrollment		
<45	84 (16.1)	426,288
45–49	51 (9.8)	101,018
50–54	75 (14.3)	84,998
55–59	90 17.2)	74,440
60–64	78 (14.9)	56,978
65–69	79 (15.1)	35,071
≥70	66 (12.6)	24,347
Race		
White	509 (97.3)	787,799
Black	14 (2.7)	15,341
State		
IA	332 (63.5)	537,252
NC	191 (36.5)	265,888
Lifetime Total Herbicide Exposure Days		
0–146 days	170 (32.5)	251,401
147–543 days	169 (32.3)	273,107
544–2453 days	184 (35.2)	278,632

[1]During the period from enrollment (1993–1997) to December 31, 2010 in NC and December 31, 2011 in Iowa.
[2]Individuals with missing ever/never exposure information or missing confounding variable information were not included in the table.

did show evidence of an exposure-response association. Lindane was the only pesticide that showed a statistically significant increasing trend in risk for NHL with both exposure metrics, for lifetime-days of lindane use the RR were = 1.0 (ref), 1.2 (0.7–1.9), 1.0 (0.6–1.7), 2.5 (1.4–4.4); p trend = 0.004 and intensity-weighted lifetime-days of use the: RR were: = 1.0 (ref), 1.3 (0.8–2.2), 1.1 (0.7–1.8), 1.8 (1.0–3.2); p trend = 0.04. DDT showed a significant trend for NHL risk with life-time days of use RR = 1.0 (ref), 1.3 (0.9–1.8), 1.1 (0.7–1.7), 1.7 (1.1–2.6); p trend = 0.02, while the intensity weighted lifetime days of use of DDT was of borderline significance: RR = 1.0 (ref), 1.2 (0.8–1.8), 1.1 (0.8–1.7), 1.6 (1.0–2.3); p trend = 0.06. The number of lifetime days of use of lindane and DDT was weakly correlated (coefficient of determination = 0.04), and the pattern of NHL risk showed little change when both were included in the model. The results for lindane adjusted for DDT were, RR = 1.0 (ref), 1.2 (0.7–2.0), 1.0 (0.5–1.8), 1.6 (0.9–3.3); p trend = 0.07 and the results for DDT adjusted for lindane were, RR = 1.0 (ref), 1.3 (0.9–2.0), 0.9 (0.6–1.6), 1.6 (0.9–2.6); p trend = 0.08).

We also evaluated pesticides by NHL sub-type. In the ever/never analyses (Table 2), permethrin was significantly associated with multiple myeloma, RR = 2.2 (1.4–3.5) and also demonstrated an exposure-response trend (RR = 1.0 (ref), 1.4 (0.8–2.7), 3.1 (1.5–6.2); p trend = 0.002) (Table 4). Similarly, there was an elevated risk of SLL/CLL/MCL with terbufos in ever/never analyses RR = 1.4 (0.97–2.0) and an exposure response trend (RR = 1.0 (ref), 1.3 (0.8–2.0), 1.6 (1.0–2.5); p trend = 0.05). For follicular lymphoma, lindane showed an elevated but non-significant association for ever use, RR = 1.7 (0.96–3.2) and a significant exposure-response association (RR = 1.0 (ref), 4.9 (1.9–12.6), 3.6 (1.4–9.5); p trend = 0.04). There were also two chemicals with evidence of exposure-response that were not associated with specific subtypes in the ever/never analyses: DDT (Dichlorodiphenyltrichloroethane) with SLL/CLL/MCL (RR = 1.0 (ref), 1.0 (0.5–1.8), 2.6 (1.3–4.8; p trend = 0.04); and diazinon with follicular lymphoma (RR = 1.0 (ref), 2.2 (0.9–5.4), 3.8 (1.2–11.4); p trend = 0.02) (Table 4).

The pattern of increased CLL/SLL/MCL risk with increased use of DDT and terbufos remained after both insecticides were placed in our model concurrently. CLL/SLL/MCL risk increased with DDT use (RR = 1.0 (ref), 0.9 (0.5–4.7); 2.4 (1.1–4.7); p trend = 0.04), and a pattern of increased CLL/SLL/MCL risk was also observed with terbufos use (RR = 1.0 (ref), 1.1 (0.6–2.1), 1.7 (0.9–3.3) p trend = 0.07), although the trend was not significant for terbufos. Similarly, the pattern of increased follicular lymphoma risk with lindane use and diazinon use remained after both insecticides were placed in our model concurrently. Follicular lymphoma risk increased with diazinon use (RR = 1.0 (ref), 4.1 (1.5–11.1); 2.5 (0.9–7.2); p trend = 0.09), and a similarly, pattern of increased follicular lymphoma risk was observed with lindane use (RR = 1.0 (ref), 1.6 (0.6–4.1), 2.6 (0.8–8.3) p trend = 0.09), although neither remained statistically significant (Table 4).

Three chemicals showed elevated risks in ever/never analyses for certain subtypes, with no apparent pattern in exposure-response analyses: metalaxyl and chlordane with SLL/CLL/MCL, RR = 1.6 (1.0–2.5) and RR = 1.4 (0.97–2.0) respectively, and methyl bromide with diffuse large B-cell lymphoma RR = 1.9 (1.1–3.3). Although there was evidence of association by subtype, and polytomous logit models indicated homogeneity across subtypes for lindane (p = 0.54), DDT (p = 0.44) and any other pesticide evaluated in this study (e.g., permethrin (p = 0.10), diazinon (p = 0.09), terbufos (p = 0.63), (last column in Table 4).

There was no evidence of confounding of the total NHL associations with either lindane or DDT. We also calculated RR for those who completed both the phase 1 enrollment and take-home questionnaires and the phase 2 questionnaire (n = 17,545) and found no meaningful difference in the RR that also included imputed exposures, although there was an increase in precision of

Table 2. Pesticides exposure (ever/never) and adjusted Relative Risk of total NHL and NHL Subtype[1].

Insecticide												
	Total NHL Cases[2]		SLL/CLL/MCL Cases[2]		Diffuse Large B-Cell Cases[2]		Follicular B-Cell Cases[2]		Other B-cell Cases[2]		Multiple Myeloma Cases[2]	
Pesticide (chemical-functional class)	Ever/Never Exposed	RR[3,4] (95% CI)	Ever/Never Exposed	RR[3,4] (95% CI)	Ever/Never Exposed	RR[3,4] (95% CI)	Ever/Never Exposed	RR[3,4] (95% CI)	Ever/Never Exposed	RR[3,4] (95% CI)	Ever/Never Exposed	RR[3,4] (95% CI)
Aldicarb	47/435	1	14/124	1.1	8/98	0.7	6/54	0.9	7/41	1.6	10/82	1.2
(carbamate-insecticide)		(0.7–1.4)		(0.6–1.8)		(0.4–1.5)		(0.3–2.2)		(0.7–3.5)		(0.6–2.2)
Carbofuran	147/317	1.1	48/86	1.2	26/78	0.8	18/39	1	13/31	0.8	31/56	1.3
(carbamate-insecticide)		(0.9–1.3)		(0.8–1.8)		(0.5–1.3)		(0.5–1.7)		(0.4–1.6)		(0.8–2.1)
Carbaryl	272/225	1	75/66	1	58/53	0.8	37/24	0.8	24/28	0.9	58/34	0.9
(carbamate-insecticide)		(0.8–1.2)		(0.7–1.5)		(0.5–1.3)		(0.5–1.3)		(0.5–1.6)		(0.6–1.4)
Chlorpyrifos	210/300	1	62/84	1	44/70	0.9	32/33	1.3	21/31	0.8	36/58	1
(organophosphate-insecticide)		(0.8–1.2)		(0.7–1.4)		(0.6–1.4)		(0.8–2.2)		(0.5–1.5)		(0.6–1.5)
Coumaphos	46/411	1.1	15/120	1.2	10/93	1	8/48	1.6	5/40	xxx	7/78	1
(organophos-phate-insecticide)		(0.8–1.5)		(0.7–2.1)		(0.5–2.1)		(0.8–3.5)				(0.1–2.1)
DDVP	55/407	1.1	13/124	0.8	10/93	1	8/48	1.3	6/39	1	12/73	1.7
(dimethyl phosphate-insecticide)		(0.8–1.5)		(0.5–1.5)		(0.5–1.9)		(0.6–2.7)		(0.4–2.4)		(0.9–3.2)
Diazinon	144/342	1	46/93	1.3	30/78	0.9	22/38	1.3	12/37	0.8	27/64	1
(organophosphorous-insecticide)		(0.8–1.3)		(0.9–1.9)		(0.6–1.4)		(0.7–2.3)		(0.4–1.6)		(0.6–1.6)
Fonofos	115/349	1.1	35/100	1.1	25/81	1.2	13/45	0.9	15/30	1.3	19/66	1.3
(organophosphorous-insecticide)		(0.9–1.4)		(0.7–1.6)		(0.7–1.9)		(0.5–1.7)		(0.7–2.5)		(0.8–2.3)
Malathion	332/163	0.9	99/43	1	72/37	0.9	46/14	1.3	30/21	0.6	61/32	0.9
(organophosphorous-insecticide)		(0.8–1.1)		(0.7–1.4)		(0.6–1.4)		(0.7–2.4)		(0.3–1.0)		(0.6–1.5)
Parathion (ethyl or methyl)	69/411	1.1	20/117	1	14/91	1	10/48	1.1	7/44	1.1	14/77	1
(organophosphorous insecticide		(0.8–1.4)		(0.7–1.4)		(0.6–1.4)		(0.8–1.5)		(0.7–1.5)		(0.8–1.5)
Permethrin (animal and crop applications)	112/363	1.1	32/106	1	18/81	0.7	18/81	1.1	9/14	0.8	20/72	**2.2**
(pyrethroid insecticide)		(0.8–1.3)		(0.6–1.5)		(0.4–1.2)		(0.6–2.0)		(0.4–1.6)		**(1.4–3.5)**
Phorate	160/325	1	53/87	1.1	31/76	0.9	20/40	0.9	19/31	0.9	26/64	1
(organophosphorous-insecticide)		(0.8–1.2)		(0.8–1.6)		(0.5–1.3)		(0.5–1.6)		(0.5–1.6)		(0.6–1.7)
Terbufos	201/267	**1.2**	64/72	1.4	42/63	1.1	31/26	1.2	26/19	1.8	32/59	1.2
(organophosphorous-insecticide)		**(1.0–1.5)**		(0.97–2.0)		(0.7–1.7)		(0.7–2.1)		(0.94–3.2)		(0.7–1.9)
Chlorinated Insecticides												
Aldrin	116/364	0.9	53/99	0.9	15/91	0.8	13/45	0.8	12/37	0.6	29/62	1.5
(chlorinated insecticide)		(0.7–1.1)		(0.6–1.4)		(0.4–1.6)		(0.4–1.6)		(0.3–1.3)		(0.9–2.5)
Chlordane	136/344	1	49/90	1.4	20/86	0.6	18/41	1.2	13/36	1	31/60	1.2
(chlorinated insecticide)		(0.8–1.3)		(0.99–2.1)		(0.4–1.0)		(0.7–2.1)		(0.7–2.0)		(0.8–1.9)
DDT	182/300	1	59/79	1.2	34/73	0.8	18/41	0.9	20/31	1.1	40/50	1.1

Table 2. Cont.

Insecticide												
	Total NHL Cases[2]		**SLL/CLL/MCL Cases**[2]		**Diffuse Large B-Cell Cases**[2]		**Follicular B-Cell Cases**[2]		**Other B-cell Cases**[2]		**Multiple Myeloma Cases**[2]	
Pesticide (chemical-functional class)	**Ever/Never Exposed**	**RR**[3,4]	**Ever/Never Exposed**	**RR**[3,4]	**Ever/Never Exposed**	**RR**[3,4]	**Ever/Never Exposed**	**RR**[3,4]	**Ever/Never Exposed**	**RR**[3,4]	**Ever/Never Exposed**	**RR**[3,4]
		(95% CI)		**(95% CI)**		**(95% CI)**		**(95% CI)**		**(95% CI)**		**(95% CI)**
(chlorinated insecticide)		(0.8–1.3)		(0.8–1.8)		(0.5–1.3)		(0.5–1.6)		(0.6–2.1)		(0.7–1.8)
Dieldrin	35/442	0.9	5/130	xxx	4/101	xxx	4/54	xxx	7/42	1	10/81	0.9
(chlorinated insecticide)		(0.6–1.2)								(0.7–2.0)		(0.5–1.4)
Heptachlor	90/384	1	33/104	1.1	10/95	1.1	9/48	1.1	13/36	0.9	17/72	1.1
(chlorinated insecticide)		(0.7–1.2)		(0.7–3.0)		(0.3–3.1)		(0.5–3.2)		(0.5–2.7))		(0.6–2.0)
Lindane	85/306	1	27/113	1.2	12/95	0.6	16/41	1.7	9/40	0.7	13/73	1.1
(chlorinated insecticide)		(0.8–1.2)		(0.6–1.5)		(0.3–1.1)		(0.96–3.2)		(0.4–1.2)		(0.5–2.0)
Toxaphene	79/397	1	21/116	0.9	14/90	0.8	9/47	1	10/40	1.1	19/73	1.1
(chlorinated insecticide)		(0.7–1.2)		(0.5–1.5)		(0.4–1.4)		(0.6–2.0)		(0.6–2.0)		(0.6–1.9)
Fungicides												
Benomyl	54/428	1.1	18/123	1.2	12/95	1.1	4/51	xxx	4/51	xxx	11/80	1.1
(carbamate fungicide)		(0.8–1.5)		(0.7–2.0)		(0.6–1.9)						(0.6–2.0)
Captan	60/406	1.1	18/118	1.1	12/91	0.9	5/51	xxx	6/39	1.1	12/76	1.2
(phthalimide fungicide)		(0.8–1.4)		(0.6–1.8)		(0.5–1.8)				(0.5–2.7)		(0.6–2.2)
Chloro-thalonil	35/474	0.8	9/135	0.9	6/107	0.5	5/60	xxx	2/50	xxx	11/84	1.2
(poly-chlorinated aromatic thalonitrile fungicide)		(0.5–1.2)		(0.4–1.9)		(0.2–1.3)						(0.6–2.3)
Maneb/	44/437	0.9	13/127	1.1	12/95	1.1	4/60	xxx	5/49	xxx	10/79	0.8
Mancozeb		(0.7–1.3)		(0.6–2.1)		(0.6–2.1)						(0.4–1.7)
(dithiocarbamate fungicide)												
Metalaxyl	108/381	1	34/106	**1.6**	27/82	1.1	10/48	0.7	10/40	0.9	21/71	0.8
(acylalanine fungicide)		(0.8–1.3)		**(1.0–2.5)**		(0.6–1.8)		(0.4–1.4)		(0.4–1.7)		(0.4–1.3)
Fumigant												
Methyl bromide	85/425	1.1	18/126	0.9	28/86	**1.9**	7/58	0.6	8/44	2.2	19/76	1
(methyl halide fumigant)		(0.9–1.5)		(0.5–1.7)		**(1.1–3.3)**		(0.2–1.4)		(0.9–5.7)		(0.6–1.8)

[1]During the period from enrollment (1993–1997) to December 31, 2010 in NC and December 31, 2011 in Iowa.
[2]Numbers of cases by NHL subtype do not sum to total number of NHL cases (n = 523) due to missing data.
[3]Adjusted RR: age (<45, 45–49, 50–54, 55–59, 60–64, 65–69, ≥70), State (NC vs. IA), Race (White vs. Black), AHS herbicides (tertiles of total herbicide use-days). Statistically significant RR and 95% confidence limits are bolded.
[4]RR was not calculated if the number of exposed cases in a pesticide-NHL subtype cell was <6 and the missing RR was marked with an XXX. Statistically significant RRs and 95% confidence limits are bolded.

Table 3. Pesticide exposure (lifetime-days & intensity weighted life-time days) and adjusted risks of total NHL incidence[1].

Insecticides						
Pesticide (chemical-functional class)	**NHL Cases[2]**	**Non-Cases[2]**	**RR[3,4] (95% CI) by Total Days of Exposure**	**NHL**	**Non-Cases**	**RR[3,4] (95% CI)**
[days of lifetime exposure for each category]				**Cases[2,]**		**Intensity-weighted days of exposure**
Aldicarb (carbamate-insecticide)						
None	238	21557	1.0 (ref)	238	21557	1.0 (ref)
Low [≤8.75]	7	633	1.1 (0.5–2.3)	6	383	1.3 (0.6–3.3))
Medium [>8.75–25.5]	5	522	0.9 (0.3–2.5)	6	853	0.9 (0.4–1.9)
High [>25.5–224.75]	5	1266	0.5 (0.2–1.3)	5	1183	0.5 (0.2–1.3)
			P trend = 0.23			P trend = 0.22
Carbofuran (carbamate-insecticide)						
None	317	36296	1.0 (ref)	317	36296	1.0 (ref)
Low [≤8.75]	63	4775	1.2 (0.9–1.6)	46	3695	1.2 (0.9–1.6)
Medium [>8.75–38.75]	32	3648	0.8 (0.6–1.2)	46	4590	1.0 (0.7–1.3)
High [>38.75–757.25]	44	4370	0.97 (0.7–1.4)	45	4477	1.0 (0.7–1.4)
			P trend = 0.69			P trend = 0.74
Carbaryl (carbamate-insecticide)						
None	128	12864	1.0 (ref)	128	12864	1.0 (ref)
Low [≤8.75]	54	4128	1.1 (0.7–1.6)	46	3962	1.0 (0.7–1.5)
Medium [8.75–55]	43	5096	0.9 (0.6–1.2)	45	4433	0.9 (0.7–1.5)
High [>56–737.5]	39	3281	1.0 (0.7–1.6)	44	4029	1.0 (0.6–1.5)
			P trend = 0.87			P trend = 0.94
Chlorpyrifos (organophosphate-insecticide)						
None	300	30393	1.0 (ref)	300	30393	1.0 (ref)
Low [≤8.75]	71	6493	1.1 (0.9–1.5)	61	6383	1.1 (0.8–1.4)
Medium [>8.75–44]	65	6892	1.1 (0.8–1.4)	60	7549	0.9 (0.7–1.2)
High [>44–767.25]	67	9380	0.8 (0.6–1.1)	60	7044	1.0 (0.7–1.3)
			P trend = 0.11			P trend = 0.85
Coumaphos (organophosphate-insecticide)						
None	411	44846	1.0 (ref)	411	44846	1.0 (ref)
Low [<8.75]	16	1510	1.0 (0.6–1.7)	15	1132	1.3 (0.8–2.1)
Medium [>8.75–38.75]	14	1076	1.2 (0.7–2.1)	14	1452	1.0 (0.6–1.6)
High [>38.75–1627.5]	13	1175	1.2 (0.7–2.0)	14	1170	1.2 (0.7–2.1)
			P for trend = 0.50			P trend = 0.48
DDVP (dimethyl phosphate-insecticide)						
None	407	44551	1.0 (ref)	407	44551	1.0 (ref)
Low [≤8.75]	19	1342	1.4 (0.9–2.1)	18	1281	1.4 (0.9–2.3)
Medium [>8.75–87.5]	17	1519	1.2 (0.7–1.9)	18	1633	1.1 (0.7–1.8)
High [>87.5–2677.5]	17	1893	0.9 (0.6–1.5)	17	1824	1.0 (0.6–1.6)
			P trend = 0.78			P trend = 0.83
Diazinon (organophosphorous-insecticide)						
None	187	17943	1.0 (ref)	187	17943	1.0 (ref)
Low [≤8.75]	28	2506	1.1 (0.7–1.6)	23	2047	1.1 (0.7–1.8)
Medium [>8.75–25]	19	1515	1.0 (0.6–1.8)	24	2246	0.9 (0.5–1.5)
High [>25–457.25]	23	1990	1.2 (0.7–1.9)	22	1708	1.3 (0.8–2.1)
			P trend = 0.52			P trend = 0.33

Table 3. Cont.

Insecticides						
Pesticide (chemical-functional class)	**NHL Cases**[2]	**Non-Cases**[2]	**RR^{3,4} (95% CI) by Total Days of Exposure**	**NHL**	**Non-Cases**	**RR^{3,4} (95% CI)**
[days of lifetime exposure for each category]				**Cases**[2,]		**Intensity-weighted days of exposure**
Fonofos (organophosphorous-insecticide)						
None	349	39570	1.0 (ref)	349	39570	1.0 (ref)
Low [≤20]	47	3812	1.3 (0.96–1.8)	37	2906	1.4 (0.97–1.9)
Medium [>20–50.75]	28	2819	1.1 (0.7–1.6)	38	3487	1.1 (0.8–1.6)
High [>50.75–369.75]	37	3385	1.1 (0.7–1.5)	36	3606	1.0 (0.7–1.4)
			P trend = 0.83			P trend = 0.87
Malathion (organophosphorous-insecticide)						
None	90	8368	1.0 (ref)	90	8368	1.0 (ref)
Low [≤8.75]	75	7284	0.97 (0.7–1.3)	60	5535	1.0 (0.7–1.4)
Medium [>8.75–38.75]	47	5779	0.7 (0.5–1.1)	59	6899	0.8 (0.6–1.1)
High [>38.75–737.5]	57	5037	0.9 (0.6–1.3)	59	5588	0.9 (0.6–1.2)
			P trend = 0.63			P trend = 0.46
Parathion (ethyl or methyl) (organophosphorous insecticide)						
None	228	21457	1.0 (ref)	228	21457	1.0 (ref)
Low [≤8.75]	9	693	1.0 (0.5–2.0)	7	612	0.9 (0.4–2.0)
Medium [>8.75–24.5]	6	351	1.4 (0.6–3.2)	8	462	1.4 (0.7–2.9)
High [>.24.5–1237.5]	6	652	0.8 (0.3–1.8)	6	621	0.8 (0.4–1.9)
			P trend = 0.64			P trend = 0.74
Permethrin (animal and crop applications)						
(pyrethroid insecticide)						
None	371	37496	1.0 (ref)	371	37496	1.0 (ref)
Low [≤8.75]	38	4315	1.1 (0.8–1.5)	33	4263	0.9 (0.6–1.3)
Medium [>8.75–50.75]	31	4611	0.8 (0.5–1.2)	33	4200	1.0 (0.7–1.4)
High [>50.75–1262.25]	33	4121	1.2 (0.8–1.7)	32	4553	1.0 (0.7–1.5)
			P trend = 0.54			P trend = 0.99
Phorate (organophosphorous-insecticide)						
None	171	16834	1.0 (ref)	171	16834	1.0 (ref)
Low [≤8.75]	27	2621	0.8 (0.5–1.2)	26	2320	0.9 (0.6–1.4)
Medium [8.75–24.5]	33	1819	1.4 (0.96–2.1)	27	1951	1.1 (0.7–1.7)
High [>24.5–224.75]	18	2246	0.6 (0.4–1.1)	25	2409	0.8 (0.5–1.3)
			P trend = 0.25			P trend = 0.44
Terbufos (organophosphorous-insecticide)						
None	267	31076	1.0 (ref)	267	31076	1.0 (ref)
Low [≤24.5]	82	8410	1.2 (0.9–1.5)	64	6895	1.1 (0.9–1.5)
Medium [>24.5–56]	54	3925	1.6 (1.2–2.1)	64	4642	1.6 (1.2–2.2)
High [>56–1627.5]	57	6080	1.1 (0.8–1.5)	63	6842	1.1 (0.8–1.5)
			P trend = 0.43			P trend = 0.44
Chlorinated Insecticides						
Aldrin (chlorinated insecticide)						
None	193	19743	1.0 (ref)	193	19743	1.0 (ref)
Low [≤8.75]	27	1613	0.9 (0.6–1.4)	20	1212	0.9 (0.6–1.4)
Medium [>8.75–24.5]	16	1002	0.8 (0.5–1.3)	20	1279	0.8 (0.5–1.3)

Table 3. Cont.

Pesticide (chemical-functional class) [days of lifetime exposure for each category]	NHL Cases[2]	Non-Cases[2]	RR[3,4] (95% CI) by Total Days of Exposure	NHL Cases[2,]	Non-Cases	RR[3,4] (95% CI) Intensity-weighted days of exposure
Insecticides						
High [>24.5–457.25]	17	903	0.9 (0.5–1.5)	19	1026	0.9 (0.6–1.5)
			P trend = 0.58			P trend = 0.74
Chlordane (chlorinated insecticide)						
None	179	19115	1.0 (ref)	179	19115	1.0 (ref)
Low [≤8.75]	47	2687	1.3 (0.97–1.9)	23	1303	1.4 (0.9–2.2)
Medium[5]	0	0	xxx	24	1747	1.0 (0.6–1.5)
High [>8.75–1600]	23	1450	1.1 (0.7–1.7)	22	1085	1.4 (0.9–2.2)
			P trend = 0.43			P trend = 0.16
DDT (chlorinated insecticide)						
None	152	18543	1.0 (ref)	152	18543	1.0 (ref)
Low [≤8.75]	43	2121	1.3 (0.9–1.8)	33	1601	1.2 (0.8–1.8)
Medium [>8.75–56]	28	1598	1.1 (0.7–1.7)	32	1760	1.1 (0.8–1.7)
High [>56–1627.5]	27	953	1.7 (1.1–2.6)	32	1305	1.6 (1.0–2.3)
			P trend = 0.02			P trend = 0.06
Dieldrin (chlorinated insecticide)						
None	235	22510	1.0 (ref)	235	22510	1.0 (ref)
Low [≤8.75]	7	472	0.7 (0.3–1.5)	5	363	0.8 (0.4–1.8)
Medium [>8.75–24.5]	8	154	2.3 (1.1–4.7)	5	106	2.2 (0.9–5.3)
High [>24.5–224.75]	2	140	0.7 (0.2–2.9)	5	298	0.8 (0.3–2.0)
			P trend = 0.47			P trend = 0.84
Heptachlor (chlorinated insecticide)						
None	205	20844	1.0 (ref)	205	20844	1.0 (ref)
Low [≤8.75]	21	1261	1.0 (0.6–1.6)	15	1110	0.8 (0.5–1.4)
Medium [>8.75–24.5]	18	679	1.5 (0.9–2.4)	16	425	2.0 (1.2–3.4)
High [>24.5–457.25]	7	600	0.7 (0.3–1.4)	14	1001	0.8 (0.5–1.4)
			P trend = 0.82			P trend = 0.88
Lindane (chlorinated insecticide)						
None	205	20375	1.0 (ref)	205	20375	1.0 (ref)
Low [≤8.75]	18	1285	1.2 (0.7–1.9)	15	976	1.3 (0.8–2.2)
Medium [>8.75–56]	13	1103	1.0 (0.6–1.7)	16	1205	1.1 (0.7–1.8)
High [>56–457.25]	14	467	2.5 (1.4–4.4)	14	673	1.8 (1.0–3.2)
			P trend = 0.004			**P trend = 0.04**
Toxaphene (chlorinated insecticide)						
None	214	20911	1.0 (ref)	214	20911	1.0 (ref)
Low [≤8.75]	14	1198	0.8 (0.5–1.4)	11	630	1.3 (0.7–2.3)
Medium [>8.75–24.5]	13	564	1.5 (0.9–2.7)	12	931	0.9 (0.5–1.6)
High [>24.5–457.25]	6	686	0.6 (0.3–1.4)	10	886	0.8 (0.4–1.5)
			P trend = 0.50			P trend = 0.38
Fungicides						
Benomyl (carbamate fungicide)						
None	219	21425	1.0 (ref)	219	21425	1.0 (ref)
Low [≤12.25]	14	896	1.7 (0.9–2.9)	9	432	2.2 (1.1–4.3)
Medium [>12.25–24.5]	4	214	2.4 (0.9–6.6)	10	732	1.7 (0.9–3.2)

Table 3. Cont.

Insecticides						
Pesticide (chemical-functional class)	**NHL Cases[2]**	**Non-Cases[2]**	**RR[3,4] (95% CI) by Total Days of Exposure**	**NHL**	**Non-Cases**	**RR[3,4] (95% CI)**
[days of lifetime exposure for each category]				**Cases[2,]**		**Intensity-weighted days of exposure**
High [>24.5–457.25]	8	834	1.0 (0.5–2.1)	7	779	0.9 (0.4–2.0)
			P trend = 0.93			P trend = 0.75
Captan (phthalimide fungicide)						
None	407	43433	1.0 (ref)	407	43433	1.0 (ref)
Low [≤0.25]	15	2334	0.8 (0.5–1.4)	15	2108	0.9 (0.6–1.5)
Medium [>0.25–12.25]	16	1004	1.5 (0.8–2.6)	15	1171	1.2 (0.7–2.2)
High [>12.25–875]	14	1823	0.8 (0.5–1.5)	14	1805	0.8 (0.5–1.5)
			P trend = 0.69			P trend = 0.52
Chlorothalonil (polychlorinated aromatic thalonitrile fungicide)						
None	474	48442	1.0 (ref)	474	48442	1.0 (ref)
Low [≤12.25]	13	1509	0.9 (0.5–1.6)	10	1800	0.6 (0.3–1.2)
Medium [>12.25–64]	9	1492	0.8 (0.4–1.6)	11	1501	0.9 (0.5–1.7)
High [>64–395.25]	9	1678	0.6 (0.3–1.3)	9	1362	0.8 (0.4–1.6)
			P trend = 0.16			PP trend = 0.52
Maneb/Mancozeb (dithiocarbamate fungicide)						
None	228	21512	1.0 (ref)	228	21512	1.0 (ref)
Low [≤7]	8	400	1.9 (0.9–3.9)	8	486	1.6 (0.8–3.3)
Medium [>7–103.25]	9	990	0.9 (0.4–1.7)	9	680	1.3 (0.6–2.6)
High [>103.25–737.5]	7	454	1.4 (0.6–2.9)	7	677	0.9 (0.4–1.9)
			P trend = 0.49			P trend = 0.78
Metalaxyl (acylalanine fungicide)						
None	209	18833	1.0 (ref)	209	18833	1.0 (ref)
Low [≤6]	16	1439	1.0 (0.6–1.8)	15	1079	1.3 (0.8–2.2)
Medium [>6–28]	15	2182	0.7 (0.4–1.3)	15	2203	0.8 (0.4–1.3)
High [>28–224.75]	13	1566	1.1 (0.6–2.1)	14	1893	0.9 (0.5–1.6)
			P trend = 0.76			P trend = 0.63
Fumigant						
Methyl bromide (methyl halide fumigant)						
None	425	45265	1.0 (ref)	425	45265	1.0 (ref)
Low [≤8]	37	2060	2.0 (1.4–2.9)	26	1680	1.8 (1.2–2.7)
Medium [>8–28]	24	3011	0.9 (0.6–1.4)	25	2501	1.1 (0.7–1.8)
High [>28–387.5]	17	2768	0.6 (0.4–1.0)	25	3571	0.8 (0.5–1.2)
			P trend = 0.04			P trend = 0.10

[1]During the period from enrollment (1993–1997) to December 31, 2010 in NC and December 31, 2011 in Iowa.
[2]Numbers of cases in columns do not sum to total number of NHL cases (n = 523) due to missing data. In the enrollment questionnaire, lifetime-days & intensity weighted life-time days of pesticide use was obtained for the insecticides: carbofuran, chlorpyrifos, coumaphos, DDVP, fonofos, permethrin and terbufos; the fungicides: captan, chlothalonil and the fumigant: methyl bromide. In the take home questionnaire lifetime-days & intensity weighted life-time days of pesticide use were obtained for the insecticides: aldicarb, carbaryl, diazinon, malathion, parathion, and phorate, the chlorinated insecticides: aldrin, chlordane, DDT, dieldrin, heptachlor, lindane, and toxaphene, the fungicides: benomyl, maneb/mancozeb and metalaxyl, therefore, numbers of NHL cases can vary among pesticides listed in the table.
[3]Adjusted RR: age (<45, 45–49, 50–54, 55–59, 60–64, 65–69, ≥70), State (NC vs. IA), Race (White vs. Black), AHS herbicides (tertiles of total herbicide use-days). Statistically significant P trends are bolded.
[4]Permethrin for animal use and crop use were combined into one category.
[5]The distribution of life-time days of chlordane exposure was clumped into two exposed groups those who with, ≤8.75 life-time days of exposure and those with >8.75 life-time days of exposure.

Table 4. Pesticide exposure (Lifetime-Days of Exposure) and adjusted risks for NHL Subtypes.

Insecticides	SLL, CLL, MCL		Diffuse Large B-cell		Follicular B-cell		Other B-cell types		Multiple Myeloma		
	RR[3,4] (95% CI)	N[2]	RR[3,4] (95% CI)	N[2]	RR[3,4] (95% CI)	N[2]	RR[3,4] (95% CI)	N[2]	RR[3,4] (95% CI)	N[2]	NHL subtype Homo-geneity Test (p-value)
Carbaryl											
None	1.0 (ref)	42	1.0 (ref)	29	1.0 (ref)	11	1.0 (ref)	14	1.0 (ref)	22	
Low	1.1 (0.6–2.2)	19	0.8 (0.4–1.6)	17	1.6 (0.6–3.9)	10	1.8 (0.7–4.3)	10	0.7 (0.3–1.4)	14	
High	0.6 (0.3–1.3)	15	1.3 (0.6–2.8)	15	2.8 (1.0–7.4)	10	0.4 (0.1–1.5)	3	1.1 (0.7–1.8)	13	
	p trend = 0.16		p trend = 0.33		p trend = 0.06		p trend = 0.63		p trend = 0.98		0.19
Carbofuran											
None	1.0 (ref)	87	1.0 (ref)	78	1.0 (ref)	39	1.0 (ref)	33	1.0 (ref)	56	
Low	1.1 (0.7–1.8)	28	0.9 (0.5–1.7)	13	1.3 (0.7–2.4)	15	0.8 (0.4–1.8)	8	1.9 ((1.1–3.3)	16	
High	1.5 (0.9–2.5)	19	0.8 (0.5–1.3)	13	0.4 (0.1–1.4)	3	0.7 (0.2–2.0)	4	0.9 (0.4–1.6)	12	
	p trend = 0.16		p trend = 0.37		p trend = 0.31		p trend = 0.46		p trend = 0.57		0.52
Chlorpyrifos											
None	1.0 (ref)	84	1.0 (ref)	70	1.0 (ref)	33	1.0 (ref)	31	1 (ref)	58	
Low	1.2 (0.8–1.8)	31	0.9 (0.6–1.5)	22	1.6 (0.9–2.9)	20	1.2 (0.6–2.2)	14	1.0 (0.6–1.8)	17	
High	0.9 (0.6–1.3)	30	1.1 (0.6–1.7)	22	1.0 (0.5–2.1)	11	0.5 (0.2–1.3)	7	0.7 (0.4–1.3)	14	
	p trend = 0.45		p trend = 0.80		p trend = 0.94		p trend = 0 13		p trend = 0.27		0.90
Coumaphos											
None	1.0 (ref)	120	1.0 (ref)	92	1.0 (ref)	48	1.0 (ref)	40	1.0 (ref)	78	
Low	1.1 (0.5–2.2)	8	0.7 (0.3–1.9)	4	2.1 (0.7–5.8)	4	xxx-	4	0.7 (0.2–2.2)	3	
High	1.5 (0.6–3.4)	6	1.6 (0.6–4.5)	4	1.4 (0.5–4.0)	4	xxx-	1	1.2 (0.4–4.0)	3	
	p trend = 0.35		p trend = 0.42		p trend = 0.47		p trend = xxx		p trend = 0.84		0.63
Diazinon											
None	1.0 (ref)	53	1.0 (ref)	40	1.0 (ref)	15	1.0 (ref)	20	1.0 (ref)	41	
Low	1.4 (0.7–2.7)	14	1.5 (0.7–3.2)	9	2.2 (0.9–5.4)	8	xxx	3	0.4 (0.1–1.2)	4	
High	1.9 (0.98–3.6)	12	1.1 (0.5–2.4)	8	3.8 (1.2–11.4)	7	xxx	2	0.5 (0.2–1.7)	3	
	p trend = 0.06		p trend = 0.72		**p trend = 0.02**		p trend = xxx		p trend = 0.35		0.09
DDVP											
None	1.0 (ref)	124	1.0 (ref)	93	1.0 (ref)	48	1.0 (ref)	39	1.0 (ref)	73	
Low	0.8 (0.4–1.9)	6	1.1 (0.4–2.7)	5	1.5 (0.6–3.9)	5	1.1 (0.4–3.7)	3	2.7 (1.2–5.8)	7	
High	0.7 (0.3–1.7)	6	0.9 (0.4–2.3)	5	1.0 (0.3–3.4)	3	0.9 (0.3–3.1)	3	1.0 (0.3–2.7)	4	
	p trend = 0.49		p trend = 0.87		p trend = 0.90		p trend = 0.91		p trend = 0.81		0.96
Fonofos											
None	1.0 (ref)	100	1.0 (ref)	81	1.0 (ref)	45	1.0 (ref)	30	1.0 (ref)	66	
Low	1.2 (0.7–2.0)	20	1.2 (0.7–2.2)	13	1.5 (0.8–3.0)	11	1.4 (0.6–3.1)	8	1.2 (0.6–2.5)	9	
High	1.0 (0.6–1.8)	15	1.2 (0.6–2.3)	11	0.3 (0.1–1.2)	2	1.1 (0.4–2.7)	6	1.4 (0.7–3.0)	9	
	p trend = 0.96		p trend = 0.65		p trend = 0.19		p trend = 0.84		p trend = 0.33		0.35
Malathion											
None	1.0 (ref)	27	1.0 (ref)	20	1.0 (ref)	6	1.0 (ref)	11	1.0 (ref)	17	
Low	0.7 (0.4–1.3)	29	0.96 (0.5–1.8)	23	1.0 (0.4–2.9)	12	1.0 (0.5–2.4)	11	1.0 (0.5–2.1)	18	
High	1.0 (0.6–1.8)	22	1.0 (0.5–2.0)	20	1.6 (0.6–4.4)	11	0.3 (0.1–0.8)	6	1.0 (0.5–2.0)	17	
Ever/Never	1.0 (0.7–1.4)		0.9 (0.6–1.4)		1.3 (0.7–2.4)		0.6 (0.3–1.0)		0.9 (0.6–1.5)		
	p trend = 0.65		p trend = 0.88		p trend = 0.25		p trend = 0.17		p trend = 0.86		0.33
Permethrin											

Table 4. Cont.

Insecticides	SLL, CLL, MCL		Diffuse Large B-cell		Follicular B-cell		Other B-cell types		Multiple Myeloma		NHL subtype Homo-geneity Test (p-value)
	RR[3,4] (95% CI)	N[2]	RR[3,4] (95% CI)	N[2]	RR[3,4] (95% CI)	N[2]	RR[3,4] (95% CI)	N[2]	RR[3,4] (95% CI)	N[2]	
None	1.0 (ref)	108	1.0 (ref)	89	1.0 (ref)	41	1.0 (ref)	38	1.0 (ref)	64	
Low	1.1 (0.6–2.0)	15	0.6 (0.3–1.2)	8	1.3 (0.6–2.7)	8	0.9 (0.3–2.7)	5	1.4 (0.8–2.7)	13	
High	0.8 (0.5–1.5)	15	1.0 (0.5–2.1)	8	1.0 (0.5–2.4)	8	0.5 (0.2–1.7)	4	3.1 (1.5–6.2)	12	
	p trend = 0.53		p trend = 0.99		p trend = 0.88		p trend = 0.28		**p trend = 0.002**		0.10
Phorate											
None	1.0 (ref)	48	1.0 (ref)	37	1.0 (ref)	20	1.0 (ref)	16	1.0 (ref)	36	
Low	1.0 (0.6–1.9)	14	1.4 (0.7–2.7)	15	1.1 (0.4–3.0)	5	0.9 (0.3–2.2)	6	0.7 (0.3–1.8)	6	
High	0.8 (0.4–1.6)	11	0.7 (0.3–2.1)	4	0.8 (0.3–2.2)	5	1.1 (0.4–3.5)	4	0.8 (0.3–2.4)	4	
	p trend = 0.51		p trend = 0.80		p trend = 0.67		p trend = 0.91		p trend = 0.73		0.77
Terbufos											
None	1.0 (ref)	72	1.0 (ref)	63	1.0 (ref)	31	1.0 (ref)	19	1.0 (ref)	59	
Low	1.3 (0.8–2.0)	32	1.2 (0.8–1.9)	29	1.6 (0.9–3.1)	15	1.8 (0.9–3.6)	17	1.1 (0.6–1.9)	12	
High	1.6 (1.0–2.5)	31	1.0 (0.5–2.0)	12	0.8 (0.4–1.7)	10	1.6 (0.7–3.9)	8	1.3 (0.7–2.7)	5	
	p trend = 0.05		p trend = 0.90		p trend = 0.48		p trend = 0.29		p trend = 0.42		0.63
Chlorinated Insecticides											
Aldrin											
None	1.0 (ref)	53	1.0 (ref)	46	1.0 (ref)	22	1.0 (ref)	20	1.0 (ref)	34	
Low	1.0 (0.5–2.0)	11	xxx	2	1.2 (0.4–3.8)	4	0.4 (0.1–1.5)	3	2.1 (0.9–4.7)	8	
High	1.0 (0.5–2.0)	10	xxx	3	0.8 (0.3–2.5)	4	1.1 (0.3–3.9)	3	1.2 (0.5–3.2)	6	
	p trend = 0.70		p trend = xxx		p trend = 0.21		p trend = 0.67		p trend = 0.40		0.98
Chlordane											
None	1.0 (ref)	48	1.0 (ref)	42	1.0 (ref)	20	1.0 (ref)	21	1.0 (ref)	32	
Low	1.8 (1.0–3.1)	16	1.0 (0.5–2.2)	8	1.7 (0.7–4.3)	6	xxx	2	1.7 (0.9–3.3)	13	
High	1.5 (0.7–3.3)	8	1.4 (0.6–3.3)	7	1.3 (0.4–4.6)	3	xxx	2	0.7 (0.2–2.2)	3	
	p trend = 0.34		p trend = 0.69		p trend = 0.70		p trend = xxx		p trend = 0.57		0.85
DDT											
None	1.0 (ref)	42	1.0 (ref)	34	1.0 (ref)	17	1.0 (ref)	16	1.0 (ref)	28	
Low	1.0 (0.5–1.8)	16	1.6 (0.4–3.1)	2	3.3 (1.4–8.1)	9	0.4 (0.3–2.5))	5	1.2 (0.6–2.6)	10	
High	2.6 (1.3–4.8)	15	1.4 (0.6–3.5)	3	1.1 (0.3–3.6)	4	2.1 (0.7–6.5)	5	0.8 (0.4–1.8)	9	
	p trend = 0.04		P trend = 0.17		p trend = 0.80		p trend = 0.64		p trend = 0.37		0.44
Heptachlor											
None	1.0 (ref)	58	1.0 (ref)	47	1.0 (ref)	24	1.0 (ref)	21	1.0 (ref)	40	
Low	1.1 (0.5–2.3)	9	xxx	3	xxx	2	xxx	3	1.3 (0.4–3.8)	4	
High	1.4 (0.7–3.0)	9	xxx	1	xxx	1	xxx	2	1.2 (0.4–3.6)	4	
	p trend = 0.16		p trend = xxx		p trend = xxx		p trend = xxx		p trend = 0.91		0.68
Lindane											
None	1.0 (ref)	57	1.0 (ref)	49	1.0 (ref)	16	1.0 (ref)	21	1.0 (ref)	43	
Low	1.2 (0.6–2.5)	10	0.6 (0.2–1.7)	4	4.9 (1.9–12.6)	6	xxx	2	xxx	3	
High	2.6 (1.2–5.6)	9	2.0 (0.6–6.5)	3	3.6 (1.4–9.5)	6	xxx	1	xxx	2	
	p trend = 0.13		p trend = 0.96		**p trend = 0.04**		p trend = xxx		p trend = xxx		0.54
Toxaphene											
None	1.0 (ref)	68	1.0 (ref)	47	1 (ref)	23	1.0 (ref)	22	1.0 (ref)	40	

Table 4. Cont.

Insecticides	SLL, CLL, MCL		Diffuse Large B-cell		Follicular B-cell		Other B-cell types		Multiple Myeloma		
	RR[3,4] (95% CI)	N[2]	RR[3,4] (95% CI)	N[2]	RR[3,4] (95% CI)	N[2]	RR[3,4] (95% CI)	N[2]	RR[3,4] (95% CI)	N[2]	NHL subtype Homo-geneity Test (p-value)
Low	0.9 (0.4–2.3)	5	1.3 (0.5–3.3)	5	xxx	2	xxx	3	0.7 (0.2–2.0)	4	
High	0.4 (0.1–1.6)	2	0.9 (0.3–3.0)	3	xxx	2	xxx	2	0.7 (0.2–2.9)	2	
	p trend = 0.08		p trend = 0.77		p trend = xxx		p trend = xxx		p trend = 0.64		0.34
Fungicides											
Captan											
None	1.0 (ref)	118	1.0 (ref)	91	1.0 (ref)	52	1.0 (ref)	39	1.0 (ref)	76	
Low	0.9 (0.4–1.9)	7	1.1 (0.5–2.4)	7	xxx	2	xxx	3	1.4 (0.5–3.4)	5	
High	1.1 (0.5–2.6)	7	0.7 (0.1–3.1)	4	xxx	1	xxx	2	1.2 (0.5–2.9)	5	
	p trend = 0.78		p trend = 0.58		p trend = xxx		p trend = xxx		p trend = 0.75		0.92
Chlorothalonil											
None	1.0 (ref)	135	1.0 (ref)	107	1.0 (ref)	60	1.0 (ref)	50	1.0 (ref)	84	
Low	0.9 (0.4–2.3)	5	1.1 (0.4–3.1)	4	xxx	3	–xxx	1	1.1 (0.4–2.8)	5	
High	1.1 (0.4–3.3)	4	0.3 (0.1–1.2)	2	xxx	2	–xxx	1	0.7 (0.6–2.3)	3	
	p trend = 0.83		p trend = 0.09		p trend = xxx		p trend = xxx		p trend = 0.56		0.76
Metalaxyl											
None	1.0 (ref)	60	1.0 (ref)	45	1.0 (ref)	25	1.0 (ref)	23	1.0 (ref)	39	
Low	2.8 (1.4–5.8)	9	1.1 (0.4–2.6)	7	xxx	3	–xxx	2	0.4 (0.1–1.1)	4	
High	1.1 (0.4–2.8)	6	1.0 (0.4–2.7)	5	xxx	2	–xxx	1	1.1 (0.4–3.2)	4	
	p trend = 0.99		p trend = 0.97		p trend = xxx		p trend = xxx		p trend = 0.87		0.92
Maneb/ Mancozeb											
None	1.0 (ref)	69	1.0 (ref)	49	1.0 (ref)	25	1.0 (ref)	26	1.0 (ref)	41	
Low	2.1 (0.7–6.0)	4	4.0 (1.4–11.6)	4	xxx	2	–xxx	0	1.0 (0.4–2.5)	5	
High	1.2 (0.3–4.0)	3	0.9 (0.3–3.1)	3	–xxx	1	–xxx	0	2.2 (0.5–9.5)	2	
	p trend = 0.84		p trend = 0.74		p trend = xxx		p trend = xxx		p trend = 0.28		0.82
Fumigant											
Methyl Bromide											
None	1.0 (ref)	126	1.0 (ref)	86	1.0 (ref)	58	1.0 (ref)	44	1.0 (ref)	76	
Low	1.1 (0.5–2.2)	9	4.0 (2.2–7.4)	15	1.4 (0.5–4.2)	4	3.6 (1.3–9.8)	5	1.0 (0.5–2.1)	8	
High	0.8 (0.4–1.8)	8	1.0 (0.5–2.1)	11	0.3 (0.1–1.1)	3	1.3 (0.3–5.0)	3	0.8 (0.4–1.8)	8	
	p trend = 0.58		p trend = 0.67		p trend = 0.08		p trend = 0.56		p trend = 0.63		0.59

[1]During the period from enrollment (1993–1997) to December 31, 2010 in NC and December 31, 2011 in Iowa.
[2]Numbers of cases in columns do not sum to total number of NHL cases (n = 523) due to missing data. Ever/never use of all 26 pesticides (table 3) do not always match with exposure-response data in table 4 because of missing data to calculate lifetime-days of use.
[3]Adjusted for age (<45, 45–49, 50–54, 55–59, 60–64, 65–69, ≥70), State (NC vs. IA), Race (White vs. Black), AHS herbicides (in tertiles of total herbicide use-days). Significant RR and 95% confidence limits are bolded.
[4]RR was not calculated if the number of exposed cases for any NHL subtype was <6 and these cells are marked XXX. Four pesticides included in Table 2 (i.e., aldicarb, benomyl, dieldrin and parathion) were not included in Table 4 because no NHL subtype included≥6 cases of a specific cell types with lifetime-days of exposure.

risk estimates (i.e., narrower confidence intervals) when we included phase 2 imputed data (n = 54,306) (data not shown). Lagging exposures by five years did not meaningfully change the association between lindane or DDT and total NHL (data not shown). The significant exposure-response trends linking use of a particular pesticide to NHL and certain NHL subtypes did not always correspond to a significant excess risk among those who ever used the same pesticide. For chemicals for which the detailed information was only asked about in the take-home questionnaire, we evaluated potential differences between the ever/never analyses based on the enrolment questionnaire and data from the same sub-set of participants who completed the exposure-response in the take-home questionnaire and found no meaningful differences in the results. We also evaluated the impact of using an updated definition of NHL; when using the original ICD-O-3 definition of NHL[19], lifetime-days of lindane use remained significantly associated with NHL risk (RR = 1.0 (ref), 1.3 (0.7–2.6), 1.2 (0.6–2.8), 2.7 (1.3–5.4), p trend = 0.006). The trend between total NHL and lifetime-days of DDT, however, was less clear and not statistically significant (RR = 1.0 (ref) 1.3 (0.9–1.8), 1.1 (0.5–2.1), 1.4 (0.8–2.6), p trend = 0.32) [Table S3 in File S1]. Carbaryl and diazinon showed non-significant trends with the older definition of NHL, but not with the newer definition used here.

Discussion

A significant exposure–response trend for total NHL was observed with increasing lifetime-days of use for two organochlorine insecticides, lindane and DDT, although RRs from ever/never comparisons were not elevated. On the other hand, terbufos use showed a significant excess risk with total NHL in ever vs. never exposed analysis, but displayed no clear exposure-response trend. Several pesticides showed significant exposure-response trends with specific NHL subtypes however, when polytomous models were used to test the difference in parametric estimates of trend among the five NHL subtypes, there was no evidence of heterogeneity in the sub-types for specific chemicals. The subtype relationships that looked particularly interesting were DDT and terbufos with the SLL/CLL/MCL subtype, lindane and diazinon with the follicular subtype, and permethrin with MM. These pesticide-NHL links should be evaluated in future studies.

Lindane (gamma-hexachlorocyclohexane) is a chlorinated hydrocarbon insecticide. Production of lindane was terminated in the United States in 1976, but imported lindane was used to treat scabies and lice infestation and for agricultural seed treatment [21] until its registration was cancelled in 2009 [22], the same year production was banned worldwide [23]. In our study, 3,410 people reporting ever using lindane (6%) prior to enrollment, 433 reported use at the phase 2 questionnaire (1%), indicating that use had dropped substantially. Oral administration of lindane has increased the incidence of liver tumors in mice and less clearly, thyroid tumors in rats [24]. Lindane produces free radicals and oxidative stress (reactive oxygen species [ROS]) [25] and has been linked with chromosomal aberrations in human peripheral lymphocytes in vitro [26].

Lindane has been linked with NHL in previous epidemiologic studies. A significant association between lindane use and NHL was observed in a pooled analysis of three population-based case-control studies conducted in the Midwestern US, with stronger relative risks observed for greater duration and intensity of use [27]. NHL was also associated with lindane use in a Canadian case-control study [28]. Lindane was significantly associated with NHL risk in an earlier report from the AHS [29]. We are not aware of any previous study that assessed the association between a NHL subtype and lindane use. The exposure-response pattern with total NHL and the follicular lymphoma subtype indicates a need for further evaluation of lindane and NHL.

DDT is an organochlorine insecticide that was used with great success to control malaria and typhus during and after World War II [29] and was widely used for crop and livestock pest control in the United States from the mid-1940s to the 1960s [30]. Its registration for crop use was cancelled in the US in 1972 [30] and banned worldwide for agricultural use in 2009, but continues to be used for disease vector control in some parts of the world [23]. In our study, 12,471 participants (23%) reported ever using DDT prior to enrollment; 12%, 8.7% and 2.3% responding to the take-home questionnaire reported their first use occurred prior to the 1960s, during the 1960s, and during the 1970s, respectively. The National Toxicology Program classifies DDT as "reasonably anticipated to be a human carcinogen" [31] and IARC classifies DDT as a "possible human carcinogen (2B)" [12], both classifications were based on experimental studies in which excess liver tumors were observed in two rodent species. Epidemiology data on the carcinogenic risk of DDT is inconsistent. NHL was not associated with use of DDT in a pooled analysis of three case-control studies in the U.S. where information on exposure was obtained from farmers by questionnaire [32]. There also was no association between the use of DDT and NHL in our study when we used an earlier definition of NHL [18], suggesting some of the inconsistency may be due to disease definition. In the large Epilymph study, no meaningful links between DDT and the risk of NHL, or diffuse large B cell lymphoma were observed, and only limited support was found for a link to CLL [33], although a case-control study of farmers in Italy suggested increased risk of NHL and CLL with DDT exposure [34]. NHL was not associated with serum levels of DDT in a prospective cohort study from the U.S. [35], but NHL was associated with the DDT-metabolite p, p'-DDE, as well as chlordane and heptachlor-related compounds (oxychlordane, heptachlor epoxide) and dieldrin, in a study with exposure measured in human adipose tissue samples [36]. In a Danish cohort, a higher risk of NHL was associated with higher prediagnostic adipose levels of DDT, cis-nonachlor, and oxychlordane [37]. In a Canadian study, analytes from six insecticides/insecticide metabolites (beta-hexachlorocyclohexane, p, p'-dichloro-DDE, hexachlorobenzene (HCB), mirex, oxychlordane and transnonachlor) were linked with a significant increased risk with NHL [38]. However, in an analysis of plasma samples from a case-control study in France, Germany and Spain, the risk of NHL did not increase with plasma levels of hexachlorobenzene, beta-hexachlorobenzene or DDE [39]. In this analysis, NHL was significantly associated with reported use of DDT, but not with the other organochlorine insecticides studied (i.e., aldrin, chlordane, dieldrin, heptachlor, toxaphene). Our findings add further support for an association between DDT and total NHL and our results on SLL/CLL/MCL are novel and should be further explored.

Permethrin is a broad-spectrum synthetic pyrethroid pesticide widely used in agriculture and in home and garden use as an insecticide and acaricide, as an insect repellant, and as a treatment to eradicate parasites such as head lice or mites responsible for scabies [40]. This synthetic pyrethroid was first registered for use in the United States in 1979 [40]. The U.S. Environmental Protection Agency classified permethrin as "likely to be carcinogenic to humans" largely based on the observed increase incidence of benign lung tumors in female mice, liver tumors in rats and liver tumors in male and female mice [41]. Permethrin was not associated with NHL overall in our study, nor in pooled case-control studies of NHL from the U.S (the NHL definition in use at the time of the study did not include MM) [42]. In our analysis, however, the risk of MM increased significantly with lifetime-days of exposure to permethrin, as had been noted in an earlier analysis of AHS data [43]. We are unaware of other studies that have found this association.

Terbufos is an organophosphate insecticide and nematicide first registered in 1974 [44]. The EPA classifies terbufos as Group E, i.e., "Evidence of Non-Carcinogenicity for Humans" [44]. We found some evidence for an association between terbufos use and NHL, particularly for the SLL/CLL/MCL subtype. NHL was not associated with terbufos in the pooled case-control studies from the

U.S. [42] but there was a non-significant association between terbufos and small cell lymphocytic lymphoma [10].

Diazinon is an organophosphate insecticide registered for a variety of uses on plants and animals in agriculture [45]. It was commonly used in household insecticide products until the EPA phased out all residential product registrations for diazinon in December 2004 [45.46]. In an earlier evaluation of diazinon in the AHS, a significant exposure-response association was observed for leukemia risk with lifetime exposure-days [47]. While there was no link between diazinon and NHL overall in this analysis, there was a statistically significant exposure-response association between diazinon and the follicular lymphoma subtype and an association with the SLL/CLL/MCL subtype that was not statistically significant. Diazinon was previously associated with NHL in pooled case-control studies from the U.S. and particularly with SLL [10].

Several other insecticides, fungicides and fumigants cited in recent reviews of the pesticide-cancer literature suggested etiological associations with total NHL [8,9], these include: oxychlordane, trans-nonachlor, and cis-nonachlor which are metabolites of chlordane; and dieldrin and toxaphene among NHL cases with t(14,18) translocations. We did not find a significant association between chlordane and total NHL nor with any NHL subtype, but we did not have information about chlordane metabolites to make a more direct comparison. Similarly we did not observe a significant association between dieldrin nor toxaphene and total NHL nor with any NHL subtypes. Mirex (1,3-cyclopentadiene), an insecticide, and hexachlorobenzene, a fungicide, were also associated with NHL risk [8,9] but we did not examine these compounds in the AHS.

This study has a number of strengths. It is a large population of farmers and commercial pesticide applicators who can provide reliable information regarding their pesticide use history [48]. Information on pesticide use and application practices was obtained prior to onset of cancer. An algorithm that incorporated several exposure determinants which predicted urinary pesticide levels was used to develop an intensity-weighted exposure metric in our study [20]. Exposure was ascertained prior to diagnosis of disease, which should eliminate the possibility of case-response bias [14]. Because of the detailed information available on pesticide use, we were able to assess the impact for the use of multiple pesticides. For example, we evaluated total pesticide use-days, and specific pesticides found to be associated with NHL or its subtypes in the AHS. We found no meaningful change in the associations with DDT, lindane, permethrin, diazinon and terbufos from such adjustments. Information on many potential NHL risk factors was available and could be controlled in the analysis.

Most epidemiological investigations of NHL prior to 2007 [17] did not include CLL and MM as part of the definition. These two subtypes made up 37% (193/523) of the NHL cases in this analysis. This is a strength of our study in that the definition of NHL used here is based on the most recent classification system [16,17] and will be relevant for comparisons with future studies. On the other hand, the inclusion of MM and CLL in the recent definition of NHL makes comparisons of our findings with earlier literature challenging, because the NHL subtypes may have different etiologies. For example, DDT was not significantly associated with NHL using the older definition, but was significantly associated with the NHL using the most recent definition of NHL because of its association with the SLL/CLL/MCL subtype (Table S1 in File S1). On the other hand, carbaryl and diazinon were associated with the old definition of NHL (although non-significantly) but not with the new definition. Lindane, however, was associated with both definitions of NHL. Lindane was significantly associated with the follicular lymphoma subtype and this subtype was included in the older and newer definition of NHL. No other pesticides were significantly associated with NHL under the old definition (Table S3 in File S1).

Although this is a large prospective study, limitations should be acknowledged. A small number of cases exposed to some specific pesticides could lead to false positive or negative findings. We also had reduced statistical power to evaluate some pesticides for total days of use and intensity-weighted days of use because some participants did not complete the phase one take-home questionnaire and the tests of homogeneity between specific pesticides and specific NHL subtypes were underpowered. Some chance associations could occur because of multiple testing, i.e., a number of pesticides, several NHL subtypes, and more than one exposure metric. Despite the generally high quality of the information on pesticide use provided by AHS participants [48,50], misclassification of pesticide exposures can occur and can have a sizeable impact on estimates of relative risk, which in a prospective cohort design would tend to produce false negative results [49].

Conclusion

Our results showed pesticides from different chemical and functional classes were associated with an excess risk of NHL and NHL subtypes, but not all members of any single class of pesticides were associated with an elevated risk of NHL or NHL subtypes, nor were all chemicals of a class included on our questionnaire. Significant pesticide associations were between total NHL and reported use of lindane and DDT. Links between DDT and terbufos and SLL/CLL/MCL, lindane and diazinon and follicular lymphoma, and permethrin and MM, although based on relatively small numbers of exposed cases, deserve further evaluation. The epidemiologic literature on NHL and these pesticides is inconsistent and although the findings from this large, prospective cohort add important information, additional studies that focus on NHL and its subtypes and specific pesticides are needed. The findings from this large, prospective cohort add important new information regarding the involvement of pesticides in the development of NHL. It provides additional information regarding specific pesticides and NHL overall and some new leads regarding possible links with NHL subtypes that deserve evaluation in future studies.

Supporting Information

File S1 This file contains Table S1, Table S2, and Table S3. Table S1, Frequency of NHL in Agricultural Health Study applicators using New (Interlymph hierarchical classification of lymphoid neoplasms) and Older Definitions (ICD-O-3). Table S2, Pesticides included in the Agricultural Health Study questionnaires by Chemical/Functional Class. Table S3, Pesticide exposure (lifetime-days) and adjusted risks of total NHL incidence (Older definition [ICD-O-3]).

Acknowledgments

Disclaimer: The findings and conclusions in this report are those of the author(s) and do not necessarily represent the views of the National Institute for Occupational Safety and Health. The United States Environmental Protection Agency through its Office of Research and Development partially funded and collaborated in the research described here under Contracts 68-D99-011 and 68-D99-012, and through Interagency Agreement DW-75-93912801-0. It has been subjected to Agency review and approved for publication.

This work was supported by the Intramural Research Program of the NIH, National Cancer Institute, Division of Cancer Epidemiology and

Genetics (Z01CP010119) and the National Institutes of Environmental Health Science (Z01ES049030). The authors have no conflicts of interest in connection with this manuscript.

Ms. Marsha Dunn and Ms. Kate Torres, (employed by Westat, Inc. Rockville, Maryland) are gratefully acknowledged for study coordination. The ongoing participation of the Agricultural Health Study participants is indispensable and sincerely appreciated.

Author Contributions

Conceived and designed the experiments: MCA DPS AB. Performed the experiments: MCA CFL KT CJH. Analyzed the data: MCA JNH CFL CJH KHB JB DWB KT DPS JAH SK GA JHL AB LEB. Contributed reagents/materials/analysis tools: MCA JB DWB CFL. Wrote the paper: MCA LEBF JNH CFL CJH KT AB DWB JHL. Designed the software: JB DWB.

References

1. Milham S (1971) Leukemia and multiple myeloma in farmers. Am J Epidemiol 94: 507–510.
2. Cantor KP (1982) Farming and mortality from non-Hodgkin's lymphoma: a case-control study. Int J Cancer 29: 239–247.
3. Blair A, Malker H, Cantor KP, Burmeister L, Wiklund K (1985) Cancer among farmers: A review. Scand J Work Environ Health 11: 397–407.
4. Pearce NE, Smith AH, Fischer DO (1985) Malignant lymphoma and multiple myeloma linked with agricultural occupation in a New Zealand cancer registration-base-study. Am J Epidemiol 121: 235–237.
5. Baris D, Silverman DT, Brown LM, Swanson GM, Hayes RB, et al. (2004) Occupation, pesticide exposure and risk of multiple myeloma. Scand J Work Environ Health 30(3): 215–222.
6. Beane Freeman LE, DeRoos AJ, Koutros S, Blair A, Ward MH, et al. (2012) Poultry and livestock exposure and cancer risk among farmers in the agricultural health study. Cancer Causes Control 23: 663–670.
7. Cordes DH, Rea DF (1991) Farming: A Hazardous Occupation. In: Health Hazards of Farming. Occupational Medicine: State of the Art Reviews. Vol6(3). Hanley & Belfus, Inc., Philadelphia, PA.
8. Alavanja M, Bonner M (2012) Occupational pesticide exposure and cancer risk. A Review. J. Toxicol Environ Health B Critic Review. 15(4): 238–263.
9. Alavanja MCR, Ross MK, Bonner MR (2013) Increased cancer burden among pesticide applicators and others due to pesticide exposure. CA, Cancer J Clin; 63(2): 120–142.
10. Waddell BL, Zahm SH, Baris D, Weisenburger DD, Holmes F, et al. (2001). Agricultural use of organophosphate pesticides and the risk of non-Hodgkin's lymphoma among male farmers (United States). Cancer Causes and Control 12: 509–517.
11. Merhi M, Raynal H, Cahuzac E, Vinson F, Cravedi JP, et al. (2007) Occupational exposure to pesticides and risk of hematopoietic cancers: meta-analysis of case-control studies. Cancer Causes Control. 18: 1209–1226.
12. IARC (1991) International Agency for Research on Cancer (IARC). Occupational Exposures in Insecticide applications and some pesticides. Lyon, France: IARC, 1991. Monographs on the Evaluation of Carcinogenic Risk to Humans, volume 53.
13. Morton LM, Slager SL, Cerhan JR, Wang SS, Vajdic CM, et al. (2014) Etiologic Heterogeneity Among NHL Subtypes: The InterLymph NHL Subtypes Project. J Natl Cancer Institute 48: 130–144.
14. Alavanja MCR, Sandler DP, McMaster SB, Zahm SH, McDonnell CJ, et al. (1996) The Agricultural Health Study. Environ Health Perspect 104: 362–369.
15. Alavanja MC, Samanic C, Dosemeci M, Lubin J, Tarone R, et al. (2003) Use of agricultural pesticides and prostate cancer risk in the Agricultural Health Study Cohort. Am J Epidemiol 157(9): 800–814.
16. SEER Program, National Cancer Institute. Available: http://seer.cancer.gov/lymphomarecode Accessed September 15, 2013.
17. Morton LM, Turner JJ, Cerhan JR, Linet MS, Treseler PA, et al. (2007) Proposed classifciation of lymphoid neoplasms for epidemiologic research from the Pathology Working Group of the International Lymphoma Epidemiology Consortium (InterLymph). Blood 110(2): 695–708.
18. Percy C, Fritz A, Ries L (2001) Conversion of neoplasms by topography and morphology from the International Classification of Disease for Oncology, second edition, to International Classification of Diseases for Oncology, 3rd ed. Cancer Statistics Branch, DCCPS, SEER Program, National Cancer Institute; 2001.
19. Heltshe SL, Lubin JH, Koutros S, Coble JB, Ji B-T, et al. (2012) Using multiple imputation to assign pesticide use for non-responders in the follow-up questionnaire in the Agricultural Health Study J. Exp Sci Environ Epidemiol 22(4): 409–416.
20. Coble J, Thomas KW, Hines CJ, Hoppin JA, Dosemeci M, et al. (2011) An updated algorithm for estimation of pesticide exposure intensity in the Agricultural Health Study. Int J Environ Res Public Health. 8(12): 4608–4622.
21. ATSDR (2005) Agency for Toxic Substances and Disease Registry. Toxicological profile for Alpha-, Beta-, Gamma- and Delta- Hexachlorcyclohexane, August, 2005. Available: http://www.atsdr.cdc.gov/Toxprofiles/tp43.pdf. Accessed 2013 Sep 15.
22. EPA (2006a) US Environmental Protection Agency (2006). Lindane; Cancellation Order December 13, 2006. Federal Register/Vol 71, number 239, page 74905.
23. Stockholm Convention Report (2009) Report of the conference of the Parties of the Stockholm Convention on Persistent Organic Pollutants on the work of its fourth meeting. Convention on Persistent Organic Pollutants. Fourth Meeting, Geneva, 4–8 May 2009. Available: http;//chm.pops.int/Portals/0/Repository/COP4/UNEP-POPS-COP.4–38.English.pdf.
24. IARC (1987) International Agency for Research on Cancer (IARC). Overall evaluation of carcinogenicity: an updating of IARC monographs volume 1 to 42. Lyon, France: IARC, 1987. Monographs on the Evaluation of Carcinogenic Risk to Humans, Supplement 7.
25. Piskac-Collier AL, Smith MA (2009) Lindane-induced generation of reactive oxygen species and depletion of glutathione do not result in necrosis in renal distal tube cells. J Toxicol and Environ Health, Part A. 72: 1160–1167.
26. Rupa DS, Reddy PP, Reddi OS (1989). Genotoxic effect of benzene hexachloride in cultured human lymphocytes. Hum Genet 83: 271–273.
27. Blair A, Cantor KP, Zahm SH (1998) Non-Hodgkin's lymphoma and agricultural use of the insecticide lindane. Am J Ind Med 33: 82–87.
28. McDuffie HH, Pahwa P, Mclaughlin JR, Spinelli JJ, Fincham S, et al. (2001) Non-Hodgkin's lymphoma and specific pesticide exposures in Men: Cross-Canada Study of Pesticides and Health. Cancer, Epidemiology, Biomarkers & Prevention 10: 1155–1163.
29. Purdue MP, Hoppin JA, Blair A, Dosemeci M, Alavanja MCR (2007) Occupational exposure to organochlorine insecticides and cancer incidence in the Agricultural Health Study. Int J Cancer. 120(3): 642–649.
30. EPA (2012) US Environmental Protection Agency 2012. DDT-A Brief History and Status. Available: http://www.epa.gov/pesticides/factsheets/chemicals/ddt-brief-history-status.htm. Accessed 2013 Sep 15.
31. NTP (2011) National Toxicology Program, Report on Carcinogen- Twelfth Edition. Available: http://ntp.niehs.nih.gov/go/roc12. Accessed 2013 Sep 15.
32. Baris D, Zahm SH, Cantor KP, Blair A (1998). Agricultural use of DDT and the risk of non-Hodgkin's lymphoma: pooled analysis of three case-control studies in the United States, Occup Environ Med 55: 522–527.
33. Cocco P, Satta G, Dubois S, Pilli C, Pillieri M, et al. (2013) Lymphoma risk and occupational exposure to pesticides: results of the Epilymph study. Occupational Environ Med 70(2): 91–98.
34. Nanni O, Amadori D, Lugaresi C, Falcini F, Scarpi E, et al. (1996) Chronic lymphocytic leukaemia and non-Hodgkin's lymphomas by histological type in farming-animal breeding workers: a population case-control study based on a priori exposure matrics Occup Environ Med 53(10): 652–657.
35. Rothman N, Cantor KP, Blair A, Bush D, Brock JW, et al. (1997) A nested case-control study of non-Hodgkin lymphoma and serum organochlorine residues. The Lancet 350: 240–244.
36. Quintana PJE, Delfino RJ, Korrick S, Ziogas A, Kutz FW, et al. (2004) Adipose tissue levels of organochlorine pesticides and chlorinated biphenyls and the risk of non-Hodgkin's lymphoma. Environ Health Perspect 112: 854–861.
37. Brauner EV, Sorensen MA, Gaudreau E, LeBlanc A, Erikson KT, et al. (2012) A prospective study of organochlorines in adipose tissue and risk of non-Hodgkin lymphoma. Environ Health Perspect. 120(1): 105–111.
38. Spinelli JJ, Ng CH, Weber JP, Connors JM, Gascoyne RD, et al. (2007) Organochlorines and risk of non-Hodgkin lymphoma. Int J Cancer. 121(12): 2767–2775.
39. Cocco P, Brennan P, Ibba A, de Sanjose Llongueras S (2008) Plasma polychlorobiphenyl and organochlorine pesticide level and risk of major lymphoma subtypes. Occup Environ Med 65: 132–140.
40. EPA 2006(b). U.S. Environmental Protection Agency. 2006. Re-registration Eligibility Decision for Permethrin: Available: http://www.epa.gov/oppsrrd1/REDs/permethrin_red.pdf. Accessed 2013 Sep 15.
41. NPIC (2012). National Pesticide Information Center. Chemicals Evaluated for Carcinogenic Potential. Office of Pesticide Programs. U.S. Environmental Protection Agency. November 2012. Available: http://npic.orst.edu/chemicals_evaluated.pdf. Accessed 2013 Sep 15.
42. De Roos AJ, Zahm SH, Cantor KP, Weisenburger DD, Holmes FF, t al. (2003) Integrative assessment of multiple pesticides as risk factors for non-Hodgkin's lymphoma among men Occup Environ Med 60: E11.
43. Rusiecki JA, Patel R, Koutros S, Beane Freeman LE, Landgren O, et al. (2009) Cancer incidence among pesticide applicators exposed to permethrin in the Agricultural Health Study. Environ Health Perspect 117: 582–586.
44. EPA 2006(b). U.S. Environmental Protection Agency. 2006. Re-registration Eligibility Decision for Terbufos: Available: http://www.epa.gov/pesticides/reregistration/REDs/terbufos_red.pdf Accessed 2013 Nov 18.
45. Environmental Protection Agency, 2004. Interim registration eligibility decision: diazinon. Available: http://www.epa.gov/pesticides/reregistration/REDs/diazinon_red.pdf. Accessed 2013 Nov 18.
46. Donalson D, Kieley T, Grube A (2002) Pesticide industry sales and usage: 1998 and 1999 market estimates. Washington, DC.: US Environmental Protection Agency, 2002 (EPA-733-R-02-001).

47. Beane Freeman LE, Bonner MR, Blair A, Hoppin JA, Sandler DP, et al. (2005). Cancer incidence among male pesticide applicators in the Agricultural Health Study cohort exposed to diazinon. Am J Epidemiol. 162: 1070–1079.
48. Blair A. Tarone R, Sandler D, Lynch CF, Rowland A, et al. (2002) Reliability of reporting on life-style and agricultural factors by a sample of participants in the Agricultural Health Study from Iowa. Epidemiology 13(1): 94–99.
49. Blair A, Thomas HT, Coble J, Sandler DP, Hines CJ, et al. (2011). Impact of pesticide exposure on misclassification on estimates of relative risks in the Agricultural Health Study. Occup Environ Med 68: 537–541.
50. Thomas KW, Dosemeci M, Coble JB, Hoppin JA, Sheldon LS, et al. (2010) Assessment of a pesticide exposure intensity algorithm in the Agricultural Health Study. J Expo Sci Environ Epidemiol 20(6): 559–569.

The Mechanism of Methylated Seed Oil on Enhancing Biological Efficacy of Topramezone on Weeds

Jinwei Zhang[1,2], Ortrud Jaeck[3], Alexander Menegat[3], Zongjian Zhang[4], Roland Gerhards[3], Hanwen Ni[1]*

1 College of Agriculture and Biotechnology, Weed Management laboratory, China Agricultural University, Beijing, China, **2** Institute of Plant Protection, Tianjin Academy of Agricultural Sciences, Tianjin, China, **3** Institute of Phytomedicine, Department of Weed Science, University of Hohenheim, Stuttgart, Germany, **4** The Office of National Pesticide R&D South Center R&D Center of Functional Additives, Central Research Institute of China Chemical Science and Technology, Beijing, China

Abstract

Methylated seed oil (MSO) is a recommended adjuvant for the newly registered herbicide topramezone in China and also in other countries of the world, but the mechanism of MSO enhancing topramezone efficacy is still not clear. Greenhouse and laboratory experiments were conducted to determine the effects of MSO on efficacy, solution property, droplet spread and evaporation, active ingredient deposition, foliar absorption and translocation of topramezone applied to giant foxtail (*Setaria faberi* Herrm.) and velvetleaf (*Abutilon theophrasti* Medic.). Experimental results showed that 0.3% MSO enhanced the efficacy of topramezone by 1.5-fold on giant foxtail and by 1.0-fold on velvetleaf. When this herbicide was mixed with MSO, its solution surface tension and leaf contact angle decreased significantly, its spread areas on weed leaf surfaces increased significantly, its wetting time was shortened on giant foxtail but not changed on velvetleaf, and less of its active ingredient crystal was observed on the treated weed leaf surfaces. MSO increased the absorption of topramezone by 68.9% for giant foxtail and by 45.9% for velvetleaf 24 hours after treatment. It also apparently promoted the translocation of this herbicide in these two weeds.

Editor: Randall P. Niedz, United States Department of Agriculture, United States of America

Funding: This research was supported by the Special Fund for Agro-scientific Research in the Public Interest (201303022) and the Deutsche Forschungsgemeinschaft, DFG (German ResearchFoundation, DFG GRK 1070, IRTG Sustainable Resource Use in North China).The funders had no role in study design, data collection and analysis, decision to publish, or preparation of the manuscript.

Competing Interests: The authors have declared that no competing interests exist.

* E-mail: hanwenni@cau.edu.cn

Introduction

Almost all herbicide formulations contain adjuvants which act as important tools to improve physical aspects of herbicide application and/or to enhance biological efficacy [1,2,3,4]. Research on adjuvant technology for agrochemicals has made good progress in recent years in part due to increased efforts by agrochemical manufacturers to ensure that the best adjuvants are used with their products for maximum performances. The efficacy of herbicide formulation can be expressed as a function of deposition, retention, absorption, translocation and phytotoxicity. Although adjuvants are not able to directly affect inherent herbicide toxicity, they can significantly alter each of the preceding terms [4]. There are various types of adjuvants with varying degrees of effectiveness on enhancing herbicide efficacy.

Methylated seed oil (MSO) is a kind of fatty acid from seed oil esterified with methyl alcohol [5]. For oil-based adjuvant, droplet spread on leaf surfaces and herbicide penetration seem to be the two predominant factors regarding the mechanism of the enhancement in herbicide efficacy [6]. Xu et al. reported that MSO could decrease the surface tension and contact angle and then increase the wetted areas of droplets on both waxy and hairy leaves [7,8]. Some reports have shown that MSO enhances the efficacy of several herbicides on certain weed species by increasing the absorption of the herbicides by weeds [9,10,11,12,13,14,15,16]. Topramezone, a hydroxyphenylpyruvate dioxygenase inhibitor, was commercially introduced in the year of 2006 [17]. When applied as a post-emergence herbicide, it controls a wide spectrum of annual grass and broadleaf weeds [18,19,20,21] and is safe to corn (*Zea mays* L.) [22,23].

In order for the optimal weed control, topramezone should be also applied with a certain adjuvant and MSO is always recommended. This herbicide has good field performance when applied with the MSO adjuvant which is made from soybean oil in China [24,25,26]. Some other reports also presented that a good efficacy could be achieved when this herbicide was tank-mixed with MSO adjuvant [27,28,29]. However, little information has been provided about the mechanism of MSO enhancing the biological efficacy of topramezone on weeds. Only Grossmann and Ehrhardt reported that a kind of MSO adjuvant Dash HC could significantly increase the foliar absorption of topramezone by plants [17].

The objective of this study was to determine the mechanism of MSO enhancing the biological efficacy of topramezone through investigating effects of MSO on solution property, droplet behavior on weed leaf surface, active ingredient absorption by weed leaves and translocation in weed plants.

Materials and Methods

Materials Used and Plant Growth Conditions

Topramezone commercial formulation was 336 g L^{-1} SC provided by BASF Co., Ltd. Adjuvant MSO (GY-HMax, methylated soybean soil) was provided by the Central Research

Institute of China Chemical Science and Technology, which is recommended by BASF as spray adjuvant for this herbicide in China. Unless indicated, the added amount of MSO was 0.3% herbicide solution (v/v).

A dicotyledonous weed velvetleaf (*Abutilon theophrasti* Medic.) and a monocotyledonous weed giant foxtail (*Setaria faberi* Herrm.) were used in this study. The weed seeds provided by Herbiseed Co., UK were pre-germinated in plastic pots (11×11×6 cm) filled with vermiculite (2–3 cm) in greenhouse (25/20±1°C day/night, additional light 122 μmol m^{-2} s^{-1} for 12 h, and 55±10% RH). After germination, velvetleaf seedlings were transplanted into 12-cm-diam plastic pots (3 plants per pot) and giant foxtail seedlings were transplanted into 7×7 cm paper pots (4 plants per pot). All pots were filled with the mixture of vermiculite: peat: clay 1:1:1 (by volume) and cultivated under the above-described condition. The plants were watered daily with tap water. No watering was done for 24 hours after herbicide application.

Efficacy Enhancement of MSO on Topramezone

At the 3- to 4-leaf stage of weeds, topramezone was applied alone or tank-mixed with MSO at 12 series doses for giant foxtail and 10 series doses for velvetleaf. The dose ranges are given in Table 1. Plants were applied with tap water as an untreated control for each weed species. Herbicide application was done with a track sprayer (Aro, Langenthal, Switzerland), which simulated a spray volume of 200 L ha^{-1} (nozzle: 8002 EVS, Teejet® Spraying Systems Co., Wheaton, IL, USA) at 3.2 kPa. The experiment was completely randomized with 3 replicates and repeated once.

Plants were harvested 3 weeks after application, dried at 80°C for 48 h and weighed. The following four-parameter Weilbull model [30] was used for dose-response curves (Equation 1):

$$Y = c + (d - c) \bullet \exp\{-\exp[b \bullet (\log x - e)]\} \qquad (1)$$

where d and c denote the upper and lower limits, respectively. e is the dose where a response half-way between the upper and lower limit (ED_{50}) is reached. b denotes the slope around the ED_{50} value.

To compare different models generated from dry biomass data of each treatment, the residual sum of squares of the regression analysis was assessed by an F-test for lack-of-fit. To compare the efficacy difference between the two treatments of topramezone applied alone and mixed with MSO, the relative potency (RP) (Equation 2) of two treatment curves was calculated based on their ED_{90} values according to Ritz's method [31].

$$RP = ED_{90}(\text{topramezone alone})/ED_{90}(\text{topramezone} + \text{MSO}) \qquad (2)$$

Surface Tension and Contact Angle

Topramezone solutions were prepared at the concentration of 0.126 g a.i. L^{-1} which corresponded to 25.2 g a.i. ha^{-1} at 200 L ha^{-1}. Solution Surface tensions of topramezone alone, topramezone with MSO and deionized water were measured with a contact angle/surface tensionmeter (JC2000C1, Shanghai Zhongchen Digital Technology Equipment Co., LTD, Shanghai, China).

The fourth leaves of velvetleaf and giant foxtail were removed at the 4- to 5-leaf stage. The leaves of giant foxtail were cut into about 2 cm-long segments. The segment of giant foxtail or the blade of velvetleaf was fixed onto a glass slide with double-sided adhesive tape. Adaxial surface was outward. Two 1 μL-droplets of the tested solutions were dropped onto one blade or segment with a microsyringe. Contact angles of the droplets were measured with the above mentioned instrument. Measurements of 10 droplets were carried out for each weed species.

Spread and Evaporation of Topramezone Droplet on Leaf Surfaces

Tested leaf samples and topramezone solutions were prepared as same as in the above-mentioned surface tension and contact angle measuring experiment. In order to distinguish droplet residual borders, 0.5 mg ml^{-1} black color (Brilliant Black BN E151, produced by Ringer Kuhlmann, Hamburg, Germany) was blended into the treatment solutions. A leaf segment of giant foxtail or a whole velvetleaf blade was fixed onto a glass slide with double-sided adhesive tape. For giant foxtail, 1 droplet of topramezone alone, topramezone with MSO and deionized water was dropped onto 1 leaf segment separately with a volumetric pipette (Eppendorf, Germany). For velvetleaf, 1 droplet of the three treatments was randomly dropped onto one blade. The droplet volume was 2 μL. Time from the moment of the droplet deposited onto the leaf surface to it completely evaporated was recorded with a stopwatch and the picture of each treated leaf was taken using a camera (SAMSUNG EX1, Samung Electronics Co., DC, South Korea). This procedure was conducted under the condition of temperature at 29±1°C and 45±10% RH. The spread area was determined by tracing the marked outline of the droplet spread on the leaf surface using the free select tool of GIMP software (version 2.8). An integrated index (λ) was used to comprehensively characterize the droplet spread and evaporation and defined as the product of the spread area and the evaporation time of a droplet on a leaf surface [7,8].

Deposition of Topramezone on Leaf Surfaces

Topramezone solutions were prepared as same as in the above-mentioned surface tension and contact angle measuring experiment. Sample plants were cultivated in the greenhouse condition as described above. One 2 μL-droplet of the prepared solution was dropped onto the adaxial side of the fourth leaf of giant foxtail or

Table 1. Topramezone doses and its recommended adjuvant preparation for giant foxtail and velvetleaf.

Weed	Treatment	Dose range g a.i. ha^{-1}	MSO* concentration %, v/v
Giant foxtail	Topramezone alone	0.394–806.4	0
	Topramezone with MSO	0.025–50.4	0.3
Velvetleaf	Topramezone	0.197–100.8	0
	Topramezone with MSO	0.197–100.8	0.3

*MSO is methylated seed oil.

velvetleaf with a volumetric pipette when these two weeds grow up to 4- to 5-leaf stage. About 1 cm-long leaf fragment treated with the herbicide solution was cut from the plants after the droplet completely evaporated and then was fixed onto a specimen holder of microscope with carbon double-side tape (NISSHIN EM. Co., Ltd. Tokyo, Japan). The deposition state of topramezone active ingredient on leaf surface was immediately observed under an environment scanning electron microscope (FEI Quanta 200, Czech Republic) at low vacuum mode at 12.5 KV. Pictures were taken at 1000×magnification.

Absorption of Topramezone by Weed Leaves

Topramezone solutions were prepared similar as in the above-mentioned surface tension and contact angle measuring experiment and the concentration was 0.504 g a.i. L^{-1}, which corresponded to 100.8 g a.i. ha^{-1} at 200 L ha^{-1}. Three 2 µL-droplets of the prepared solutions were dropped onto the adaxial side of the flag leaf of each giant foxtail plant (at flowering stage) and the fourth leaf of velvetleaf plant (at 4- to 5-leaf stage) with a volumetric pipette. Every treatment was repeated 20 times for giant foxtail and 12 times for velvetleaf. Total leaf sample weight was around 1–2 g. Treated leaves were harvested at 2, 6, 24 and 48 hours after application (HAT). All harvested leaves were firstly washed with 70 ml deionised water for 15- to 20-s and then rinsed with 30 ml deionised water for 5- to 10-s. After being washed, the sample leaves of each treatment were immersed into 25 ml chloroform for 5-to 10-s to dissolve cuticle waxes and to remove herbicide retained on the cuticle referencing Beckett and Stoller's method [32].

After being washed by chloroform, the leaves were cut into small pieces and put into a 50-ml centrifuge tube. Acetonitrile solution of 15 ml including 2% formic acid v/v was added to the tube. The plant sample was ground with an Ultra-Turrax (IKA®-Werke GmbH & Co. KG, Staufen, Germany) for around 2 min. Sodium chloride of 1 g and anhydrous magnesium sulfate of 4 g were added to the ground sample. Then the tube was capped and immediately vortexed vigorously for 1.5 min. The sample stood for 2 h at room temperature to extract the herbicide completely from the leaf sample and the water could be absorbed completely by anhydrous magnesium sulfate. Next, the tube was centrifuged for 5 min at 4000 rpm and then 1.5 ml of the upper layer was accurately removed using a volumetric pipette and transferred to a new 2-ml centrifuge tube. Primary secondary amine of 35 mg, graphitized carbon black of 15 mg and anhydrous magnesium sulfate of 150 mg were added to this tube. The tube was then vortexed vigorously for 1.5 min and then centrifuged for 3 min at 8000 rpm. Finally the resulting aqueous (top) layer was filtered through 0.22- µm nylon syringe filters. The extracted solution of 10 µL was injected into the UPLC-MS/MS system. The UPLC-MS/MS analytical conditions were according to Li's method [33]. Peak area data of topramezone alone and mixed with adjuvant MSO treatments at different HATs were compared.

Translocation of Topramezone in Weeds Plants

The fourth leaf of both giant foxtail and velvetleaf plants were coated with aluminum paper completely when these two weeds grew up to 4-leaf stage. Topramezone alone or mixed with MSO was applied at the rate of 25.2 g a.i. ha^{-1} on whole tested plants with the track sprayer mentioned above. When the droplets were dried on the leaf surfaces, around 10–15 min after herbicide application, the coated aluminum paper was taken off. Maximum quantum efficiency (Fv/Fm) of PS II of the fourth leaf was measured with an IMAGING-PAM M-Series Chlorophyll Fluorometer (Heinz Walz GmbH, Germany) before herbicide application and 2, 3, 4 and 5 days after treatment. Fv/Fm of PS II indicated the ratio of variable fluorescence (Fm − Fo) and maximum fluorescence (Fm), where Fo is dark fluorescence yield. Ten individual plants were measured for each treatment. For determination of Fo, plants were dark adapted for 30 min prior to the measurement. All measurements were conducted in a dark room under green illumination to avoid other photosynthetic active radiation except that emitted by the IMAGING PAM light source. After dark adaption, the fourth leaf of each plant was illuminated with a light saturation pulse of 580 µM m^{-2} s^{-1} and a wavelength of 450 nm for Fv/Fm determination. While measuring the Fv/Fm value, chlorophyll fluorescence images were taken using a CCD camera with a 680 nm filter.

To determine that the Fv/Fm value changes as a function of the natural logarithm of date after herbicide treatment, Fv/Fm was calculated according to the following equation 3 referencing the model of Rasmussen et al. [34,35]:

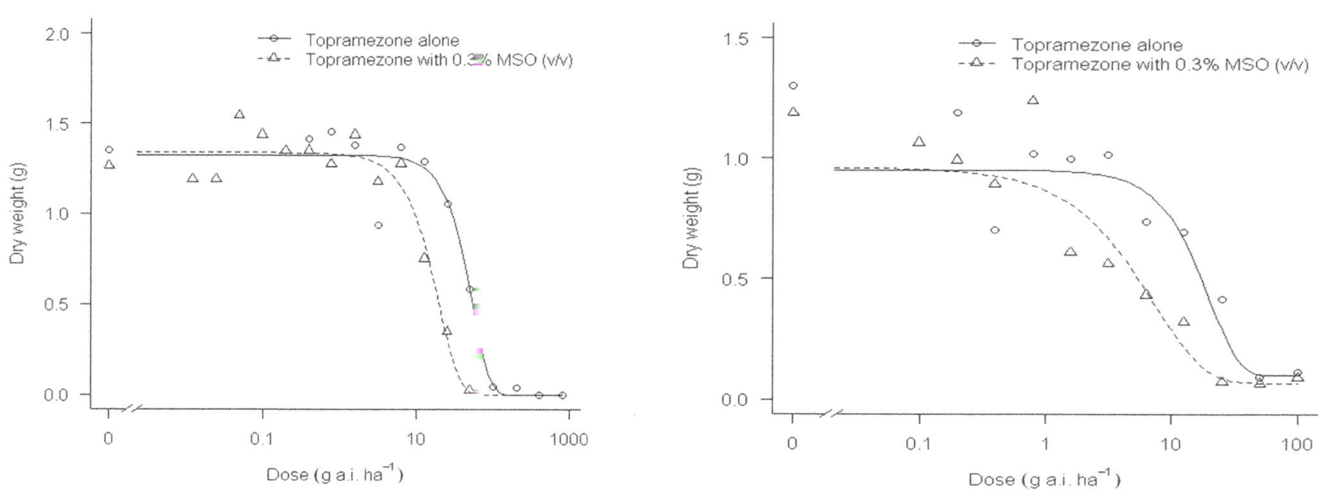

Figure 1. Dose-response curve of giant foxtail and velvetleaf on topramezone alone and tank-mixed with MSO. The left is the result of giant foxtail and the right is the result of velvetleaf. MSO means methylated soybean oil served as an adjuvant.

Table 2. Effect of MSO on the surface tension of topramezone solution.

Treatment	Surface tension [a]
	mN/m
Topramezone with MSO*	33.4±1.1 b
Topramezone alone	65.5±1.4 a
Water	67.2±0.4 a

*MSO is methylated seed oil; [a] Values (means ± SE) followed by same letters are not significantly (P≤0.05) different.

Table 3. Effect of MSO on topramezone droplet contact angle on the leaf surfaces of giant foxtail and velvetleaf.

Weed	Treatment	Contact angle [a]
		o
Giant foxtail	Topramezone with MSO*	101.1±3.0 b
	Topramezone alone	124.9±3.1 a
	Water	124.8±4.0 a
Velvetleaf	Topramezone with MSO	48.6±4.2 b
	Topramezone alone	66.5±3.0 a
	Water	65.4±4.6 a

*MSO is methylated seed oil; [a] Values (means ± SE) followed by same letters are not significantly (P≤0.05) different.

$$Fv/Fm = (Fv/Fm)_0 \bullet \exp[-c \bullet \ln(d+1)] \qquad (3)$$

where $(Fv/Fm)_C$ is the estimated Fv/Fm value in untreated plant leaves before herbicide application, c is the slope parameter, d is the date after herbicide treatment.

The 95%-Confidence intervals (95%-CI) of parameter c were calculated and difference significance between different treatments was judged according to the 95%-CIs overlap with each other or not.

The surface tension and contact angle, spread and evaporation experiment were repeated once. Data of two repeated experiments were combined to analyze because there was no interaction between two experiments. Data were subjected to ANOVA and means were separated by Fisher's protected LSD test at the 5% level of probability. Absorption and translocation experiments were also conducted twice and data combined to be analyzed. All analysis and graphs were done with the statistical software R (R version 2.15.2) [36].

Results and Discussion

Effect of MSO on Topramezone Efficacy

The four-parameter Weilbull model fitted well for both velvetleaf (P = 0.59) and giant foxtail (P = 0.48) curves (Fig 1). After topramezone was tank-mixed with MSO, its biological efficacy was enhanced significantly. For giant foxtail, the ED_{90} value of treatment topramezone with MSO was only 33.53 g a.i. ha^{-1}, while the ED_{90} value of treatment topramezone alone was 85.06 g a.i. ha^{-1}. The former treatment was 1.5-fold more effective than the latter treatment. For velvetleaf, the ED_{90} value of treatment topramezone with MSO was only 15.84 g a.i. ha^{-1}, while the ED_{90} value of treatment topramezone alone was 31.56 g a.i. ha^{-1}. The former treatment was 1.0-fold more effective than the latter treatment.

It should be noticed that MSO enhances herbicide efficacy on weeds and also increases phytotoxicity risk on crops. However, topramezone is highly selective between corn and weed species [17] and it has low risk on the crop when applied with MSO. Some field experimental results demonstrated that this herbicide was still highly safe to corn when it was tank-mixed with MSO [24,25,26].

Effect of MSO on Surface Tension and Contact Angle of Topramezone Solution

There was no difference between surface tension of topramezone alone solution and water control treatment. After topramezone solution was mixed with MSO, its surface tension and contact angle on weed leaves decreased significantly (Table 2, 3). The surface tension decreased around 50%. This result is similar to the results of Xu et al., which demonstrated that MSO decreased the surface tension and contact angle of distilled water on the waxy leaves [7]. The decrease of the contact angle was higher on velvetleaf (26.9%) than on giant foxtail (19.0%).

Surface tension and contact angle are two important factors in adjuvant research [37]. Surface tension is determined only by the physical-chemical property of an adjuvant, but contact angle is the result of interaction between droplet and target surface. For adjuvant at normal application rates, the lower the surface tension

Table 4. Effect of MSO on spread and evaporation of topramezone on leaf surface of giant foxtail and velvetleaf.

Weed	Treatment	Spread area [a]	Evaporation time [a]	λ[a]
		cm^2	Min	$min \times cm^2$
Giant foxtail	Topramezone with MSO*	0.27±0.03 a	14.66±1.90 b	3.65±0.34 a
	Topramezone alone	0.02±0.00 b	32.42±1.97 a	0.62±0.08 b
	Water	0.01±0.00 b	30.47±2.51 a	0.36±0.05 b
Velvetleaf	Topramezone with MSO	0.08±0.01 a	17.48±0.84 a	1.41±0.20 a
	Topramezone alone	0.05±0.01 b	19.63±1.16 a	0.94±0.12 b
	Water	0.03±0.01 b	17.77±0.61 a	0.55±0.12 b

*MSO is methylated seed oil; [a] Values (means ± SE) in the same column followed by same letters are not significantly (P≤0.05) different. λ means the product of the spread area and the evaporation time.

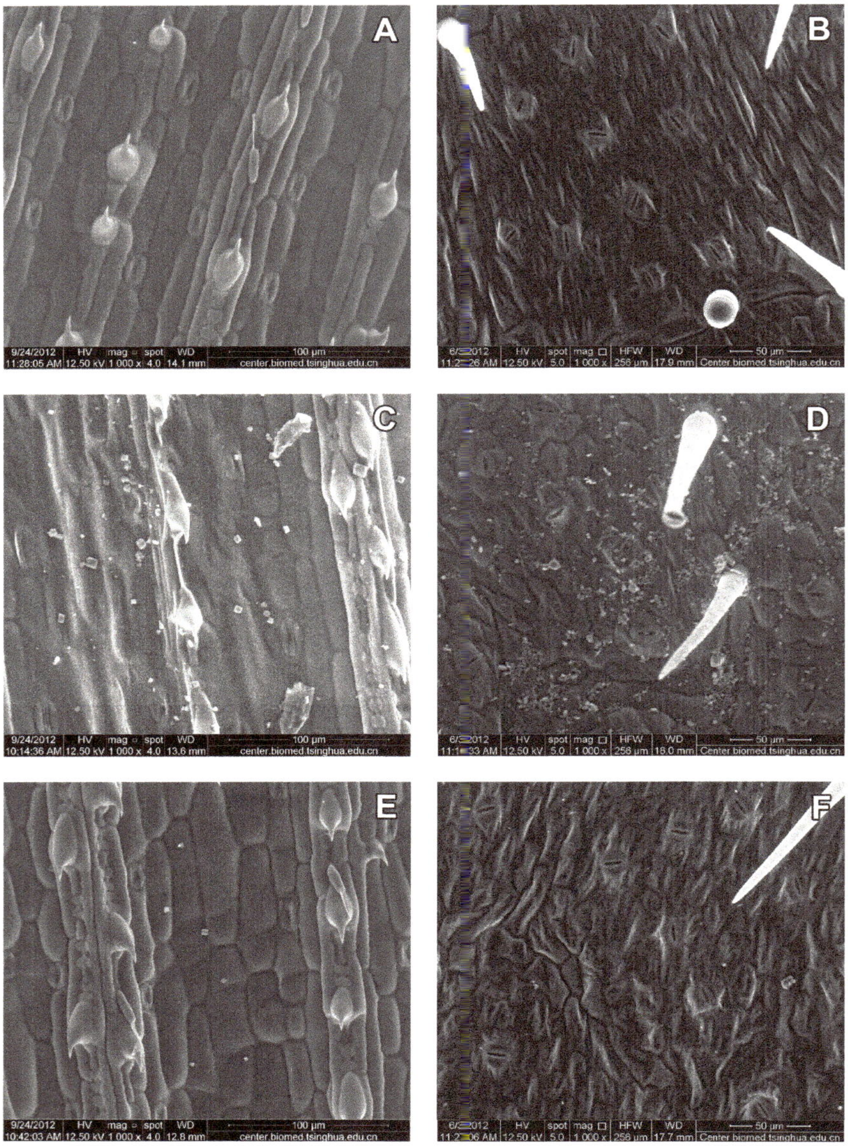

Figure 2. Active ingredient depositions of topramezone on the adaxial sides of giant foxtail and velvetleaf leaves. A, C and E are giant foxtail leaves, and B, D and F are velvetleaf leaves. The upper two (A and B) means blank control, the middle (C and D) means treated with topramezone alone and the lower two (E and F) means treated with topramezone mixed with 0.3% MSO (v/v); topramezone dose is 25.2 g a.i. ha^{-1}; control means treated with deionised water; pictures were photographed at 1000× magnification using an environment scanning electron microscope at low vacuum mode at 12.5 kV.

is, the lower the contact angle is on target surface [37]. A leaf was considered "wettable" if the water contact angle was less than 90 ^{0} and "non-wettable" if the water contact angle was greater than 90 ^{0} [38,39]. According to this rule, velvetleaf could be classified as "wettable" while giant foxtail as "non-wettable". Our results showed that MSO could decrease the contact angle on these two kinds of leaves, but decreasing degree was larger on the "wettable" leaf surface (velvetleaf) than on the "non-wettable" leaf surface (giant foxtail).

Effect of MSO on Spread and Evaporation of Topramezone Droplet on Leaf Surfaces

The droplet of topramezone alone solution almost did not spread on the surface of giant foxtail and velvetleaf leaves till complete evaporation. However, the spreading area increased significantly on the leaves of these two weeds when the solution was mixed with MSO at the concentration 0.3% v/v. After topramezone solution was mixed with MSO, the evaporation time of the droplets shortened significantly on giant foxtail, but did not on velvetleaf. The index λ of treatment topramezone with MSO was significantly higher than those of treatments topramezone alone and water control on both weed species (Table 4).

After adjuvant MSO was mixed, increasing spread areas of the droplets is due to the decrease of their contact angles on weed leaves. Larger spread area on giant foxtail than on velvetleaf is probably due to the waxier giant foxtail leaf surface than that of velvetleaf. Sanyal et al. reported that the wax content of velvetleaf was 7.4 μg cm^{-2} and that of green foxtail was 19.1 μg cm^{-2}

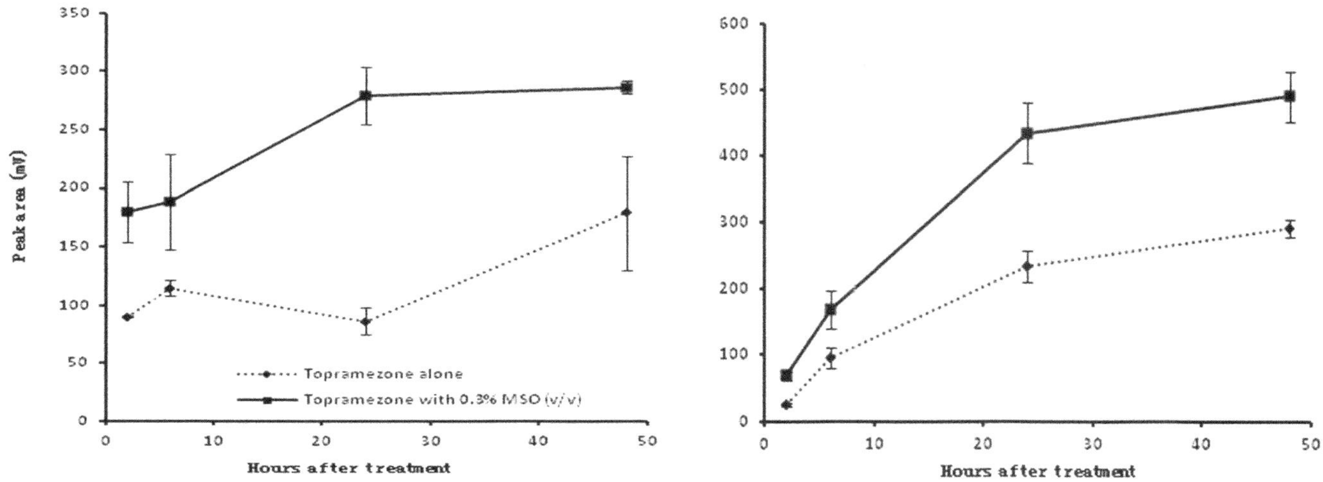

Figure 3. Topramezone absorption of giant foxtail and velvetleaf after treatment. The left is the result of giant foxtail and the right is the result of velvetleaf. Topramezone dose is 100.8 g a.i. ha^{-1}.

[40,41]. In general, a larger droplet spread area and extending liquid state increases pesticide uptake in the target plants. However, these two desirable conditions are contradictory because a larger wetted area always occurred with less evaporation time. Xu et al. introduced the integrated indexλto evaluate droplet spread and resistance to drying [7,8]. This index is more reasonable to characterize the ability of an adjuvant in increasing spread and decreasing evaporation. In our study, we found the same phenomena of larger spread area but less evaporation time after mixing MSO. However, the indexλvalue of treatment topramezone with MSO was significantly higher than those of treatments topramezone alone and water control on both velvetleaf and giant foxtail. The result was similar to the results of Xu et al., in which MSO was good at increasing indexλof distilled water on both waxy and hairy leaf surfaces [7,8].

Effect of MSO on Deposition of Topramezone on Leaf Surfaces

The leaf surface of giant foxtail was relatively more flat and waxy while that of velvetleaf was more hairy (Fig 2). A lot of crystals of active topramezone ingredient were observed on the leaf surface of giant foxtail or velvetleaf treated with topramezone alone. However, only a few crystals were observed on the leaf surfaces of giant foxtail and velvetleaf treated by topramezone mixed with MSO. The results indicated that MSO decreased topramezone crystallization on both weed leaf surfaces.

Due to nearly no spread of droplets of topramezone applied alone, less active ingredient penetrated into the leaf. A lot of active ingredient was deposited and then crystalized on the leaf surface after water evaporation. While in the treatment of topramezone

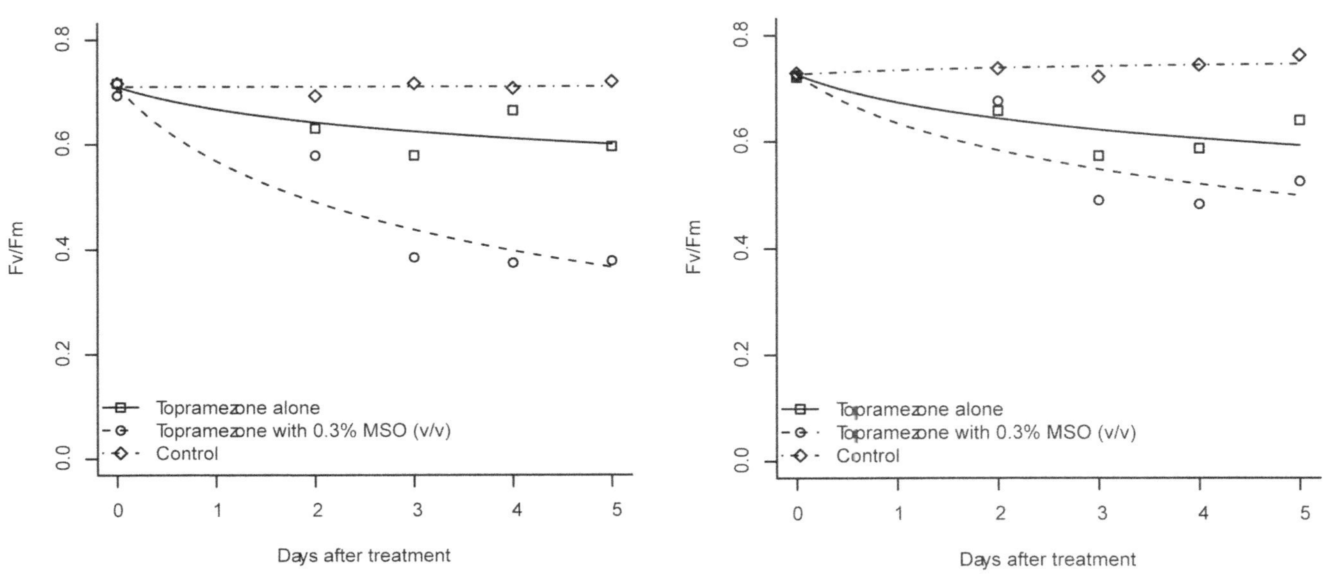

Figure 4. PS II photochemistry maximum quantum efficiency (Fv/Fm) of the leaf of giant foxtail and velvetleaf at different time after herbicide application. The left is the result of giant foxtail and the right is the result of velvetleaf. Topramezone dose is 25.2 g a.i. ha^{-1}; control means treated with tap water.

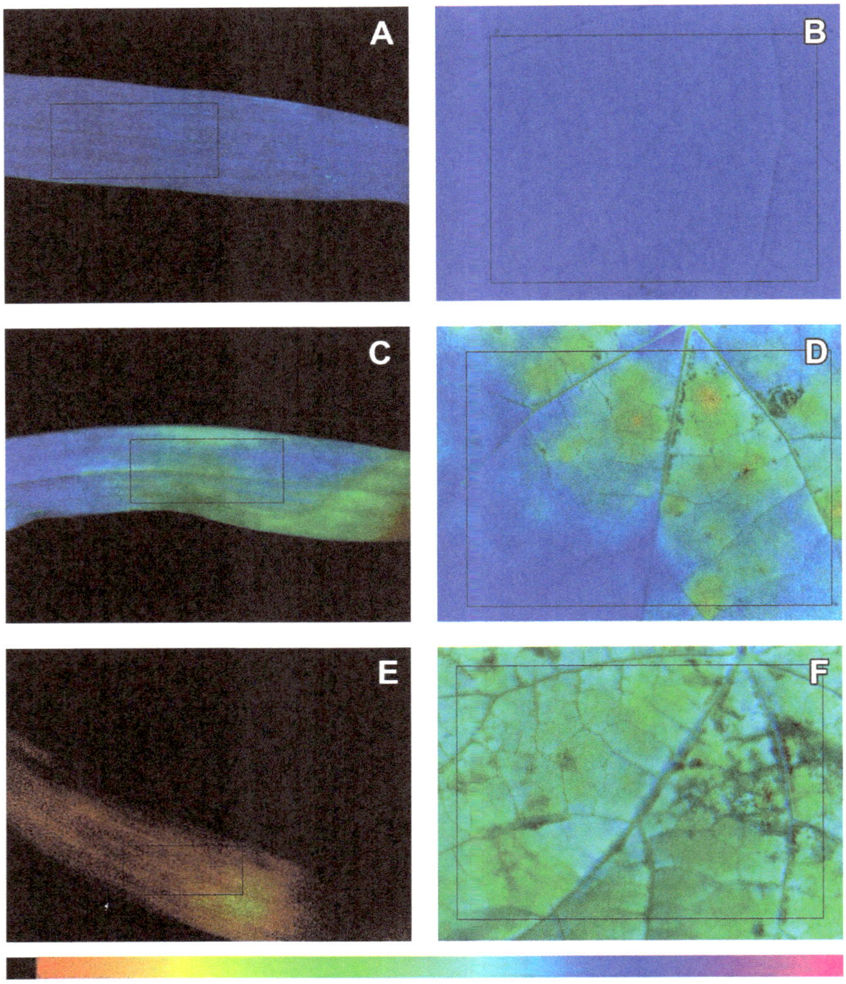

Figure 5. Chlorophyll fluorescence images of leaves of giant foxtail and velvetleaf treated by different solutions 5 days after treatment. A, C and E are giant foxtail leaves, and B, D and F are velvetleaf leaves; the upper two (A and B) means blank control, the middle (C and D) means treated with topramezone alone and the lower two (E and F) means treated with topramezone mixed with 0.3% MSO (v/v); Topramezone dose is 25.2 g a.i. ha^{-1}; control means treated with tap water. Maximum quantum efficiency is reduced more when the color turns from blue to red (black).

mixed with MSO, the droplet spread significantly larger compared with the treatment of topramezone applied alone, so less active ingredient was crystalized as solid state.

Effect of MSO on Absorption of Topramezone by Leaves

Topramezone absorption by giant foxtail and velvetleaf leaves increased from 2 HAT to 48 HAT whether the herbicide was applied alone or with MSO. However, compared with the treatment of topramezone applied alone, the absorption amount was larger when it was mixed with MSO at any sampling times after herbicide application for both weed species. Especially at 24 HAT, the absorption amount of topramezone increased by 68.9% for giant foxtail and by 45.9% for velvetleaf (Fig 3). The result indicated that MSO increased topramezone absorption by both weed leaves.

Grossmann and Ehrhardt found that the adjuvant Dash HC (a kind of MSO) increased the foliar uptake of giant foxtail, sorghum (*Sorghum bicolor*), black nightshade (*Solanum nigrum*) and corn [17], which was similar to our result. Less herbicide is absorbed by the leaf if the herbicide is in a solid rather than in a liquid form [42]. MSO reduced the crystallization of active topramezone ingredient, and then resulted in increasing absorption when toramezone was mixed with MSO.

Effect of MSO on Translocation of Topramezone in the Plants

Lack-of-fit tests showed that equation 3 well described the change of the Fv/Fm value as time ($P>0.05$). The Fv/Fm values of the coated leaves declined exponentially with time either the plants treated by topramezone alone or by topramezone mixed with MSO. However, the declining speed of the treatment of topramezone with MSO was faster than that of the treatment of topramezone alone (Fig 4). In the case of giant foxtail, 95%-Cl of the curve slope for topramezone with MSO was −0.1532 to −0.0974, while that of topramezone alone was −0.0649 to −0.0091, which meant that they were statistically different with each other (the 95%-CIs did not overlap with each other). In the case of velvetleaf, the slope of the treatment of topramezone with MSO was steeper than that of the treatment of topramezone alone, but their slope 95%-CIs slightly overlapped. The former was −0.1077 to −0.0497, and the latter was −0.0738 to 0.0159. The chlorophyll fluorescence image of the giant foxtail leaf treated by topramezone with MSO was yellow to black while that of the leaf treated by topramezone alone was only light green; the image of the velvetleaf leaf treated by topramezone with MSO was green while that of the leaf treated by topramezone alone was only light green (Fig 5). These results indicated that the active ingredient of topramezone was translocated more and faster from treated leaves to untreated leaves (coated leaves) when this herbicide was mixed with MSO than applied alone, which resulted in more severe and faster photosynthesis inhibition of the untreated leaves.

There is a strong correlation between the translocation and the absorption of a herbicide in the plants [14,16,43]. In this study, MSO significantly increased absorption and translocation of topramezone for giant foxtail, but it significantly increased only absorption of this herbicide but not translocation for velvetleaf. This phenomenon needs further study in the future.

For agriculture, the enhancement of herbicide biological efficacy by the addition of adjuvant can contribute to reducing herbicide application rates [44]. The performance of MSO adjuvants is dependent upon the source and composition of the oil, the nature of the herbicide and treated plant species. It has been demonstrated clearly that appropriate combinations of herbicides with adjuvants can greatly enhance the rate and efficiency of herbicides delivery to target sites, and thus the ultimate activity [2,6]. Our results showed that the combination of methyleated soybean oil adjuvant and topramezone enhanced the herbicide being delivered to the target sites and then provided a really good efficacy on weeds. This kind MSO is also used as an adjuvant for some other herbicides on weed control. In addition, this adjuvant is minimum health risk to humans and the environment because it comes from soybean and has good biodegradation ability.

Acknowledgments

The authors thank Dr. Yongquan Zheng, Mr. Fengshou Dong, and Dr. Yuanbo Li from the Institute of Plant Protection, Chinese Academy of Agricultural Sciences, Key Laboratory of Pesticide Chemistry and Application for their technical assistance in the experiment of topramezone absorption, and also Dr. Victor Rueda Ayala from the Department of Weed Science, Institute for Phytomedicine, University of Hohenheim for his statistical assistance.

Author Contributions

Conceived and designed the experiments: JZ HN ZZ. Performed the experiments: JZ OJ AM. Analyzed the data: JZ OJ AM. Contributed reagents/materials/analysis tools: ZZ RG. Wrote the paper: JZ.

References

1. Green JM, Foy CL (2000) Adjuvants: test design, interpretation, and presentation of results. Weed Technol 14: 819–825.
2. Kirkwood RC (1993) Use and mode of action of adjuvants for herbicides: a review of some current work. Pestic Sci 38: 93–102.
3. Penner D (2000) Activator adjuvants. Weed Technol 14: 785–791.
4. Zabkiewicz JA (2000) Adjuvants and herbicidal efficacy- present status and future prospects. Weed Res 40: 139–149.
5. Miller P, Westra P (1996) Herbicide surfactants and adjuvants. Bulletin No. 0.599, Crop Series. Colorado State University Cooperative Extension, Fort Collins, Co.
6. Gauvrit C, Cabanne F (1993) Oils for weed control: uses and mode of action. Pestic Sci 37: 147–153.
7. Xu L, Zhu H, Ozkan HE, Bagley WE, Derksen RC, et al. (2010) Adjuvant effects on evaporation time and wetted area of droplets on waxy leaves. Transactions of the ASABE 53: 13–20.
8. Xu L, Zhu H, Ozkan HE, Bagley WE, Krause CR (2011) Droplet evaporation and spread on waxy and hairy leaves associated with type and concentration of adjuvants. Pest Manag Sci 67: 842–851.
9. Bukun B, Lindenmayer RB, Nissen SJ, Westra P, Shaner DL, et al. (2010) Absorption and Translocation of aminocyclopyrachlor and aminocyclopyrachlor-methyl ester in canada thistle (*Cirsium arvense*). Weed Sci 58: 96–102.
10. Bunting JA, Sprague CL, Riechers DE (2004) Absorption and activity of foramsulfuron in giant foxtail (*Setaria faberi*) and woolly cupgrass (*Eriochloa villosa*) with various adjuvants. Weed Sci 52: 513–517.
11. Enloe SF, Nissen SJ, Westra P (1999) Absorption, fate, and soil activity of quinclorac in field bindweed (*Convolvulus arvensis*). Weed Sci 47: 136–142.
12. Hart SE, Kells JJ, Penner D (1992) Influence of adjuvants on the efficacy, absorption and spray retention of primisulfuron. Weed Technol 6:592–598.
13. Pester TA, Nissen SJ, Westra P (2001) Absorption, translocation, and metabolism of imazamox in jointed goatgrass and feralrye. Weed Sci 49:607–612.
14. Sharma SD, Singh M (2000) Optimizing foliar activity of glyphosate on *Bidens frondosa* and *Panicum maximum* with different adjuvant types. Weed Res 40: 523–533.
15. Thompson WM, Nissen SJ, Masters RA (1996) Adjuvant effects on imazethapyr, 2,4-D and picloram absorption by leafy spurge (*Euphorbia esula*). Weed Sci 44: 468–475.
16. Young BG, Hart SE (1998) Optimizing foliar activity of isoxaflutole on giant foxtail (*Seteria faberi*) with various adjuvants. Weed Sci 46: 397–402.
17. Grossmann K, Ehrhardt T (2007) On the mechanism of action and selectivity of the corn herbicide topramezone: a new inhibitor of 4-hydroxyphenylpyruvate dioxygenase. Pest Manag Sci 63: 429–439.
18. Bollman J, Boerboom C, Becker R, Fritz V (2007) New weed control options for sweet corn. In: Proc Of the 2007 Wisconsin fertilizer, aglime and pest management conference. pp. 216–221.
19. Porter RM, Vaculin PD, Orr JE, Immaraju JA, O'Neal WB (2005) Topramezone: a new active for postemergence weed control in corn. In: Proceedings of the North Central Weed Science Society. pp. 60.
20. Ransom CV, Ishida JK (2005) Weed control and crop response with Option® and Impact® herbicides in furrow-irrigated field corn. In: Malheur experiment station annual report 2005, Oregon State University, Special report. pp. 27–32.
21. Schonhammer A, Freitag J, Koch H (2006) Topramezone eineuer Herbizid-wirkstoff zur hochselektiven Hirse-und Unkrautbekampfung im Mais. J Plant Dis Prot 23: 1023–1031.
22. Soltani N, Sikkema PH, Zandstra J, O' Sullivan J, Robinson DE (2007) Response of eight sweet maize (*Zea mays* L.) hybrids to topramezone. Hort Sci 42: 110–112.
23. Gitsopoulos TK, Melidis V, Evgenidis G (2010) Response of maize (*Zea mays* L.) to post-emergence applications of topramezone. Crop Prot 29: 1091–1093.
24. Zhou X, Zhu J, Chen Q, Zheng S, Tu M, et al. (2010) Efficacy of 33.6% topramezone SC in weed control in maize field. Phytomedicine 23: 41–44 (in Chinese).
25. Zheng L, Lv Y, Ni H (2011) Efficacy comparison of four post-emergence herbicides in weed control in corn. Agrochemicals 50: 597–613 (in Chinese with English abstract)
26. Zhang J, Zheng L, Jäck O, Yan D, Zhang Z, et al. (2013) Efficacy of four post-emergence herbicides applied at reduced doses on weeds in summer maize (*Zea mays* L.) fields in North China Plain. Crop Prot 52: 26–32.
27. Senesac A (2007). New herbicide for sweet corn. Crop Talk 3: 7.
28. Young BG, Zollinger RK, Bernards ML (2007) Variability of tembotrione efficacy as influenced by commercial adjuvant products. In: 62 of the North Central Weed Science Society Proceedings. 141.
29. Zollinger R, Ries JL (2006) Comparing mesotrione, tembotrione, and topramezone. In: 61 of the 2006 North Central Weed Science Society Proceedings. 114.
30. Streibig JC, Rudemo M, Jensen JE (1993) Dose-response curves and statistical methods. In: Streibig JC, Kudsk P, Herbicide Bioassays. Boca Raton: CRC Press. pp. 29–55
31. Ritz C, Cedergreen N, Jensen JE, Streibig JC (2006) Relative potency in nonsimilar dose-response curves. Weed Sci 54: 407–412.
32. Beckett TH, Stoller EW (1991) Effects of methyl ammonium and urea ammonium nitrate uptake of thifensulfuron in velvetleaf (*Abutilon theophrasti*). Weed Sci 41: 218–224.
33. Li Y, Dong F, Liu X, Xu J, Li J, et al. (2011) Miniaturized liquid-liquid extraction coupled with ultra-performance liquid chromatography/tandem mass spectrometry for determination of topramezone in soil, corn, wheat, and water. Anal Bioanal Chem 400: 3097–3107.
34. Rasmussen J, Bibby BM, Schou AP (2008) Investigating the selectivity of weed harrowing with new methods. Weed Res 48: 523–532.
35. Rasmussen J, Nielsen HH, Gundersen H (2009) Tolerance and selectivity of cereal species and cultivars to postemergence weed harrowing. Weed Sci 57(3): 338–345.
36. R Development Core Team (2011). R: A language and environment for statistical computing. R Foundation for Statistical Computing, Vienna, Austria. ISBN 3-900051-07-0, Available: http://www.R-project.org/.
37. Sun J, Foy CL (1998) Physico-chemical properties of several commercial organosilicones, their blends, and selected other adjuvants. In: Nalewaja JD, Goss GR, Tann RS, editors. Pesticide formulations and application systems:

eighteenth volume, ASTM STP 1347, American Society for Testing and Materials, West Conshohocken, PA. pp. 281–293.

38. Cape JN (1983) Contact angles of water droplets on needles of Scots pine (*Pinus sylvestris*) growing in polluted atmospheres. New Phytol 93: 293–299.
39. Bhushan B, Jung YC (2008) Wetting, adhesion, and friction of superhydrophobic and hydrophilic leaves and fabricated micro/nanopatterned surfaces. J. Physics: Condensed Matter 20 225010: 1–24.
40. Sanyal D, Bhowmik PC, Reddy KN (2006a) Leaf characteristics and surfactants affect primisulfuron droplet spread in three broadleaf weeds. Weed Sci 54: 16–22.
41. Sanyal D, Bhowmik PC, Reddy KN (2006b) Influence of leaf surface micromorphology, wax content, and surfactant on primisulfuron droplet spread on barnyardgrass (*Echinochloa crus-galli*) and green foxtail (*Setaria viridis*). Weed Sci 54: 627–633.
42. Hess FD, Falk RH (1990) Herbicide deposition on leaf surface. Weed Sci 38: 280–288.
43. Dodds DM, Reynolds DB, Massey JH, Smith MC, Koger CH (2007) Effect of adjuvants and urea ammonium nitrate on bispyribac efficacy, absorption, and translocation in barnyardgrass (*Echinochloa crus-galli*). II. absorption and translocation. Weed Sci 55: 406–411.
44. Blackshaw RE, O'Donovan JT, Harker KN, Clayton GW, Stougaard RN (2006) Reduced herbicide doses in field crops: A review. Weed Biol Manag 6: 10–17.

Permissions

All chapters in this book were first published in PLOS ONE, by The Public Library of Science; hereby published with permission under the Creative Commons Attribution License or equivalent. Every chapter published in this book has been scrutinized by our experts. Their significance has been extensively debated. The topics covered herein carry significant findings which will fuel the growth of the discipline. They may even be implemented as practical applications or may be referred to as a beginning point for another development.

The contributors of this book come from diverse backgrounds, making this book a truly international effort. This book will bring forth new frontiers with its revolutionizing research information and detailed analysis of the nascent developments around the world.

We would like to thank all the contributing authors for lending their expertise to make the book truly unique. They have played a crucial role in the development of this book. Without their invaluable contributions this book wouldn't have been possible. They have made vital efforts to compile up to date information on the varied aspects of this subject to make this book a valuable addition to the collection of many professionals and students.

This book was conceptualized with the vision of imparting up-to-date information and advanced data in this field. To ensure the same, a matchless editorial board was set up. Every individual on the board went through rigorous rounds of assessment to prove their worth. After which they invested a large part of their time researching and compiling the most relevant data for our readers.

The editorial board has been involved in producing this book since its inception. They have spent rigorous hours researching and exploring the diverse topics which have resulted in the successful publishing of this book. They have passed on their knowledge of decades through this book. To expedite this challenging task, the publisher supported the team at every step. A small team of assistant editors was also appointed to further simplify the editing procedure and attain best results for the readers.

Apart from the editorial board, the designing team has also invested a significant amount of their time in understanding the subject and creating the most relevant covers. They scrutinized every image to scout for the most suitable representation of the subject and create an appropriate cover for the book.

The publishing team has been an ardent support to the editorial, designing and production team. Their endless efforts to recruit the best for this project, has resulted in the accomplishment of this book. They are a veteran in the field of academics and their pool of knowledge is as vast as their experience in printing. Their expertise and guidance has proved useful at every step. Their uncompromising quality standards have made this book an exceptional effort. Their encouragement from time to time has been an inspiration for everyone.

The publisher and the editorial board hope that this book will prove to be a valuable piece of knowledge for researchers, students, practitioners and scholars across the globe.

List of Contributors

Liping Ke, RuiE Liu, Bijue Chu, Xiushuang Yu, Yuqiang Sun, Xiaofei Cheng and Huizhong Wang
College of Life and Environmental Science, Hangzhou Normal University, Hangzhou, Zhejiang, China

Jie Sun
College of Agronomy, Shihezi University, Shihezi, Xinjiang, China

Brian Jones
Faculty of Agriculture, Food and Natural Resources, The University of Sydney, Sydney, Australia

Gang Pan and Shuijin Zhu
College of Agriculture and Biotechnology, Zhejiang University, Hangzhou, Zhejiang, China

Nina Cedergreen
Department of Plant and Environmental Sciences, University of Copenhagen, Frederiksberg, Denmark

Arnaud Duhoux and Christophe Délye
INRA, UMR1347 Agroécologie, Dijon, France

Shiv Shankhar Kaundun, Sarah-Jane Hutchings, Richard P. Dale and Eddie McIndoe
Syngenta, Jealott's Hill International Research Centre, Biological Sciences, Bracknell, United Kingom

Guo-Qi Chen and Sheng Qiang
Weed Research Laboratory, Nanjing Agricultural University; Nanjing, China

Yun-He He
Weed Research Laboratory, Nanjing Agricultural University; Nanjing, China
Department of Landscape Architecture, Zhejiang Agricultural and Forestry University, Hangzhou, China

Wu-Chien Chien, Chi-Ming Chu, Senyeong Kao, Sui-Lung Su and Ching-Huang Lai
Department of Public Health, National Defense Medical Center, No. 161, Section 6, Neihu District, Taipei City, Taiwan, Republic of China

Chi-Hsiang Chung
Graduate Institute of Life Sciences, National Defense Medical Center, No. 161, Section 6, Neihu District, Taipei City, Taiwan, Republic of China

Jouni J. K. Jaakkola
Center for Environmental and Respiratory Health Research, Institute of Health Sciences, University of Oulu, Oulu, Finland

Natália Corniani, Edivaldo D. Velini and Ferdinando M. L. Silva
São Paulo State University, Faculty of Agronomic Sciences, Botucatu, SP, Brazil

N. P. Dhammika Nanayakkara
National Center for Natural Products Research, School of Pharmacy, University of Mississippi, University, MS, United States of America

Matthias Witschel
BASF SE, GVA/HC-B009, Ludwigshafen, Germany

Franck E. Dayan
USDA-ARS Natural Products Utilization Research Unit, University, MS, United States of America

Jodie S. Holt, Shana R. Welles, Sylvia M. Heredia, Alejandra Martinez-Berdeja, Kai T. Palenscar, Lynn C. Sweet and Norman C. Ellstrand
Department of Botany and Plant Sciences, University of California Riverside, Riverside, California, United States of America

Katia Silvera
Smithsonian Tropical Research Institute, Balboa, Ancon, Republic of Panama

Ian M. Heap
International Survey of Herbicide Resistant Weeds, Corvallis, Oregon, United States of America

Jessica K. Leet, Casey D. Lindberg, Luke A. Bassett, Gregory M. Isales, Krystle L. Yozzo, Tara D. Raftery and David C. Volz
Department of Environmental Health Sciences, Arnold School of Public Health, University of South Carolina, Columbia, South Carolina, United States of America

Marianne Cousin, Jean-Luc Brunet and Luc P. Belzunces
INRA, UR 406 Abeilles & Environnement, Laboratoire de Toxicologie Environnementale, Avignon, France

Elaine Silva-Zacarin
Universidade Federal de São Carlos, Laboratory of Structural and Functional Biology, Sorocaba, São Paulo State, Brazil

André Kretzschmar
INRA, UR Biostatistiques et Processus Spatiaux, Avignon, France

Mohamed El Maataoui
Université d'Avignon, Pôle Agrosciences, Avignon, France

Floriane Larras, Agnès Bouchez, Frédéric Rimet and Bernard Montuelle
Institut National de la Recherche Agronomique, UMR 0042, Carrtel, Thonon, France

Xiangyang Yuan, Na Ning, Yinyuan Wen, Shuqi Dong, Meiqiang Yin, Meijun Guo, Binqiang Wang, Lei Feng and Pingyi Guo
Laboratory of Crop Chemical Regulation and Chemical Weed Control, Agronomy College, Shanxi Agricultural University, Taigu, the People's Republic of China

Liguang Zhang
Laboratory of Crop Chemical Regulation and Chemical Weed Control, Agronomy College, Shanxi Agricultural University, Taigu, the People's Republic of China
Institute of Crop Sciences, Shanxi Academy of Agricultural Sciences, Taiyuan, the People's Republic of China

Xia Yang, Xin-Yan Yu and Yong-Feng Li
Institute of Plant Protection, Jiangsu Academy of Agricultural Sciences, Nanjing, China

Adam S. Davis
United States Department of Agriculture/Agricultural Research Service, Global Change and Photosynthesis Research Unit, Urbana, Illinois, United States of America

Jason D. Hill
Department of Bioproducts and Biosystems Engineering, University of Minnesota, St. Paul, Minnesota, United States of America

Craig A. Chase
Leopold Center for Sustainable Agriculture, Iowa State University, Ames, Iowa, United States of America

Ann M. Johanns
Department of Economics, Iowa State University Extension and Outreach, Osage, Iowa, United States of America

Matt Liebman
Department of Agronomy, Iowa State University, Ames, Iowa, United States of America

Jing Li, Hui Yu, Fengzhen Zhang, Chaoyang Lin, Jianhua Gao, Jun Fang, Xiahui Ding, Zhicheng Shen and Xiaoli Xu
State Key Laboratory of Rice Biology, Institute of Insect Sciences, School of Agriculture and Biotechnology, Zhejiang University, Hangzhou, China

Collin W. Ahrens and Carol A. Auer
Agricultural Biotechnology Lab, Department of Plant Science, University of Connecticut, Storrs, Connecticut, United States of America

Kesiane M. Costa and Izaque S. Maciel
Postgraduate Program in Medicine and Health Sciences, Pontifical Catholic University of Rio Grande do Sul, Porto Alegre/RS, Brazil

Luiza W. Kist
Postgraduate Program in Medicine and Health Sciences, Pontifical Catholic University of Rio Grande do Sul, Porto Alegre/RS, Brazil
Laboratory of Genomics and Molecular Biology, Faculty of Biosciences, Pontifical Catholic University of Rio Grande do Sul, Porto Alegre/RS, Brazil

Maria M. Campos
Postgraduate Program in Medicine and Health Sciences, Pontifical Catholic University of Rio Grande do Sul, Porto Alegre/RS, Brazil
Institute of Toxicology and Pharmacology, Pontifical Catholic University of Rio Grande do Sul, Porto Alegre/RS, Brazil
Faculty of Dentistry, Pontifical Catholic University of Rio Grande do Sul, Porto Alegre/RS, Brazil

Maurício R. Bogo
Postgraduate Program in Medicine and Health Sciences, Pontifical Catholic University of Rio Grande do Sul, Porto Alegre/RS, Brazil
Laboratory of Genomics and Molecular Biology, Faculty of Biosciences, Pontifical Catholic University of Rio Grande do Sul, Porto Alegre/RS, Brazil
Postgraduate Program in Cellular and Molecular Biology, Pontifical Catholic University of Rio Grande do Sul, Porto Alegre/RS, Brazil

Wilhelm Klümper and Matin Qaim
Department of Agricultural Economics and Rural Development, Georg-August-University of Goettingen, Goettingen, Germany

Luhua Zhang, Haiwei Chen, Yanlan Li, Yanan Li, Defu Chen and Xiwen Chen
Laboratory of Molecular Genetics, College of Life Sciences, Nankai University, Tianjin, China

Shengjun Wang, Jinping Su and Xuejun Liu
Tianjin Crop Research Institute, Tianjin, China

Shweta Sharma
Biochemistry Division, CSIR-Central Drug Research Institute, Lucknow, Uttar Pradesh, India

Jayanta Sarkar and Sudhir Sinha
Biochemistry Division, CSIR-Central Drug Research Institute, Lucknow, Uttar Pradesh, India
Academy of Scientific and Innovative Research, New Delhi, India

Chandana Haldar
Department of Zoology, Banaras Hindu University, Varanasi, Uttar Pradesh, India

Jie Jin, Mengxia Yu, Chao Hu, Li Ye, Lili Xie, Jin Jin and Feifei Chen
Department of Hematology, the First Affiliated Hospital of Zhejiang University, Hangzhou, People's Republic of China
Institute of Hematology, Zhejiang University School of Medicine, Hangzhou, People's Republic of China

Hongyan Tong
Department of Hematology, the First Affiliated Hospital of Zhejiang University, Hangzhou, People's Republic of China
Institute of Hematology, Zhejiang University School of Medicine, Hangzhou, People's Republic of China
Myelodysplastic syndromes diagnosis and therapy center, Zhejiang University School of Medicine, Hangzhou, People's Republic of China

Arno Cimadom, Elisabet Zöttl and Sabine Tebbich
Department of Behavioural Biology, University of Vienna, Vienna, Austria

Angel Ulloa, Birgit Fessl and Francesca Cunninghame
Charles Darwin Foundation, Puerto Ayora, Santa Cruz Island, Galápagos, Ecuador

Patrick Meidl
Institute for Science and Technology, Klosterneuburg, Austria

Markus Zöttl
Department of Zoology, University of Cambridge, Cambridge, United Kingdom

Erwin Nemeth and Michael Dvorak
BirdLife Austria, Vienna, Austria

Leila Priscila Peters, Giselle Carvalho, Paula Fabiane Martins, Manuella Nó brega Dourado, Milca Bartz Vilhena and Ricardo Antunes Azevedo
Departamento de Genética, Escola Superior de Agricultura Luiz de Queiroz, Universidade de São Paulo, Piracicaba, Brazil

Marcos Pileggi
Departamento de Biologia Estrutural, Molecular e Genética, Universidade Estadual de Ponta Grossa, Ponta Grossa, Brazil

Michael C. R. Alavanja, Jonathan N. Hofmann, Kathryn H. Barry, Stella Koutros, Gabriella Andreotti, Jay H. Lubin, Aaron Blair and Laura E. Beane Freeman
Division of Cancer Epidemiology and Genetics, National Cancer Institute, Rockville, Maryland, United States of America

Charles F. Lynch
College of Public Health, University of Iowa, Iowa City, Iowa, United States of America

Cynthia J. Hines
National Institute for Occupational Safety and Health, Cincinnati, Ohio, United States of America

Joseph Barker and Dennis W. Buckman
IMS, Inc, Calverton, Maryland, United States of America

Kent Thomas
National Exposure Research Laboratory, U.S. Environmental Protection Agency, Research Triangle Park, North Carolina, United States of America

Dale P. Sandler and Jane A. Hoppin
Epidemiology Branch, National Institute for Environmental Health Sciences, Research Triangle Park, North Carolina, United States of America

Jinwei Zhang
College of Agriculture and Biotechnology, Weed Management laboratory, China Agricultural University, Beijing, China
Institute of Plant Protection, Tianjin Academy of Agricultural Sciences, Tianjin, China
Ortrud Jaeck, Alexander Menegat and Roland Gerhards
Institute of Phytomedicine, Department of Weed Science, University of Hohenheim, Stuttgart, Germany

Zongjian Zhang
The Office of National Pesticide R&D South Center R&D Center of Functional Additives, Central Research Institute of China Chemical Science and Technology, Beijing, China

Hanwen Ni
College of Agriculture and Biotechnology, Weed Management laboratory, China Agricultural University, Beijing, China

Index